SHAPE AND FORM IN PLANTS AND FUNGI

Frontispiece

Facing page Use of β-glucuronidase (GUS) as a reporter to study complex developmental interactions in *Pyrenopeziza brassicae* by light microscopy. (a) Mycelium of a GUS transformant of *P. brassicae* NH10 (NH10pNOM102/1) crossed with wild-type JH26, 7 days after crossing stained with magenta-β-D-glucuronide (scale bar 25 μm). (b) Mycelium of NH10pNOM102/18 interacting with wild-type JH26 mycelium after 4 days on CMM medium and stained with X-gluc (scale bar 1 mm). (c) Crushed apothecia from a cross of *P. brassicae* NH10pNOM102/18 and wild-type JH26 and stained with X-gluc (scale bar 250 μm). (d) Septate ascospores resulting from a cross of *P. brassicae* NH10pNOM102/18 and wild-type JH26 stained with X-gluc (scale bar 25 μm). From Ashby & Johnstone, Ch. 20, this volume.

Reverse facing page **a** Low-power micrograph of a 4 × 1 mm jerusalem artichoke tuber pith explant after 7 days of incubation on a xylogenic concentration of auxin, cleared and stained with safranin as in Dalessandro (1973). **b** Negative print of an aerial photograph of a fossilized creek system. Silt in a former marshland drainage system shows up light (dark in this negative) against dark fenland soil (light in this negative). **c** Painting by Pery Burge. **d** Micrograph of the cleared and stained result of an experiment in which an auxin gradient and a cytokinin counter-gradient were set up across a thin sheet of jerusalem artichoke tissue. The source of auxin was at the left, the source of cytokinin far to the right, of the region of the sheet depicted here. From Hanke & Green, Ch. 11, this volume.

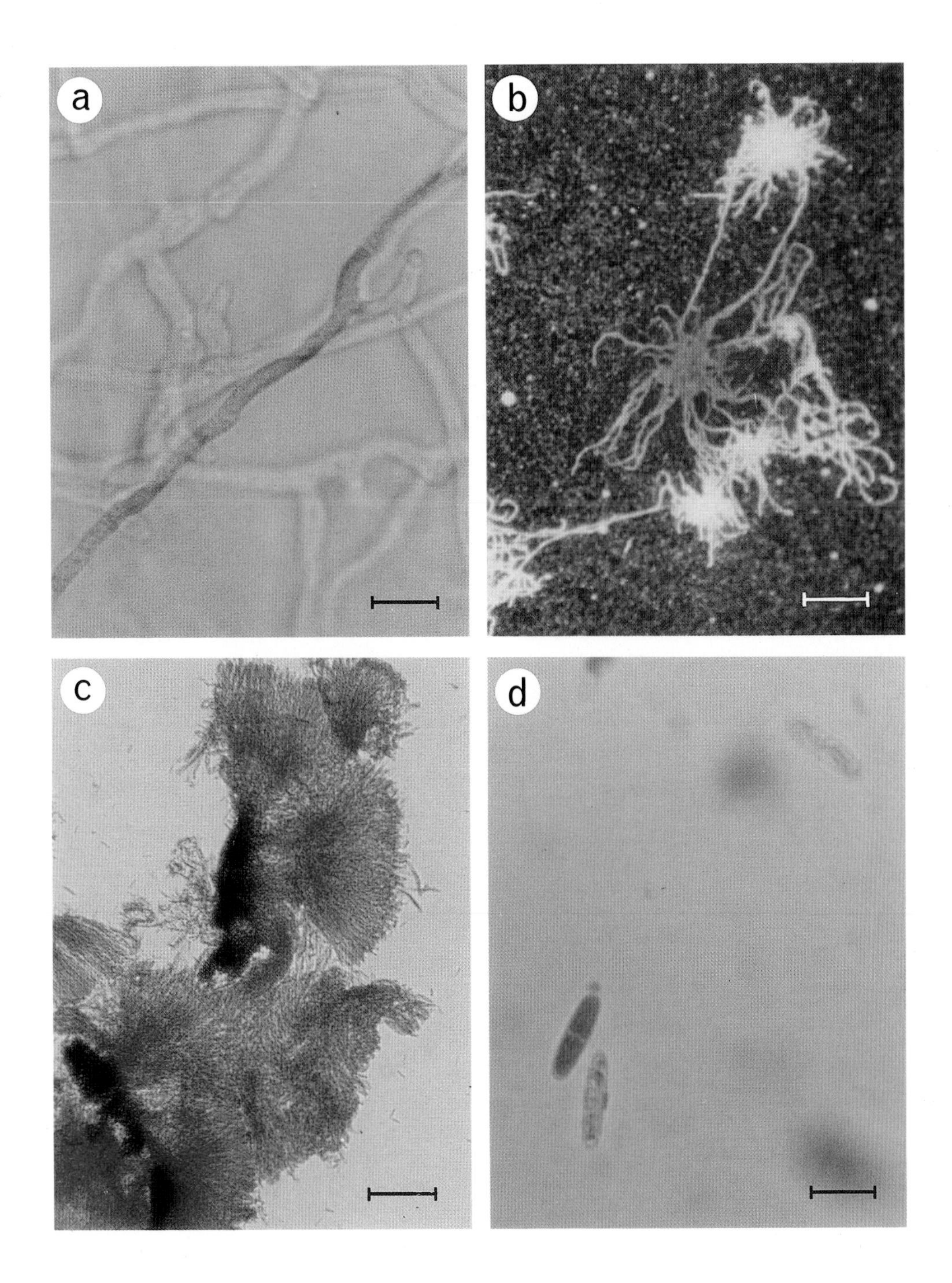
a
b
c
d

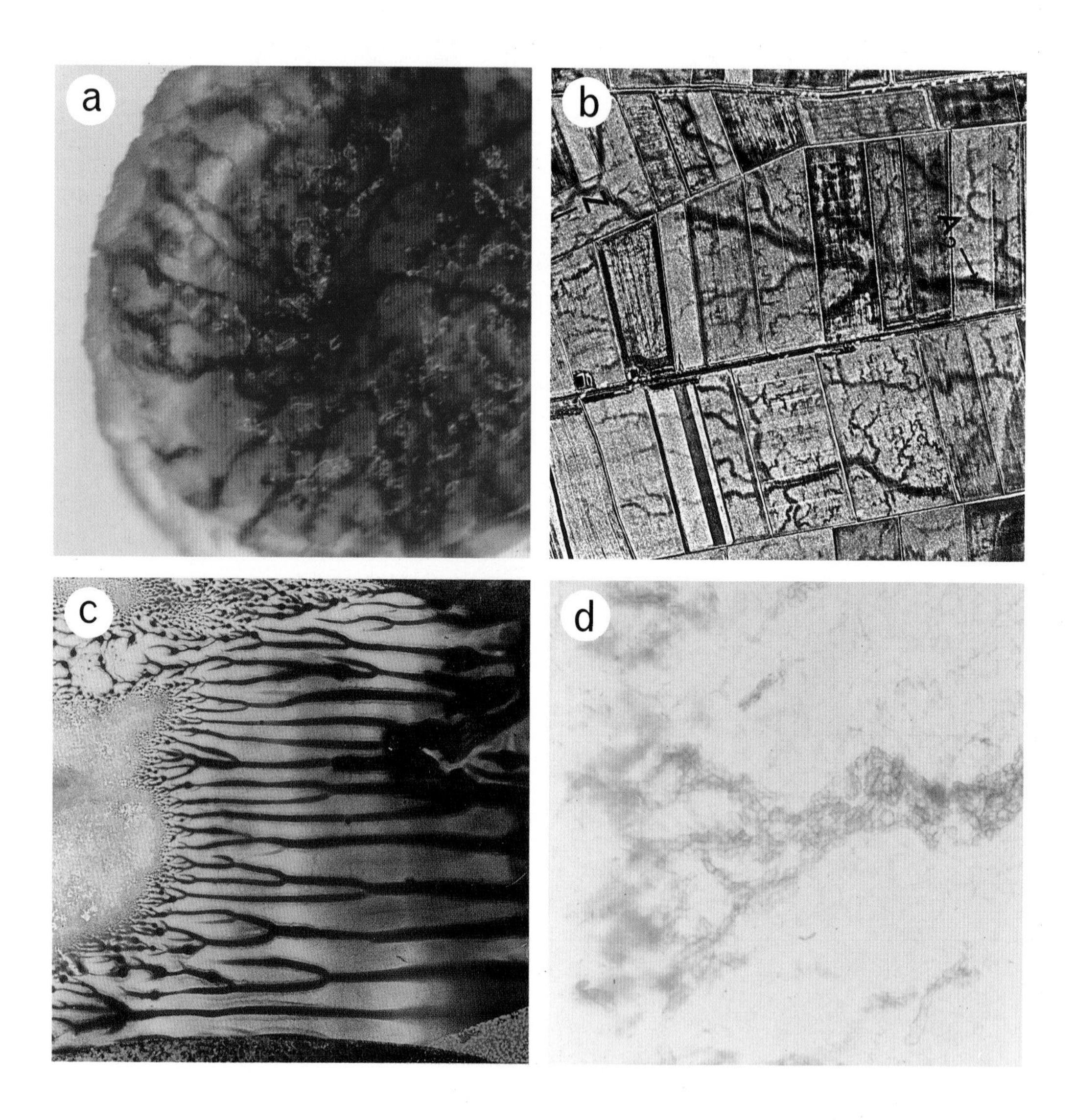
a
b
c
d

LINNEAN SOCIETY SYMPOSIUM SERIES | NUMBER 16

SHAPE AND FORM IN PLANTS AND FUNGI

edited by
D.S. Ingram and A. Hudson

Papers presented at a
Linnean Society Regional Meeting
held at the Royal Botanic Garden
Edinburgh.

Published for the Linnean Society of London
by Academic Press

ACADEMIC PRESS
Harcourt Brace & Company, Publishers
London San Diego New York Boston
Sydney Tokyo Toronto

ACADEMIC PRESS LIMITED
24/28 Oval Road
London NW1 7DX

United States edition published by
ACADEMIC PRESS INC.
San Diego, CA 92101

A catalogue record is available from the
British Library

ISBN 0-12-371035-9

This book is printed on acid-free paper

Typeset by Photo·graphics, Honiton, Devon
Printed in Great Britain by The Bath Press, Avon

Contents

Contributors

A.M. Ashby, Department of Plant Sciences, University of Cambridge, Downing Street, Cambridge CB2 3EA, U.K.

P.W. Barlow, Department of Agricultural Sciences, University of Bristol, AFRC Institute of Arable Crops Research, Long Ashton Research Station, Bristol BS18 9AF, U.K.

R.M. Bateman, Departments of Earth and Plant Sciences, Oxford University, Parks Road, Oxford OX1 3PR, U.K. Present address: Royal Botanic Garden, Edinburgh EH3 5LR, U.K. and Royal Museum of Scotland, Chambers Street, Edinburgh, EH1 1JF, U.K.

A.D. Bell, School of Biological Sciences, University of Wales, Bangor, Gwynedd LL57 2UW, U.K.

B.L. Burtt, Royal Botanic Garden, Edinburgh EH3 5LR, U.K.

P. Carol, Cambridge Laboratory, John Innes Centre, Norwich NR4 7UH, U.K.

E. Coen (and D. Bradley, R. Elliott, R. Simon, J. Romero, S. Hantke, S. Doyle, M. Mooney, D. Luo, P. McSteen, L. Copsey, C. Robinson, R. Carpenter), John Innes Institute, John Innes Centre, Norwich NR4 7UH, U.K.

J.W. Crawford, Scottish Crop Research Institute, Dundee DD2 5DA, U.K.

W.A. DiMichele, Department of Palaeobiology, N.M.N.H., Smithsonian Institution, Washington, DC 20560, U.S.A.

L. Dolan, John Innes Institute, John Innes Centre, Norwich NR4 7UH, U.K. and Plant Science Institute, Department of Biology, University of Pennsylvania, Philadelphia, PA. 19104, U.S.A.

D. Edwards, Department of Geology, University of Wales, P.O. Box 914, Cardiff CF1 3YE, U.K.

G.W. Gooday, Department of Cell and Molecular Biology, University of Aberdeen, Marischal College, Aberdeen AB9 1AS, U.K.

B.C. Goodwin, Department of Biology, The Open University, Walton Hall, Milton Keynes MK7 6AA, U.K.

N.A.R. Gow, Department of Cell and Molecular Biology, University of Aberdeen, Marischal College, Aberden AB9 1AS, U.K.

S.J. Green, Department of Plant Sciences, University of Cambridge, Downing Street, Cambridge CB2 3EA, U.K.

G.S. Griffith, School of Biological Sciences, University of Bath, Claverton Down, Bath BA2 7AY, U.K.

D.E. Hanke, Department of Plant Sciences, University of Cambridge, Downing Street, Cambridge CB2 3EA, U.K.

N. Harberd, Cambridge Laboratory, John Innes Centre, Norwich NR4 7UJ, U.K.

A. Hay, Royal Botanic Gardens, Mrs Macquarie's Road, Sydney 2000, Australia.

A. Hudson, Institute of Cell and Molecular Biology, University of Edinburgh, King's Buildings, Mayfield Road, Edinburgh EH9 3JH, U.K.

D.S. Ingram, Royal Botanic Garden, Edinburgh EH3 5LR, U.K.

K. Johnstone, Department of Plant Sciences, University of Cambridge, Downing Street, Cambridge CB2 3EA, U.K.

M. Knight, Institute of Cell and Molecular Biology, University of Edinburgh, King's Buildings, Mayfield Road, Edinburgh EH9 3JH, U.K.

P. Linstead, John Innes Institute, John Innes Centre, Norwich NR4 7UH, U.K.

D.J. Mabberley, Department of Plant Sciences, Oxford University, South Parks Road, Oxford OX1 3RB, U.K.

D.G. Mann, The Royal Botanic Garden, Edinburgh EH3 5LR, U.K.

D. Moore, School of Biological Sciences, Stopford Building, The University, Manchester M13 9PT, U.K.

R.J. Pankhurst, Royal Botanic Garden, Edinburgh EH3 5LR, U.K.

J. Peng, Cambridge Laboratory, John Innes Centre, Norwich NR4 7UJ, U.K.

R.S. Poethig, Plant Science Institute, Department of Biology, University of Pennsylvania, Philadelphia, PA. 19104, U.S.A.

A.D.M. Rayner, School of Biological Sciences, University of Bath, Claverton Down, Bath BA2 7AY, U.K.

N.D. Read, Institute of Cell and Molecular Biology, University of Edinburgh, King's Buildings, Mayfield Road, Edinburgh EH9 3JH, U.K.

K. Ritz, Scottish Crop Research Institute, Dundee DD2 5DA, U.K.

K. Roberts, John Innes Institute, John Innes Centre, Norwich NR4 7UH, U.K.

A.J. Trewavas, Institute of Cell and Molecular Biology, University of Edinburgh, King's Buildings, Mayfield Road, Edinburgh EH9 3JH, U.K.

R. Watling, Royal Botanic Garden, Edinburgh EH3 5LR, U.K.

H.G. Wildman, Department of Natural Products Discovery, Glaxo Group Research Limited, Greenford Road, Greenford, Middlesex UB6 0HE, U.K.

Preface

D'Arcy Wentworth Thompson, author of *On Growth and Form* (Cambridge University Press, 2nd edition, 1942), was born at No. 3 Brandon Street, Edinburgh, little more than a stone's throw from the Royal Botanic Garden. This fact did not lead directly to the decision to hold a Linnean Society Regional Symposium Meeting on *Shape and Form in Plants and Fungi* at the Garden in September 1992, but most, perhaps all, of those who took part must have at some time been influenced by D'Arcy Thompson's ideas and thinking. I like to think that he was present in spirit at the Symposium and that he, like me, enjoyed the stimulating and scholarly debate engendered by the speakers whose contributions are recorded in this volume.

Two additional contributions have since been added to widen the scope of the book: that by Bateman and DiMichele (Ch. 4), on saltational evolution of form in vascular plants; and that by Hudson (Ch. 13), which provides a synopsis of recent research on genes controlling flower development in plants, to supplement the summary of the lecture of Coen *et al.* (Ch. 13), who were unable to provide a complete manuscript on this important topic.

Shape may be regarded as the external contour of an organism, while *Form* describes the three-dimensional arrangement of the component parts. These are distinct yet interdependent concepts, and both shape and form are the consequence of *Growth and Development.* The speakers at the Symposium were chosen for their ability to address these matters from a variety of standpoints: descriptive, philosophical, morphological, taxonomic, evolutionary, physiological, genetic, molecular-biological and mathematical. Their contributions should not be seen as an attempt to cover the entire field of study, but rather as a collection of essays on a general theme. Taken individually or together, however, I am sure that they will stimulate as much new thought and intellectual excitement for the reader as they did for the participants at the Symposium. The strength of the collection lies, I believe, in its diversity of approach and philosophy rather than in its uniformity.

Each reader will distil different conclusions from close analysis of the essays. For me, some important themes were: the crucial significance of accurate observation and an understanding of perception in the study of shape and form; the creative tension between the holistic and reductionist approaches to analysis of these parameters and the interdependence of one approach on the other; the paucity of information on the internal messages and forces that link the genome and the phenotype; the probable significance of calcium in this context; the power of molecular-genetic and mathematical/computer analysis as tools for the study of shape and form; and the potential of research with unicellular plants and with fungi, where the relative simplicity of the systems for study and the increasing availability of the appropriate tools for analysis will surely soon lead to major advances in understanding.

If I could make one plea for the future, however, it would be that in the excitement to apply the techniques of cell and molecular biology to the analysis of shape and form in plants and fungi, the importance of understanding the whole organism and its place in the environment is not forgotten.

D.S. Ingram
Royal Botanic Garden Edinburgh
September 1993

Acknowledgements

DSI wishes to thank Dr Andrew Hudson for his help in devising the programme and in editing this volume, Mrs Rose Clement, whose organization and planning skills made the Symposium possible, Mrs Helen Hoy, Mrs Marisa Main and other staff of the Royal Botanic Garden Edinburgh for their assistance in the practical business of running the Symposium, the Gatsby Charitable Foundation for the provision of travel grants for students and, last but not least, the Linnean Society of London and the Trustees of the Royal Botanic Garden Edinburgh for their joint sponsorship of the Symposium.

Part I

PLANTS

CHAPTER

1

Generative explanations of plant form

B.C. GOODWIN

CONTENTS

Abstract

Explaining plant form requires some description of the morphogenetic field that generates adult morphology, including the mechanical forces that contribute to three-dimensional structure. A model is described for morphogenesis in the giant unicellular green alga, *Acetabularia acetabulum*, which presents a simple but non-trivial example of pattern formation in a single cell. On the basis of experimental work and using the dynamics of calcium–cytoskeleton interaction together with wall growth, this model simulates the basic morphogenetic sequence of the developing organism and provides possible explanations of the structures observed, including transiently produced whorls of sterile bracts that serve no apparent function. The characteristics of the process suggest that the morphogenetic trajectory may be robust in the sense that many parameters (genes) can be altered and the same basic form is produced. The role of genes and genetic programmes in morphogenesis is discussed, and the argument for robust morphogenesis is extended to the case of phyllotaxis in higher plants. This has evolutionary implications, which are considered.

INTRODUCTION

The message to be transmitted in this chapter is by no means original, but it does need to be reiterated periodically. Goethe expressed it in his insight into plant form as dynamic transformation (metamorphosis) through his concept of the 'primal plant' (Urpflanze). His belief was that this allowed him to 'go on forever inventing plants

Shape and Form in Plants and Fungi
ISBN 0–12–371035–9

and know that their existence is *logical*; that is to say, if they do not actually exist, they could, for they are not the shadowy phantoms of a vain imagination, but possess *an inner necessity and truth*.' (cf. Hegge, 1987). The issue concerns understanding and explanation in science at the level of invariant generative properties that embody this 'inner necessity and truth'. In contemporary biology there is a very strong tendency to seek explanations of organismic properties at two quite different levels: an internal level of gene activities and an external level of natural selection. While both of these have a role to play in the study of biological form, neither provides us with the type of explanation that we seek in science. The problem will be illustrated by particular example taken from work on plant morphogenesis, and then generalized to the broader picture of the relationships between development and evolution.

MORPHOGENESIS

The example used to illustrate the message is morphogenesis in the giant unicellular alga, *Acetabularia acetabulum*, whose life cycle is shown in Fig. 1. What follows has been published elsewhere (Goodwin, 1990; Goodwin & Brière, 1992) so the description will be kept to a minimum. The habitat of *Acetabularia* is the shallow waters around the shores of the Mediterranean. Isogametes fuse to produce a roughly spherical zygote which breaks symmetry, producing a growing stalk and a branching rhizoid

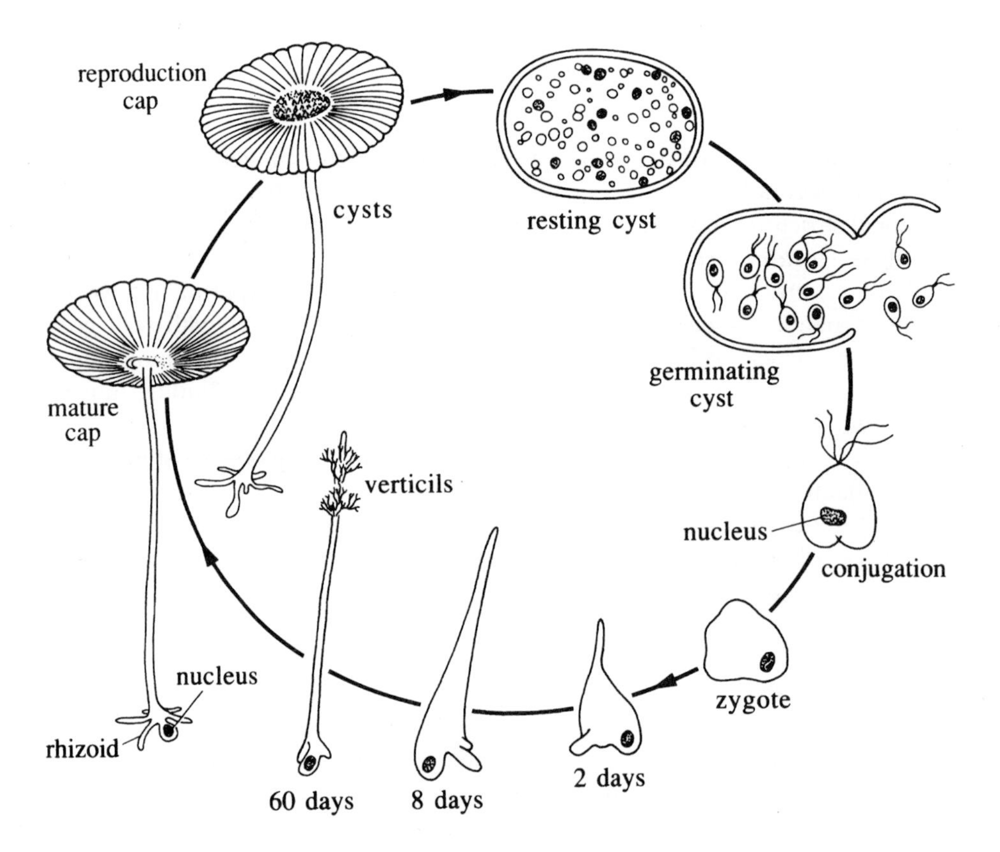

Figure 1 The life cycle of *Acetabularia acetabulum*.

that anchors the alga to the substratum and houses the nucleus in one of its branches. When the stalk reaches a length of about 1–1.5 cm after several weeks of growth, a ring of small bumps arises around the tip, growing into a vertical or whorl of bracts that branch successively as they grow. The tip renews its growth from the centre of the whorl and, after a few days during which the stalk grows several millimetres, another whorl is produced. This process repeats itself until a cap primordium is generated (Fig. 2), a structure with the same circular symmetry as a whorl but consisting of many rays joined together in a disc-shaped structure, the cap. This grows radially, the whorls drop off, resulting in the morphology of the adult: a rhizoid (still containing the nucleus), a stalk 3–5 cm in length and about 0.5 mm in diameter, and a cap whose diameter is approximately 0.5 cm: a giant differentiated cell (Fig. 3).

The alga is also capable of regenerating a cap after a cut through the stalk. It does so following the same sequence of shape changes as in normal growth. After the cut has healed with the production of a new cell wall, a tip emerges, grows, produces a series of whorls and then generates a cap. The regenerate with the rhizoid is, after the whorls have dropped off, indistinguishable from the original mature alga. In this chapter the dynamic processes responsible for the characteristic sequence of shape changes during normal development and regeneration are discussed.

CALCIUM–CYTOSKELETON DYNAMICS

The importance of calcium in morphogenesis was first suggested by studies showing that changing the concentration of this ion in the sea water in which the algae develop

Figure 2 Formation of a cap primordium of *Acetabularia acetabulum*, after three whorls.

Figure 3 Morphology of the mature cell of *Acetabularia acetabulum*, showing rhizoid, stalk and cap. The whorls have dropped off.

causes dramatic changes of morphology (Goodwin & Pateromichelakis, 1979; Goodwin *et al.*, 1983). These changes could be duplicated by adding to the sea water ions (Co^{2+} or La^{3+}) which block calcium channels, so the effects of reduced calcium are not simply on the mechanical properties of the cell wall but extend inside the cell. The wavelength of the whorl pattern may also be systematically altered by changing the calcium concentration in the medium (Goodwin *et al.*, 1987; Harrison & Hillier, 1985). Calcium is known to have significant effects on the mechanical state of the cytoskeleton (Menzel & Elsner-Menzel, 1989; Menzel *et al.*, 1992; Shelanski, 1989; Williamson, 1984). It changes the viscosity and elastic modulus of the cytoplasm by influencing the state of polymerization of actin and tubulin, and by activating actomyosin contraction and enzymes such as gelsolin which cut actin filaments. Some of these influences are shown schematically in Fig. 4.

For morphogenesis to occur it is necessary to have a medium in which spatial patterns are generated spontaneously. The most likely candidate for this in a cell is certainly the cytoplasm. It was established by Turing (1952) that coupled biochemical reactions combined with diffusion can produce spatially non-uniform patterns of reactants which he called morphogens. These reaction–diffusion systems have been

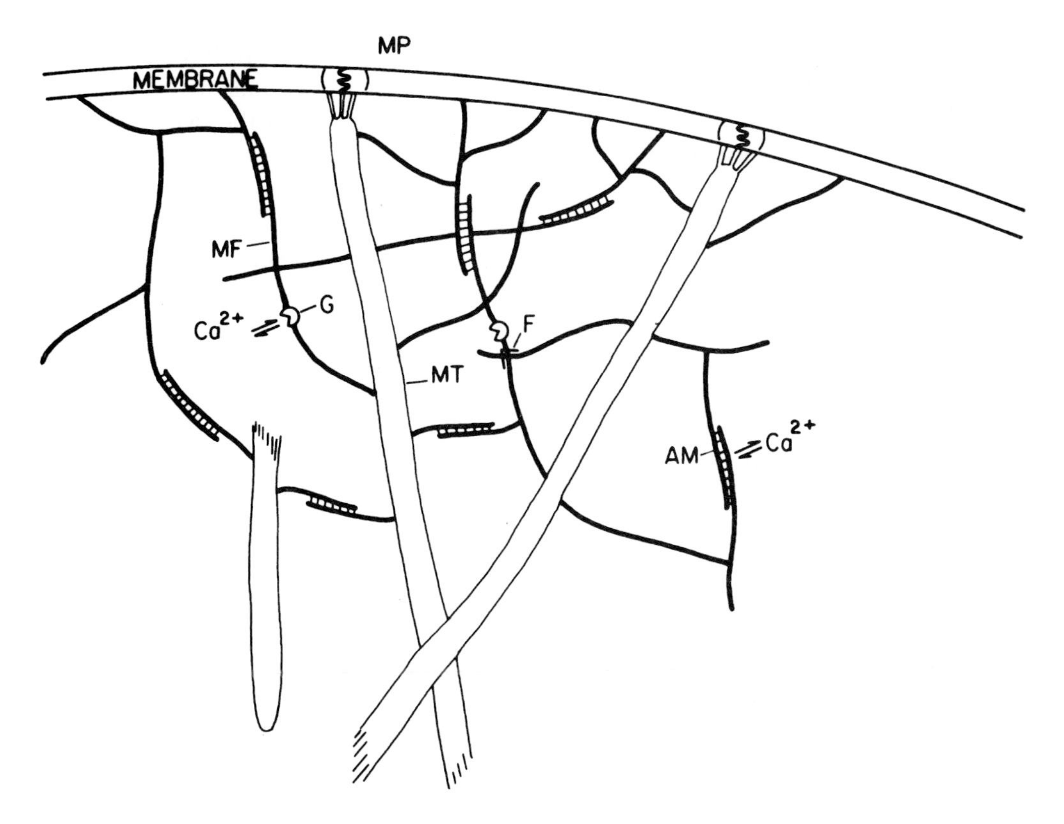

Figure 4 Schematic diagram showing structural compounds of the cytoskeleton. AM, actomycin; F, fimbrin; G, gelsolin; MF, microfilament; MP, membrane protein; MT, microtubule.

extensively used by Meinhardt (1982) and by Murray (1989) to model morphogenetic processes in a variety of organisms, and they have been considered by Harrison and Hillier (1985) and by Goodwin *et al.* (1985). However, the morphogenetic effects of calcium and its influence on the cytoskeleton suggested that this system itself might play the role of primary pattern generator. The first step in explaining this possibility was to derive equations that describe the mechanical properties of the cytoskeleton, the cytoplasmic regulation of calcium, and their interaction. This was done by Goodwin and Trainor (1985) and the coupled equations were shown to have the property of spontaneous bifurcation for particular ranges of the parameters. Within this range, spatial patterns of cytosolic free calcium and mechanical strain in the cytoplasm develop from random perturbation of the system from a spatially uniform initial condition. The reason for this behaviour lies in certain basic properties of the calcium–cytoskeleton interaction, which will now be described.

Cytosolic free calcium is regulated in eukaryotic cells at concentrations of 100 nM or so by plasmalemma pumps, by sequestration mechanisms involving the endoplasmic reticulum, vesicles or vacuoles, and by binding to cytoplasmic proteins and chelating agents such as calcitonin and calmodulin. Studies of actin gels have shown that as calcium rises above 100 nM it induces gel breakdown and solation by activation of enzymes such as gelsolin. At higher concentrations calcium initiates contraction of actomyosin filaments so that the cytoplasm becomes more resistant to deformation (Nossal, 1988). At calcium concentrations above about 5 μM depolymerization of filaments and microtubules, and the progressive action of gelsolin, have the consequence that the cytoplasm becomes progressively solated. A qualitative description of this behaviour in terms of changes in the elastic modulus of the cytoplasm as a function of calcium is shown in Fig. 5. This describes how calcium affects the mechanical state of the cytoplasm. It may be deduced that there is also a reciprocal action of the mechanical state of the cytoskeleton on free calcium concentration. It is assumed that strain or deformation of the cytoplasm results in release of calcium from the bound or sequestered state to free ions. Therefore, regions that happen to have elevated strain will also have elevated free calcium levels. But increased free calcium causes gel breakdown and solation. This results locally in more strain (deformation) since the cytoplasm is assumed to be under tension, hence there is further calcium release. The result is a positive feedback loop in the regions of calcium concentration where the slope of the elastic modulus curve as a function of calcium is negative (see Fig. 5): a local, random increase of calcium above the steady-state level initiates a run-away calcium release and increase of cytoplasmic strain. This, however, is stabilized by the effects of diffusion which tends to reduce the calcium gradients and also by the opposing effects of calcium on actomyosin contraction, which increases the elastic modulus and so decreases the strain (region of positive slope, Fig. 5).

The argument also works in reverse: where calcium levels are decreased the strain is also reduced since the cytoplasm is more gel-like (higher elastic modulus) and so free calcium will be bound or sequestered, decreasing it still further. In terms of reaction–diffusion dynamics, calcium plays the role of short-range activator while mechanical strain is like the long-range inhibitor. The result of the interactions is that spatial patterns of calcium concentration and strain can arise spontaneously from initially uniform conditions when the equation parameters are in the bifurcation range. The model is qualitatively similar to that discussed by Oster and Odell (1983).

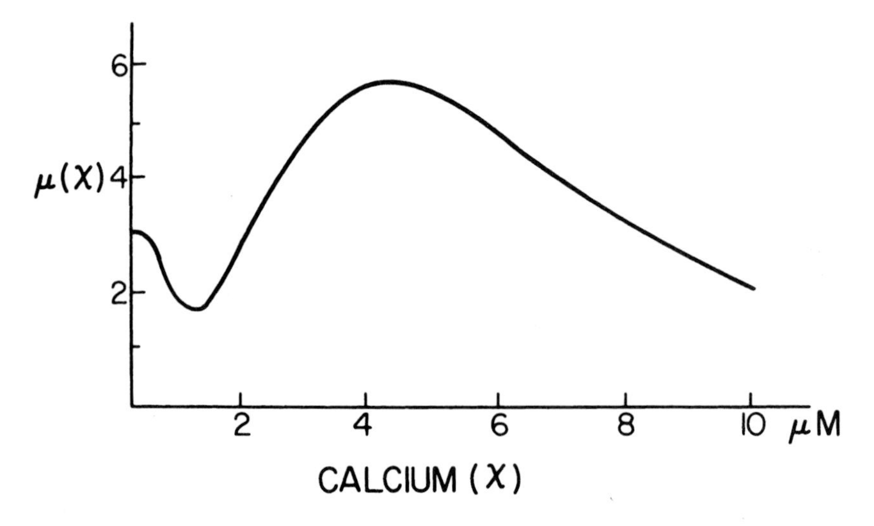

Figure 5 Variation of elastic modulus with free Ca^{2+} concentration (micromolar) in *Acetabularia ace- tabulum*. Qualitative relations only.

CELL WALL DYNAMICS

The properties of the cytoplasm just described give it the characteristics of an excitable medium which can spontaneously generate spatial patterns, both stationary and dynamic, i.e. propagating waves. This is sufficient to initiate pattern; but morphogenesis involves changes of geometry. In the case of *Acetabularia* and many other developing organisms, morphogenesis is linked to growth, so the cell wall must undergo localized changes of shape together with elongation. The wall is described in the model as a purely elastic shell (about 2 μm thick) whose state changes as a function of strain in the underlying cytoplasm, which is a thin shell about 10 μm thick closely apposed to the wall, the plasmalemma separating them. This functional coupling is assumed to arise via strain-activated pumps in the plasmalemma that cause the wall to soften by excretion of protons or hydrolases. The large vacuole in the centre of the cell is an osmotic organelle, separated from the cytoplasmic shell by another membrane, the tonoplast. The vacuole exerts a pressure that is resisted by the wall. Patterns of strain in the cytoplasm are thus reflected in the elastic modulus of the wall, which undergoes elastic deformations as a result of the outward-directed osmotic pressure. A growth process was introduced into the model whereby new wall material was added wherever wall strain exceeded a threshold value so that elastic deformations led to plastic changes, in accordance with experimental evidence (Cleland, 1971; Green *et al.*, 1971). Growth of the cytoplasm was coupled to wall growth, while vacuolar pressure remained constant. These relations are described mathematically in Brière and Goodwin (1988).

SIMULATIONS OF MORPHOGENESIS IN *ACETABULARIA*

The calcium–cytogel and cell wall equations were used for a finite-element simulation of growth and morphogenesis. Parameters were adjusted so that the calcium–cytoskeleton equations were in the bifurcation range, making spatial patterns possible. The shape that developed depended on the characteristic wavelength of the pattern, and also on the parameters that describe wall growth. Simulations started with uniform initial conditions, on a dome representing a regenerating apex (Fig. 6). All variables start off spatially uniform (flat) and spontaneously develop a pattern, calcium rising to a maximum at the tip, as does the gel strain. The result is that the wall softens in this region and there is an elastic deformation. A three-dimensional view of this is shown in Fig. 7: a tip is produced.

This is the first stage of the regenerative process. A characteristic feature of whorl formation is the flattening of the conical tip just prior to the appearance of the ring of hair primordia that initiates a whorl. This was something that was not understood; the model gave an explanation. As growth occurs at the tip, plastic changes of geometry follow the elastic deformations, and there is an interesting interaction between the shape generated and the dynamic behaviour of the calcium–cytogel system. After an initial stage of growth, the calcium gradient with the maximum at the tip becomes unstable and transforms into an annulus, the maximum level of calcium occurring away from the tip. The region of maximum cytogel strain also changes in a similar manner (Fig. 8). Wall softening is now greatest in this annular region, resulting also in maximal wall curvature proximal to the tip, with consequent flat-

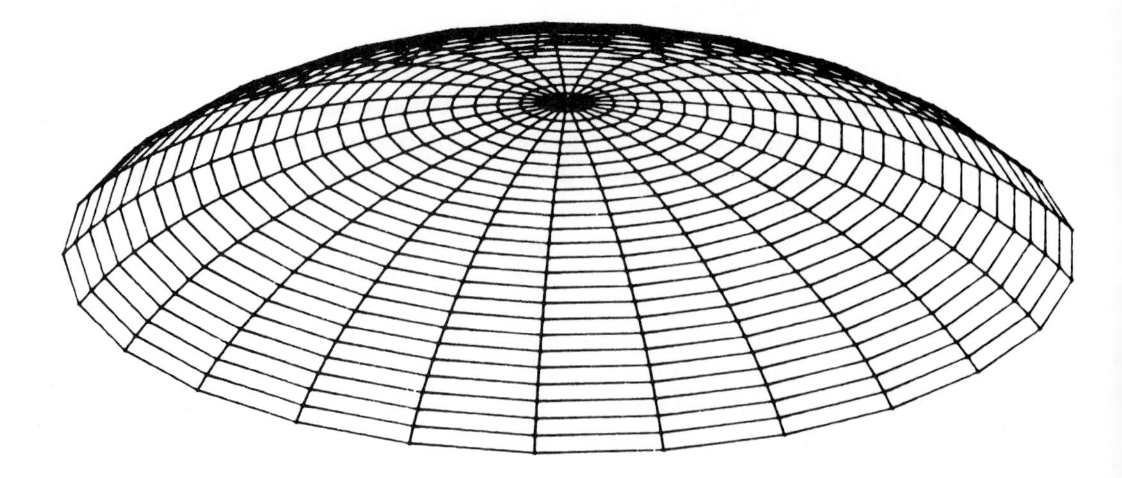

Figure 6 Computer simulation of regeneration in *Acetabularia acetabulum* in which the cytoplasm and the cell wall are described as shells made up of finite elements which obey equations describing their dynamics as mechanochemical or elastic media, respectively.

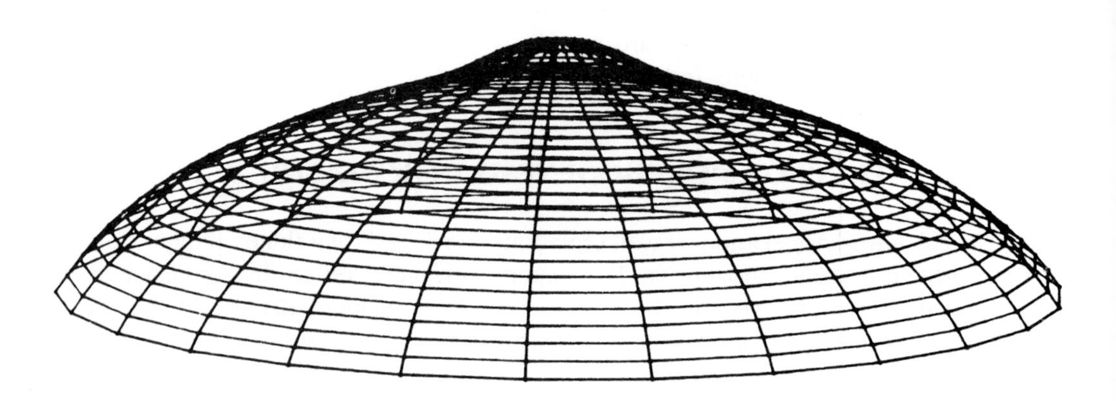

Figure 7 Tip initiation in the model for regeneration in *Acetabularia acetabulum*, resulting from spontaneous symmetry breaking of the cytogel–calcium dynamic, gradient formation with maxima of strain and calcium at the tip, and wall softening as a result of interaction between cytoplasm and the wall.

tening of the tip itself, where the elastic modulus of the wall is now larger than in the annular region.

A calcium annulus of this type was perturbed to see if it could spontaneously generate a pattern similar to that of a whorl. It did so, producing a ring of peaks of calcium that can be interpreted as the initiator of the whorl pattern. Unfortunately the finite-element programme is not yet sufficiently robust to allow the study of the growth of such small elements, breaking the axial symmetry of the growing tip. This work is in progress. But the sequence of pattern changes observed, namely gradient and tip formation, elongation, annulus formation and tip flattening, and the bifurcation of an

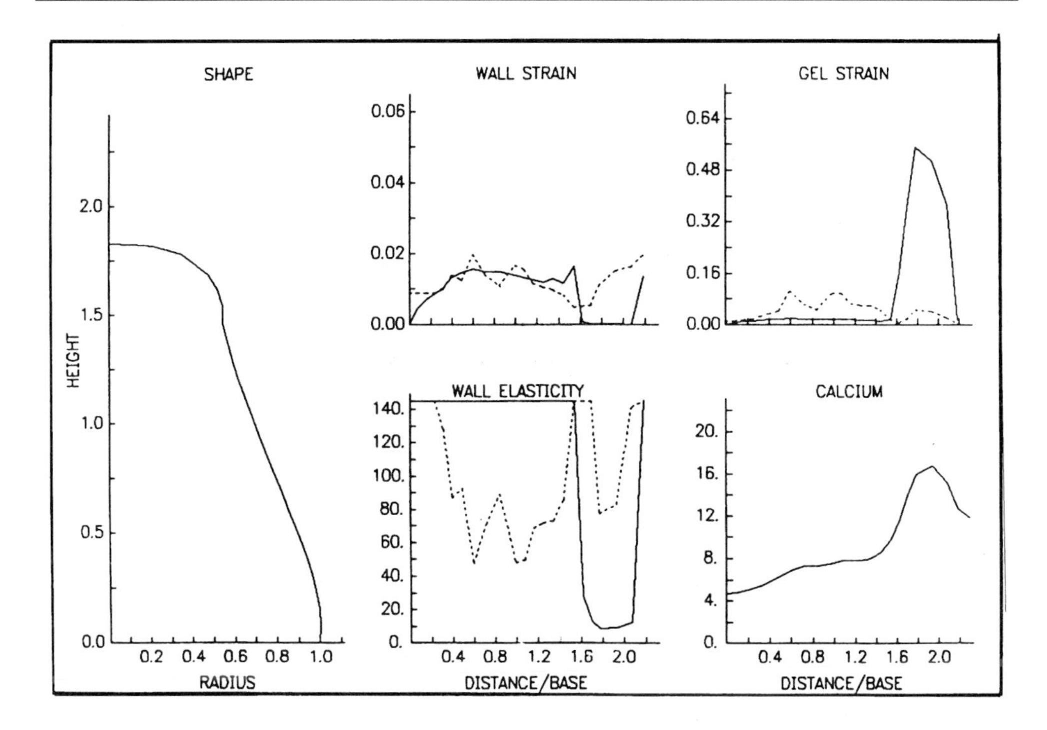

Figure 8 Section through the mesh describing the regenerating tip of *Acetabularia acetabulum*, showing shape, wall strain and elastic modulus, cytogel strain and free calcium concentration as a function of distance from base (origin, O, on the abscissa of the graphed variables, radius 1 in the curve showing shape). An annulus of calcium forms spontaneously after a period of growth, resulting in tip flattening.

annular pattern to a ring of calcium peaks, provides a very natural dynamic explanation of a basic morphogenetic sequence.

As the apex elongates the annulus itself becomes unstable and intermittently collapses back to a gradient with a maximum at the tip; an annulus then reforms and grows in amplitude. This occurs with a rather irregular frequency, suggestive of the somewhat irregular frequency of successive whorl initiations as the alga grows. The simulations have so far failed to produce a cap. This requires another pattern which has been approached, but never achieved, in which lateral growth exceeds longitudinal. The conditions for cap formation appear to require more anisotropy in the strain field than is currently in the model.

What is the role of genes in this process? They specify parameter values so that viable species exist in a range that allows the morphogenetic trajectory to unfold. It is not yet known how large this domain is in parameter space. But the ease with which suitable parameter values were found suggests that it is extensive – a large basin of attraction seems to exist. It is known that most of the parameters can be changed without significantly affecting behaviour, so the morphogenetic sequence is quite robust. It may be speculated that it is generic in this space – a structurally stable trajectory. If this is the case, then there should be many species of algae in the taxon containing *Acetabularia* that all share the same basic morphology. This is indeed the case: the Dasycladales are an ancient group of giant unicellular green algae most of which are known from fossil remains of calcified cell walls that extend back at least to the Cambrian era, 570 million years ago. These, together with the 20 or more living species, are described in a recent monograph by Berger and Kaever (1992). All species have a rhizoid and an axis with bracts, most of them produced in whorls. In the majority of species the bracts are the gametangia (where the cysts and gametes form) and no caps are produced, so *Acetabularia* belongs in this respect to a minority class. However, it continues to produce whorls despite their sterility and apparent lack of function, being transient structures that are shed. It is known also that the algae can grow perfectly well without whorls, whose production can be prevented by simply reducing the calcium concentration in the sea water (Goodwin *et al.*, 1983), and these whorl-less stalks can then make caps and produce cysts simply by restoring calcium to normal levels. So why does *Acetabularia* produce whorls of sterile bracts?

An obvious answer is that these structures result from the generic dynamics of morphogenesis in this group of organisms, all of which share the same basic organization. They are produced not because they are useful but because they are natural. This is the way forms in nature are explained – the shape of a diamond crystal, vortices in liquids, elliptical trajectories of planetary movement about the sun. These are the natural, robust forms generated by particular dynamic processes. In biology the notions of function or adaptation or natural selection are often used to explain organismic form, but these ideas address only the question why a structure persists. They fail to explain why the structure exists, i.e. what kind of dynamic process makes the form possible as one of its stable solutions. This gets at the generative roots of the problem, as described also by Newman and Comper (1990). Eighty-six years ago D'Arcy Thompson (1917) made precisely the same point in the introduction to his volumes *On Growth and Form*: 'So long and so far as "fortuitous variation" and the "survival of the fittest" remain ingrained as fundamental and satisfactory hypotheses in the philosophy of biology, so long will these satisfactory and specious causes tend to stay severe and diligent enquiry .... to the great arrest of future discovery.'

Is there a genetic programme that directs the morphogenetic sequence in *Acetabularia*? That genes are involved in the process there is no doubt – they set the values of the parameters. How many genes are involved is not known, but certainly every one of the 26 parameters must involve the contributions of many genes. For example, the effective diffusion constant of free calcium in the cytoplasm, one of the sensitive parameters that affects the dynamics, is dependent upon the concentration and infinity constants of any protein that binds calcium such a calmodulin, calcitonin and any analogue of troponin involved in the cytoskeleton. Concentration regulation of proteins involves other genes, so there could be as many as ten genes involved in specifying this one parameter. The restoring force of the cytoplasm or the elastic modulus of the cell wall depends on numerous gene products that contribute to the mechanical properties of these cellular components. Several hundred genes could probably be accounted for in the model, all contributing to parameter values. Not one of these needs to be known in detail, since generic properties are being sought – classes of process, types of form that can be generated. Of course, the more detailed information there is available, the more accurate the model. But it is not constructed from a detailed knowledge of gene activites of the type that is the goal of, say, the human genome project, which promises to reveal all about human development and disease, among other things (see Strohman (1992) for an incisive critique of these claims). The model is constructed from knowledge about the physics, chemistry, biology and mathematics of relevant processes described at the appropriate level, which in the case of *Acetabularia* morphogenesis is the macroscopic field of the cytoplasm and the cell wall, The equations are field equations describing patterns in space; they are not defined by a genetic programme, which addresses the problem at the wrong level for explaining morphogenesis (cf. Goodwin 1985, 1990 for more detailed analyses of this approach to development). In general, it is clear that gene activities change in systematic ways during development. What this means is that the dynamics is hierarchical, 'parameters' themselves becoming variables on particular time-scales and resulting in the hierarchy of processes in space and time that characterize development (Mittenthal & Baskin, 1992). The first-order model indicates that moving boundary processes of a relatively simple kind can generate morphogenetic sequences that capture basic features of the natural process. More complex patterns of morphogenesis seem likely to have equivalently robust dynamic properties (Goodwin *et al.*, 1993; Goodwin, 1993), defining a category of generic process that underlies biological forms.

Genes can be expected to contribute to the stabilization of generic morphogenetic states, with spatially ordered transcription patterns that not only reinforce natural patterns but also modify the characters of the elements in distinctive ways. Examples of the latter are the homoeotic genes that have been identified in *Arabidopsis* (Bowman & Meyerowitz, 1991) and the way these and other genes have spatial distributions that correlate with the particular forms of the components of the flower (sepals, petals, stamens, carpels), as described by Coen and Carpenter (1992). These genes do not generate the whorls of elements in the flower; they modify their structure, and the spatial patterns of gene products result in a transformation series of the whorls. These are very interesting and important results. But the problem of plant form clearly cannot be solved at the level of gene products. How genes act in modifying shape will become clear only when there becomes available a generative model that identifies the actual forces at work in morphogenesis, what types of shape can be produced, and what influences are effective in transforming one shape into another.

THE EVOLUTIONARY PERSPECTIVE

The consequences of the point of view outlined above for the study of biological form in an evolutionary context are of some interest. It is a distinct possibility that all conserved features of organismic structure represent generic properties of morphogenetic dynamics. Consider, for example, the patterns of leaves on higher plants, the problem of phyllotaxis. Leaves are generated by the growing tip of a branch (the meristem) according to well-defined rules, as shown in Fig. 9. The whorled pattern, in which two, three or more leaves are produced at each node, is actually a particular form of the pattern which *Acetabularia* itself generates in its whorls of bracts, although the meristem is a multicellular structure in contrast to the unicellular algal tip, despite their similarity of dimension (100–200 μm in diameter). It seems perfectly reasonable to propose that these patterns of phyllotaxis are the primary stable modes generated by the meristem as a morphogenetic field, obeying particular equations involving cytoplasmic state, mechanical stresses, growth and planes of cell division. This is precisely the hypothesis being explored by Green (1987, 1989) in his interesting studies of leaf phyllotaxis. The fact that a majority of plant species (>80%) have spiral phyllotaxis could then be explained by this trajectory having the largest basin

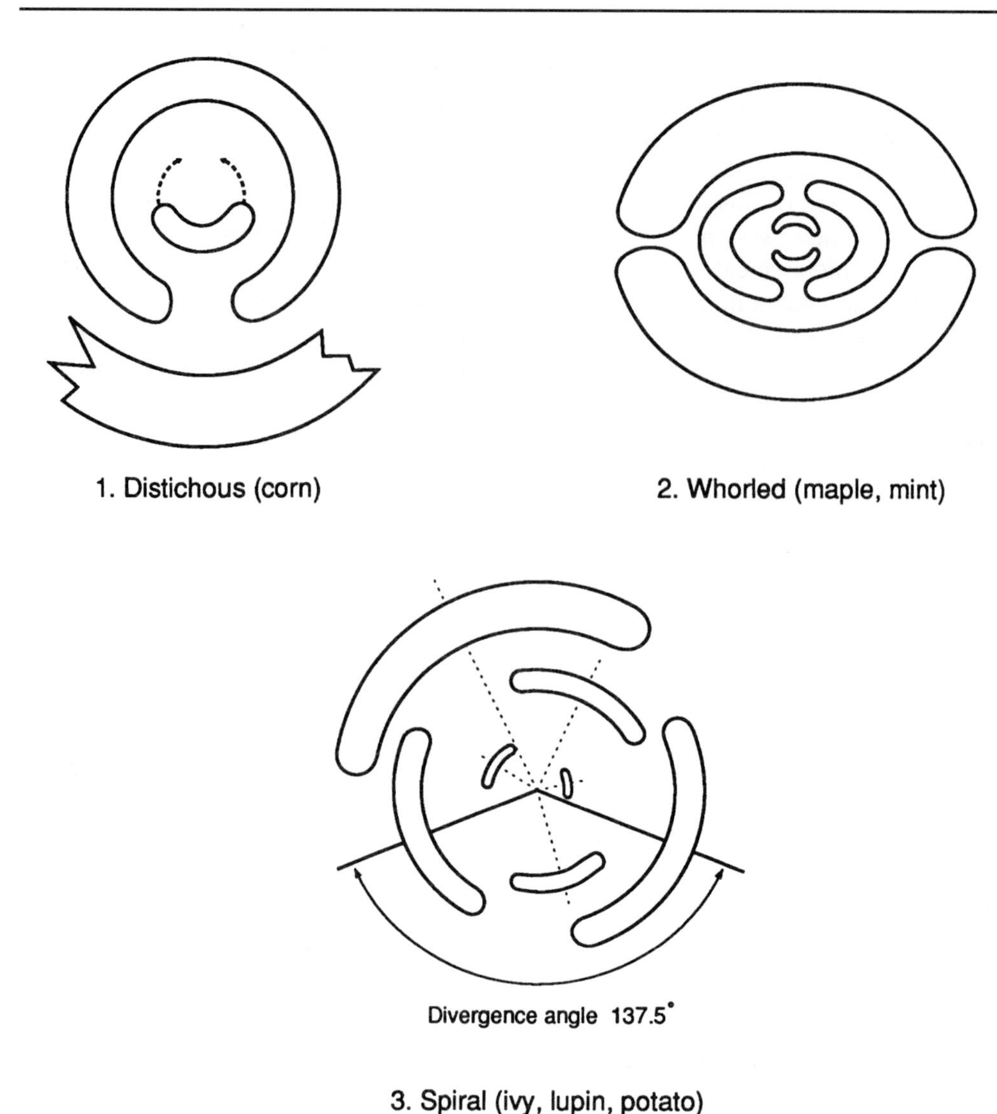

Figure 9 The three basic types of leaf phyllotaxis in angiosperms.

of attraction in the morphogenetic space of the meristem. All three types of phyllotaxis work perfectly well, as they must for survival. But the differential abundance of the different forms may be due primarily to the sizes of the basins of attraction in morphospace, rather than differential selection due to different degrees of fitness of the patterns. There is no conflict between these different aspects of stability; in fact, they belong together in a dynamic analysis of overall life-cycle stability, which includes the generative process that produces the organism in the first place.

There are many other examples of conserved characters in evolution such as the body plans that define the different phyla, and structures such as tetrapod limbs whose basic form is preserved despite a great diversity of expression serving a variety of functions in different species. The concept of homology, describing invariant features of diverse structures, expresses the conserved nature of the biological forms that are the foundation of rational systematics, taxonomy. There are now morphogenetic models that suggest what the invariant generative properties may be that underlie these forms, such as tetrapod limbs (Oster *et al.*, 1988; Shubin & Alberch, 1986) or the coiled shells of gastropods (Raup & Michelson, 1965). A model has been presented by Ho (1990) that provides a rational taxonomy of teratologies observed in *Drosophila* in terms of developmental regularities. The objective of these and other studies is to provide explanations of biological forms in terms of generative dynamics of morphogenetic fields. It is the generic states of these fields that represent the stable forms available to life, hence the types of organism that can arise during evolution.

REFERENCES

BERGER, S. & KAEVER, M.J., 1992. *Dasycladales: An Illustrated Monograph of a Fascinating Algal Order.* Stuttgart, Germany: Thieme.

BOWMAN, J.L. & MEYEROWITZ, E.M., 1991. Genetic control of pattern formation during flower development in *Arabidopsis.* In G.I. Jenkins & W. Schuch (eds), *Molecular Biology of Plant Development,* pp. 89–115. Symposium of the Society of Experimental Biology XLV. Cambridge.

BRIÈRE, C. & GOODWIN, B.C., 1988. Geometry and dynamics of tip morphogenesis in *Acetabularia. Journal of Theoretical Biology, 131:* 461–475.

CLELAND, R., 1971. Cell wall extension. *Annual Review of Plant Physiology, 22:* 197–222.

COEN, E. & CARPENTER, R. (1992). The power behind the flower. *New Scientist, 25:* 24–27.

GOODWIN, B.C., 1985. What are the causes of morphogenesis? *BioEssays, 3:* 32–35.

GOODWIN, B.C., 1990. Structuralism in biology, In *Science Progress,* vol. 74, pp. 227–244. Oxford, UK: Blackwell.

GOODWIN, B.C., 1993. Development as a robust natural process. In W.D. Stein & F. Varela (eds), *Thinking About Biology.* Santa Fe Institute Studies in the Sciences of Complexity. Reading, MA, USA: Addison-Wesley.

GOODWIN, B.C. & BRIÈRE, C., 1992. A mathematical model of cytoskeletal dynamics and morphogenesis in *Acetabularia.* In D. Menzel (ed.), *The Cytoskeleton of the Algae,* pp. 219–238. Boca Raton, FL, USA: CRC Press.

GOODWIN, B.C. & PATEROMICHELAKIS, S., 1979. The role of electrical fields, ions, and the cortex in the morphogenesis of *Acetabularia. Planta, 145:* 427–435.

GOODWIN, B.C. & TRAINOR, L.E.H., 1985. Tip and whorl morphogenesis in *Acetabularia* by calcium-regulated strain fields. *Journal of Theoretical Biology, 117:* 79–106.

GOODWIN, B.C., SKELTON, J.C. & KIRK-BELL, S.M., 1983. Control of regeneration and morphogenesis by divalent actions in *Acetabularia mediterranea. Planta, 157:* 1–7.

GOODWIN, B.C., MURRAY, J.D. & BALDWIN, D., 1985. Calcium: the elusive morphogen? In *Acetabularia 1984,* S. Bonnotto, F. Cinelli & R. Billiau (eds)

GOODWIN, B.C., BRIÈRE, C. & O'SHEA, P.S., 1987. Mechanisms underlying the formation of spatial structure in cells. In R.K. Poole & A.P.T. Trinci (eds), *Spatial Organization in Eukaryotic Microbes,* pp. 1–9. Oxford: IRL Press.

GOODWIN, B.C.. KAUFFMAN, S.A. & MURRAY, J.D., 1993. Is morphogenesis and intrinsically robust process? *Journal of Theoretical Biology 163:* 123–148.

GREEN, P.B., 1987. Inheritance of pattern: analysis from phenotype to gene, *American Zoology, 27:* 657–673.

GREEN, P.B. 1989. Shoot morphogenesis, vegetative through floral, from a biophysical perspective. In E. Lord & C. Barner (eds), *Plant Reproduction: From Floral Induction to Pollination,* Am. Soc. Plant Physiol. Symp. Series, vol. 1, pp. 58–75.

GREEN, P.B., ERICKSON, R.O. & BUGGY, J., 1971. Metabolic and physical control of cell elongation rate. In vitro studies in *Nitella, Plant Physiology, 47:* 423–430.

HARRISON, L.G. & HILLIER, N.A., 1985. Quantitative control of *Acetabularia* morphogenesis by extracellular calcium: a test of kinetic theory. *Journal of Theoretical Biology, 114:* 177–192.
HARRISON, L.G., GRAHAM, K.T. & LAKOWSKI, B.C., 1988. Calcium localisation during *Acetabularia* whorl formation: evidence supporting a two-stage hierarchical mechanism. *Development, 104:* 255–262.
HEGGE, H., 1987. Theory of science in the light of Goethe's science of nature. In F. Amrine, F.J. Zucked & H. Wheeler (eds), *Goethe and the Sciences: a Reappraisal,* pp. 195–218. Boston, USA: D. Reidel.
HO. M.W., 1990. An exercise in rational taxonomy. *Journal of Theoretical Biology, 147:* 43–57.
MEINHARDT, H., 1982. *Models of Biological Pattern Formation.* London, UK: Academic Press.
MENZEL, D. & ELSNER-MENZEL, C., 1989. Induction of actin-based contraction in the siphonous green alga *Acetabularia (Chlorophycea)* by locally restricted calcium influx. *Botanica Acta, 102:* 164–171.
MENZEL, D., JONITZ, H. & ELSNER-MENZEL, C., 1992. The cytoskeleton in the life cycle of *Acetabularia* and other related species of dasyclad green algae. In D. Menzel (ed.), *The Cytoskeleton of the Algae,* ch. 11. Boca Raton, FL, USA: CRC Press.
MITTENTHAL, J.E. & BASKIN, A.B., 1992. *The Principles of Organization in Organisms,* vol. XIII, Santa Fe Institute Studies in the Sciences of Complexity. Reading, MA, USA: Addison-Wesley.
MURRAY, J.D., 1989. *Mathematical Biology.* Berlin: Springer-Verlag.
NEWMAN, S.A. & COMPER, W.D., 1990. 'Generic' physical mechanisms of morphogenesis and pattern formation. *Development, 110:* 1–18.
NOSSAL, R., 1988. On the elasticity of cytoskeletal networks. *Biophysics Journal, 53:* 349–359.
OSTER, G.F. & ODELL, G.M., 1983. The mechanochemistry of cytogels. In A.E. Bishop (ed.), *Fronts, Interfaces and Patterns,* Amsterdam, The Netherlands: North Holland, Elsevier Science Division.
OSTER, G.F., SHUBIN, N., MURRAY, J.D. & ALBERCH, P., 1988. Evolution and morphogenetic rules: the shape of the vertebrate limb in ontogeny and phylogeny. *Evolution, 42:* 862–884.
RAUP, D.M. & MICHELSON, A., 1965. Theoretical morphology of the coiled shell. *Science, 147:* 1294–1295.
SHELANSKI, M.L., 1989. Intracellular ionic calcium and the cytoskeleton in living cells. *Annals of the New York Academy of Sciences, 568:* 121.
SHUBIN, N.H. & ALBERCH, P., 1986. A morphogenetic approach to the origin and basic organization of the tetrapod limb. *Evolutionary Biology, 20:* 319–387.
STROHMAN, R.C., 1992. Ancient genomes, wise bodies, unhealthy people: limits of a genetic paradigm in biology and medicine. Wellness Lecture Series, UC Berkeley, U.S.A.
THOMPSON, D'ARCY W., 1917. *On Growth and Form.* Cambridge, UK: Cambridge University Press.
TURING, A.M., 1952. The chemical basis of morphogenesis. *Philosophical Transactions of the Royal Society, B, 237:* 37–72.
WILLIAMSON, R.E., 1984. Calcium and the plant cytoskeleton. *Plant Cell Environment, 7:* 431.

CHAPTER

2

The origins of shape and form in diatoms: the interplay between morphogenetic studies and systematics

DAVID G. MANN

CONTENTS

Abstract

The silicified cell walls of diatoms exhibit an enormous variety of shape and form, which demands explanation in terms of adaptation, evolution and morphogenesis. This chapter examines the ontogeny of diatom shape and form, and the extent to which shape and form are inherent, inherited, produced *de novo* or imposed. Correct interpretation of morphological variation is critical in systematic and evolutionary studies and hence, given the importance of diatoms in aquatic ecosystems world-wide and the enormous number of diatom species, there is an urgent need for a better understanding of diatom morphogenesis, which can only be achieved by a multidisciplinary approach.

Shape and Form in Plants and Fungi
ISBN 0-12-371035-9

INTRODUCTION

The diatoms are an extremely abundant and diverse group of aquatic and semiterrestrial plants, characterized by a unique type of cell wall impregnated with silica. Ecologically and geochemically diatoms are of immense significance, because of the part they play in the silicon and carbon cycles; it has been estimated, for instance, that they may account for 20–25% of global primary production (Werner, 1977). Fossil diatoms are used as stratigraphic markers for tertiary sediments, both marine and freshwater, while their importance for palaeoecological reconstruction (either using indicator species or transfer functions calculated from assemblages) grows year by year (Battarbee *et al.*, 1990; Davis *et al.*, 1985; Fritz *et al.*, 1991). In cell biology and biochemistry, diatoms have provided valuable experimental systems for studying mitosis and cell division (Pickett-Heaps, 1991) and silicon metabolism (Volcani, 1978).

For all these different types of study, taxonomy, that 'most pervasive of all branches of biology' (NERC, 1992), is essential. And taxonomy depends to a considerable extent on an analysis and understanding of shape and form.

Altogether, there are probably between 10^5 and 10^6 species of diatoms, of which only a tenth or less have yet been described. These numbers are guesses, based on the number of species recognized at present, the ease with which new species can still be discovered using traditional criteria for distinguishing species in diatoms, and the few in-depth studies of variation patterns in relation to breeding, which suggest that the traditional species concept in diatoms is far too broad (Mann, 1989a).

Classifications of diatoms produced between 1850 and 1970 relied principally on the shape and form, symmetry and patterning of the cell wall. Diatom genera and higher taxa were often separated by differences in the overall shape and symmetry of the cell wall, while species distinctions were generally based on differences in the arrangement and spacing of ribs, pores and other wall ornamentation. Since scanning electron microscopes became widely available, it has been possible to study the cell wall in ever greater detail. The extra data gained, together with information concerning sexual reproduction and the structure of the protoplast, such as the configuration of the plastids and nucleus and how this changes during the cell cycle (e.g. Mann, 1989b; Mann & Stickle, 1991, 1993), have shown that the traditional classifications are quite good, but not good enough. In particular, contradictions have been found between the relationships suggested by shape and symmetry on the one hand, and by the detailed structure of the cell wall and protoplast on the other. Thus, for example, *Cymbella* Ag. and *Amphora* Ehrenb. both have elongate valves that are asymmetrical about the apical plane (Figs 5, 6); both have often, therefore, been included in the same family or subfamily (e.g. in the Cymbellaceae by Patrick & Reimer, 1975; in the Gomphocymbelloideae by Hustedt, 1930) but separated from bilaterally symmetrical diatoms, i.e. diatoms with 'naviculoid symmetry' (Figs 2, 7, 8) such as *Navicula* Bory *sensu lato* (placed in the Naviculaceae by Patrick & Reimer, 1975, and in the Naviculoideae by Hustedt, 1930). Recent investigations, however, have shown that *Navicula* is very heterogeneous, that some groups previously included within *Navicula*, such as *Placoneis* Mereschk. (Fig. 7) have a very similar valve structure to *Cymbella* and its close ally *Encyonema* Kütz. (Fig. 5), and that *Amphora* contains some species (Fig. 6) that, except in their symmetry, are very similar indeed to *Navicula sensu stricto* (Fig. 8). As a result, in a recent reclassification of the diatom genera (Round *et al.*, 1990), diatoms of widely different symmetries are often classified together in

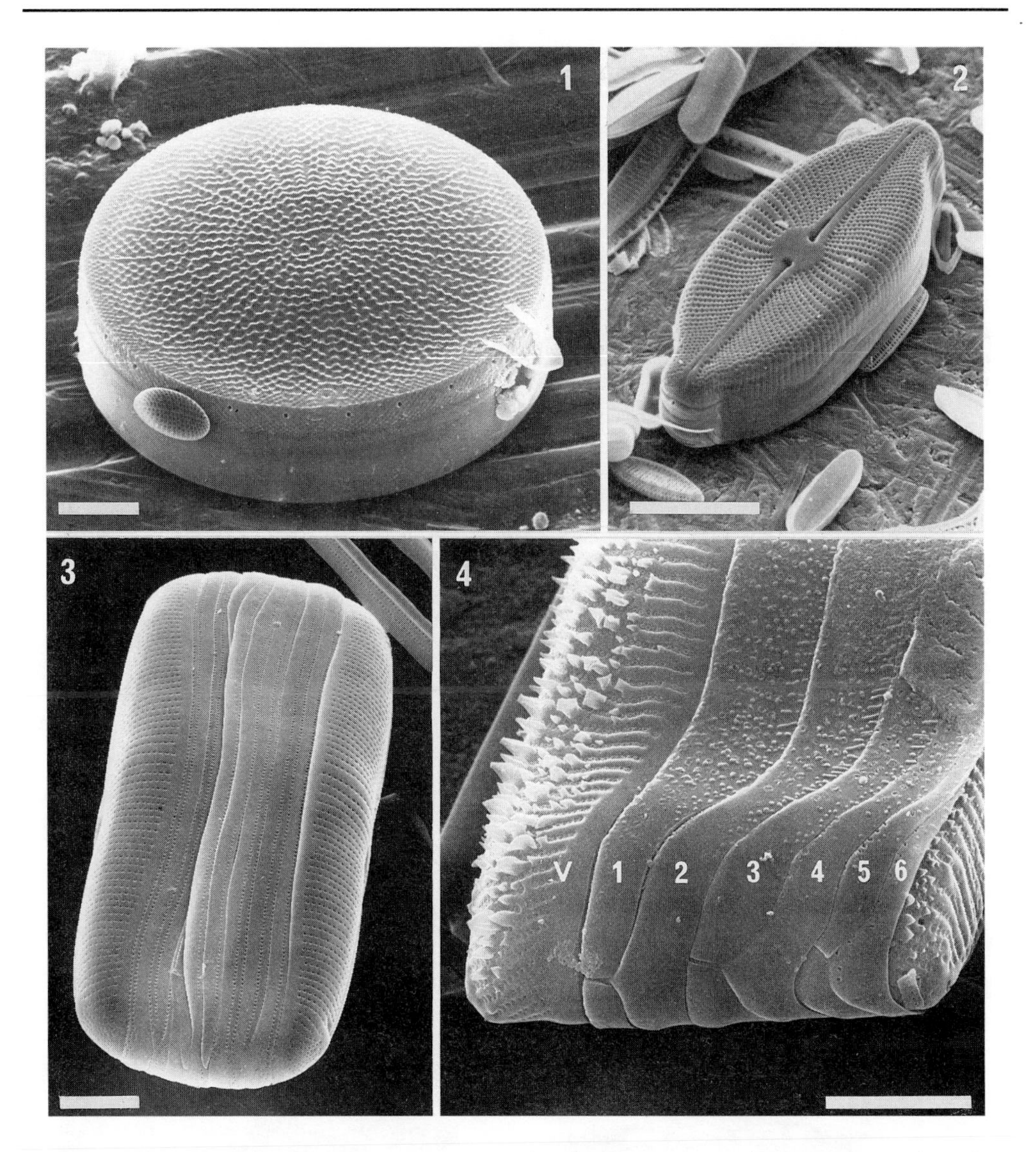

Figures 1–4 Diatom frustules, SEM. **1** *Actinocyclus* Ehrenb.: a centric diatom with circular valves and a shallow girdle. Scale bar 20 μm. **2** *Cosmioneis* Mann & Stickle: an elongate pennate diatom, with a pair of raphe slits along the centre of each valve. Scale bar 10 μm. **3** *Parlibellus* Cox: side (girdle) view, showing the girdle bands lying between the two valves (left and right). Scale bar 5 μm. **4** *Eunotia* Ehrenb.: the end of a frustule. The epitheca lies to the left, consisting of a valve (V) and six girdle bands (1–6), which are all split rings; the open ends of bands 1, 3 and 5 are visible, while those of bands 2, 4 and 6 lie at the opposite end of the frustule (not shown). Scale bar 5 μm.

the same family, while diatoms with similar shapes or symmetries are sometimes placed in different orders. The implication is clearly that shape and symmetry have been subject to a great deal of convergent and parallel evolution.

There have been some attempts to account for this parallelism in terms of adaptation to habitat (Medlin, 1991). However, while there is a clear correlation between certain types of cell shape and the mode of life of the diatom – heteropolar, wedge-

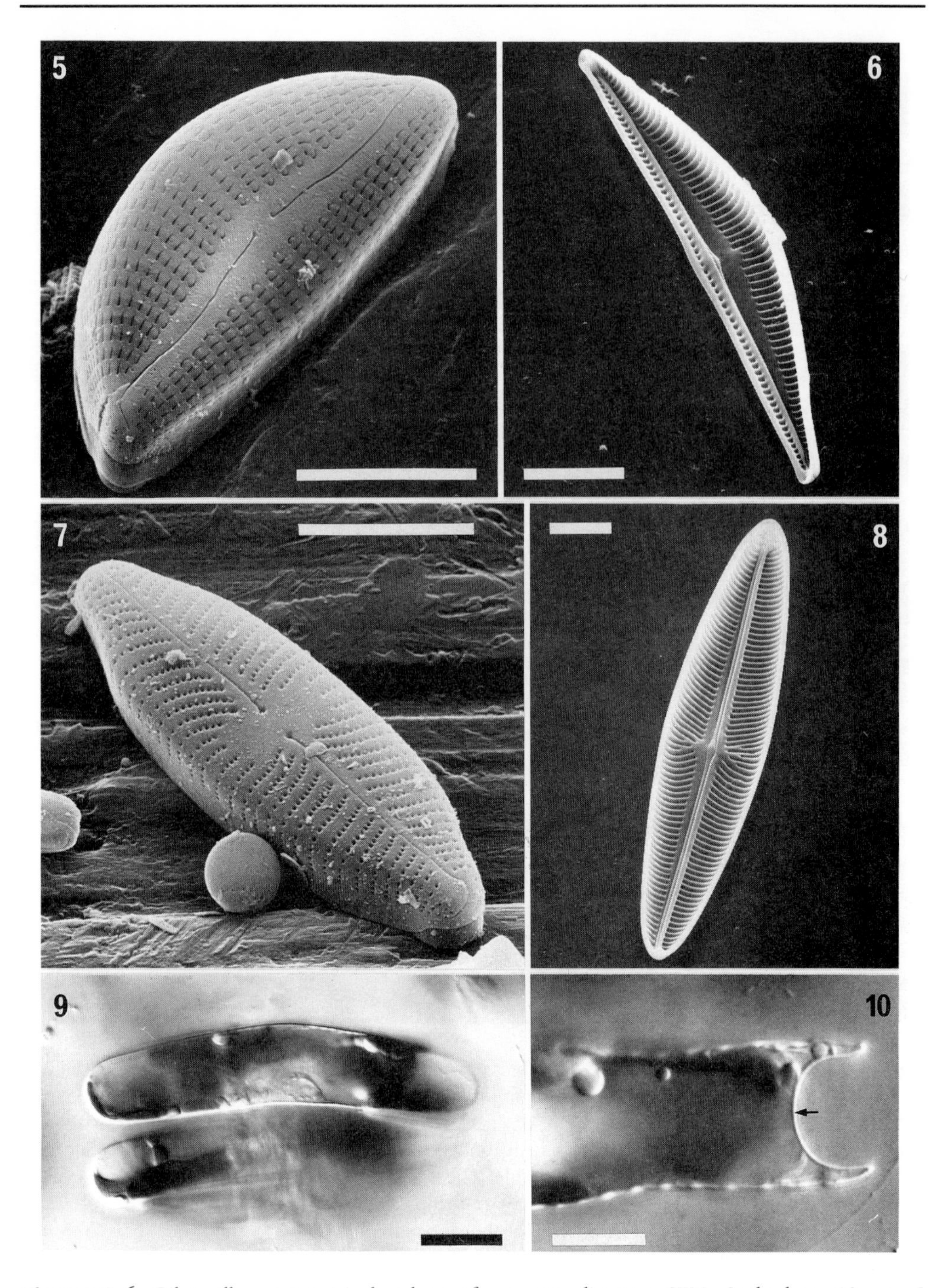

Figures 5, 6 Bilaterally asymmetrical valves of pennate diatoms, SEM. Scale bars 10 μm. **5** *Encyonema* (formerly within *Cymbella*), exterior. **6** *Seminavis* D.G. Mann (formerly within *Amphora*), interior. **Figures 7, 8** Symmetrical valves, SEM. Scale bars 10 μm. **7** *Placoneis* (formerly within *Navicula*), exterior. **8** *Navicula*, interior. **Figure 9** Curved auxospore of *Amphora*, LM; asymmetrical valves like those shown in Figs 5 & 6 can be formed in such auxospores (see text and Figs 29, 30). Scale bar 20 μm. **Figure 10** Plasmolysed auxospore of *Caloneis ventricosa*, showing inversion of the soft tip of the auxospore (arrow; compare Figs 9, 26–28). Scale bar 10 μm.

shaped cells, for example, are characteristic of diatoms that live attached to rocks, plants or other substrata by a mucilage stalk secreted from the narrower end of the cell – so far, there have been few or no rigorous demonstrations of the selective value of a particular shape or symmetry. As so often happens with attempts to distinguish evolutionary or ecological 'strategies', the plausibility of the argument that a particular shape is an adaptation to a particular environment is undermined by the fact that other shapes also occur in the same environment and seem to be equally adaptive. Very little is known about the subtleties of niche separation and the adaptive trade-offs that enable the coexistence of different species with different morphologies.

At present, although it is by no means completely futile to speculate about the functional significance of diatom shapes and symmetries, it is more instructive to look at how the shapes and symmetries are produced during ontogeny. This provides a sound basis for understanding the variation pattern, which is a prerequisite for good classification and phylogenetic reconstruction. Ontogenetic studies help in two main ways. First, they help determine which shapes and structures are truly homologous and which are only superficially alike, by providing a more detailed basis for comparison. Second, they show what developmental changes would be necessary for the evolution of one morphology from another.

THE STRUCTURE OF THE DIATOM CELL WALL

Each diatom is completely surrounded by a wall, consisting of organic components and silica, beneath which is the plasmalemma (Fig. 14).

The diatom cell wall consists of several siliceous elements, coated by organic material, and often also a discrete layer of polysaccharide, the diatotepic layer, which lies between the siliceous elements and the plasmalemma (von Stosch, 1981). The silica part of each cell wall, i.e. the part that survives treatment with the strong acids or other oxidizing agents used to clean diatoms for microscopy, is called the frustule. This is composed of two overlapping halves or thecae (Fig. 14), one of which (the epitheca) is slightly larger than the other (the hypotheca). Each theca consists of a large end piece, called a valve (Figs 1, 2) and a number of smaller strips or hoops of silica, called girdle bands (Figs 3, 4). Within a theca, the valve overlaps the first girdle band and the girdle bands overlap each other, from the valve outwards (Figs 3, 4, 14).

Immediately after mitosis, the plasmalemma usually invaginates to form a cleavage furrow, which brings about cytokinesis, so that there are two protoplasts within the parental cell wall (Fig. 14). A new valve is then produced within each daughter cell beneath the plasmalemma of the cleavage furrow. All the frustule elements are formed within the cell in special silica deposition vesicles, which are always intimately associated with the plasmalemma (e.g. Fig. 14). Once each element is complete, it is secreted from the cell through exocytosis, and integrated into the cell wall. New valves are formed immediately after cytokinesis, while girdle bands are formed at various times during the cell cycle, depending upon the species. The addition of new girdle bands allows the two overlapping thecae to move apart to accommodate growth, which is thus unidirectional, as in the fission yeast *Schizosaccharomyces* Lindner (see Mitchison, 1971); growth can be continuous throughout the cell cycle or may take place in one to several discrete steps (Olson *et al.*, 1986).

The valves (and often the girdle bands too) are porous and in most cases are constructed from ribs, between which are rows of pores (Figs 1–8, 11–13, 42–45). Valve formation is sequential, the ribs extending out from a circular or elongate pattern centre (Mann, 1984a; Pickett-Heaps *et al.*, 1990). The simple rib and pore structure can be obscured or modified, however, by silica added late in the development of the valve. Thus, in some diatoms, systems of hexagonal or rectangular chambers are superimposed on the basic rib framework, while in others the spaces between the ribs are filled in to form areas of plain silica.

SETTING THE PROBLEM

Figures 11–13 show three frustules, all of which would traditionally be classified as belonging to *Sellaphora* (= *Navicula*) *pupula* (Kütz.) Mereschk. The frustules are elongate and more or less symmetrical side to side and end to end. Each valve is flat-topped with downturned sides and has a central raphe system containing two long slits, which are involved in cell motility. Extending out from the raphe system are transverse ribs and lines of pores (striae), although in the centre there is a broad, plain area, where the ribs and pores are restricted to the edge of the valve. The patterns of ribs and pores differ subtly among the three frustules – for example, the striae are almost parallel in Fig. 12 and slightly more radial in Fig. 11. However, the most obvious difference between the frustules is in their outline: one is lanceolate with broadly rostrate ends (Fig. 11), another is linear-lanceolate (Fig. 12), while the

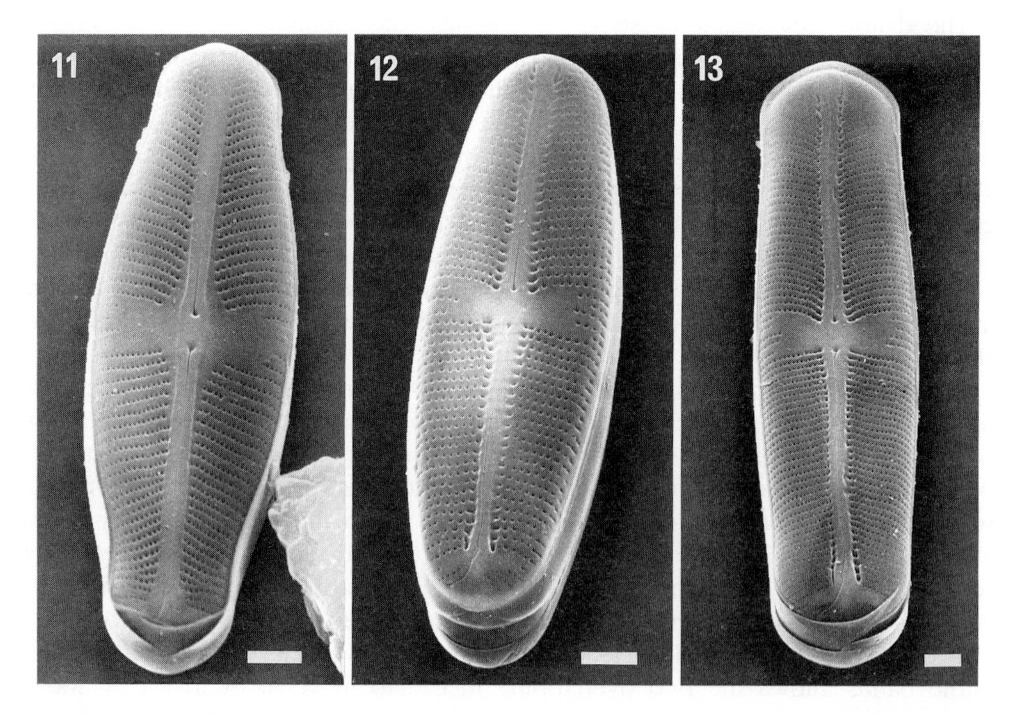

Figures 11–13 *Sellaphora pupula*, SEM: three different frustules, with subtly different striation patterns and outlines, representing three reproductively isolated populations (gamodemes). Scale bars 2 μm.

third is almost linear (Fig. 13). And the poles of the valve are rounded in Fig. 12, but slightly acute in Figs 11 and 13.

The fact that all three frustules would usually be identified as belonging to the same species (following e.g. Hustedt, 1930; Krammer & Lange-Bertalot, 1986) implies, of course, that the differences between them have little significance, but this is not so. Observations similar to those reported by Mann (1984b, 1989a) show that the three frustules belong to three different reproductively isolated gamodemes – effectively biological species. Each gamodeme varies in shape and size during its life cycle, but the multidimensional pattern of variation does not overlap the variation shown by either of the other gamodemes. There are also many other gamodemes within *Sellaphora pupula*, each again with its own distinct combination of shape, size and pattern.

Thus, the problems to be solved are difficult indeed. Not only is it necessary to understand how some diatoms come to be elongate (Fig. 2) rather than round (Fig. 1), some bilaterally symmetrical (Figs 7, 8) but others asymmetrical (Figs 5, 6), some flat-topped (Figs 42, 44) but others domed or undulate (Figs 43, 45); it also has to be explained how very subtle differences in outline are produced and maintained. In addition, sophisticated means must be found to describe, analyse and document the differences, although these will not be discussed here.

THE GENESIS OF SHAPE AND FORM: SOME GENERAL OBSERVATIONS

Before considering diatoms in particular, it is worth considering how shape and form are produced in biological systems in general. Shape and form can be inherent to biological material; inherited; produced *de novo* as a result of developmental processes controlled more or less directly by a genetic programme; or imposed by other cells, other organisms, or the environment.

Inherent shape and form are shown by the crystals or crystallites produced by various organisms. The coccolithophorids (a group of unicellular algae belonging to the division Haptophyta = Prymnesiophyta) synthesize organic scales that become calcified through the deposition of calcite or aragonite. Plates and rods of calcium carbonate are produced with a crystalline structure (e.g. see Manton & Oates, 1975; Manton *et al.*, 1976) and the forms they take are undoubtedly related to their chemical composition. Within cells too there are structures that appear to owe their form to a crystalline structure. The pyrenoids of algal cells, now convincingly shown to be a metabolic compartment of the chloroplast responsible for carbon fixation and containing RuBisCO (ribulose bisphosphate carboxylase-oxidase) and RuBisCO activase (McKay & Gibbs, 1991; McKay *et al.*, 1991), sometimes adopt regular geometrical forms, apparently because the pyrenoid matrix is crystalline (Holdsworth, 1968; Taylor, 1972). Thus, in the diatom genus *Sellaphora* Mereschk. the pyrenoids are often tetrahedral (Mann, 1984c, 1989b). Diatom cell walls, on the other hand, do not owe their shape and form to the inherent properties of silica; unlike the calcium carbonate deposited by coccolithophorids, diatom silica is not crystalline, except perhaps within submicroscopic domains (Pickett-Heaps *et al.*, 1990).

In most unicellular organisms, shape and form seem to be generated *de novo*, as part of an organized programme of cytomorphogenesis involving the cytoskeleton and/or wall synthesis. Usually this process is not much affected by other cells, since morphogenesis is completed when the cells have already separated, after cell division. Thus, placoderm desmids such as *Micrasterias* Ag. produce their new, often elaborately lobed semicells with only minimal contact between the daughter cells; the final morphology is built up by controlled softening and synthesis of wall material in each daughter cell, almost completely independently of what happens in the other (Kallio & Lehtonen, 1981; Kiermayer, 1981). Diatoms are quite different.

PHYSICAL INHERITANCE OF SHAPE IN DIATOMS

Diatoms have the remarkable property that shape is directly inherited; the shape of one cell physically determines the shape of its progeny, within certain limits that will be discussed. When a diatom cell divides mitotically, the new valves and girdle bands are produced inside the parental cell wall. Thus, since the new valves and girdle bands form directly beneath the plasmalemma of the newly cleaved cell (Fig. 14), their outline must conform to the internal contours of the girdle, which is basically a cylinder capped by the two valves. So, the immediate answer to the question, why does a particular frustule have a particular shape, is that its parent had that shape.

But this answer is not complete, since the new valves almost always differ in shape from the parental valves (Figs 19, 21, 23), albeit slightly, except in the special case that both are circular (Figs 1, 42, 43). This difference is not random but follows simple rules (Geitler, 1932): the new valves are more squat and more simply rounded. During vegetative growth and division, populations of diatoms with long, parallel-sided valves give rise to populations with short, elliptical valves (Figs 19, 21, 23). Rostrate or capitate apices tend to be lost (Round *et al.*, 1990, fig. 54).

There is no evidence to suggest that this involves any action on the part of the cell. A tendency towards the spherical is inevitable in any turgid cell with a homogeneous, flexible cell wall (Thompson, 1942). The girdle bands (or at least those bands beneath which the new valves form) are arranged so as to produce a more or less homogeneous cylinder and are generally thin and flexible. Furthermore, since they are often open at one end (split rings: Fig. 4), it is likely that the girdle can also stretch slightly, through shear between one girdle band and the next, although individual bands appear to be almost inextensible. The girdle can therefore be deformed by excess internal pressure and will tend to adopt the shape giving minimum surface area for the volume enclosed.

It is particularly interesting to examine the shape changes that occur during the life cycle in diatoms like *Gomphonema* Ehrenb. and *Rhoicosphenia* Grun. (Fig. 21), where the girdle is not a right cylinder (i.e. the ends, the valves, are not at right angles to the axis of the cylinder): at the 'head pole' the girdle is much wider than at the 'foot pole', making the cells wedge-shaped in side view (Fig. 20A). Because it is wider at one pole than the other, the girdle will not be deformed equally at the two poles by turgor pressure, but will expand outwards more near the head pole (Fig. 20B; contrast Fig. 18B). The point of maximum width thus moves a little closer to the head pole (relative to the centre) with each cell division (Fig. 21 and see also Geitler, 1932, fig. 19; Hohn, 1959; Kociolek & Stoermer, 1988; Mann, 1982a). In *Rh. curvata* (Kütz.) Grun. and some species of *Gomphonema* the valves are initially almost isopolar (Hohn, 1959; Mann, 1982a) but become distinctly heteropolar with time (Fig. 21).

Similarly, a difference in the width of the girdle between the two sides of a frustule

(Fig. 22A) will lead to unequal distortion of the girdle (Fig. 22B), and hence to the development or accentuation of lateral asymmetry in the valves.

The extent of shape change will clearly depend on the pressure differential across the wall, the particular characteristics of the girdle in the diatom being considered and the exact position of the new valves within the girdle when they are formed (see also Crawford, 1981). Thus the more flexible the girdle, or the greater the pressure

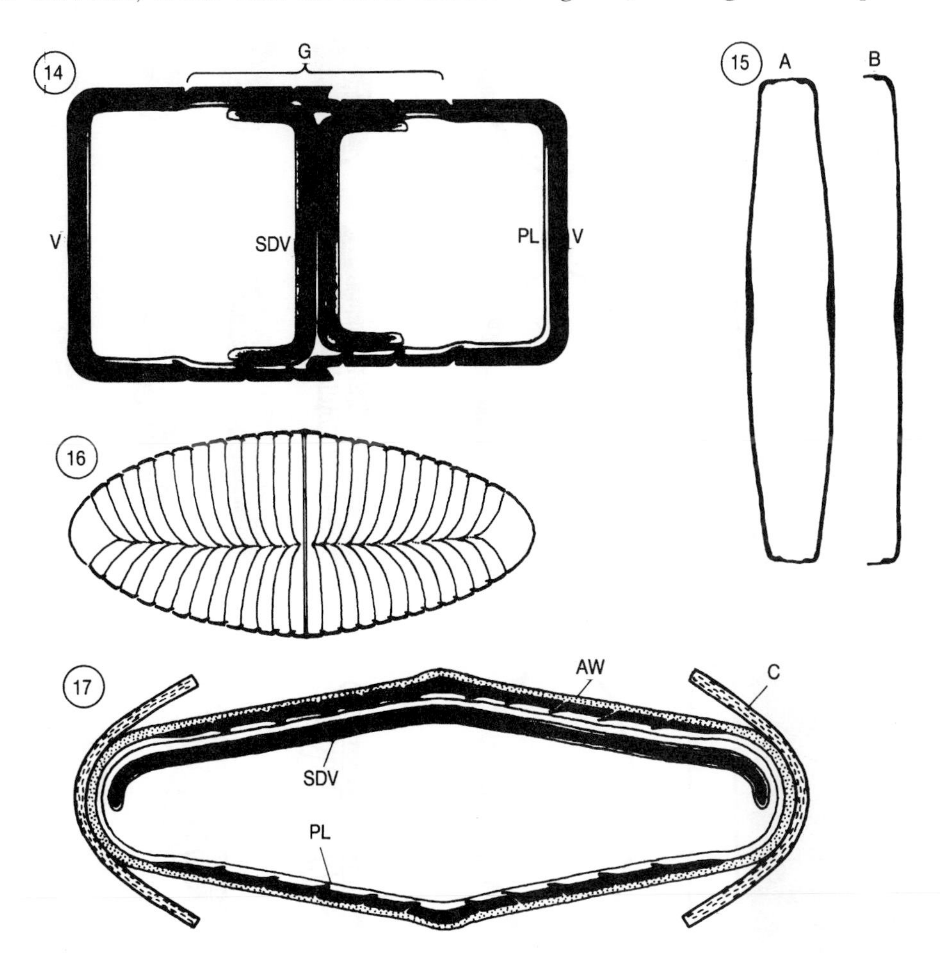

Figure 14 Diagrammatic section through a recently divided diatom cell. At the ends are two valves (V). Between is the girdle (G), consisting here of two overlapping sets of three girdle bands, which appear in section, top and bottom. Beneath the wall is the plasmalemma (PL). In the centre, new valves are being formed within the daughter cells, in silica deposition vesicles (SDV).
Figure 15 *Craticula cuspidata*, girdle view. A. Frustules of a live, turgid cell, with arched valves. B. An isolated valve, with flat top (A & B are about 100 μm long).
Figure 16 The auxospore casing of *Lyrella*, showing the perizonial bands. The central band is a complete hoop, whereas the others are split rings, their ends aligned to form a suture.
Figure 17 Diagrammatic section of an expanded auxospore. At the tips are two organic caps (C), representing the remains of the zygote wall. Beneath these is the auxospore wall, consisting of organic layers (stippled) and a series of perizonial bands (black, in section), which girdle the auxospore like the girdle bands of vegetative cells (Fig. 14), except that they overlap from the centre outwards. Beneath the auxospore wall is the plasmalemma (PL); within the auxospore the first initial valve is being formed within a silica deposition vesicle (SDV).

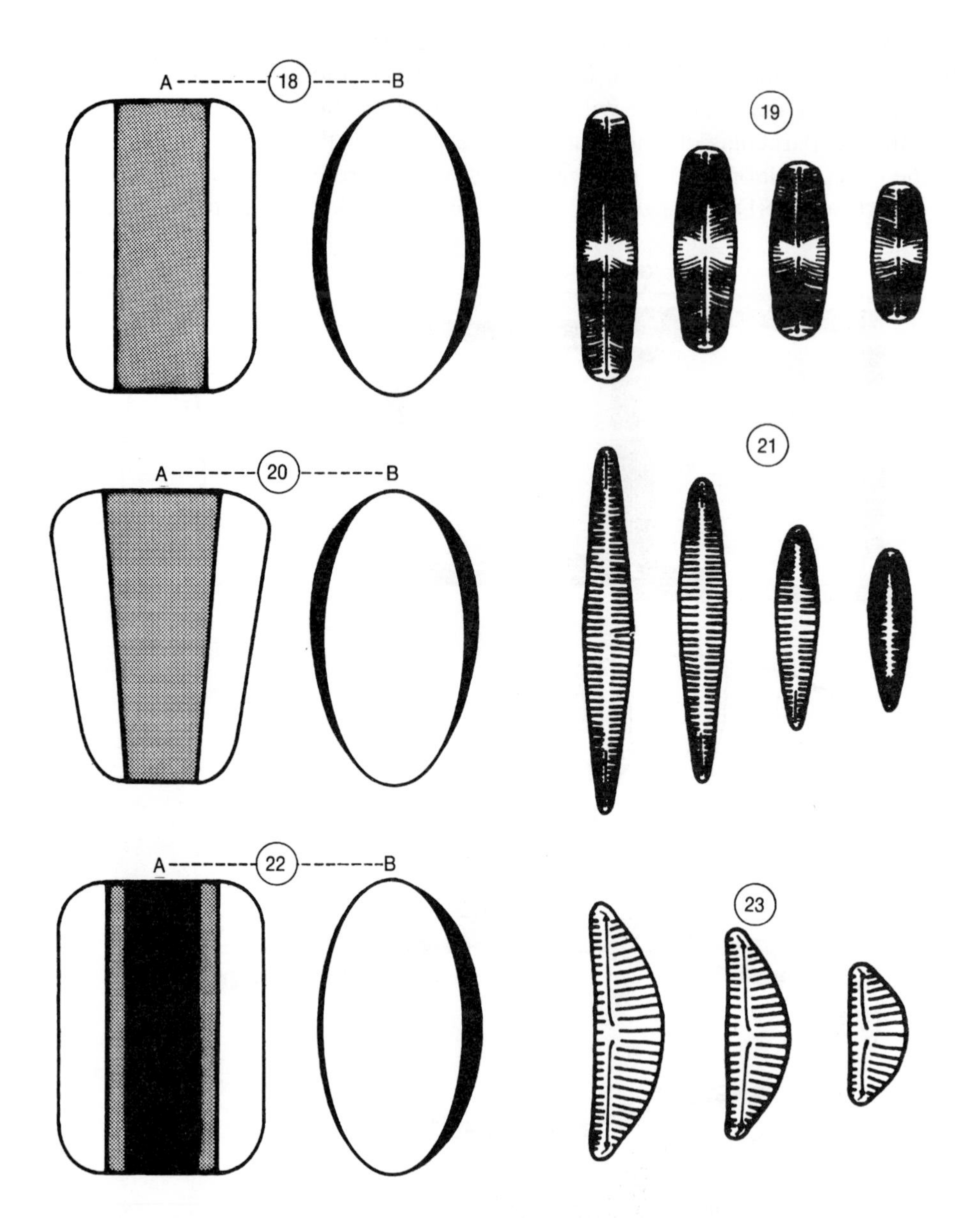

Figures 18–23 Distortion of the girdle and shape change in elongate diatoms. 18A, 20A & 22A show cells in girdle view, with the girdle shaded. 18B, 20B & 22B show the same cells in valve view; the inner ellipse represents the valve itself, while the black shape indicates (and exaggerates) the outward distortion of the girdle caused by a turgid cell. Figs 19, 21 & 23 show examples of shape change in diatoms with girdle structures as in Figs 18, 20 & 22. **18** Girdle a right cylinder, undergoing symmetrical, central distortion. **19** Shape change in *Sellaphora*: valves become shorter and more squat but maintain the same symmetry. **20** Girdle wider at one end than the other, undergoing greater distortion near its wider end. **21** Shape change in *Rhoicosphenia*: the widest point of the valve moves closer to one pole as size decreases. **22** Girdle wider on one side (lighter shading) than the other and therefore undergoing asymmetrical lateral distortion. **23** Shape change in *Encyonema*: asymmetry becomes exaggerated as size decreases.

differential, the more the shape can be expected to change per cell division. Likewise, the longer the girdle cylinder and the further away from the parental valves that the new valves and girdle bands are formed, the greater the potential for differences in shape between parent and offspring (see Crawford, 1981).

The amorphous silica of the diatom frustule is probably more flexible than is often assumed. Thus, for example, the valves of *Craticula cuspidata* (Kütz.) D.G. Mann (a freshwater diatom) are flat when first formed during cell division, and they return to being flat if the cell is killed or if individual valves are isolated after cleaning with oxidizing acids (Figs 15B, 44; Mann & Stickle, 1991), but they are strongly arched in living cells, as a result of turgor pressure (Fig. 15A). Measurements by Übeleis (1957) indicate an excess internal hydrostatic pressure of 0.4–0.8 MPa or more in this species.

As yet, there appear to be no quantitative data on the flexibility and extensibility of the diatom cell wall. In time, however, it should be possible to model shape change in diatoms from a·knowledge of the thickness and characteristics of the various parts of the cell wall and the pressure differential between cell and environment. Meanwhile, a qualitative understanding of how changes in shape are brought about is important if the taxonomist is to be able to interpret variation patterns correctly. Fortunately, the shape–size trajectories of closely related diatoms do not often intersect, so that it is possible to identify the valves belonging to each race or species, regardless of which stage in the life cycle is observed.

Taxonomists should remember, however, that any alteration in the pressure differential across the cell wall, as might occur for instance in a habitat subject to marked changes in salinity, may affect the course of shape change and bring about long-term effects, even if conditions later return to their initial states. Slight but consistent differences between the shapes exhibited by two allopatric diatom populations, or by two sympatric populations, may reflect genotypic differentiation, as in Figs 11–13, but do not guarantee it.

In addition, irregularities in cell division, whether caused by osmotic stress, damage to the frustule by parasites or grazers, or some failure of the cytokinetic machinery, can produce effects on shape that may remain detectable for many generations. Some diatoms seem to be more susceptible to this than others. The cells of *Cocconeis* Ehrenb., for example, usually have elliptical valves, but not uncommonly they develop indentations, which can be useful for detecting cell lineages *in vivo*, on the leaves of aquatic angiosperms or bryophytes where *Cocconeis* often lives.

But the manner in which diatom cells divide can also act as a homoeostatic mechanism tending to restore cell shape after perturbation. For instance, in a diatom where the girdle is similar in structure and depth on either side of the frustule (Fig. 18), the stresses exerted on the girdle by the turgid cell within will always tend to correct the shape towards bilateral symmetry. This must surely be how, in genera like *Meridion* Ag. and *Synedra* Ehrenb. (Geitler, 1940a, b), the irregular and sometimes bizarre cells produced immediately after sexual reproduction give rise to symmetrical cells.

Even though there is some flexibility (and very limited extensibility) in the diatom cell wall, the mechanism of cell division in diatoms (involving the formation of new valves and girdle bands within the confines of the parent cell wall) usually causes an absolute decline in the mean cross-sectional area of the girdle cylinder within each diatom population (Figs 19, 21, 23); at each cell division, one daughter cell is almost exactly the same size as the parent, while the other is slightly smaller (Fig. 14). The process of size diminution continues for months or years in nature (Jewson, 1992;

Mann, 1988), but is eventually interrupted by a growth phase, and it is this that must be considered next.

THE GENERATION OF SHAPE *DE NOVO* IN DIATOMS DURING AUXOSPORULATION

Size is restored in diatoms by a process called auxosporulation, which is usually accompanied by meiosis and sexual reproduction (diatoms are diplonts). During auxosporulation the old cell walls are sloughed off and an auxospore (usually a zygote) is produced, which expands and then constructs a new frustule with two initial valves (Figs 17, 26, 32–37). The details of how the auxospore is produced vary greatly but need not be considered here. What is important is that auxosporulation is a process where shape and symmetry are generated *de novo*. Initially, the auxospore is spherical or ellipsoidal (Figs 24, 26A). Sometimes the auxospore wall is more or less homogeneous in its properties and expansion is isometric, in which case the initial valves will be circular (Round *et al.*, 1990, fig. 63). In many diatoms, however, expansion is not isometric: parts of the auxospore wall become selectively hardened during expansion, through the incorporation of strips or bands of silica (Figs 16, 17, 25, 27, 28, 35, 36; Round *et al.*, 1990), and expansion is thus restricted to particular areas of the auxospore. Shape is built up sequentially, through a form of tip growth (Figs 24, 26–28).

Figures 24–37 show how different shapes are produced in a variety of bipolar diatoms. In each case the auxospore is initially spherical or ellipsoidal, with an organic wall. The first step in expansion in the examples shown (although this is not true of all diatoms) is the formation of a new organic wall beneath the first, the rupture of the first organic wall (which usually persists as caps over the auxospore poles), and the formation of a silica band (the primary transverse perizonial band), which encircles the equator of the auxospore (Figs 16, 17, 25, 27–29, 35). The expanding auxospore remains completely surrounded by its own organic wall but its tips remain plastic and are driven outwards by the hydrostatic pressure within the auxospore; they can be inverted by strong plasmolysis (Fig. 10). As the auxospore grows, more

---

Figures 24, 25 *Neidium* auxospores (the auxospore in Fig. 25 was about 90 μm long). **24** Auxospore expansion: outline traced at 00.00, 23.25, 24.22, 26.34 (dashed) and 29.17 h (solid). **25** Section through auxospore: the perizonial bands (in section and dashed) are formed just within the margins of the silicified caps (C), so that all bands have the same diameter and the expanded auxospore is parallel-sided.

Figures 26–29 *Craticula*. **26** Spherical zygote (A; diameter about 35 μm) and partly expanded auxospore, with organic caps (C); two plastids (stippled) and two nuclei (dashed circles) are present in each auxospore. **27** Expanding auxospore: the soft, deformable apical dome of the auxospore is shown at two times (pale-shaded semicircles); the darker-shaded region indicates the section of new auxospore created through tip growth and subsequent hardening, through the formation of perizonial bands (shown in section and dashed or dotted). **28** Partially expanded, tapering auxospore, showing the perizonium, together with the organic caps over the ends of the auxospore (see Fig. 26B). **29** Curved auxospore (e.g. of *Amphora*).

Figures 30, 31 Curved auxospores with the positions of the initial valves indicated by shading and in section (below). **30** Initial valves formed on either side of the plane of curvature. **31** Initial valves produced at right angles to the plane of curvature.

perizonial bands are added on either side of the primary bands (Figs 17, 27, 28, 35, 36; see also Mann, 1982b; Round *et al.*, 1990; von Stosch, 1982), each overlapping the next (Fig. 17); the structure of the perizonium is thus like that of the vegetative cell (Fig. 14), but reversed. Depending on the rate of expansion relative to the addition of new bands, the expanded auxospore may be parallel-sided (*Pinnularia* Ehrenb.: Fig. 32), lanceolate (*Petroneis* Stickle & Mann, *Lyrella* Karayeva: Figs 16, 34; Mann & Stickle, 1993) or rhombic (*Stauroneis* Ehrenb.: Fig. 33, or *Craticula* Grun.: Mann & Stickle, 1991); it may have a swollen central portion (*Caloneis ventricosa* (Ehrenb.)

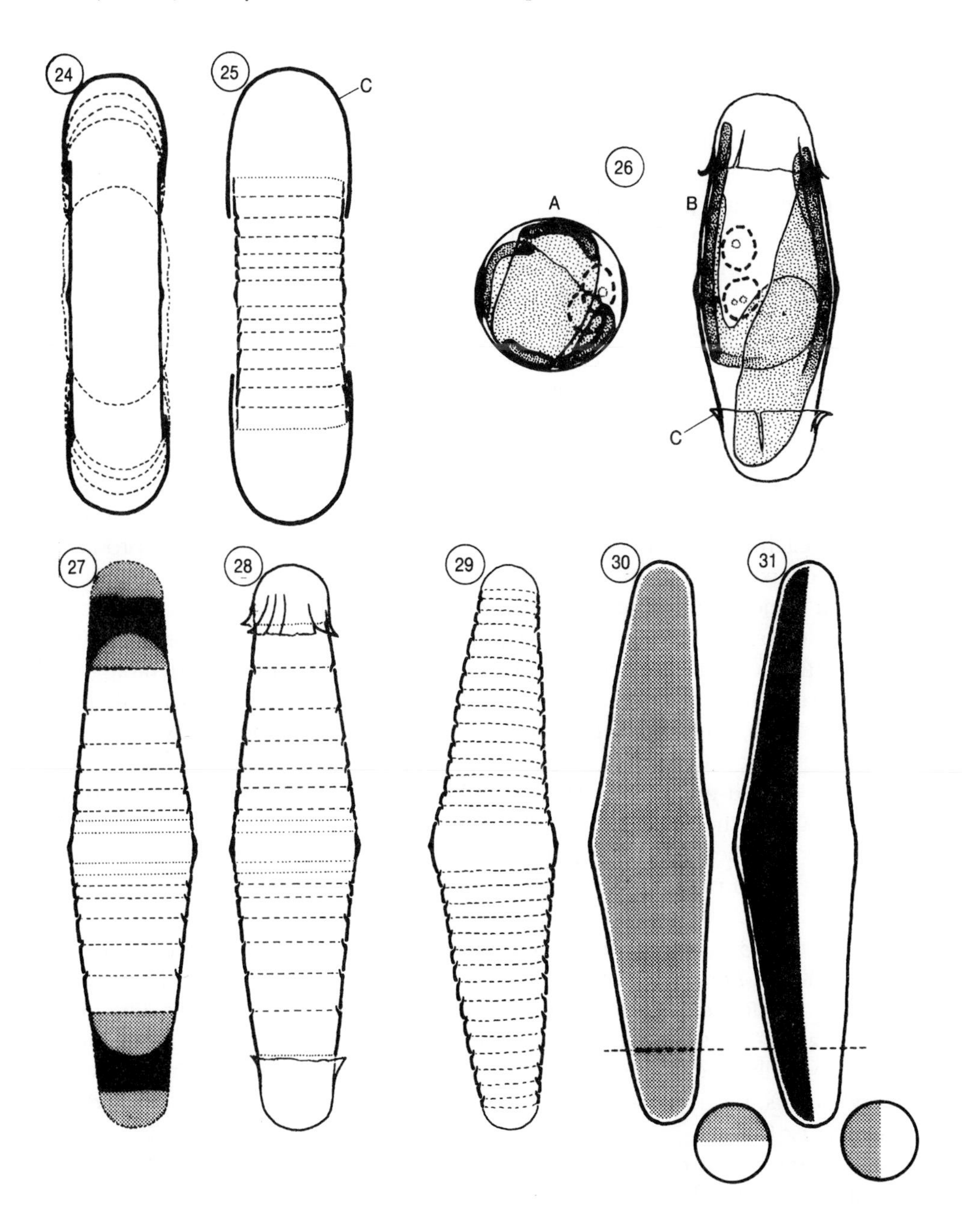

Meister: Fig. 35, Mann, 1989c), and so on. Auxospores are not always straight – some develop a slight or marked curvature (Figs 9, 29; Mann, 1982b) – and it also appears that development can be heteropolar, one end developing a different shape from the other (Cholnoky, 1929). In all cases, the development of the auxospore is constant within a species or race and is clearly under tight control.

Here, then, is the development of shape *de novo*, depending on a delicate balance between wall hardening and expansion. If auxospores are stressed osmotically (Fig. 10) or otherwise during expansion, morphogenesis is disturbed, producing abnormal initial cells. This is not uncommon in nature. However, although the immediate descendants of the initial cells inherit the abnormalities, they are rarely evident in later generations, for the reasons discussed in the previous section.

Neidium Pfitzer (Figs 24, 25; Mann, 1984d) shows a slight but interesting variation on the usual pattern of development. Like *Pinnularia* and many other pennate diatoms, *Neidium* produces linear auxospores (Fig. 25), but it does so in a different way. Instead of maintaining a dynamic balance between apical expansion and subapical hardening, via the formation of the perizonium, *Neidium* controls the width of the auxospore poles during expansion by enclosing them within silicified caps, formed around the ellipsoidal zygote before expansion (Figs 24, 25).

The shape of the expanded auxospore is not necessarily passed on to the vegetative cells. Two extra factors must be taken into account. First, the protoplast of the auxospore sometimes contracts away from the auxospore wall just before the formation of the initial frustule, modifying the shape produced during auxospore expansion (Fig. 37). This will only affect the shapes of cells in succeeding generations if the contraction involves the sides of the auxospore where the new girdle forms. In *Lyrella* auxospores, there is a marked contraction away from the auxospore wall before the formation of each valve, but only on the sides where the initial valves are produced; the lanceolate auxospores therefore give rise to lanceolate initial cells (Mann & Stickle, 1993). But in some *Navicula* species, such as *N. ulvacea* (Berk. *ex* Kütz.) Van Heurck (this species does not belong to *Navicula sensu stricto*) or *N. cryptocephala* Kütz., contraction affects the girdle region of the initial cell, resulting in a considerable modification to shape. In *N. ulvacea* a lanceolate auxospore gives rise to a virtually linear initial cell (Fig. 37), while in *N. cryptocephala* a centrally expanded auxospore produces a linear-lanceolate initial cell (Geitler, 1968).

Second, in curved auxospores (Figs 9, 29), since the curvature is always in one plane, the shape of the initial valves will depend on where they are produced. If the pervalvar axis (i.e. the axis of the girdle cylinder, passing through both valves) of the initial cell lies in the plane of curvature of the auxospore, then the initial valves will be symmetrical but unlike; one will be convex and the other concave when seen from the side (Fig. 31). This 'achnanthoid symmetry' occurs in *Achnanthes* Bory, *Achnanthidium* Kütz., *Rhoicosphenia*, *Rhoikoneis* Grun., *Campylopyxis* Medlin and some other genera (Round *et al.*, 1990). If, on the other hand, the pervalvar axis of the initial cell is at right angles to the plane of curvature of the auxospore, then the initial valves will be alike but curved – laterally asymmetrical (Fig. 30). Such 'cymbelloid symmetry' occurs, for instance, in *Cymbella*, *Encyonema* (Fig. 5), *Amphora* and *Epithemia* Bréb *ex* Kütz.

How does any of this help the systematist? First, it shows that the same final shape can be produced in different ways. A linear diatom, for example, can be produced through balanced tip growth and wall silicification during the expansion of the auxo-

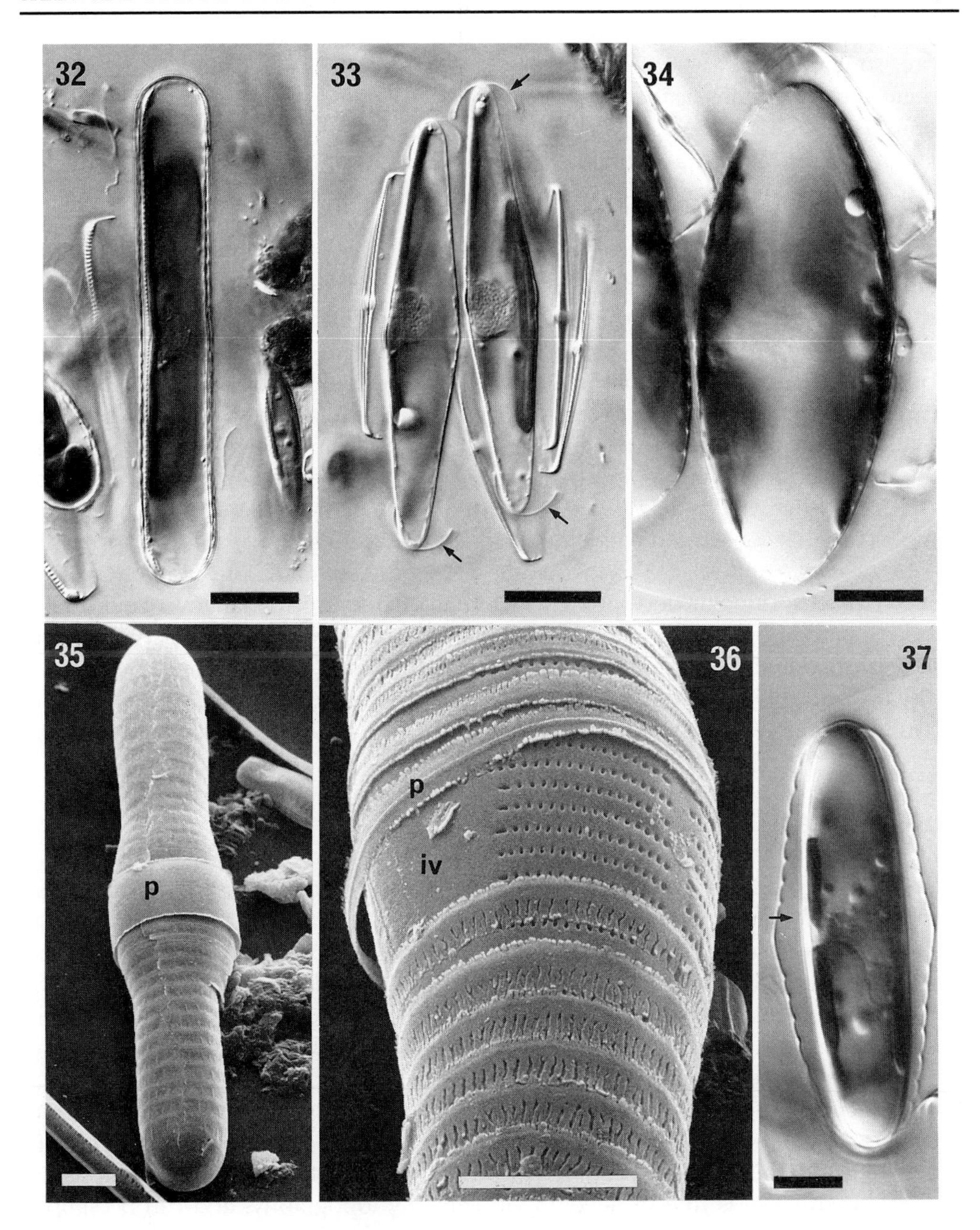

Figures 32–37 Auxospores of pennate diatoms. LM (except Figs 35, 36). **32** *Pinnularia*: linear outline. Scale bar 20 μm. **33** *Stauroneis*: almost rhombic outline. Scale bar 20 μm. **34** *Petroneis*: elliptical outline. Scale bar 20 μm. **35** *Caloneis* auxospore, SEM. The auxospore bears a central expansion. Note the wider primary perizonial band (p). Scale bar 10 μm. **36** *Rhoicosphenia*: centre of a partially disrupted auxospore casing, showing perizonial bands overlying the initial valve (iv). The central perizonial band (p) is little wider than the other bands in this species. Scale bar 5 μm. **37** *Navicula ulvacea*: expanded auxospore containing the initial cell. Note that during the formation of the initial valves, the cell has contracted away from the perizonium (e.g. arrow), so that a linear initial cell is formed within a lanceolate auxospore. Scale bar 10 μm.

spore, so that the auxospore diameter is constant from centre to pole. Or it can be produced by direct physical control of auxospore diameter by silica caps, as in *Neidium* or *Biremis* Mann & Cox (Mann, in press). Or it can be produced secondarily, through contraction of the protoplast within the auxospore after expansion is complete, whatever the shape of the auxospore itself (lanceolate, centrally expanded, etc.). Thus, similar shapes are not necessarily equivalent ontogenetically and so not necessarily homologous.

Second, it can be predicted that some shape transitions may be brought about more easily during evolution than others. For instance, in diatoms that produce curved auxospores, the change from cymbelloid to achnanthoid (or naviculoid) symmetry appears to be relatively simply achieved, by reorientation of the initial cell with respect to the auxospore (see above and Figs 30, 31). Hence in some cases saltatory evolution of shape can be expected to occur, since a minor developmental change can produce profound effects on the form of the adult. Cladistic or molecular studies may reveal whether this has indeed occurred. Other changes are inherently more complex. The change from cymbelloid symmetry (Figs 5, 6, 23) to the heteropolarity of *Gomphonema* and related forms (cf Fig. 21) requires a minimum of two developmental changes: isopolar development of the auxospore must change to heteropolar development; and either the auxospore must become straight or the pervalvar axis of the initial cell must be shifted through 90° with respect to the plane of curvature of the auxospore. Effectively, then, different shape changes require different weighting in taxonomic analysis.

THE MUTUAL INFLUENCE OF SIBLING CELLS

The mode of cell division in diatoms not only ensures that the parent cell passes on its shape, with slight modification, to its progeny, it also means that in many cases the sibling cells directly affect each other's morphology. This is because the protoplasts of sibling cells often remain appressed during part or all of valve formation (Figs 14, 38), so that any bulges formed on one sibling must correspond to equal depressions formed in the other (Figs 38, 39); most diatoms with interactive division have flat-topped valves (Figs 42, 44).

The phenomenon of 'complementarity' has been discussed by Mann (1984a) and Round *et al.* (1990). It is not universal. In some diatoms cell division is 'non-interactive' (Fig. 40): the protoplasts contract away from each other during cytokinesis. In this case each valve is free to develop virtually any topography, and can be simply domed (Figs 40, 43) or have an extremely elaborate morphology (Figs 41, 45). Complex topographies are built up sequentially as the valve expands out from the pattern centre. This is done through a combination of controlled expansion and contraction of the protoplast together with wall hardening, as the new valves are deposited and thickened (the study of *Chaetoceros* Ehrenb. by von Stosch *et al.*, 1973, is particularly instructive). The cytoskeleton is probably also involved, supporting and moulding the form of the valve until it is mature (Pickett-Heaps *et al.*, 1990).

Non-interactive division is rare in freshwater diatoms, except in some large species of *Surirella* Turp., e.g. *S. robusta* Ehrenb. (Pickett-Heaps *et al.*, 1988, 1990): the metabolic costs of reducing the volume of the daughter cells while the new valves are being produced, presumably through active expulsion of water to the hypotonic

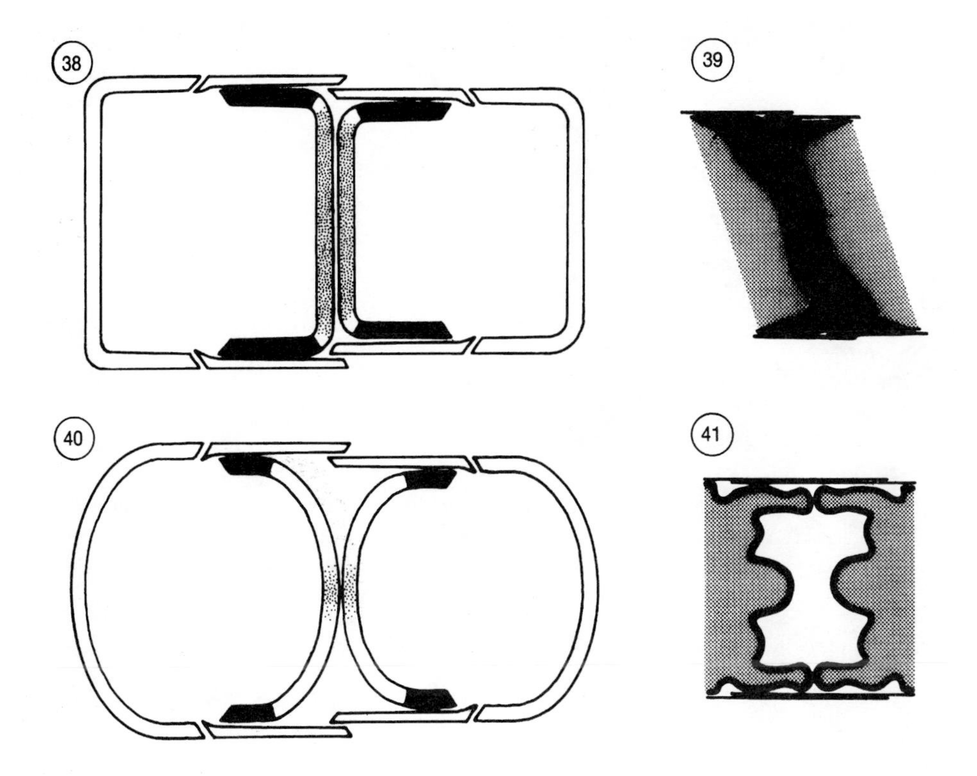

Figure 38 Diagrammatic section through a diatom exhibiting interactive division. The parts of the new valves moulded by the parental girdle are shown in black; those moulded by the other daughter cell (and thus showing complementarity) are shown stippled.

Figure 39 *Tryblionella*: sibling valves within the parental girdle, showing complementary ridges and depressions.

Figure 40 Diagrammatic section through a diatom exhibiting non-interactive division (shading as in Fig. 39).

Figure 41 Biddulphioid centric diatom: sibling valves within the parental girdle. Valve features can develop virtually without restraint within the space between sibling protoplasts after cell division.

medium, must usually outweigh any selective advantages there might be for each of the extra morphologies possible with non-interactive division. This appears to have been a Bauplan constraint during diatom evolution, so that freshwater diatoms are generally simpler in form than their marine counterparts (Round *et al.*, 1990, show valve shape and form in about 250 genera of freshwater and marine diatoms). It is interesting to note that *Skeletonema subsalsum* (A. Cleve) Bethge switches between interactive and non-interactive division, according to the salinity of the medium (Paasche *et al.*, 1975). The valve morphology changes accordingly: when division is interactive the valves have a flat top (the protoplasts are pressed against each other during valve formation) but, with higher salinity, non-interactive division occurs and the valves are rounded.

In a significant minority of interactive diatoms, the interface between the sibling cells is thrown into a series of sinusoidal waves – in one dimension or two – so that the valves are undulate. A simple case is that of *Tryblionella* W. Smith (Fig. 39), where the valve is transversely undulate, with one wavelength between the two margins.

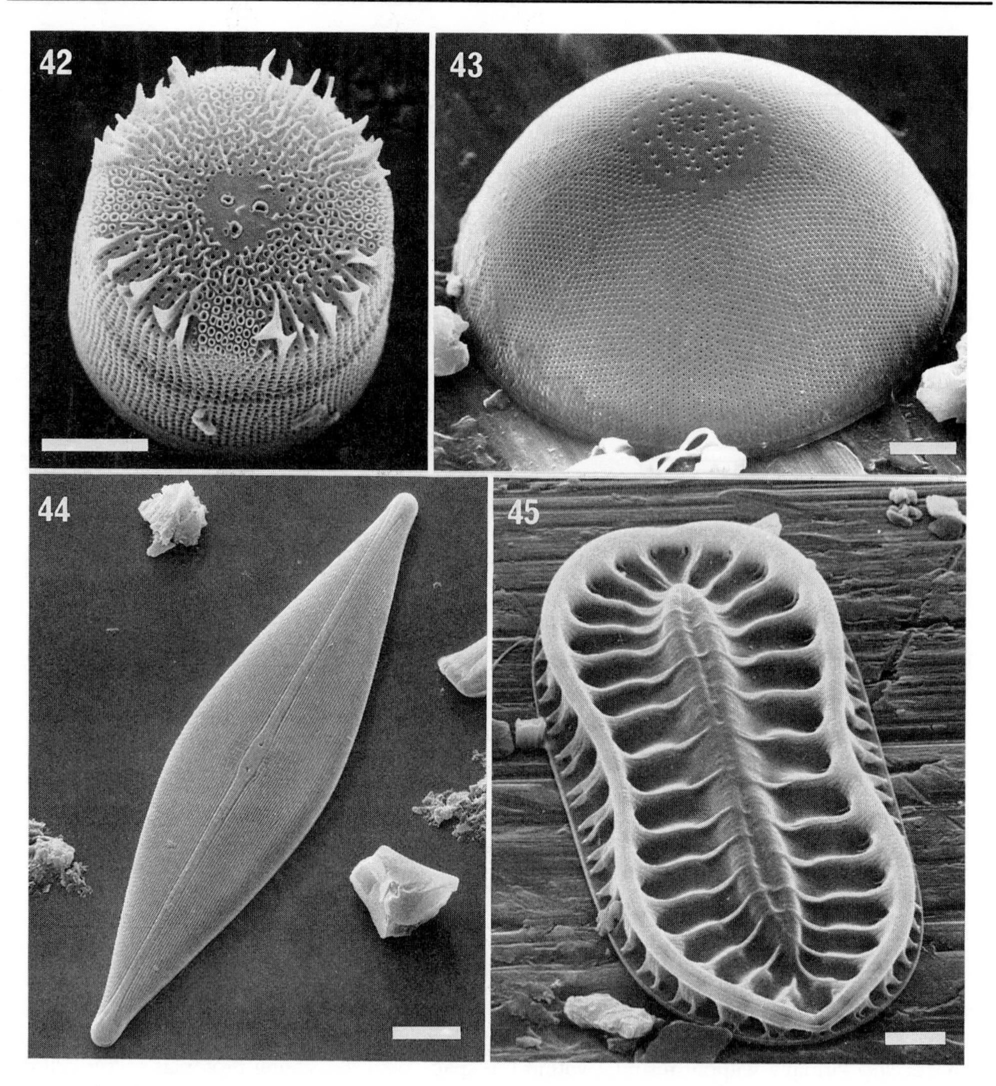

Figures 42–45 Diatom valves, showing morphologies developed during interactive (Figs 42, 44) and non-interactive (Figs 43, 45) division. Note the simple flat-topped valves formed during interactive division. The two interactive diatoms are from freshwater; the other two are marine. Scale bars 10 μm. **42** *Orthoseira* Thwaites. **43** *Hyalodiscus* Ehrenb. **44** *Craticula*. **45** *Surirella*.

In *Cymatopleura* W. Smith, the valve is longitudinally undulate (see Mann, 1987; Round *et al.*, 1990 for illustrations); here there are several waves between one pole of the valve and the other, and their amplitude often varies systematically along the valve. In some centric diatoms more complex waveforms are produced. *Stephanodiscus* Ehrenb. species often have two types of valve, representing complementary two-dimensional waveforms, one with a central dome and a concentric submarginal groove, the other with a central depression and a submarginal elevation (Round *et al.*, 1990). *Actinoptychus* Ehrenb. and *Glorioptychus* Hanna species have more complex tangential undulations, with alternately raised and depressed radial sectors (Round *et al.*, 1990).

Where division is interactive, the production of a bulge by one daughter cell will automatically produce a depression in the other, so that in theory at least, each valve might merely be the negative of the other, mirroring its morphogenesis. However, the orderliness and fairly constant wavelengths of the sinusoidal undulations mentioned above suggest that the sibling cells act as a coordinated pair. If their plasmalemmas are firmly bonded to each other across their interface, then local contraction of the protoplasm (by formation of a gel?) in one cell, but not in the other, could cause bending of the interface, in the same way that a bimetallic strip bends when heated. A series of undulations would thus represent standing waves in the concentration of some chemical that modulates contraction, which might in turn reflect periodic variation in the magnitude and direction of flux of this chemical across the two plasmalemmas: calcium would seem to be the obvious candidate for the messenger.

FINAL REMARKS

From the discussion and examples, it should be clear that the diatom taxonomist must understand the development of shape and form to interpret variation correctly. However, the taxonomist may be ill-equipped (technically and intellectually) to investigate morphogenesis. It is to be hoped, therefore, that more molecular biologists, biochemists and computer scientists can be persuaded to study diatoms, since otherwise taxonomists will be crippled as they attempt to grapple with the 10^5 or 10^6 species there may be, to produce a natural classification. For the morphogeneticist interested in explaining the genesis of shape, form and pattern in individual cells, diatoms offer the greatest challenges within the plant kingdom; perhaps this in itself will be enough to generate interest. This chapter has concentrated on the development of shape and form, but the origins of the ribs and pores on the valves (Figs 1–8, 11–13, 42–45) are equally or more perplexing. After cytokinesis, beneath the plasmalemma of the cleavage furrow, a silicon deposition vesicle appears and grows and within it an intricate pattern of ribs and pores develops, all in a few tens of minutes (Pickett-Heaps *et al.*, 1990). The patterns, although very diverse, have certain features in common and some are tantalizingly like those that can be produced by reaction–diffusion systems (Lacalli, 1981; Meinhardt, 1981; Meinhardt & Klingler, 1987).

But the relationship between taxonomist and morphogeneticist should not be like that between the dogs and the children in the parable (Mark 7, v. 28), taking crumbs that happen to fall from the dinner table. Taxonomists can and must play a much more important part in morphogenetic studies. Explanations of shape and form can only be developed through multidisciplinary approaches, involving the full battery of techniques and tools now available. It is therefore impractical to investigate morphogenesis in more than a handful of carefully chosen 'experimental organisms'. The taxonomist is therefore vital, since he is in the best position both to define the essential features of the Bauplan that any *general* model of morphogenesis must explain, and to test each morphogenetic model for its ability to account for *particular* aspects of the bewildering diversity of shape and form that exists.

ACKNOWLEDGEMENTS

Aspects of this research have been supported by SERC grants GR B/0936 and C/68484, and by a grant from the Royal Society towards the purchase of a photomicroscope.

Thanks are also due to Debbie White for preparing the photographs and Stephen Droop for discussion and support.

REFERENCES

BATTARBEE, R.W., MASON, J., RENBERG, I. & TALLING, J.F., 1990. *Palaeolimnology and Lake Acidification.* London: The Royal Society.

CHOLNOKY, B. von, 1929. Beiträge zur Kenntnis der Auxosporenbildung. *Archiv für Protistenkunde, 68:* 471–502.

CRAWFORD, R.M., 1981. Some considerations of size reduction in diatom cell walls: 253–261. In R. Ross (ed.), *Proceedings of the 6th Symposium on Recent and Fossil Diatoms,* Koenigstein: O. Koeltz.

DAVIS, R.B., ANDERSON, D.S. & BERGE, F., 1985. Palaeolimnological evidence that lake acidification is accompanied by loss of organic matter. *Nature, 316:* 436–438.

FRITZ, S.C., JUGGINS, S., BATTARBEE, R.W. & ENGSTROM, D.R., 1991. A diatom-based transfer function for salinity, water-level, and climate reconstruction. *Nature, 352:* 706–708.

GEITLER, L., 1932. Der Formwechsel der pennaten Diatomeen. *Archiv für Protistenkunde, 78,* 1–226.

GEITLER, L., 1940a. Die Auxosporenbildung von *Meridion circulare. Archiv für Protistenkunde, 93:* 288–294.

GEITLER, L., 1940b. Gameten- und Auxosporenbildung von *Synedra ulna* im Vergleich mit anderen pennaten Diatomeen. *Planta, 30:* 551–566.

GEITLER, L., 1968. Auxosporenbildungen bei einigen pennaten Diatomeen und *Nitzschia flexoides* n. sp. in der Gallerte von *Ophrydium versatile. Österreichische botanische Zeitschrift, 115:* 482–490.

HOHN, M.H., 1959. Variability in three species of *Gomphonema* (Bacillariophyceae) undergoing auxospore formation. *Notulae Naturae 316:* 1–7.

HOLDSWORTH, R.H., 1968. The presence of a crystalline matrix of pyrenoids of the diatom *Achnanthes brevipes. Journal of Cell Biology, 37:* 831–837.

HUSTEDT, F., 1930. Bacillariophyta (Diatomeae). In A. Pascher (ed.), *Die Süsswasser-Flora Mitteleuropas,* vol. 10, 2nd edn. Jena: G. Fischer.

JEWSON, D.H., 1992. Size reduction, reproductive strategy and the life cycle of a centric diatom. *Philosophical Transactions of the Royal Society of London, B 336:* 191–213.

KALLIO, P. & LEHTONEN, J., 1981. Nuclear control of morphogenesis in *Micrasterias.* In O. Kiermayer (ed.), *Cytomorphogenesis in Plants,* pp. 191–213. Vienna: Springer-Verlag.

KIERMAYER, O., 1981. Cytoplasmic basis of morphogenesis in *Micrasterias.* In O. Kiermayer (ed.), *Cytomorphogenesis in Plants,* pp. 147–189. Vienna: Springer-Verlag.

KOCIOLEK, J.P. & STOERMER, E.F., 1988. Taxonomy and systematic position of the *Gomphoneis quadripunctata* species complex. *Diatom Research, 3:* 95–108.

KRAMMER, K. & LANGE-BERTALOT, H., 1986. Bacillariophyceae. 1. Teil: Naviculaceae. In H. Ettl, J. Gerloff, H. Heynig & D. Mollenhauer (eds), *Süsswasserflora von Mitteleuropa,* vol. 7/1. Stuttgart: G. Fischer Verlag.

LACALLI, T.C., 1981. Dissipative structures and morphogenetic pattern in unicellular algae. *Philosophical Transactions of the Royal Society of London, B, 294:* 547–588.

MANN, D.G., 1982a. Structure, life history and systematics of *Rhoicosphenia* (Bacillariophyta) I. The vegetative cell of *Rh. curvata. Journal of Phycology, 18:* 162–176.

MANN, D.G., 1982b. Structure, life history and systematics of *Rhoicosphenia* (Bacillariophyta) II. Auxospore formation and perizonium structure of *Rh. curvata. Journal of Phycology, 18:* 264–274.

MANN, D.G., 1984a. An ontogenetic approach to diatom systematics. In D.G. Mann (ed.), *Proceedings of the 7th International Diatom Symposium,* pp. 113–144. Koenigstein: O. Koeltz.

MANN, D.G., 1984b. Observations on copulation in *Navicula pupula* and *Amphora ovalis* in relation to the nature of diatom species. *Annals of Botany, 54:* 429–438.

MANN, D.G., 1984c. Protoplast rotation, cell division and frustule symmetry in the diatom *Navicula bacillum*. *Annals of Botany, 53:* 295–302.

MANN, D.G., 1984d. Auxospore formation and development in *Neidium* (Bacillariophyta). *British Phycological Journal, 19:* 319–331.

MANN, D.G., 1987. Sexual reproduction in *Cymatopleura*. *Diatom Research, 2:* 97–112.

MANN, D.G., 1988. Why didn't Lund see sex in *Asterionella*? A discussion of the diatom life cycle in nature. In F.E. Round (ed.), *Algae and the Aquatic Environment*, pp. 383–412. Bristol: Biopress.

MANN, D.G., 1989a. The species concept in diatoms: evidence for morphologically distinct, sympatric gamodemes in four epipelic species. *Plant Systematics and Evolution, 164:* 215–237.

MANN, D.G., 1989b. The diatom genus *Sellaphora*: separation from *Navicula*. *British Phycological Journal, 24:* 1–20.

MANN, D.G., 1989c. On auxospore formation in *Caloneis* and the nature of *Amphiraphia* (Bacillariophyta). *Plant Systematics and Evolution, 163:* 43–52.

MANN, D.G., 1993. Patterns of sexual reproduction in diatoms. *Hydrobiologia, 269/270:* 11–20.

MANN, D.G. & STICKLE, A.J., 1991. The genus *Craticula*. *Diatom Research, 6:* 79–107.

MANN, D.G. & STICKLE, A.J., 1993. Life history and systematics of *Lyrella*. *Nova Hedwigia, Beiheft 106:* 43–70.

MANTON, I. & OATES, K., 1975. Fine-structural observations on *Papposphaera* Tangen from the southern hemisphere and on *Pappomonas* gen. nov. from South Africa and Greenland. *British Phycological Journal, 10:* 93–109.

MANTON, I., SUTHERLAND, J. & McCULLY, M., 1976. Fine structural observations on coccolithophorids from South Alaska in the genera *Papposphaera* Tangen and *Pappomonas* Manton and Oates. *British Phycological Journal, 11:* 225–238.

McKAY, R.M.L. & GIBBS, S.P., 1991. Composition and function of pyrenoids: cytochemical and immunochemical approaches. *Canadian Journal of Botany, 69:* 1040–1050.

McKAY, R.M.L., GIBBS, S.P. & VAUGHN, K.C., 1991. RuBisCo activase is present in the pyrenoid of green algae. *Protoplasma, 162:* 38–45.

MEDLIN, L.K., 1991. Evidence for parallel evolution of frustule shape in two lines of pennate diatoms from the epiphyton. *Diatom Research, 6:* 109–124.

MEINHARDT, H., 1982. *Models of Biological Pattern Formation.* London: Academic Press.

MEINHARDT, H. & KLINGLER, M., 1987. A model for pattern formation on the shells of molluscs. *Journal of Theoretical Biology, 126:* 63–89.

MITCHISON, J.M., 1971. *The Biology of the Cell Cycle.* Cambridge: Cambridge University Press.

NERC, 1992. Evolution and biodiversity – the new taxonomy. Report of a Committee set up by the Natural Environment Research Council and Chaired by Professor J.R. Krebs, FRS. Natural Environment Research Council, U.K.

OLSON, R.J., WATRAS, C. & CHISHOLM, S.W., 1986. Patterns of individual cell growth in marine centric diatoms. *Journal of General Microbiology, 132:* 1197–1204.

PAASCHE, E., JOHANNSON, S. & EVENSEN, D.L., 1975. An effect of osmotic pressure on the valve morphology of the diatom *Skeletonema subsalsum* (A. Cleve) Bethge. *Phycologia, 14:* 205–211.

PATRICK, R. & REIMER, C.W., 1975. The diatoms of the United States, vol. 2, part 1. *Monographs. Academy of Natural Sciences of Philadelphia, 13:* 1–213.

PICKETT-HEAPS, J.D., 1991. Cell division in diatoms. *International Review of Cytology, 128:* 63–108.

PICKETT-HEAPS, J.D., COHN, S., SCHMID, A.M.-M. & TIPPIT, D.H., 1988. Valve morphogenesis in *Surirella*. *Journal of Phycology, 24:* 35–49.

PICKETT-HEAPS, J.D., SCHMID, A.M.-M. & EDGAR, L.A., 1990. The cell biology of diatom valve formation. In F.E. Round & D.J. Chapman (eds), *Progress in Phycological Research*, vol. 7, pp. 1–168. Biopress, Bristol.

ROUND, F.E., CRAWFORD, R.M. & MANN, D.G., 1990. *The Diatoms. Morphology and Biology of the Genera.* Cambridge: Cambridge University Press.

STOSCH, H.A. von, 1981. Structural and histochemical observations on the organic layers of the diatom cell wall. In R. Ross (ed.), *Proceedings of the 6th Symposium on Recent and Fossil Diatoms*, pp. 231–252. Koenigstein: O. Koeltz.

STOSCH, H.A. von, 1982. On auxospore envelopes in diatoms. *Bacillaria, 5:* 127–156.

STOSCH, H.A. von, THEIL, G. & KOWALLIK, K.V., 1973. Entwicklungsgeschichtliche Untersuchungen an zentrischen Diatomeen. V. Bau und Lebenszyklus von *Chaetoceros didymum*, mit Beobachtungen über einige andere Arten der Gattung. *Helgoländer wissenschaftlicher Meeresuntersuchungen, 25:* 384–445.

TAYLOR, D.L., 1972. Ultrastructure of *Cocconeis diminuta* Pantocsek. *Archiv für Mikrobiologie, 81:* 136–145.

THOMPSON, d'A.W., 1942. *On Growth and Form.* Cambridge: Cambridge University Press.

ÜBELEIS, I., 1957. Osmotischer Wert, Zucker- und Harnstoffpermeabilität einiger Diatomeen. *Sitzungsberichte der Akademie der Wissenschaften in Wien, mathem.-naturw. Kl., Abt. I, 166:* 395–433.

VOLCANI, B.E., 1978. Role of silicon in diatom metabolism and silicification. In G. Bendz & I. Lundquist (eds), *Biochemistry of Silicon and Related Problems*, pp. 177–204. New York: Plenum.

WERNER, D., ed., 1977. *The Biology of Diatoms (Botanical Monographs 13).* Oxford: Blackwell Scientific.

CHAPTER

3

Towards an understanding of pattern and process in the growth of early vascular plants

D. EDWARDS

CONTENTS

Abstract

Branching patterns are described from the four major groups of early vascular plants, the Rhyniophytina, Trimerophytina, Zosterophyllophytina and Lycophytina. In the oldest examples of the Rhyniophytina and rhyniophytoids, isotomous branching whether planar or three-dimensional, was associated with determinate growth imposed by terminal sporangia, with the latter produced synchronously on all branches. Such branching was frequently accompanied by spiralling of cells and twisting of axes and sporangia. In *Rhynia gwynne-vaughanii*, the only axial fossil in which the apical meristems are preserved, the terminal sporangia were shed and further upward growth achieved by laterals. The Zosterophyllophytina exhibit the greatest range in branching forms, showing, in particular, variation in frequency of branching (e.g. production of K + H forms), and pronounced overtopping producing pseudomonopodial systems with extended periods of spore production. Circinate vernation, characteristic of examples with scattered lateral sporangia and often planar branching systems, is described at the cellular level in *Trichopherophyton* and analogues sought

Shape and Form in Plants and Fungi
ISBN 0–12–371035–9

in the processes recorded in extant fern megaphylls. Various kinds of anomalous branching including subaxillary branches are described and the functions of such systems debated. Problems of the recognition of roots in plants of axial organization are discussed, and the presumed lycophyte *Drepanophycus* used to exemplify probable shoot-borne roots early in the evolution of vascular plants.

INTRODUCTION

What follows links some observations on the branching patterns in Silurian and Lower Devonian vascular plants, with anatomical and morphological data relevant to deliberation as to how they might have grown. The approach is thus descriptive, not exhaustive, idiosyncratic, and somewhat simplistic. Any inferences are so tentative that they are hardly sufficiently respectable to form the basis for hypothesis. The reader is referred to Niklas (1978) for a more quantitative assessment, including the modelling of branching systems and to Rothwell (in press) for a discussion on the importance of branching pattern in the definition of major clades of embryophytes.

This account centres on the earliest vascular plants as they were probably the first component of terrestrial vegetation with erect branching sporophytic systems, although not the earliest colonizers of subaerial substrates (Edwards & Selden, 1992; Gray, 1985a, b). However, with the exception of some microphyllous lycophyte-like forms, these sporophytes were branching, axial vegetative systems, normally naked, sometimes spiny. The term 'axis' is preferred to stem, root, or shoot as it conveys ignorance on the organographic status of the normally isolated fragments of extremely simple plants, where anatomy is not preserved and information on organ initiation is lacking. Its use also emphasizes the difficulties, indeed the undesirability, of employing predominantly angiosperm-orientated terminology and concepts for these very early vascular plants.

Perhaps the most striking implication in terms of development relates to the apex itself. Its lack of leaf and bud primordia hampers direct analogy with extant plants, and raises questions as to the sources of plant growth regulators involved in differentiation. That similar systems to those operating today already existed in the early vascular plant plexus, is evidenced by somewhat limited information on the morphology of the apex and pattern of vascular differentiation (e.g. Hueber, 1992; Kidston & Lang, 1920b) in the lycophyte-like forms.

The axial plants have been grouped together in the Psilophyta (e.g. Høeg, 1967), but Banks (1975) recognized three major evolutionary lines (designated as the subdivisions Rhyniophytina, Zosterophyllophytina and Trimerophytina) in the complex (Table 1), and suggested that taxa not easily fitted into these subdivisions might 'express the vigour of evolutionary activity during Early Devonian times'. More recent research has confirmed this, revealing even greater disparity in these early vascular plants such that their classification remains in flux (Kenrick & Crane, 1991; Li & Edwards, 1992). The microphyllous forms (including *Baragwanathia* and *Drepanophycus*) have recently been placed in the Lycophytina, although the spatial relationship between sporangium and sporophyll remains conjectural (see for example Gensel, 1992, and Hueber, 1992).

In the context of plant growth the imperfections of the fossil record are particularly daunting. Early vascular plants are preserved mainly as compressions and permin-

Table 1 Characteristics of major groups of early vascular plants (based mainly on Banks, 1975)

Rhyniophytina
Leafless homosporous plants of axial organization with isotomous and anisotomous and rarely pseudomonopodial branching, and terminal sporangia including fusiform, spherical and discoidal forms. Xylem terete in cross-section. ?Centrarch.
e.g. *Rhynia gwynne-vaughanii, Cooksonia pertoni, Uskiella spargens*

Zosterophyllophytina
Leafless homosporous plants of axial organization with isotomous, anisotomous and pronounced pseudomonopodial branching. Lateral sporangia, splitting into two valves, aggregated into spikes or distributed on distal parts of the plant. Xylem elliptical or terete in section. Exarch.
e.g. *Gosslingia breconensis, Trichopherophyton teuchansii, Tarella trowenii*

Trimerophytina
Leafless homosporous plants of axial organization with isotomous, anisotomous and pronounced pseudomonopodial branching. Elongate sporangia terminate branched specialized lateral branch systems and appear as trusses. Xylem terete in section (possibly lobed). Centrarch. e.g. *Psilophyton dawsonii*

Lycophytina
Drepanophycales (sometimes called prelycophytes).
Herbaceous, homosporous plants with unbranched microphylls. Xylem deeply lobed in section. Exarch or slightly mesarch. Sporangial position equivocal: cauline or axillary.
e.g. *Asteroxylon mackiei, Drepanophycus spinaeformis*

eralizations. Thus, shape and form may seem easily circumscribed, but fossils, particularly permineralized ones, are rare, and for compressions the compaction process tends to obliterate any three-dimensional organization. Furthermore, the fossils are fragmentary: most consist of rather featureless axial systems on which sporangial characteristics and their position, together with xylem anatomy and in some instances branch patterns, are used for identification (Banks, 1975; see Table 1). Thus there is heavy reliance on features of the mature reproducing sporophyte. In considering sterile systems, assignation to taxa and recognition of position on the growing plant are conjectural. Further, in that for compression fossils, preservation may result from possession of cuticle and structural tissues, both containing relatively resilient biopolymers, apical regions may have low fossilization potential. Even in pyrite and calcium carbonate permineralizations, parenchyma is not present, and once again it is from the Aberdeenshire Rhynie Chert silicified plants that information on growing points may be derived (Hueber, 1992. pp. 484, 486; Kidston & Lang, 1920a).

BRANCHING PATTERNS

In axial plants, two major types of branching pattern, both produced by dichotomy of the apex, can be distinguished. In the first, division is strictly isotomous with daughter branches repeating the process, and with products of one division diverging at more or less the same angle from the parent axis (Fig. 1C). In anisotomous branching (sometimes termed pseudomonopodial), unequal division of the shoot apex was succeeded by more extensive, ± 'vertical' growth of the wider daughter branch, which on dividing again in a similar manner appears as a 'main axis' bearing 'laterals'. Such overtopping is particularly evident where successive daughter branches are at 180° (at least in compression fossils) and the main axis appears zig-zagged to varying degrees as it retains evidence of this sequential branching (Fig. 1G). The lateral branches may repeat this process or branch isotomously. In certain overtopped systems the main axis shows no deviations from the vertical and bears laterals spirally or alternately, with again variation in the branching pattern of the latter (Fig. 1F).

Further variation in anisotomous branching is produced by frequency/proximity of branching events and the presence of aborted, arrested or dormant axes. These are considered in more detail later, as will be the three-dimensional nature of the systems. In the early lycophytes, isotomous and anisotomous systems with well-defined laterals are recorded, but branching tends to be less regular and more infrequent. Since such branching sporophytes point to the existence of prolonged apical meristematic activity the very limited information available on the apex itself will be presented first.

Apical meristems

In their classic descriptions of Rhynie Chert plants, Kidston and Lang (1920b) illustrated three apices which they attributed to *Rhynia gwynne-vaughanii*, because they were associated with a cluster of mature axes of that species. The most complete longitudinal section (Fig. 2) has a semicircular distal outline and is composed of closely packed 'cells', whose outlines are difficult to define as each consists of a central dense golden-brown region surrounded by a lighter (less granular) layer. Within the former is sometimes seen an even darker, almost circular area of relative size similar to that of a nucleus in meristems of extant vascular plants. The outermost layer has isodiametric or, more normally, rectangular cells with anticlinal long axes (Fig. 3). It has a uniform appearance over the summit, but looks less well-organized, with tangentially expanded cells, over the flanks. There is no evidence for an apical cell or for cell linkages, the bean-shaped area tissue (*c.* 220 μm deep) below the dermal layer being composed of more or less isodiametric cells with no intercellular spaces, which become less distinct but more elongate proximally (Figs 2, 6). Close to this apex is another, fortuitously cut in paradermal section (Fig. 5), with 'packets' of four cells looking remarkably similar to those in the *Osmunda* shoot tip (Bierhorst, 1977), although in the extant plant they ultimately result from the activity of an apical cell.

In the longitudinally sectioned apex (although possibly not medial), there is no pronounced zonation or indications of a procambium in the one millimetre of axis with preserved cells. A slight distortion on the left-hand side of the apical dome (Figs 2, 4), with possible periclinal divisions close to the surface, is suggestive of an apex in early stages of division. This may be a preservational feature but there is some resemblance to sections through the dividing apex of the rhizophore of *Selaginella martensii* (Webster & Steeves, 1967). This organ may well provide the closest living analogue to the naked fossil apex, particularly when it branches isotomously before initiation of the root cap (see for example Webster & Steeves, 1967, Pl. 11, figure 9). During the process segregation of two groups of meristematic cells occurs, before

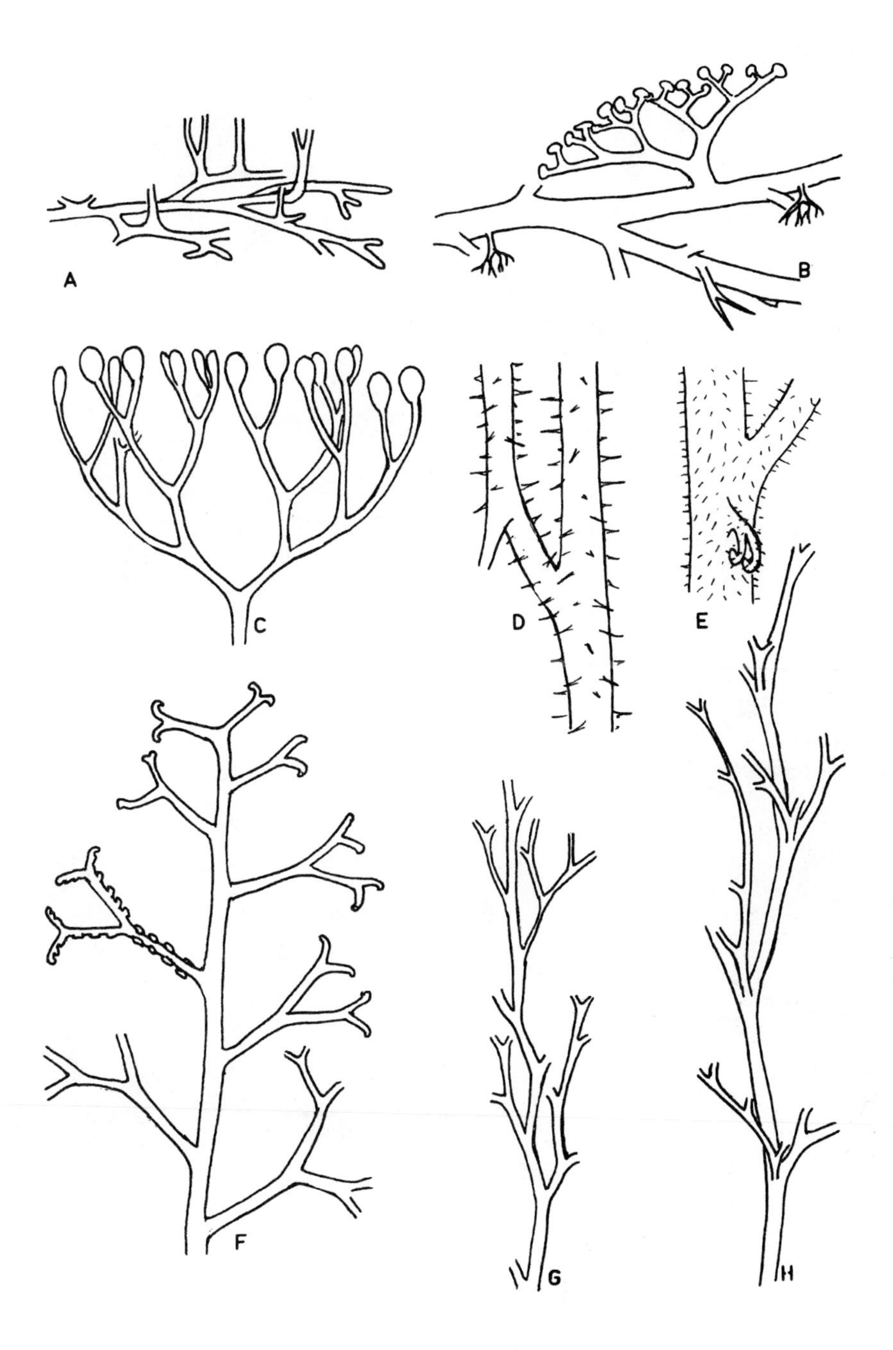

Figure 1 Branching patterns in early land plants with axial organization. **A** K- and H-branching (*Zosterophyllum*: after Gerrienne, 1988). **B** Isotomous and anisotomous branching with lateral root-like branches (*Hsüa*: after Li, 1982). **C** Isotomous with terminal sporangia at one level (*Uskiella*: after Shute & Edwards, 1989). **D** Anomalous branching in *Sawdonia ornata* (after Rayner, 1983). **E** Subaxillary branching (*Deheubarthia*: after Edwards *et al.*, 1989). **F** Anisotomous with main axis and laterals in a planar system (*Thrinkophyton*: after Kenrick & Edwards, 1988 with sporangia added to one lateral branch system). **G** Anisotomous with overtopping producing zig-zag (*Gosslingia breconensis*). **H** Subaxillary branching (*Anisophyton gothani*: after Hass & Remy, 1986 with spines present on all branches omitted).

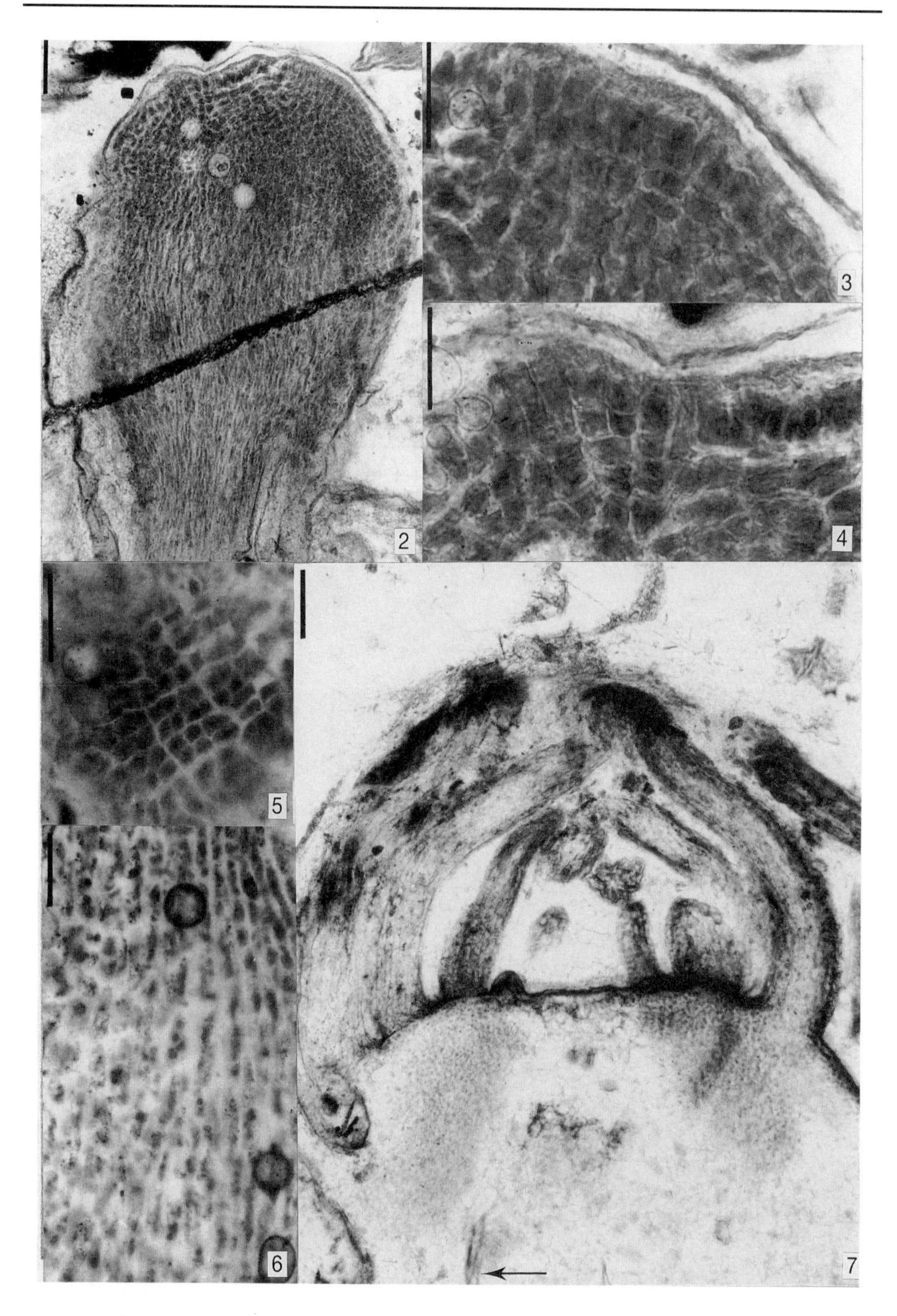

Figures 2–6 Sections through axial apical regions in *Rhynia gwynne-vaughanii* (?Siegenian, Rhynie Chert, Aberdeenshire). **2** Longitudinal section apex initials. Hunterian Museum (H) 2425. Scale bar 100 μm. **3,4** Magnification of apical initials in Fig. 2. Note the possible initiation of branching in Fig. 4. Scale bars 50 μm. **5** Transverse section apical initials, H 2425. Scale bar 50 μm. **6** Longitudinal section apex showing undifferentiated but elongate cells. H 2426. Scale bar 100 μm.

Figure 7 Longitudinal section *Asteroxylon mackiei* with leaf primordia. Arrow indicates beginnings of cauline vascular tissue differentiation, USNM 446314 (by kind permission of Dr F.M. Hueber). Scale bar 100 μm.

there is any external evidence for division. (The distinction of two meristematic areas in a morphologically undivided apex may result in the differentiation of two procambial strands proximally, and thus may explain why, in many isotomously dividing fossil axes, branching of the vascular strand occurs some distance below bifurcation of the axis itself, a feature also noted in *Selaginella*.)

A prominent and unusual feature of the fossil apex is the presence of a well-developed cuticle that completely covers the apex (Fig. 2) and delimits the axis proximally when cells are no longer preserved, and where it resembles the undulating cuticle characteristic of a more mature axis with poor preservation. Thus, it is unfortunately impossible to observe cell maturation or expansion–elongation in such apices. In longitudinal sections through mature axes of *R. gwynne-vaughanii*, cortical cells are much wider than those in the apex, but they are rarely much longer than wide, suggesting further cell division during a putative extension phase. The only elongate cells at maturity are those of the vascular system, a condition that more or less holds for most of the Rhynie Chert plants.

In contrast, the epidermal cells of many rhyniophytoids and of the erect branches of *Horneophyton* (Hass, 1991) are extremely long ($< 1000\ \mu m$ in *Horneophyton*), suggestive of considerable extension growth but little cell division. In the hypodermal region, typically of thick-walled cells in rhyniophytoids, cells are long, but more difficult to quantify. The elongate epidermal cells frequently demonstrate spiralling. A Silurian example shows a clockwise spiral (as viewed from the apex) of similar steepness in all branches (Figs 8, 9). Stomata are present, but of low density – a further characteristic of these systems. Although the tip appears rounded and complete, the actual apex is directed into the stub. In *Tortilicaulis* (Figs 10, 11, 13) the spiralling extends into the terminal sporangium and may be involved in longitudinal dehiscence into two valves. The functional significance of such spiralling and gross twisting (Fig. 14) in vegetative axes is less easy to explain. Mechanistically it may be related to the unevenly thickened cells noted in the hypodermis of some of these plants.

In mature axes of extant plants, collapsed protoxylem elements are evidence for extension growth. That it is very difficult to give accurate descriptions of protoxylem wall thickenings in early vascular plants (Edwards, 1993) – this tissue being identified by the relatively small size of elements in transverse section and preservation – may perhaps be considered evidence that extension did occur. By comparison with the maturation pattern associated with crozier uncurling in living ferns, the undifferentiated xylem strand in the straight part of the *Trichopherophyton* tip (see later, Fig. 16) may indicate that there was further extension in this region. The actual apex of *Trichopherophyton* shows no individual cells (Fig. 12). The slightly clearer lateral meristem will be described later (Fig. 16).

Lycophytes

Hueber (1992) has recently described a longitudinal section through a shoot tip of silicified *Asteroxylon mackiei* (Fig. 7). The apex itself is flat with several, somewhat indistinct, apical initials and surrounded by a number of leaf primordia. Below the apical initials is a region of 'central mother cells' and, more proximally, traces leading to the leaves and the beginnings of delimitation of the zoned cortex. Cauline xylem differentiation is apparent even further down the stem. Hueber also illustrated a procambial strand within the leaf itself, but as tracheidal thickenings have not been observed beyond the base of a mature leaf, xylem presumably never became differentiated. The cauline apical organization was compared favourably with that in extant *Lycopodium reflexum*, and considered evidence for secure relationship of *Asteroxylon*, sometimes considered a prelycophyte (see for example Gensel, 1992; Hueber, 1992), with the lycophytes. Kidston and Lang (1920a) described transverse sections from just above (Fig. 23) and below a stem apex (Figs 21, 22). The former has spirally arranged leaves in transverse section and the latter, proximal to leaf insertion, shows

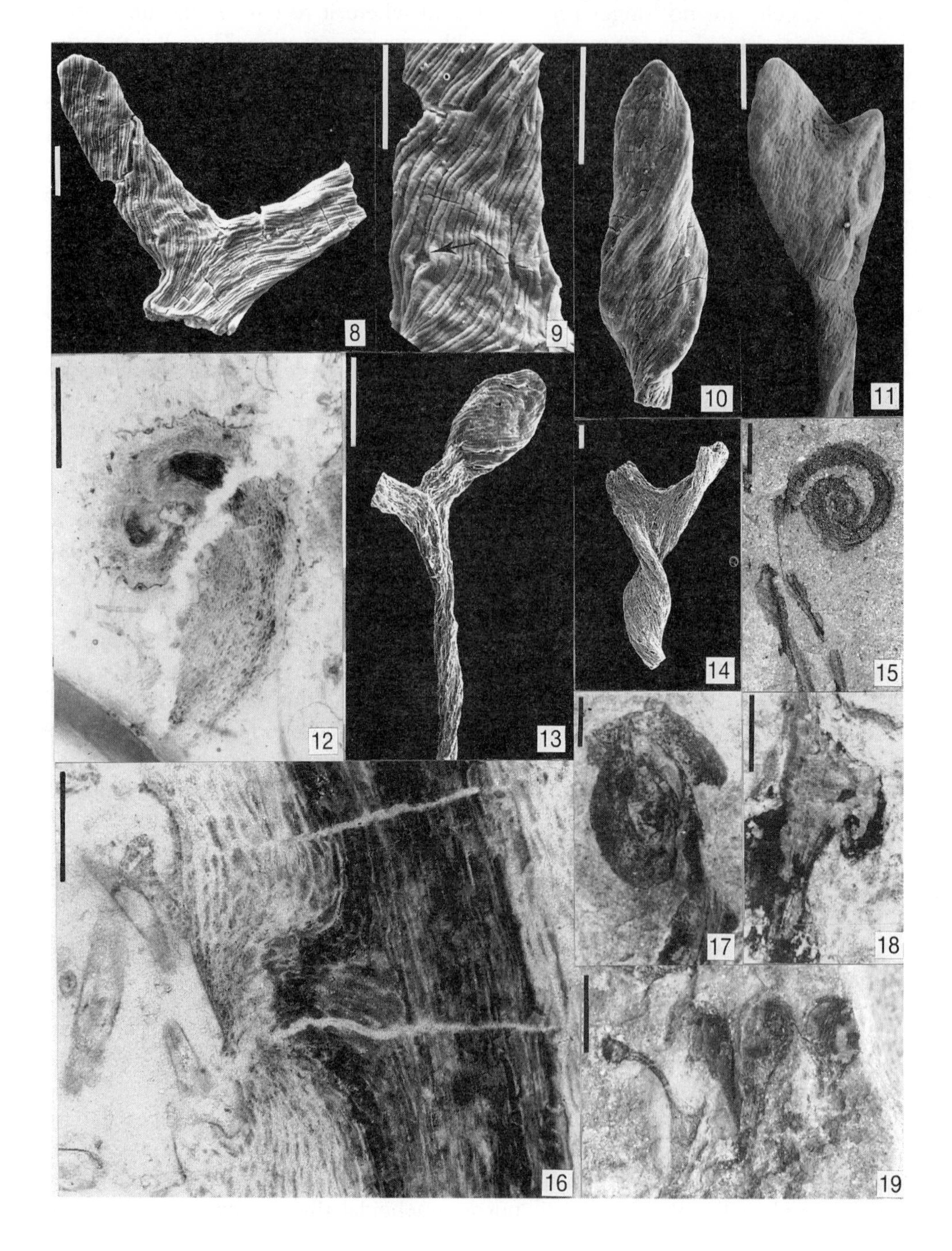

differentiation of protoxylem at the poles of the actinostele, with metaxylem less well preserved and lacking differentiated wall thickenings (Figs 21, 22). Spiral and annular thickenings characterize the narrow protoxylem elements. Kidston and Lang also illustrated sections with undifferentiated cauline vascular tissue distinguished by its denser, but normally non-cellular appearance, and others in various stages of cortical maturity (Fig. 22), but with only protoxylem differentiated in the vascular tissues.

Isotomous branching

This produces the very orderly, symmetrical and simple branch systems thought characteristic of the presumed earliest vascular plants as exemplified by *Cooksonia.* While Lower Devonian *C. pertoni* has been demonstrated vascular (Edwards *et al.*, 1992), Silurian representatives, including the earliest from Irish Wenlock (Upper Silurian) sediments (Edwards *et al.*, 1983), lack information on vascular tissues in their isotomous axes. This situation also holds for a large number of taxa (termed informally rhyniophytoid) with variously shaped sporangia, some bearing spines, terminating smooth, usually isotomously, branching axes from the Upper Silurian and Lower Devonian (e.g. Edwards & Davies, 1990; Fanning *et al.*, 1992).

Silurian sterile axes are also usually isotomous. In the few that show overtopping, there is little difference in axis diameter in the daughter branches. The oldest sterile axes (*Eohostinella* of unknown affinity) from Llandovery sediments are reported as 'dichotomous' but branch points are not well illustrated (Schopf *et al.*, 1966).

The most completely preserved fertile specimens (e.g. *Steganotheca striata, C. pertoni, C. caledonica, Uskiella spargens*) demonstrate synchrony of branching (in that axes are of similar lengths between successive branch points) and in production of terminal sporangia (i.e. more or less all at the same level; see Figs 1C and 19 for *Uskiella* (Shute & Edwards, 1989)). A similar pattern is noted in isotomizing laterals, e.g. in Emsian *Hsüa robusta,* possibly an advanced rhyniophyte (Li, 1992). Thrice dichotomizing lateral branches are inserted spirally on presumed vertical axes and these terminate in ± reniform to spherical bivalved sporangia of *C. caledonica* or

---

Figure 8 Scanning electron micrograph of bifurcating vegetative tip showing spiralling of elongate cells, CR/LL 1. Silurian (Ludlow), Shropshire. Scale bar 200 μm.

Figure 9 Close up of Fig. 8.

Figures 10, 11, 13 Lower Devonian sporangia of *Tortilicaulis* showing twisting. Scale bars 200 μm. **10** NMW 93. 98G. 15 **11** Bifurcating sporangium, NMW 93. 98G. 13. **13** Sporangium with twisting branching axis, NMW 93. 98.G. 14.

Figures 12, 16 Longitudinal section hooked apex of *Trichopherophyton teuchansii* (Siegenian Rhynie Chert). **12** Entire specimen showing procambial cylinder distally, AGL 93/9. Scale bar 1 μm. **16** Lateral meristem with procambial strand. Note sections through precociously developed unicellular hairs, AGL 93/4. Scale bar 200 μm.

Figure 14 Scanning electron micrograph of fragmentary Silurian axis showing gross twisting. CR/LL 2. Scale bar 200 μm.

Figure 15 *Gosslingia breconensis.* (Siegenian, S. Wales). Circinate tip with loose coiling. V. 26575. Scale bar 2 μm.

Figures 17, 18 *Tarella trowenii* (Siegenian, S. Wales). Scale bars 2 μm. **17** Tightly coiled tip in broad axis. NMW85.18G.3. **18** Lateral 'hook'. NMW85.18G.10.

Figure 19 *Uskiella spargens* (Siegenian, S. Wales). Four of six terminal sporangia at the same level. V.63095. Scale bar 2 mm.

Zosterophyllum type (Fig. 1B). Less-branched laterals are reported in *Junggaria spinosa* from the Late Silurian of north-west China (Cai *et al.*, 1993).

Whether or not there were changes in the *plane* of division at successive isotomies is difficult to detect in compression fossils. In some cases where overlapping axes and sporangia give the impression of a bushy three-dimensional system (e.g. *C. bohemica* – Schweitzer, 1983; *C. caledonica* – Edwards, 1970a), but in others where branching angles are wider (e.g. *Steganotheca striata* – Edwards, 1970a), there is the possibility that the system was planar. The expectation of three-dimensionality in the archetypal land plants as hypothesized in the Telome theory may colour interpretations of the fossils. However, planar branching in an upright system may well be untenable on mechanical grounds (although characteristic of compound megaphylls). Isotomous branching predominates in the earliest vascular plants (rhyniophytes: rhyniophytoids), but is also found in some more complex members of the group and forms part of the branching system in zosterophyllophytes and trimerophytes.

The conversion of vegetative apices to sporangium production would have involved changes in the planes and type of cell division, sometimes associated with gross changes in shape and size, and would have eventually led to the cessation of growth of that particular axis. Hueber (1992), in considering sporangium ontogeny in relation to evolution of major lineages of early vascular plants, distinguished two major groups within the rhyniophytes s.1. based on sporangial shape: viz. fusiform as in *Salopella* (and eventually the Trimerophytina, etc.) and globose or reniform as in *Cooksonia* (and eventually the Zosterophyllophytina). Both forms are envisaged as a continuation of the subtending axis, with the wall of the sporangia continuous with that of the axial cortex (e.g. in *Psilophyton dawsonii*, Banks *et al.*, 1975), but were suggested to have differed in the number of initials involved in sporangial initiation. Thus, the more robust and morphologically distinct globular sporangia were compared with a eusporangium in which a number of initials are involved in initiation, while the development of less modified fusiform sporangia was postulated to show greater similarity with the leptosporangium in its most primitive form (Bierhorst, 1971: 206).

Hueber did not explicitly extend this hypothesis into the nature of the vegetative apex in the two types. In the examples presented here there is no evidence of an apical cell in the presumed *Rhynia* apex (Figs 2–5) with its fusiform sporangia, or in the lateral meristem in *Trichopherophyton* (Fig. 16), where preservation is poorer, and whether the initials were of a branch or sporangium cannot be determined. Work in Cardiff on Lower Devonian rhyniophytoids emphasizes that the fusiform sporangium is a continuation of the vegetative axis. Thus in *Tortilicaulis* the spiralling epidermal cells are seen in both (and this is associated with predetermined longitudinal dehiscence; Fig. 13), while the bifurcating sporangia show that the timing of sporangial initiation could coincide with various stages in the branching of an axis (Fig. 11).

The limitations imposed by determinate growth are obvious in the earliest members where all apices would be converted from vegetative to reproductive growth (e.g. Edwards & Davies, 1990), and escape from this constraint by overtopping and pseudomonopodial growth must have evolved a number of times in separate groups. In *Rhynia gwynne-vaughanii* a different 'solution' is apparent. Accurate reconstructions of the gross morphology of Rhynie Chert plants demands meticulous analysis of serial transverse sections together with fortuitous longitudinal sections. Using such techniques Edwards (1980) was able to demonstrate that *R. gwynne-vaughanii* was a more complex plant than is evident from the original reconstruction by Kidston and Lang (1921). Some branching in the erect vegetative system is isotomous, but the axes bear hemispherical projections, some of which have developed into lateral branches. The terminal sporangium is separated from the subtending axis by a dark region interpreted as an abscission zone (Fig. 20), and distally directed lateral branches arise just below this zone. Edwards suggested that the entire sporangium was shed and that the lateral branch continued upward growth. Whether or not the final dichotomy produced a short axis with a sporangium and a latent branch, or whether the latter arose *de novo* is conjectural, although the branching angle may favour the latter. Thus, an individual axis may have continued growing after sporangial production, although its extent and the fate of the lateral remains unknown. More recently, the shedding (or isolation) of complete sporangia has been recorded in *Huvenia* and *Stockmansella* (united with *Rhynia* in the possession of S-type tracheids: see Kenrick *et al.*, 1991).

Anisotomous branching

The earliest fertile example is the *Junggaria spinosa* (≡ *Cooksonella*) of north-west China mentioned earlier (Cai *et al.*, 1993) from the latest Silurian (Pridoli) at a time when isotomous branching predominated in Laurussia. Marked overtopping with fertile lateral systems (e.g. *Renalia* sp., Fanning *et al.*, 1992) is recorded in the early Devonian. In both taxa the sporangia too are advanced in that there is marginal modification for dehiscence. The Trimerophytina are usually far more complexly organized vegetatively, showing combinations of isotomous and anisotomous branching with further variation produced by changes in frequency of branching and dormant, arrested or aborted branches. The most thoroughly investigated is *Psilophyton dawsonii* (Banks *et al.*, 1975), reconstructed from calcium carbonate permineralizations and the uncovering (dégagement) of compressions. Branching is isotomous in the widest main axes. In vegetative systems, subsequent branching is in a close spiral and then by further successive dichotomies with the ultimate 'branchlets' considered possible precursors of leaves (Banks, 1980). At higher levels, fertile lateral systems are inserted alternately (a system described as pseudomonopodial by Banks) while subsequent branching is anomalous because in two closely spaced dichotomies there is abortion of the central axis so that just a stump remains between the remaining two. Further isotomous branching produces trusses of 64–128 sporangia. The overall shape of the primary xylem and that of the protoxylem predicts some of these branching events (Banks, 1980).

Regular branching patterns are more difficult to discern in *P. crenulatum* (Doran, 1980). Trifurcations (produced by double dichotomies) are recorded in the spiny axes as well as isotomous and anisotomous branching, the first producing clustering at the base of vertical, often fertile, branches and also close to the ultimately forking axes with recurved tips. Circinate coiling, a feature usually associated with the zosterophylls, is recorded in some lateral primordia. Doran compared some of these branching patterns in rhizomatous and upright systems with those of Carboniferous ferns in the Anachoropteridaceae and Botryopteridaceae.

Branching in the Zosterophyllophytina

The Zosterophyllophytina with lateral rather than terminal sporangia is the largest group of early vascular plants. Much of its variation derives from positioning of sporangia and branching patterns, which encompass all the forms already described and which can be related to increased sporing capacity and duration of sporing, as well as to increased stature. In *Zosterophyllum* itself, the oldest member, erect and sparsely branched axes are attached to a well-defined plagiotropic basal axial system in which branching is described as K- or H-shaped (Fig. 1A). This shape depends on the proximity of a second isotomy to the first, with normally similar axial width in all branch products. Although originally reconstructed as part of the aerial system or basal region with one branch directed downwards and one up (Walton, 1964), Gensel *et al.* (1969) interpreted similar patterns in *Kaulangiophyton* as horizontal systems with some branches upwardly directed. A similar position is evident in Gerrienne's recent reconstruction for *Z. deciduum* (Gerrienne, 1988; Fig. 1A), and also in the Chinese *Discalis* (Hao, 1989).

In the remaining zosterophylls the branching erect aerial system is anisotomous, and the main axis usually straight or slightly zig-zagged (Fig. 1F–H). The pattern may be repeated in the lateral branch systems, with the most distal branches isotomous. All three branching types may be seen in the same specimen of *Gosslingia breconensis.* In *Sawdonia ornata* the laterals curve upwards to lie parallel to the main axis (Fig. 1D), while *Thrinkophyton* shows regular branching and insertion of sporangia (Fig. 1F).

In all these examples, major branching is believed to be in one plane. Evidence comes from the non-superpositioning of laterals in specimens preserved on one bedding plane and from permineralized *Sawdonia* (Rayner, 1983) where the whole planar system is at an angle to bedding. In the various papers relating to these plants, authors have attempted to explain the adaptive significance of this apparently unstable system, e.g. in terms of light harvesting. It is also possible that this was a developmental constraint, that the apex was just not programmed for changes in 'caulitaxy', particularly if the plant also possessed a creeping system that favoured planar branching. It may well be related to circinate vernation (see later). It appears not to have been a major disadvantage because *Sawdonia ornata* is one of the very few Early Devonian plants to have persisted into Late Devonian times.

It is perhaps no coincidence that in addition to this morphological bilateral symmetry the cross-sectional shape of the xylem is elliptical, and sporangia are arranged in one or two rows. For recent discussion on the zosterophylls, see Niklas and Banks (1990), Lyon and Edwards (1991), Gensel (1992) and Hueber (1992).

Another conspicuous feature of these zosterophylls lacking strobili, and also *Zosterophyllum divaricatum*, is circinate vernation. Lateral axes may terminate in small hooks (e.g. *Thrinkophyton* (Fig. 1F), *Koniora*) or be more extensively coiled either loosely (e.g. *Gosslingia* (Fig. 15) and the earliest indeterminate example from the Gedinnian) or tightly (e.g. *Tarella*, Fig. 17). In most cases the youngest sporangia are some distance below the coiled tips, although in *Tarella* a single sporangium has been recorded just proximal to the coiled region. As Lyon and Edwards (1991) noted, such coiling would have offered protection from desiccation and possible injury to the young apex, although its planar nature would have left lateral sides of the axes exposed. In this respect it would be useful to know exactly where sporangia are initiated, i.e. whether at the growing apex as seems most likely when compared with strobili (see earlier) or from lateral meristems on mature stems. Pursuing a putative protective function it is perhaps relevant that isolated apices of *S. ornata* are particularly spiny and in *Trichopherophyton teuchansii* the unicellular spines are precociously developed in the hooked apex (Figs 12, 16). This Rhynie Chert species offers the best opportunity for detailed anatomical investigations of circinate vernation. Only one such tip, fortuitously sectioned longitudinally, has been found (Lyon & Edwards, 1991; Fig. 12) and is described in detail here for the first time.

The specimen is about 3 mm long and the cut end about 1.3 mm in diameter, and the apex, which is directed distally as a result of the almost complete coil, is about 0.6 mm, although the cells of the meristem itself cannot be distinguished. Indeed, apart from in the vascular system and the unicellular spines, individual cells are represented by darker, slightly fuzzy areas and are thus in marked contrast to the relatively larger cells of the ground tissue systems in mature axes where very clearly defined cell walls surround clear areas. The most prominent feature of the longitudinal section is the undifferentiated vascular system – a central cylinder of more darkly coloured elongate cells showing no tracheidal thickenings or differentiation into zones and occupying about 40% of the diameter of the stem (Fig. 16). This procambium has a lateral branch which ends abruptly below a superficial meristematic region (Fig. 16), whose surface is more or less flush with that of the rest of the axis, but differs in the darker appearance of the cells and their lack of hairs. Well-developed hairs occur immediately adjacent to the meristem in this section, and in the equivalent region to the meristem in preceding and succeeding peels. The meristem faces and is just below the hooked main apex, but whether it is a sporangial or lateral branch initial cannot be distinguished. As the limited evidence for branching in *Trichopherophyton* suggests it was isotomous, possibly pseudomonopodial, a sporangial destination seems more likely, although the massive procambial strand, almost as wide as that of the main axis, is perhaps more appropriate to an axis (but see earlier). In either case, it shows that protection of lateral meristem was achieved by its position and more importantly, in this taxon, by the surrounding hairs. The dimensions of these hairs, including those of the subtending epidermal cells, are similar to those on the mature axes, even though the remaining epidermal cells in the area cannot be so distinguished and measured. If the lateral meristem is indeed part of the axial system it shows that lateral branching was not achieved by gross 'cleavage' of the apex, that the branch primordium was differentiated in the same plane as the coiled tip thus contributing to a planar branching system.

In extant plants, circinate vernation characterizes, with very few exceptions (e.g. the leaves of *Drosera binata*), the leaves of ferns, including Carboniferous representatives. From the work of Steeves and colleagues comes detailed anatomical information on how the crozier is formed and how it uncoils in *Osmunda*. However, the fern leaf is a determinate organ and the zosterophyll aerial branching system, at least where sporangia are not organized into a strobilus, was presumably open. Thus, while the manner of crozier formation (by increased number of cells on the outer side of curve) and its uncoiling (by mainly adaxial cell division and then elongation on the inside) may have been similar, uncoiling in the fern is preceded by cessation of all apical activity. Indeed Bower (1884) noted in *Pilularia*, the fern system most superficially similar to the ancient axial one, that the activity of the two-sided wedged-shaped apical cell ceases after elongation of the distal part of the leaf (similar perhaps to that part of the leaf lacking pinnules in *Osmunda*) and before the uncoiling. Ferns with indeterminate growth (e.g. *Lygodium*) may thus be more usefully compared, but it has not been possible to find any detailed information on their growing apices. In *Stromatopteris*, whose organs are described as 'ill-defined' (Bierhorst, 1969, 1973), for the leaves themselves (although their derivation may be anomalous), subsequent development appears to conform to that in *Osmunda* but this was not specifically stated.

In the Lower Devonian fossils, curvature in the apices ranges from a recurved tip, as in *Trichopherophyton* (essentially a hook of one turn) to those with at least two complete coils (e.g. *Tarella*, Fig. 17). Some taxa show only the first type (e.g. *Thrinkophyton*) while others have both (e.g. *Tarella*). In *Tarella* there are two plus turns in the larger axes and hooks in the short broad lateral structures described earlier and in narrower axes. This suggests changes in the nature of the tips during the growth of the plant and possibly a combination of determinate growth (genetically constrained) with some further constraint imposed by the metabolic state of the vegetative system (see also Hueber, 1992). Because of the size of the aerial system compared with that of a fern frond, it is possible, but considered unlikely, that the entire system could have developed from a single crozier borne on some kind of rhizome. However, the fern megaphyll does show how a bilaterally symmetrical planar system can be derived from a coiled structure with lateral appendages in more or less one row.

Returning to the *Tarella* example, maintenance of the two-turn arrangement during growth of the wider axes would have required continual accommodation of the new innermost tissues by the enveloping ones as well as eventual straightening. Again the *Osmunda* data may provide the mechanistic evidence, although there have been insufficient comparative studies on extant ferns to estimate variation in development. There is, for example, evidence suggesting that curvature is not always achieved by increase in cell numbers on the longer side. In *Cibotum* and *Alsophila*, ground tissue cells in this region just prior to uncoiling are distinctly longer, although this may be related to the uncoiling process itself (see Steeves, 1963). Unfortunately in *Trichopherophyton*, although abaxial and adaxial sides superficially appear different with the impression of wider cells on the abaxial, such differences cannot be quantified, particularly in relation to cell length and numbers.

Considering tissue differentiation, that in *Osmunda* may well seem a modern analogue, particularly for those zosterophylls with a hypodermal zone of elongate cells with thickened walls. Briggs and Steeves (1959) recognized a hypodermal zone of cells within half a turn of the apex. This zone elongated while ground tissue cells remained short as the back of the crozier extended, with lignification occurring below the elongation zone. The earliest tracheids, invariably helical, were also found within half a turn of the apex, but were destroyed during uncoiling and were replaced by further tracheidal initials during this phase: subsequent maturation was completed after cessation of elongation. No such cortical zonation was noted in *Trichopherophyton*, which is an atypical zosterophyll in that it lacks a sterome (Lyon & Edwards, 1991). Further, no differentiated elements could be detected towards the edge of the procambial strand as might be anticipated in exarch xylem, although the darkened nature of the region may well obscure them. It is, however, noteworthy that maturation of metaxylem tracheids has still not occurred some 3 mm below the apex, but initiation of the procambial strand itself occurs very close, within 160 μm of the apical meristem and about 40 μm below the lateral one.

Anomalous branching

Under this heading are included lateral structures whose position and appearance suggest deviation from the precise and predictable branching sequences just discussed, which are usually interpreted as the equivalent of aerial shoot systems. Such lateral branches are broadly divided into two groups depending on whether or not they are regular in positioning.

In compression fossils, many of the zosterophylls possess a small raised area or depression called an axillary tubercle or subaxillary branch just below a branch point. Where preserved in permineralizations, it projects at right angles to the planar branching system, probably on the same side in successive branches (*Gosslingia*: Edwards, 1970b). In *Deheubarthia*, development/dégagement in this region revealed a downwardly diverted branch, covered with spines as are the aerial branches, and ending in a tight coil (Fig. 1E). Forking was occasionally observed (Edwards *et al.*, 1989). In contrast, in *Crenaticaulis verriculosus* the presumed gravitropic branch lacked the rows of enations characteristic of the main axis (Banks & Davis, 1969). Upwardly directed branching axes occurring in this position in *Anisophyton gothani* are spiny (Fig. 1H), as is the main axial system, while in *A. potonei* there are normally small hooks in this position (Haas & Remy, 1986). Such lateral structures were termed Angular-Organe, based on similarities with the angular meristems developing into rhizophores in *Selaginella* (e.g. Webster, 1992). Anatomical studies involving serial sections through the branch point show that the subaxillary branch is an integral part of the aerial branching system. In *Gosslingia* the elliptical xylem strand characteristic of a 'main' axis becomes more elongate, separates off a large terete strand, the subaxillary branch trace, and then a much more poorly defined and smaller trace to the lateral branch. The terete strand turns abruptly through 90° and enters a very short branch. In *Deheubarthia*, the trace to the subaxillary branch is also relatively large, but here the main axis strand becomes J-shaped at branching and a smaller lateral branch trace separates off *before* that to the subaxillary branch, a reversal of the sequence apparent in the external morphology.

Haas and Remy emphasized that their angular organs reflected the marked dorsiventrality of the axial system, again making comparison with *Selaginella*. The morphological status of the latter has been much debated (see Gifford & Foster, 1989; Webster, 1992) and its plasticity, seen for example in the development of leaves under experimental conditions (e.g. Wochok & Sussex, 1976), is perhaps paralleled in the Lower Devonian plants. Non-apical branching of a different, but still regular, kind is seen in *Sawdonia* where Rayner identified a small rounded slightly hooked projection lacking spines on the abaxial surface of a lateral branch just above its inception (2–3 diameters beyond the main axis). Such hemispherical projections are replaced by smooth downwardly directed axes (Fig. 1D), sometimes branched. Rayner (1983) speculated as to whether they were dormant lateral branches with retained capacity for further growth, and thus a possible adaptation for colonization of an unstable environment. A similar function was suggested for pronounced lateral outgrowths in *Tarella* (Edwards & Kenrick, 1986), which are not regularly spaced and are of two major kinds, short hooked or circinate structures with wide bases (Fig. 18) and no tapering, and more elongate forms sometimes tapering and almost incomplete distally. Unfortunately, orientation of these structures with respect to the main branching system could not be determined and hence they may have functioned as rooting structures or dormant buds capable of further growth upwards.

Hemispherical projections characterize a number of taxa of diverse affinity including *Zosterophyllum*, *Renalia*, *Rhynia* and *Sawdonia* and may be the functional equivalents of dormant buds, or abscissed lateral branches. In *Rhynia gwynne-vaughanii*, those near the base of the aerial system bear tufts of rhizoids. Perhaps the most bizarre collection of such appendages occur in similar positions in *Bitelaria clubjanski* (Incertae sedis) in which clavate structures are interpreted as aborted branches while those with rounded apices are considered dormant (Johnson & Gensel, 1992). Root-like structures are rarely reported in Lower Devonian fossils (but see *Psilophyton dawsonii*, Banks *et al.*, 1975), a notable exception being the regularly isotomous branching narrow axes located just above the much larger, fertile lateral systems in *Hsüa robusta* (Li, 1982, 1992; Fig. 1B). They are apparently exogenous in origin with

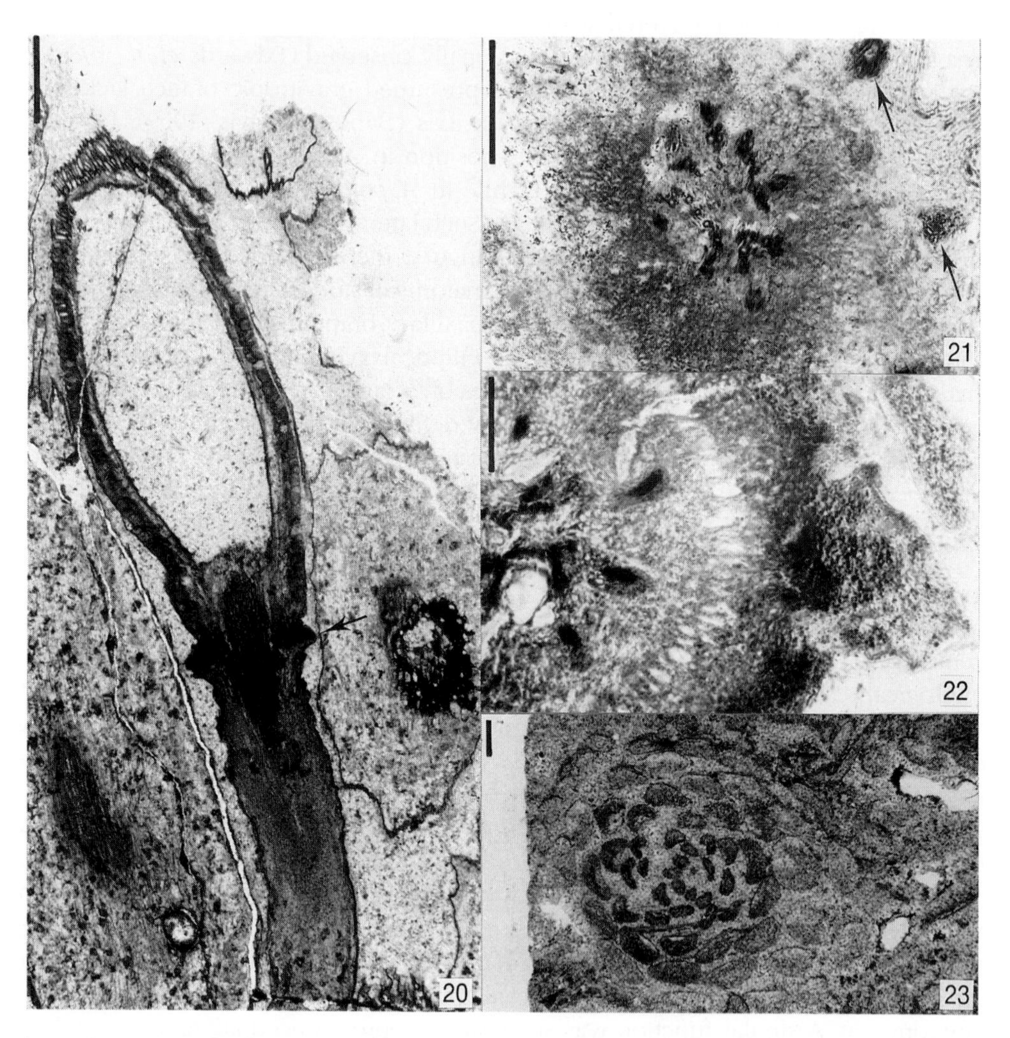

Figure 20 Longitudinal section *Rhynia gwynne-vaughanii* showing empty sporangium and abscission zone (arrowed). H470–50. Scale bar 500 μm.

Figures 21–23 Transverse section stem apices of *Asteroxylon mackiei*. Scale bars 500 μm. **21** Central region of stem showing undifferentiated metaxylem. Arrowed are traces to leaves. H2547. **22** Stem with undifferentiated metaxylem but differentiated cortex and leaf base. H2553. **23** Section above apex showing leaf arrangement. H2554.

their vascular tissues in clear connection with those of the main branching system. They occur on both creeping and erect axes.

ROOTS AND EARLY LAND PLANTS – OR 'WHAT'S IN A NAME?'

Discussions on the functions of the anomalous branching systems described earlier inevitably overlap with speculation on their organographic nature and invoke attempts to 'shoehorn' such systems into the primary dichotomy of shoot and root – a concept, associated traditionally with the bipolar angiosperm embryo, that has dominated vascular plant morphology (see, for example, Groff & Kaplan, 1988). The pteridophyte system is better referred to Goebbel's homorhizic category that encompasses plants lacking a bipolar embryo, all of whose roots are shoot-borne, and for which the appellation 'adventitious' may be inappropriate. However, the earliest fossil members (often combined with extant Psilotales) are traditionally considered rootless, and indeed where sufficiently completely preserved are best considered as a shoot system with an upright aerial component concerned with photosynthesis and reproduction, and a basal, usually horizontal, component concerned with anchorage and absorption (e.g. *Rhynia gwynne-vaughanii*). Such plants thus represent a stage before the evolution of the root/shoot dichotomy, and within such a complex might well be detected the initial stages in the evolution of roots. However, their recognition as endogenously produced, naked, gravitropic, absorbing systems in the anomalous branching of usually axial fragments of mature plants is a daunting and (bearing in mind the *Selaginella* rhizophore controversy) probably fruitless exercise. But it cannot be denied that some of these lateral systems, e.g. in *Hsüa*, look remarkably similar to roots, even though in the Chinese plant their positioning on the presumed erect, reproducing axial systems is puzzling. Even where anatomy is preserved, as in the Rhynie Chert, critical information may be missing (e.g. apices with root caps, endogenous origins) or ambiguous (e.g. *Trichopherophyton* with little anatomical distinction, particularly in the exarch xylem, between axes with spines and those with rhizoids or possibly root hairs).

The fossil lycophytes provide some evidence for shoot-borne roots. Rayner (1984) recorded less regular branching in naked axes attached to wider more characteristically branching stems of *Drepanophycus*, both with and without leaves. The latter occurrences are more frequent. Rayner interpreted the lateral systems as roots comparable to the stem-borne examples in extant Lycopodiales and cited evidence from branching pattern, relative size and their positioning both on the plant and in the surrounding sediment. Their oblique orientation to the bedding planes on which the stems are preserved and their penetration of several centimetres of sediment suggested that the plants were preserved *in situ*. Since many of the stems are leafless, Rayner postulated that they too were subterranean. However, Schweitzer (1980) had earlier published a reconstruction with an apparently surface-colonizing rhizomatous system bearing deltoid-shaped outgrowths (contrasting with the awl-like leaves of the orthotropic stems) and tufts of frequently branching roots. For *Asteroxylon mackiei*, Kidston and Lang (1920b) described a transition region with scale-like outgrowths lacking leaf-traces between 'leafy' shoots with stellate xylem and smooth cylindrical axes with cylindrical xylem. The latter system was interpreted as a subterranean rhizome, but with 'behaviour' very root-like, both in position and physiologically, although rhizoids have not been recorded in *Asteroxylon*. In calling them rhizomes, the authors were emphasizing their exogenous origin, although in other aerial leafy axes, they illustrated a unique example of anatomically preserved endogenous branching in a stem also distinguished by a clearly demarcated layer, interpreted as endodermous, around the vascular tissue. The fate of the lateral branch is not indicated.

Kidston and Lang reconstructed the transition zone with some confidence from a number of specimens rather than a series of sections from a single plant, and in their line-drawing reconstruction of a whole plant, placed it near the base of aerial axes in continuity with horizontal leafless rhizomes bearing more branched 'root-like' systems. Thus for both the axial plants and early lycophytes, there is the inference that the sporophyte may have exhibited unipolar growth with initial rhizomatous phases, and further that the young sporophyte continued to be dependent on the gametophyte at least during the early stages of its development (Rothwell, in press). Such plants are characteristic of Rothwell's Psilotioid grade which also encompasses extant Psilopsida, Lycopodiales and Ophioglossales. Rothwell also distinguished a Cooksonioid grade for plants characterized by determinate growth and a persistent gametophyte (as in bryophytes), although there is no direct evidence for the latter. Indeed it would be a less radical change to concepts on early land plant life cycles to consider Cooksonioid and Psilotioid grades as a continuum characterized by changes in sporophytic branching patterns. Nevertheless, Rothwell's hypothesis involving gametophytes is a fascinating one, and to some extent supported by recent data on Devonian gametophytes (e.g. Remy *et al.*, 1980; Remy & Hass, 1991; Schweitzer, 1983). The detection of gametophytes of rhyniophytoids in the Silurian and of the early bryophytes becomes an even greater challenge.

Deliberations such as these demand not only the continuing requirement for the collection of new material, but also a reassessment of existing data. In asking new questions of the fossil record, we can go a little way to not only reconstructing the morphology and anatomy of early land plants but to resurrecting them as growing and metabolizing organisms.

REFERENCES

BANKS, H.P., 1975. Reclassification of Psilophyta. *Taxon, 24:* 401–413.

BANKS, H.P., 1980. The role of *Psilophyton* in the evolution of vascular plants. *Review of Palaeobotany and Palynology, 29:* 165–176.

BANKS, H.P. & DAVIS, M.R., 1969; *Crenaticaulis*, a new genus of Devonian plants allied to *Zosterophyllum*, and its bearing on the classification of early land plants. *American Journal of Botany, 56:* 436–449.

BANKS, H.P., LECLERCQ, S. & HUEBER, F.M., 1975. Anatomy and morphology of *Psilophyton dawsonii*, sp.n. from the late Lower Devonian of Quebec (Gaspé), and Ontario, Canada. *Palaeontographica Americana, 8:* 77–127.

BIERHORST, D.W., 1969. On *Stromatopteris* and its ill-defined organs. *American Journal of Botany, 56:* 166–174.

BIERHORST, D.W., 1971. *Morphology of Vascular Plants.* New York: Macmillan.

BIERHORST, D.W., 1973. Non-appendicular fronds in the Filicales. *Botanical Journal of the Linnean Society, 67 (Suppl. 1):* 45–57.

BIERHORST, D.W., 1977. On the stem apex, leaf initiation and early leaf ontogeny in filicalean ferns. *American Journal of Botany, 64:* 125–152.

BOWER, F.O., 1884. On the comparative morphology of the leaf in the vascular cryptogams and gymnosperms. *Philosophical Transactions of the Royal Society, 175:* 565–615.

BRIGGS, W.R. & STEEVES, T.A., 1959. Morphogenetic studies on *Osmunda cinnamomea* L. – the mechanism of crozier uncoiling. *Phytomorphology, 9:* 134–147.

CAI, C.-Y., DOU, Y.-W., & EDWARDS, D., 1993. New observations on a Pridoli plant assemblage from north Xinjiang, northwest China, with comments on its evolutionary and palaeogeographical significance. *Geological Magazine, 130:* 155–170.

DORAN, J.B., 1980. A new species of *Psilophyton* from the Lower Devonian of northern New Brunswick, Canada. *Canadian Journal of Botany, 58:* 2241–2262.

EDWARDS, D., 1970a. Fertile Rhyniophytina from the Lower Devonian of Britain. *Palaeontology, 13:* 451–461.

EDWARDS, D., 1970b. Further observations on the Lower Devonian plant, *Gosslingia breconensis* Heard. *Philosophical Transactions of the Royal Society of London, B, 258:* 225–243.

EDWARDS, D. 1993. Cells and tissues in the vegetative sporophytes of early land plants. *New Phytologist, 125*: 225–247.

EDWARDS, D. & DAVIES, M.S., 1990. Interpretations of early land plant radiations: 'facile adaptationist guesswork' or reasoned speculation? In P.D. Taylor & G.P. Larwood (eds), *Major Evolutionary Radiations*, pp. 351–376. The Systematics Association, Special volume no. 42. Oxford: Clarendon.

EDWARDS, D. & KENRICK, P., 1986. A new zosterophyll from the Lower Devonian of Wales. *Botanical Journal of the Linnean Society, 92:* 269–283.

EDWARDS, D. & SELDEN, P.A., 1992. The development of early terrestrial ecosystems. *Botanical Journal of Scotland, 46:* 337–366.

EDWARDS, D., FEEHAN, J. & SMITH, D.G., 1983. A late Wenlock flora from Co. Tipperary, Ireland. *Botanical Journal of the Linnean Society, 86:* 19–36.

EDWARDS, D., KENRICK, P. & CARLUCCIO, L.M., 1989. A reconsideration of cf. *Psilophyton princeps* (Croft and Lang, 1942), a zosterophyll widespread in the Lower Old Red Sandstone of South Wales. *Botanical Journal of the Linnean Society, 100:* 293–318.

EDWARDS, D., DAVIES, K.L. & AXE, L., 1992. A vascular conducting strand in the early land plant *Cooksonia*. *Nature, 357:* 683–685.

EDWARDS, D.S., 1980. Evidence for the sporophytic status of the Lower Devonian plant *Rhynia gwynne-vaughanii* Kidston and Lang. *Review of Palaeobotany and Palynology, 29:* 177–188.

FANNING, U., EDWARDS, D. & RICHARDSON, J.B., 1992. A diverse assemblage of early land plants from the Lower Devonian of the Welsh Borderland. *Botanical Journal of the Linnean Society, 109:* 161–188.

GENSEL, P.G., 1992. Phylogenetic relationships of the zosterophylls and lycopsids: evidence from morphololgy, paleoecology and cladistic methods of inference. *Annals of the Missouri Botanical Garden, 79:* 450–473.

GENSEL, P., KASPER, A. & ANDREWS, H.N., 1969. *Kaulangiophyton*, a new genus of plants from the Devonian of Maine. *Bulletin of the Torrey Botanical Club, 96:* 265–276.

GERRIENNE, P., 1988. Early Devonian plant remains from Marchin (North of Dinant Synclinorium, Belgium), I. *Zosterophyllum deciduum* sp. nov. *Review of Palaeobotany and Palynology, 55:* 317–335.

GIFFORD, E.M. & FOSTER, A.S., 1989. *Morphology and Evolution of Vascular Plants.* New York: W.H. Freeman.

GRAY, J., 1985a. The microfossil record of early land plants: advances in understanding of early terrestrialization, 1970–1984. *Philosophical Transactions of the Royal Society of London, B, 309:* 167–195.

GRAY, J., 1985b. Ordovician-Silurian land plants: the interdependence of ecology and evolution. In M.G. Bassett & J.D. Lawson (eds), *Autecology of Silurian Organisms.* Special Papers in Palaeontology, vol. 32, pp. 281–294. London: The Palaeontological Association.

GROFF, P.A. & KAPLAN, D.R., 1988. The relation of root systems to shoot systems in vascular plants. *Botanical Review, 54:* 387–422.

HAO SHOU-GANG, 1989. A new zosterophyll from the Lower Devonian (Siegenian) of Yunnan, China. *Review of Palaeobotany and Palynology, 57:* 155–171.

HASS, H., 1991. Die Epidermis von *Horneophyton lignieri* (Kidston & Lang) Barghoorn & Darrah. *Neues Jahrbuch für Geologie und Paläontologie Abhandlungen, 183:* 61–85.

HASS, H. & REMY, W., 1986. Angular-Organe-Ausdrück einer wenig beachteten morphogenetischen Strategie devonischer Pflanzen. *Argumenta Palaeobotanica, 7:* 155–171.

HØEG, O.A., 1967. Ordre *Incertae Sedis* des Barinophytales. In E. Boureau (ed.), *Traité de Paléobotanique*, vol. 2, pp. 400–433. Paris: Masson et Cie.

HUEBER, F.M., 1992. Thoughts on the early lycopsids and zosterophylls. *Annals of the Missouri Botanical Garden, 79:* 474–499.

JOHNSON, N.G. & GENSEL, P.G., 1992. A re-interpretation of the Early land plant, *Bitelaria* Istchenko and Istchenko, 1979, based on new material from New Brunswick, Canada. *Review of Palaeobotany and Palynology, 74:* 109–139.

KENRICK, P. & CRANE, P.R., 1991. Water-conducting cells in early fossil land plants: implications for the early tracheophytes. *Botanical Gazette, 152:* 335–356.

KENRICK, P. & EDWARDS, D., 1988. A new zosterophyll from a recently discovered exposure of the Lower Devonian Senni Beds in Dyfed, Wales. *Botanical Journal of the Linnean Society, 98:* 97–115.

KENRICK, P., REMY, W. & CRANE, P.R., 1991. The structure of the water-conducting cells in the enigmatic early land plants *Stockmansella langii* Fairon-Demaret, *Huvenia kleui* Hass et Remy and *Sciadophyton* sp. Remy *et al.*, 1980. *Argumenta Palaeobotanica, 8:* 179–191.

KIDSTON, R. & LANG, W.H., 1920a. On Old Red Sandstone plants showing structure, from the Rhynie Chert Bed, Aberdeenshire. Part II. Additional notes of *Rhynia Gwynne-Vaughani*, Kidston and Lang; with descriptions of *Rhynia major*, n.sp. and *Hornea Lignieri*, n.g., n.sp. *Transactions of the Royal Society of Edinburgh, 52:* 603–627.

KIDSTON, R. & LANG, W.H., 1920b. On Old Red Sandstone plants showing structure, from the Rhynie Chert Bed, Aberdeenshire. Part III. *Asteroxylon Mackiei*, Kidston and Lang. *Transactions of the Royal Society of Edinburgh, 52:* 643–680.

KIDSTON, R. & LANG, W.H., 1921. On Old Red Sandstone plants showing structure, from the Rhynie Chert Bed, Aberdeenshire. Part IV. Restorations of the vascular cryptogams, and discussion of their bearing on the general morphology of the Pteridophyta and the origin of the organisation of land plants. *Transactions of the Royal Society of Edinburgh, 52:* 831–854.

LI CHENG-SEN, 1982. *Hsüa robusta*, a new land plant from the Lower Devonian of Yunnan, China. *Acta Phytotaxonomica Sinica, 20:* 331–342.

LI CHENG-SEN, 1992. *Hsüa robusta*, an early Devonian plant from Yunnan Province, China and its bearing on some structures of early land plants. *Review of Palaeobotany and Palynology, 71:* 121–147.

LI CHENG-SEN & EDWARDS, D., 1992. A new genus of early land plants with novel strobilar construction from the Lower Devonian Posongchong Formation, Yunnan Province, China. *Palaeontology, 35:* 257–272.

LYON, A.G. & EDWARDS, D., 1991. The first zosterophyll from the Lower Devonian Rhynie Chert, Aberdeenshire. *Transactions of the Royal Society of Edinburgh: Earth Sciences, 82:* 323–332.

NIKLAS, K.J., 1978. Branching patterns and mechanical design in Palaeozoic plants: a theoretic assessment. *Annals of Botany, 42:* 33–39.

NIKLAS, K.J. & BANKS, H.P., 1990. A reevaluation of the Zosterophyllophytina with comments on the origin of lycopods. *American Journal of Botany, 77:* 274–283.

RAYNER, R.J., 1983. New observations on *Sawdonia ornata* from Scotland. *Transactions of the Royal Society of Edinburgh: Earth Sciences, 74:* 79–93.

RAYNER, R.J., 1984. New finds of *Drepanophycus spinaeformis* Göppert from the Lower Devonian of Scotland. *Transactions of the Royal Society of Edinburgh: Earth Sciences, 75:* 353–363.

REMY, W. & HASS, H., 1991. Gametophyten und Sporophyten im Unterdevon-Fakten und Spekulationen. *Argumenta Palaeobotanica, 8:* 193–223.

REMY, W., SCHULTKA, S., HASS, H. & FRANZMEYER, F., 1980. *Sciadophyton* Steinmann – ein Gametophyt aus dem Ems. *Argumenta Palaeobotanica, 6:* 73–94.

ROTHWELL, G.W. The fossil history of branching: implications for the phylogeny of land plants. In: Hoch, D.C. and Stephenson, A.G. (eds), *Experimental and Molecular Approaches in Plant Biosystematics.* Monographs in Systematic Botany, Missouri Botanical Garden (in press).

SCHOPF, J.M., MENCHER, E., BOUCOT, A.J. & ANDREWS, H.N., 1966. Erect plants in the early Silurian of Maine. *U.S. Geological Survey. Professional Paper, 550-D:* D69–D75.

SCHWEITZER, H.-J., 1980. Über *Drepanophycus spinaeformis* Göppert. *Bonner Paläobotanische Mitteilungen, 7:* 1–29.

SCHWEITZER, H.-J., 1983. Die Unterdevonflora des Rheinlandes. *Palaeontographica, Abteilung B, 189:* 1–138.

SHUTE, C. & EDWARDS, D., 1989. A new rhyniopsid with novel sporangium organization from the Lower Devonian of South Wales. *Botanical Journal of the Linnean Society, 100:* 111–137.

STEEVES, T.A., 1963. Morphogenetic studies of fern leaves. *Journal of the Linnean Society (Botany), 58:* 401–415.

WALTON, J., 1964. On the morphology of *Zosterophyllum* and some early land plants. *Phytomorphology, 14:* 155–160.

WEBSTER, T.R., 1992. Development problems in *Selaginella* (Selaginellacea) in an evolutionary concept. *Annals of the Missouri Botanical Garden, 79:* 632–647.

WEBSTER, T.R. & STEEVES, T.A., 1967. Developmental morphology of the root of *Selaginella martensii* Spring. *Canadian Journal of Botany, 45:* 395–404.

WOCHOK, Z.S. & SUSSEX, I.M., 1976. Redetermination of cultured root tips to leafy shoots in *Selaginella willdenovii. Plant Science Letters, 6:* 185–192.

CHAPTER

4

Saltational evolution of form in vascular plants: a neoGoldschmidtian synthesis

RICHARD M. BATEMAN & WILLIAM A. DIMICHELE

CONTENTS

Shape and Form in Plants and Fungi
ISBN 0–12–371035–9

Abstract

Saltational evolution, a much abused term, is here narrowly defined as a genetic modification that is expressed as a profound phenotypic change across a single generation and results in a potentially independent evolutionary lineage (prospecies: 'hopeful monster' of Richard Goldschmidt). Dichotomous saltation is driven by mutation within a single ancestral lineage, and can result not only in instantaneous speciation but also in the simultaneous origin of a supraspecific taxon. Reticulate saltation is driven by allopolyploidy and thus incorporates genes of two ancestral lineages; it results in speciation only. Several exceptionally rapid but multigenerational evolutionary mechanisms are collectively termed parasaltational. Saltational evolutionary mechanisms probably generated many vascular plant species and most higher taxa. Hypotheses of saltation can be falsified using cladograms, which also provide an essential context for interpretations of evolutionary process as well as pattern – here illustrated using studies of evolutionary developmental change in architectures of fossil lycopsids and living angiosperms.

The neoGoldschmidtian synthesis advocated here accepts Goldschmidt's concept of speciation across a single generation but rejects his preferred causal mechanism of large-scale mutations and his requirement for competitively high levels of fitness in the monsters. Rather, we postulate that vast numbers of hopeful monsters are continuously generated by mutation of key homoeotic genes that control ontogeny via morphogens ('D-genes' of Wallace Arthur). The fitness of hopeful monsters is inevitably too low to survive competition-mediated selection – their establishment requires temporary release from selection in unoccupied niches. The prospecies can then be honed to competitive fitness by gradual reintroduction to neodarwinian selection.

When viewed backward through geological time, niches become less well-defined and more often vacant, causing a corresponding increase in the probability of successful establishment of hopeful monsters. Hence, saltation was most important among the earliest land plants, explaining the Siluro-Devonian origins of all class-level taxa. Although D-genes are similar at the molecular level across the biotic kingdoms, their phenotypic expression differs between higher plants and higher animals; this reflects highly contrasting modes of growth, notably the localization of plant growth in numerous meristems and consequent continuous, largely iterative development. More importantly, the sessile life-style of plants renders competition indirect and environmentally mediated; thus, vectorial selection is a far less profound cause of evolution in plants than in animals. Plants enjoy much greater latitude for non-lethal experimentation in form by saltation. Future advances in the study of evolutionary mechanisms will require cladograms that use phenotypically expressed genes as characters, rather than static morphology or cryptic base pairs, thereby allowing reciprocal illumination between phenotype and genotype.

INTRODUCTION

Despite continual undercurrents of dissent, the neodarwinian model of competitively driven selection has dominated evolutionary theory for several decades. Critics of neodarwinism have advocated a wide range of alternative or supplementary theories and thus have never coalesced into a coordinated opposition. Widespread discord continues among proponents of competing evolutionary theories, driven by two fallacious arguments. Firstly, contrasting hypotheses tend to be regarded as mutually exclusive; just as there is undoubtedly 'one true tree' of evolutionary relationships among all species, it is assumed that there must also be one true underlying mechanism that is transcendent over all others. Secondly, any one theory must be generally applicable to all taxa. Even when originally advanced as taxon-specific, an attractive theory is rapidly coopted by increasing numbers of sympathetic biologists studying other taxa. The generalization process usually requires progressive abandonment of facets of the original theory as additional organisms are shoe-horned into place; indeed, the relative resilience of Darwin's (1859) original formulation of adaptation by competitive selection can be attributed to the exceptional range of invertebrate, vertebrate and plant groups studied by this remarkable natural historian before finalizing his evolutionary hypothesis (e.g. Desmond & Moore, 1991).

However, the neodarwinian synthesis was fashioned largely by biologists and palaeobiologists studying only one of the five to seven widely recognized kingdoms (Whittaker, 1969; Field *et al.,* 1988; Margulis & Schwartz, 1988; Fernholm *et al.*, 1989; Gouy & Li, 1989; Woese *et al.*, 1990; Grant & Horn, 1992; Wainwright *et al.*, 1993), namely the Animalia (Haldane, 1932; Simpson, 1944; Dobzhansky, 1951; Mayr, 1963; Maynard Smith, 1972; Ayala & Valentine, 1979; Dawkins, 1986, 1989; Futuyma, 1986, 1987; Ridley, 1992). Emphasis on an even narrower and more highly derived clade, the Vertebrata, has further weakened the validity of generalizing the neodarwinian paradigm across the far more extensive tree of life (cf. Fernholm *et al.*, 1989; Woese *et al.*, 1990). Certainly, the remarkable repertoire of the most prominent living evolutionary essayist, S.J. Gould, is notable for a dearth of botanical examples. Reviews of evolutionary mechanisms in the kingdoms Plantae (e.g. Bell, 1992; Cronquist, 1968; Grant, 1971; Knoll, 1984; Knoll *et al.*, 1984; Knoll & Niklas, 1987; Raven *et al.*, 1986; Stace, 1989; Stebbins, 1950, 1971, 1974; Takhtajan, 1969, 1980, 1992) and Fungi (e.g. Rayner *et al.*, 1987) are far fewer and have coopted with minor modifications theories that are essentially zoocentric.

Occasional attempts to challenge (or at least supplement) neodarwinism have experienced similarly divergent fates according to the kingdom under scrutiny. Contributions by zoologists (e.g. Goldschmidt, 1940; Gould & Lewontin, 1979) have often prompted heated debate, whereas those from botanists (e.g. Croizat, 1962; Lewis, 1962; Sattler, 1986; Traverse, 1988; van Steenis, 1969, 1976) have received little attention.

Several Victorian naturalists (e.g. Bateson, 1894) advocated various saltational mechanisms; indeed, following his catalytic voyage on *HMS Beagle*, Darwin himself flirted with saltation (Desmond & Moore, 1991, p. 225). Perhaps the most notorious advocacy of saltation was *The Material Basis of Evolution*, an uncompromising antidarwinian critique first published in 1940 by Berkeley developmental zoogeneticist Richard Goldschmidt. Most notably, Goldschmidt argued that 'systemic mutations' (large-scale chromosomal rearrangements) altered early developmental trajectories to generate 'hopeful monsters' – teratological lineages of radically different phenotypes. By chance, some hopeful monsters possessed high levels of fitness that enabled their persistence as new lineages of great evolutionary significance. Burdened thus with an inviable underlying mechanism and an unreasonable assertion of competitively high fitness, aggressive proponents of the neodarwinian *New Synthesis* easily – but wrongly – discredited Goldschmidt's entire saltational paradigm (cf. Dawkins, 1986; Dietrich, 1992; Gould, 1977a, b). Occasional attempts to resurrect aspects of Goldschmidt's paradigm (Arthur, 1984, 1988; Bateman, in press; Dietrich, 1992; DiMichele *et al.*, 1989; Gould, 1982; Schindewolf, 1950; Stidd, 1987; Valentine, 1980; van Steenis, 1976; Waddington, 1957) had little impact on the scientific community.

We believe that saltation is supradarwinian rather than antidarwinian. Here, we argue that saltation has been a stronger driving force of evolution among higher plants than higher animals, and among extinct rather than extant species. We also emphasize the importance of a phylogenetic context for the interpretation of evolutionary processes as well as patterns, focusing on developmental constraints to vascular plant architecture.

DEFINING SALTATIONAL EVOLUTION

Core mechanisms: saltation *sensu stricto*

Although few evolutionary biologists have provided explicit definitions of saltational (or 'saltatory') evolution, it is clear that overall the term has been used to encompass a wide range of often conflicting concepts (cf. Dawkins, 1986). Below we examine potential criteria for defining and recognizing saltation.

Cause of evolutionary change

By definition, any evolutionary event must involve changes in both genotype and phenotype *sensu lato* (i.e. *sensu* Dawkins, 1982). Hence, we immediately exclude from further consideration all non-genetic contributors to phenotype – ecophenotypy and ontogeny, significant causes of phenotypic variation within most plant species. This highlights a subtle difference between the concept of a teratos – an individual possessing a radically different morphology from its immediate ancestor(s) irrespective of the underlying causal mechanism – and Goldschmidt's concept of a 'hopeful monster', where a genetic cause of the morphological discontinuity is assumed and non-heritable causes (for example, an environmental perturbation of a plant meristem) are specifically excluded.

Discontinuity of evolutionary change

As noted by van Steenis (1976) and Dawkins (1986), almost any biological phenomenon is discontinuous if viewed in sufficient detail. Any point mutation (or, for that matter, any meiotic division with recombination) is a quantum change in genotype at the scale of base-pair sequences, and cannot in itself define saltation.

Magnitude of evolutionary change

If the mere presence of an evolutionary discontinuity is insufficient, an objective measure of the magnitude of the discontinuity is desirable in order to define a quantitative threshold that must be surpassed for a change to qualify as saltation. It also becomes important to specify whether the evolutionary change should be measured genotypically or phenotypically, as recent studies have conclusively refuted Goldschmidt's (1940) assertion that the magnitude of genetic change is reflected in the magnitude of its phenotypic expression. We would argue that hopeful monsters are best quantified phenotypically, provided that an underlying genetic cause of the phenotypic change has been documented.

Directionality of evolutionary change

There is no requirement that a saltational change should be progressive in the sense of increasing overall complexity or fitness. Indeed, current evidence suggests that saltational events which suppress developmental genes and consequently reduce morphological complexity are more common than saltational events which increase overall complexity. In either case, we believe that decreased initial fitness is inevitable, and that saltation therefore lacks vectorial properties (i.e. predictable directionality). Admittedly, not all current saltationists are willing to relinquish the concept of a spontaneously fit monster.

Rate of evolutionary change

Some authors (e.g. Ayala & Valentine, 1979) defined saltation as a period when the temporal rate of evolution (change/time) is substantially greater than the long-term average within the lineage; this criterion underlay the original concept of 'punctuated equilibria' (Eldredge & Gould, 1972). We concur with the assertion of Dawkins (1986), Lemen & Freeman (1989) and Gould & Eldridge (1993) that mere changes in evolutionary rate are not true saltation, and that punctuationist patterns are more remarkable for periods of apparent evolutionary stasis than for periods of rapid evolutionary change. We believe that saltation is better defined by generation time than absolute time: a saltational change must occur across a single generation. This crucial and controversial defining criterion excludes several contrasting mechanisms of rapid evolutionary change that we collectively term 'parasaltational' (see later).

Reproductive isolation caused by the evolutionary change

Hopeful monsters resulting from a high-magnitude phenotypic change will have a much greater likelihood of retaining their novel phenotype if they become isolated from introgression with the parental population, particularly if the isolation reflects intrinsic properties of the monsters rather than mere *ad hoc* spatial separation (i.e. sympatry rather than allopatry). However, new lineages can become established even in the absence of reproductive isolation (Arthur, 1984). This criterion is therefore ancillary to, rather than inherent in, saltation.

Demographic entity affected by the evolutionary change

Bateman & Denholm (1989) argued that there is a recognizable and definable demographic hierarchy: individual organisms > populations > infraspecific taxa > species. (Here, population describes a geographically and ecologically restricted aggregate of conspecific organisms: ecotopodemes *sensu* Gilmour & Heslop-Harrison, 1954; avatars *sensu* Damuth, 1985; metapopulations *sensu* Levin, 1993.) Saltation can temporarily compress this hierarchy. A single hopeful monster is by definition highly geographically and ecologically restricted and can also be reproductively isolated, thereby meeting the criteria of a biological species *sensu* Mayr (1963). However, it is unlikely to be awarded specific rank by a practising taxonomist – indeed, it is unlikely to even be examined by a practising taxonomist, given the relative rarity of both hopeful monsters and taxonomists. In practice, a taxonomic species is obliged to prove its historical tenacity, by establishing a sizeable population that persists through many generations. Most lineages resulting from saltation fail to survive beyond a single generation, and few exceed 10^1–10^3 generations; these ephemeral entities are better described as prospecies. It should be emphasized that there is no intrinsic biological distinction between prospecies and taxonomic species; they can only be distinguished retrospectively, on the basis of the far greater temporal continuity and spatial extent of the latter.

Degree of genetic novelty generated during the evolutionary change

Mutation, the underlying cause of hopeful monsters, is by definition the generation of genetic novelty. However, other modes of genotypic change rely on the mixing of pre-existing genes from two species (hybridization) or on the duplication of pre-existing genes in one species (autopolyploidy). Mixing and duplication are combined and fixed in allopolyploidy (Stace, 1989, 1993; Stebbins, 1971; Thompson & Lumaret, 1992). Hybridization *per se* does not generate a new evolutionary lineage across a single generation and hence fails to qualify as saltation. In contrast, the genetic isolation that generally (though not inevitably) follows the doubled karyotype in allopolyploidy successfully generates a new saltational lineage.

Thus, two distinct modes of saltation are evident. In dichotomous saltation, hopeful monsters originate by mutation; one new daughter lineage diverges from the ancestral lineage, thereby forming a dichotomous pattern that can in theory be resolved cladistically (though note that this scenario requires persistence of the parental species beyond the speciation event, in contravention of cladistic principles). In reticulate saltation, allopolyploidy combines elements from two ancestral lineages; the resulting reticulate pattern cannot be adequately accommodated in a dichotomous cladogram. The general absence of mutation in reticulate saltation restricts the potential range of phenotypic innovation, so that speciation events are less likely to coincide with the origins of supraspecific taxa than is the case in dichotomous saltation (cf. Arthur, 1984; Stace, 1989, 1993; van Steenis, 1976). Reticulate saltation remains important, however, as it is more likely to lead to long-term lineage establishment.

Definitions

Taken together, the above arguments lead to the following formal definitions of key terms relating to instantaneous speciation:

Saltation: a genetic modification that is expressed as a profound phenotypic change across a single generation and results in a potentially independent evolutionary lineage (for the present, we prefer to evade our responsibility to quantify 'profound').

Dichotomous saltation: saltation driven by mutation within a single ancestral lineage.

Reticulate saltation: saltation driven by allopolyploidy and thus incorporating genes of two ancestral lineages.

Teratos (or *teras*): an individual showing a profound phenotypic change from its parent(s) irrespective of the underlying cause (plural *terata*).

Hopeful monster: an individual showing a profound phenotypical change from its parent(s) that demonstrably reflects a genetic modification.

Prospecies: a recently evolved lineage possessing the essential properties of a taxonomic (putatively biological) species but yet to achieve acceptable levels of abundance and historical continuity (longevity).

Ancillary mechanisms: parasaltation

The narrowness of the above definition of saltation excludes several under-explored evolutionary mechanisms capable of causing speciation events that are greatly accelerated but not instantaneous.

Firstly, the stringent requirement for both genotypic and phenotypic change across a single generation excludes most mutations of recessive alleles; here, the genotypic change can only be expressed in the F_1 generation in rare cases where a recessive mutation in a germ cell precursor is followed by self-fertilization involving two mutant gametes (Arthur, 1984). Hybridization *per se* is also excluded (cf. Abbott, 1992).

Specifying instantaneous speciation also rules out scenarios that focus on populations of small effective sizes, typically due to reduction induced by various forms of stress or by a vicariance event leading to allopatry (Levinton, 1988). The neutral theory (Kimura, 1983) states that random sampling effects alone can lead to allele fixation or extinction in small populations, largely independent of selective advantage. Various reformulations of Wright's (1932, 1968) shifting balance theory (Carson, 1985; Lande, 1986; Levin, 1970, 1993; Lewis, 1962, 1966, 1969; Templeton, 1982) predict that random genetic drift in small populations can temporarily override selective pressures on alleles, thereby allowing populations to cross valleys on the adaptive landscape to the slopes of another peak, which is then climbed by classic neodarwinian selection. Drift is expressed most profoundly when it disrupts and destabilizes developmental homoeostasis (Levin, 1970). Although most such populations fail, this process provides an opportunity to reorganize the developmental programming under conditions of low intraspecific competition and high physical stress ('catastrophic selection' *sensu* Lewis, 1962, 1966, 1969; see also Carson, 1985). An alternative formulation allows populations to rapidly cross adaptive valleys for some characters as a correlated response to selection imposed directly on other characters (Price *et al.*, 1993).

Shifting balance scenarios are consistent with evolutionary patterns that were termed punctuated equilibria by Eldredge & Gould (1972) – long periods of stasis followed by brief periods of rapid phenotypic change. Vermeij (1987) extended the ecological component of these scenarios, arguing that periods of stasis reflect neodarwinian processes and are punctuated by ecosystem disruptions that locally reduce selection pressure, species diversity and population sizes. Each such disruption allows escalation – a brief interval of intense competition to fill the vacated niches that increases the fitness of the competitors. Many of these observations apply equally well to populations that are very small not because they have declined but because they have just evolved by saltation – we will return to them later. In contrast, other explanations of punctuational patterns focus on selection among species (Gould, 1986; Gould & Eldredge, 1977, 1993; Levinton, 1988; Stanley, 1975) or even among clades (Williams, 1992). Requiring differential survival of lineages, these hypotheses lie well outside the bounds of saltation *sensu stricto*.

CLADISTIC TESTS OF SALTATIONAL HYPOTHESES

The dissemination of cladistic methods during the 1980s has elevated phylogeny reconstruction to a cornerstone of modern evolutionary biology. Although the methodology is becoming increasingly complex, the basic principles of cladistics remain simple (e.g. Doyle, 1993; Farris, 1983; Forey *et al.*, 1992; Patterson, 1982; Wiley, 1981; Wiley *et al.*, 1991). Each species selected for comparison is coded for a predetermined range of characters, each potentially informative character possessing two or more states among the chosen range of species. In the most common protocol, alternative states for each character are designated primitive and derived (polarized) by selecting one or more of the species as outgroup(s) (e.g. Maddison *et al.,* 1984). Parsimony is then used to filter, from among the many possible repeatedly dichotomous arrangements (topologies) of the species, one or more topologies that require the smallest number of state transitions (steps) among the characters coded in the primary matrix. The resulting distribution of character-state transitions across the chosen most-parsimonious topology can (and in our opinion should) be interpreted as an explicit statement of evolutionary pattern. However, if a bistate character changes more than once on the topology (Fig. 1A), cannot be scored for at least one species (Fig. 1B), or must be scored as two or more states for at least one species (Fig. 1C), the precise position of the state transition can be ambiguous. A particular optimization algorithm is then selected to yield an arbitrary solution to each ambiguity (e.g. Swofford, 1993).

The resulting preferred most-parsimonious cladogram, replete with optimized character-state transitions, is open to interpretation in terms of evolutionary process. However, many cladists deliberately ignore this opportunity. The few mechanistic interpretations attempted to date have focused on adaptation (Coddington, 1988; Donoghue, 1989; Greene, 1986; Harvey & Pagel, 1991; Maddison, 1990), and require three important codicils. (1) Many traits are adaptive (contribute to the fitness of the organism) but far fewer are adaptations that evolved via natural selection to fulfil a specific function. This key distinction is often overlooked. (2) Morphological cladistic analyses by definition employ 'form' as characters, but rarely include explicit functions

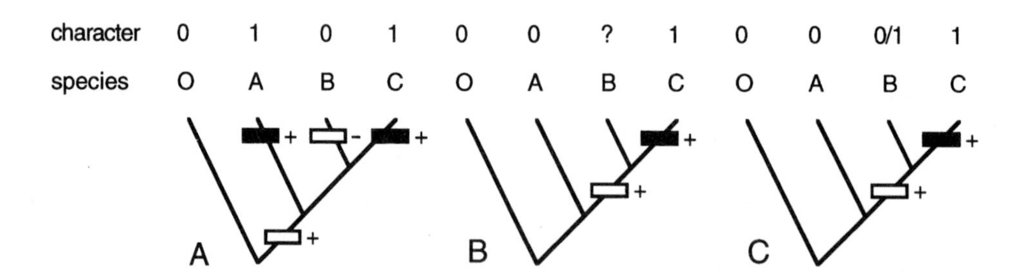

Figure 1 Three causes of ambiguity in the positions of character-state transitions on cladograms. **A** Character conflict. Because it is incongruent with other characters (not shown), this character is depicted as homoplastic. It undergoes two state changes: two origins under Deltran optimization (black boxes), and an origin and a loss under Acctran optimization (white boxes) (Swofford, 1993). **B** Missing value. Inability to score species B for this character results in arbitrary insertion of 0 under Deltran, yielding a late transition, or 1 under Acctran, yielding an earlier transition. **C** Polymorphism. An analogous situation to (B), but caused by the presence of both the primitive and the derived states in species B. Again, 0 is preferred under Deltran but 1 under Acctran.

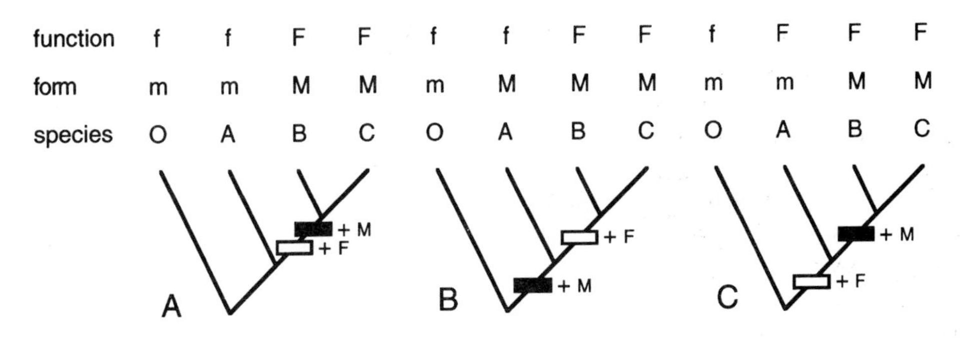

Figure 2 Cladograms as tests of adaptation and exaptation, illustrating the relative positions of transitions in a morphological character (m>M, black boxes) and its postulated function (f>F, white boxes). **A** Transitions in form and function coincide, a pattern consistent with a causal relationship involving selection-mediated adaptation. **B** Evolution of form precedes function, a pattern consistent with exaptation; the character did not evolve for its present function, but was later coopted. **C** Evolution of function precedes form, a pattern implying no significant correlation between the two observations. Adapted from Donoghue, 1989, fig. 3.

(Lauder, 1990). Fortunately, this is not a serious handicap to intepretation, as particular functions can be plotted on a cladogram *a posteriori* (mapped). (3) When attempting to infer evolutionary process from cladistic pattern, it is only possible to state that a particular process is consistent with a particular pattern. Demonstrating such a correlation requires additional biological data not coded into the original cladistic matrix.

To be consistent with a hypothesis of adaptation, a particular form and function must evolve on the same branch of the cladogram (Fig. 2A: Donoghue, 1989). A form appearing below the postulated function on the cladogram (Fig. 2B) is consistent with a hypothesis of exaptation; the form either (1) evolved non-adaptively, or (2) evolved adaptively but for a different function, only later acquiring its present function (Gould & Vrba, 1982). A form appearing above the putative function on the cladogram (Fig. 2C) refutes the hypothesis of positive correlation. We reiterate that arrangements of form and function consistent with adaptation or exaptation are not positive proof of such hypotheses; rather, the value of the cladograms is negative, allowing falsification of postulated correlations.

We argue that this same logic can be extended to allow cladistic tests of non-adaptive saltational hypotheses. These similarly rely on falsification, but the emphasis switches from demonstrating the simultaneous origin of a character state and its presumed adaptive function to demonstrating the simultaneous origin of several character states. This in turn focuses attention on long branches – those supported by several character-state transitions – and requires a literal (and thus controversial) interpretation of the cladogram as an evolutionary history. In this scenario, potentially developmentally correlated characters changing simultaneously on the cladogram are assumed to have changed simultaneously during evolution, most probably as the direct or indirect consequence of a single mutational event. In other words, saltation is regarded as the null hypothesis to explain certain types of long branch. (However, note that a long branch is consistent with any evolutionary process. Note also that long branches tend to be mutually attractive during tree-building, often yielding incorrect topologies: Hendy & Penny, 1989.)

Saltational hypotheses are tested most effectively by coding further species and adding them to the original data matrix. In the simple hypothetical example presented in Fig. 3, an outgroup species (O) has been used to root a cladogram of three ingroup species (A–C). Most of the resulting branches are short, each being supported by only one character-state transition, but the longer terminal branch subtending species C is supported by four potentially correlated character-state transitions (C1–C4). It is consistent with (though by no means proof of) the hypothesis that all four character-state transitions reflect the suppression in species C of a D-gene that controlled early ontogeny in the more primitive species O, A and B. It allowed the repeated dichotomy of apical meristems in both the rooting (C1) and shooting (C2) axial systems. Once this ability was lost, only one reproductive cone could be generated (C3) at the end of the life history of the individual; the plant therefore became monocarpic (C4).

This initial saltational hypothesis was then tested by coding three additional species (D–F) that are closely related to species A–C. Analysis of the new expanded matrix could in theory alter the previous perception of the relationships of species A–C, though in Fig. 3 the original topology is supplemented rather than overturned by the addition of species D–F. Species D is attached to a branch irrelevant to the saltational hypothesis. Species E is attached to the relevant branch, but fails to dissociate the putatively correlated character states. However, species F is intercalated between the derived state of character 1 (which it possesses) and those of characters 2–4 (which it lacks). Following this reanalysis, suppression of meristematic division is perceived as two separate evolutionary events – the first in the root system and the second in the shoot system – though the remainder of the original saltational hypothesis pertaining to the shoot system (C2–C4) remains viable. Of course, this remains susceptible to refutation if yet more species can be added to the analysis of the clade, or if

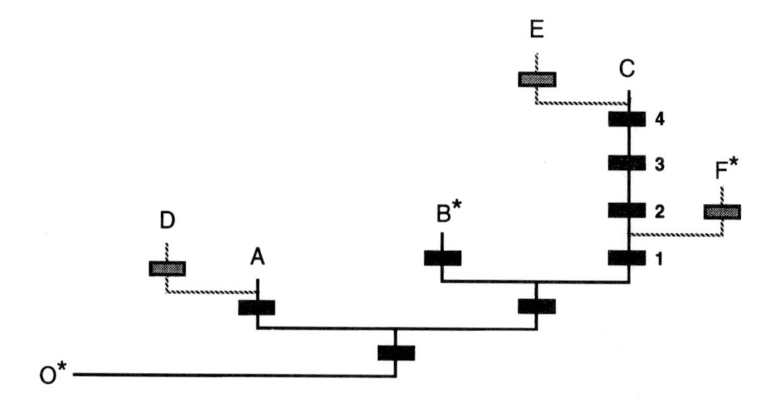

Figure 3 Cladograms as tests of saltation. The initial hypothetical analysis of outgroup species O and ingroup species A–C yields a single fully resolved topology. The relatively long branch subtending species C is supported by transitions in four potentially developmentally correlated characters: 1, suppression of dichotomy in root apical meristems; 2, suppression of dichotomy in shoot apical meristems; 3, switch from multiple to single reproductive cones; 4, switch from polycarpy to monocarpy. Simultaneous change in these four characters by saltation is assumed as the null hypothesis, and then tested by reanalysing the matrix with three additional ingroup species (labelled D–F). Species D is attached to an irrelevant branch. Although species E is connected to the relevant branch, it fails to dissociate the four correlated character-states. However, addition of species F separates character 1 from characters 2–4. This implies that the state transition occurred earlier in character 1 and thereby falsifies that part of the saltational hypothesis. Note that the fossil status of three of the species (asterisked) has no bearing on their analytical performance.

subsequent ontogenetic observations prove incongruent with the hypothesis of developmental linkage.

As in the tests of adaptation and exaptation, the cladistic test of saltation relies primarily on falsification rather than proof. Nonetheless, the probability of a correct inference of saltation will be greatest if (1) all known species of the clade, extant and extinct, have been included in the analysis, and (2) a developmental correlation can be demonstrated among the simultaneously changing characters that is consistent with transformation of a single developmental gene. Thus, the cladogram prompts re-examination of relevant characters in the context of functional and developmental integration. This requires direct observation of ontogeny and study of underlying genetic controls within a comparative framework. Even given this information, it is difficult to demonstrate that integration of the characters accompanied speciation, particularly if a considerable period of time separates speciation and observation.

LIFE AS A VASCULAR PLANT

Before further discussing plant evolution, it is desirable to review briefly what Darley (1990) graphically termed 'the essence of plantness'. The life history of an individual can be divided into the economic phenomena that govern the everyday activities of the plant within a particular generation, and the more temporally restricted reproductive phases that allow the transfer of genetic information to the next generation (Eldredge, 1989).

Physiology and ecology

In economic terms, plants are autotrophic primary producers. Able to synthesize energy-rich carbohydrates from carbon dioxide using solar radiation, they are basal to almost all food chains (the most notable exception is deep-sea hydrothermal vents). Plants also concentrate water, minerals and atmospheric oxygen. It should be emphasized that plants inherited these abilities from less derived groups. For example, photosynthesis characterizes many unicellular eukaryotes of the Protoctista and the prokaryotic cyanobacteria of the Monera. Plants first acquired chloroplasts as symbiotic cyanobacteria; earlier, a common ancestor of plants and animals had acquired mitochondria as symbiotic bacteria (e.g. Fernholm *et al.*, 1989; Knoll, 1992; Margulis, 1993; Margulis & Schwartz, 1988; Woese *et al.*, 1990). Multicellularity, and thus tissue differentiation, evolved independently in plants and animals.

Much of the subsequent evolution of plants has improved physiological efficiency. Planar megaphyllous leaves borne in Fibonacci phyllotaxis increase incident light, while their cuticle and stomata enhance water conservation and gaseous exchange. Specialized vascular tissues allow rapid transport of nutrients and hormones around the plant body, driven by passive transpiration tension in the acropetally conducting xylem but an active process in the basipetally conducting phloem. Xylem transport helps to maintain the turgor pressure that is the primary support mechanism of plant tissues. Turgor is generated in cell vacuoles, which are constrained by the primary cellulose walls of cells that are mutually adhesive (Wiebe, 1978). Later in ontogeny this 'hydraulic skeleton' becomes inadequate and secondary cell walls are laid down, providing structural reinforcement at the expense of suppressing the potential for further cell division or expansion. Roots anchor the plant to the substrate and increase surface area for absorption of water and nutrients, thereby confining the organism to a sedentary, sessile life-style. Thus, motility in vascular land plants is passive and largely restricted to aquatic species, in contrast with the active motility of flagellate photosynthetic protoctistans (e.g. *Euglena*) and chlorophyte algae (e.g. *Volvox*). It should be noted that the sessile life-style, like photosynthesis, probably preceded the conquest of the land by plants (cf. DiMichele *et al.*, 1992; Kenrick & Crane, 1991; Selden & Edwards, 1989).

Meristems and architecture

Another major factor constraining the evolution of form in plants is the restriction of all cell division and most cell expansion to a small proportion of the plant body, in regions that are termed meristems. These permit indeterminate ('open') growth throughout the life history of most vascular plant species. Early land plants possessed only axial apical meristems that allowed elongation of shoots and roots, but these subsequently diversified to generate more specialized shoot meristems for the production of leaves and sporophylls, and lateral meristems that enabled secondary growth (vascular cambium). This led to greater numbers of plant organs possessing greater degrees of physiological independence; in some species most organs are dispensable, provided that at least one indeterminate meristem is retained by the plant (Bazzaz & Carlson, 1979; White, 1979). Best known for allowing organ abscission during ontogeny, this property also confers remarkable resilience to physical damage and the opportunity for various modes of meristem-mediated asexual reproduction. The more advanced vascular plant clades evolved embryonic bipolarity of meristems to generate distinct shoot and root axial systems (Bateman, in press). Architectural flexibility in plants is largely confined to variations in the type, number, spatial arrangements and branching patterns of meristems (e.g. Hallé *et al.*, 1978).

Darley (1990) convincingly argued that the relatively complex growth patterns of higher plants are largely a substitute for motility, maximizing potential responses to environmental perturbations within this severe economic constraint. Most interactions among plants, both infraspecific and interspecific, are indirect – they are mediated by the environment. Sophisticated hormonal systems remain essential to regulate growth and metabolism (Cheplick, 1989; Poethig, 1990), but chemical interactions among plants are uncommon (exceptions include certain phytotoxins and 'phytopheromones'). Physicochemical interactions, notably mycorrhizal associations, are more common. Some species have saprophytic or parasitic life histories. Other interspecific relationships are purely physical, such as epiphytism and lianescent strangulation. Moving on from economic to reproductive interactions, sexual reproduction requires an intermediate environmental or animate vector for dissemination of at least one gender of gamete and often of the resulting zygote. Such disseminules involve periods of low metabolic activity and thus allow a resilience to environmental catastrophes greater than that observed in higher animals (e.g. Boulter *et al.*, 1988). Furthermore, plants lack the complex central nervous systems and consequent interactive behaviours of higher animals; for example, choice of mate is purely physiological and generally *ad hoc* within the population. Sadly, intelligent vegetables remain the preserve of science ficition (e.g. Wyndham, 1951).

Thus, plants compete primarily with the environment rather than with each other, and where they do interact they do so unconsciously – key points that we will develop later.

PLANT DEVELOPMENT AND ARCHITECTURE

Intraspecific description

The largest scale and thus most obvious aspect of the shape and form of vascular plants is their overall architecture. Body shape and size are among the most biologically significant aspects of a plant, as they strongly influence life history and ecological role (e.g. Hallé *et al.*, 1978; Phillips & DiMichele, 1992). We define architecture as the idealized, genetically determined morphology of a plant. Architecture is subject to varying degrees of non-genetic, ecophenotypic modification to generate the observed phenotypic growth habit: the ultimate form of a plant as expressed in its physiognomy (Bateman & DiMichele, 1991; Gottlieb, 1984, 1986; Hallé *et al.*, 1978; Meeuse, 1986; Mosbrugger, 1990; Tomlinson, 1982, 1987). Other classifications of form are more nebulous, such as the distinction among tree (large bodied, non-recumbent), shrub (medium bodied, non-recumbent), pseudoherb (small bodied or recumbent, woody), and herb (small bodied or recumbent, non-woody) (Bateman & DiMichele, 1991; Bateman *et al.*, 1992).

The plant body is a composite of modular constructional units generated by various types of meristem (a metapopulation *sensu* White, 1979; see also Buss, 1987; Harper *et al.*, 1986; Tomlinson, 1982). Appendages borne on axes are generally produced by determinate meristems as repeatable (and often dispensable) units (Hallé, 1986; Hardwick, 1986). However, evolution can blur the axis–appendage distinction. For example, the megaphyllous leaves of ferns and seed plants are positionally appendicular but share a developmental programme with axes; leaves of many early tracheophytes resemble dichotomous axial branching systems and were termed telomes and telome–mesome aggregates (Stewart & Rothwell, 1993; Stidd, 1987; Zimmermann, 1959). A plant is characterized primarily by the number of types of such modules present and by their spatial arrangements. Variations in the size, shape and relative numbers of such modules are less profound (Bateman, in press).

Moreover, the ultimate expression of the genotypically mediated body form of an individual plant is strongly influenced by chance factors such as disturbance, predation and proximity to limiting resources (Tomlinson, 1982). Plant form permits opportunistic responses to local conditions and events; such flexibility is crucial, given their inability to relocate when subjected to stress. Thus, genetically determined architectures incur a strong stochastic overprint, perturbing the idealized form.

Interspecific comparison: evolutionary developmental change

Conceptual aspects of comparative development have long been dominated by zoologists (e.g. Alberch *et al.*, 1979; Arthur, 1984, 1988; de Beer, 1940; Gould, 1977a; Haeckel, 1868; Kauffman, 1993; Kluge & Strauss, 1985; McKinney, 1988; McKinney & McNamara, 1991; Raff & Kaufman, 1983; Riedl, 1979; von Baer, 1828; Wake, 1989). Neobotanical applications have been relatively few (e.g. Gottlieb & Jain, 1988; Guerrant, 1982; Iltis, 1983; Kellogg, 1990; Sattler, 1988, 1993; Takhtajan, 1972; Weston, 1988; Williams *et al.*, 1990) and palaeobotanical applications even fewer (Bateman, in press;

Bateman & DiMichele, 1991; Bateman *et al.*, 1992; Doyle, 1978; Meyen, 1988; Rothwell, 1987; Stidd, 1980). Much is owed to Gould (1977a) for rigorously defining a set of descriptive terms, Alberch *et al.* (1979) for translating those definitions into semi-quantitative plots of shape against time (or against its crude proxy, size), and Fink (1982) for developing cladistic tests of hypotheses of developmental transitions.

Broadly (and somewhat superficially), changes in development between ancestor and descendant can be attributed to heterotopy – a spatial (positional) change in the expression of a trait – and heterochrony – a temporal change in the expression of a trait. Changes in the relative times of onset or offset of growth, or of the rate of morphological development, can be used to define six end-member modes of heterochrony that can be assigned to two main categories: relative to the ancestor, the descendant shows more morphological change if peramorphic and less morphological change if paedomorphic (Alberch *et al.*, 1979). Allometric changes that alter size but not shape (giantism and dwarfism) lie outside the formal definition of heterochrony (Alberch *et al.*, 1979; Gould, 1977a). Bateman (in press) argued that changes in whole-organism traits such as the timing of onset of sexual maturity and the number of definable developmental stages in the ancestral and descendant ontogenies should also be excluded from heterochrony *sensu stricto*. He preferred to coin the broader collective term evolutionary developmental change for these phenomena, and suggested that architectural comparisons among plants are best made at the level of definable organs rather than entire organisms.

Terminology developed to describe changes in the number or nature of developmental stages in compared ontogenies also presents difficulties. Each change can be an addition, deletion or substitution, which can be terminal or non-terminal in the ontogenetic sequence (O'Grady, 1985). Unfortunately, a change in a single gene can affect more than one developmental stage, either by pleiotropism, epigenesis, or mere cascade effects; parity between cause and effect cannot be assumed. Another problem is presented by morphological simplification; whether due to heterochronic paedomorphism or stage deletion, this can undermine cladistic analyses (Bateman, in press; Bateman *et al.*, 1992).

To summarize, comparative plant development suffers from the same conceptual deficiencies as descriptive plant development. When torn from their zoological roots in order to be applied to higher plants, and when shoe-horned into a cladistic framework, those concepts clearly require reappraisal (Bateman, in press; Sattler, 1988, 1993). Such revisions are now underway.

CASE STUDIES OF SALTATION IN PLANT ARCHITECTURE

The following case studies were selected from among the few cladistic analyses of vascular plants performed to date within a single taxonomic order. The range of appropriate studies was further restricted by several *a priori* requirements for the nature of the data matrix (cf. Bateman, in press). Studies coding species rather than infraspecific or supraspecific taxa were preferred. The matrix should lack both polymorphic coding (i.e. each character of each species should be represented by a single character state) and large tracts of missing values. Characters should be dominantly morphological and some should describe overall architecture. Interestingly, architectural characters were omitted from a surprisingly large proportion of morphological cladistic studies. On the rare occasions when they were included they proved unusually homoplastic, thereby decreasing the resolution of the analysis (e.g. Bateman, in press; Bateman *et al.*, 1992; Funk, 1982; Hill & Camus, 1986). Recognition of their problematic nature led to their deliberate *a priori* omission from many studies.

The following case studies were selected from among the few analyses that conform to the above criteria. The chosen clades differ greatly in taxonomic affinities and in the relative proportions of living and fossil species. These brief summaries focus on relevant aspects of the interpretations; further analytical details should be sought in the original publications.

Montanoa (Asteraceae: Angiospermales)

Funk (1982) performed a morphological cladistic analysis of 25 exant species of the predominantly tropical American asteracean genus *Montanoa*. Most were small-bodied shrubs or lianas, but five species were woody trees reaching about 30 m. Mapping of habitat preferences across the cladogram showed that the five arboreous (tree-sized) species differed from the remainder in favouring high-altitude cloud forests; moreover, three of the five species were studied karyotypically and all proved to be high-level polyploids. Had chromosome counts been obtained from all 25 species analysed, precise correlations and co-occurrences could have been calculated between karyotype, habitat and morphological characters such as body size (Harvey & Pagel, 1991). Even with the present incomplete evidence, the association of the three parameters is clear-cut. Funk (1982) and Funk & Brooks (1990) argued that in the Asteraceae polyploidy frequently promotes arboreousness; this in turn leads to unusually inefficient vascular conductance; and this in turn confines composite trees to perpetually humid habitats such as cloud forests. Interestingly, a similar pattern has been documented in *Lobelia*, which shows repeated origins of woody tetraploid and hexaploid species from putative ancestors that are both diploid and herbaceous (Knox *et al.*, 1993).

The *Montanoa* cladogram (Fig. 4) shows that four independent origins of arboreousness are required to explain the phylogenetic positions of the five tree species (unlike the lycopsid example below, no evolutionary transitions from tree to shrub are required). Together, these observations are consistent with frequent generation of arboreous prospecies by *ad hoc* polyploidy events among the diploid, low-altitude species of *Montanoa*. Only polyploid seeds – embryonic hopeful monsters – that fortuitously reach cloud forest habitats have any chance of establishing evolutionarily significant populations (differential habitat preferences render the polyploids allopatric relative to their ancestors). Although most polyploid *Montanoa* seeds presumably fail even in appropriate habitats, at least four of the saltationally generated prospecies achieved historical continuity, and one ((3) on Fig. 4) apparently underwent at least one additional speciation event without further change in ploidy. This illustrates another important point. Stace (1989, 1993) estimated that perhaps the majority of all living angiosperm species are polyploid. However, this does not mean that half the extant species owe their origin to polyploidy events; given that speciation occurs frequently among polyploids, far fewer ploidy changes are required to explain their evolution.

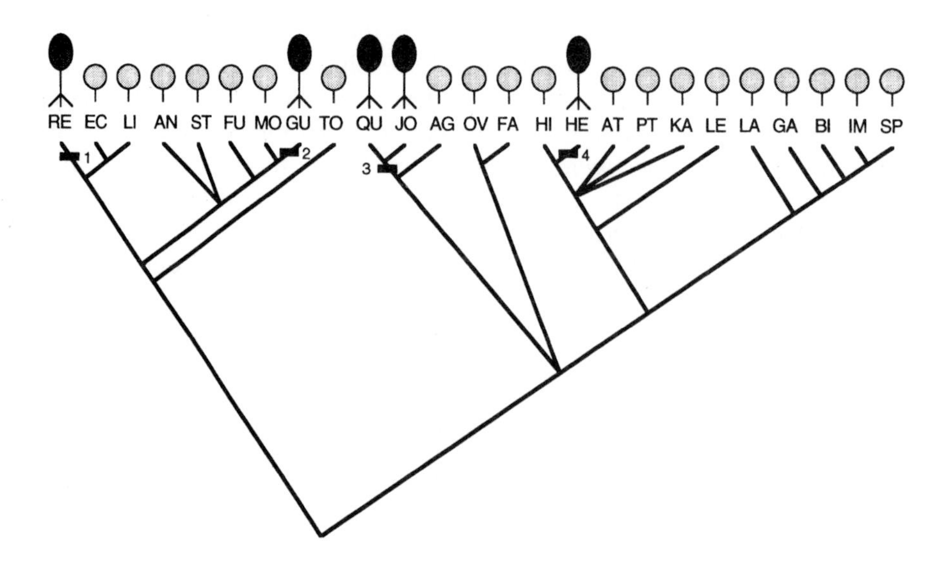

Figure 4 Incompletely resolved preferred most-parsimonious cladogram of 25 extant species of the composite genus *Montanoa*, showing four independent origins (1–4) of large-bodied trees from smaller-bodied shrubs and lianas. The origin of arboreousness coincides with polyploidy events (reticulate saltation) and with ecological transitions from low- to high-altitude tropical forest habitats. Species as follows: RE, *M. revealii*; EC, *M. echinacea*; LI, *M. liebmannii*; AN, *M. andersonii*; ST, *M. standleyi*; FU, *M. frutescens*; MO, *M. mollissima*; GU, *M. guatemalensis*; TO, *M. tormentosa*; QU, *M. quadrangularis*; JO, *M. josei*; AG, *M. angulata*; OV, *M. ovalifolia*; FA, *M. fragrans*; HI, *M. hibiscifolia*; HE, *M. hexagona*; AT, *M. atriplicifolia*; PT, *M. pteropoda*; KA, *M. karwinskii*; LE, *M. leucantha*; LA, *M. laskowskii*; GA, *M. grandiflora*; BI, *M. bipinnatifida*; IM, *M. imbricata*; SP, *M. speciosa*. Data from Funk, 1982; figure adapted from Funk & Brooks, 1990, fig. 102.

Rhizomorphales (Lycopsida)

The rhizomorphic lycopsids are the most derived portion of the lycophyte clade (lycopsids plus zosterophyllopsids), which is sister group to the remainder of the eutracheophytes (e.g. Crane, 1990). They evolved independently many of the features generally regarded as characterizing the early seed plants, including bipolar embryonic growth, secondary growth producing wood and periderm, the arboreous growth form, and indehiscent integumented megasporangia (Bateman, in press; Bateman *et al.*, 1992; Phillips & DiMichele, 1992). Observable anatomical differences highlight the non-homology of these features between the two clades. Appendages are comparatively poorly developed in the rhizomorphic lycopsids; microphylls, sporophylls and rootlets are all supplied by a single, narrow vascular strand. More significantly, their bipolar growth exhibited much stronger developmental parallelism between the rooting and shooting axial systems than that documented in seed plants. Also, the apical meristems followed an animal-like developmental pattern of closed, determinate growth. Consequently, individual growth modules experienced high degrees of physiological independence and thus were unusually well defined. This in turn meant that the idealized genotypic architecture was unusually faithfully reproduced in the actual growth habitat observed in the phenotype (Bateman & DiMichele, 1991).

A detailed experimental cladistic analysis was performed on the group by Bateman *et al.* (1992), and its implications for evolutionary developmental change in architecture were explored in detail by Bateman (in press). Sixteen adequately reconstructed fossil species, comprising ten genera, were included; they were scored for 12 binary architectural characters. A simplified version of the preferred most-parsimonious tree, reduced to generic level, is reproduced in Fig. 5. This reveals a tremendous range of body size and architectural plans; only one triplet and one pair of genera possess identically coded architectures, and of these only the most derived pair is depicted as monophyletic. However the phylogenetic relationships and character-state changes are interpreted, the architectural history of the group appears anarchic. On several occasions similar architectures evolved in parallel. Trees alternate with far smaller-bodied genera, and isotomous, exclusively terminal crown-branching alternates with anisotomous, at least partially non-terminal lateral branching. Loss of organs is an iterative theme, occurring as morphological simplification via the suite of heterochronic phenomena that are collectively termed paedomorphosis.

Bateman (in press) argued that the remarkable frequency and radical nature of these architectural transitions, combined with the determinate modular growth and presumably relatively simple genetic control of development, render this an especially

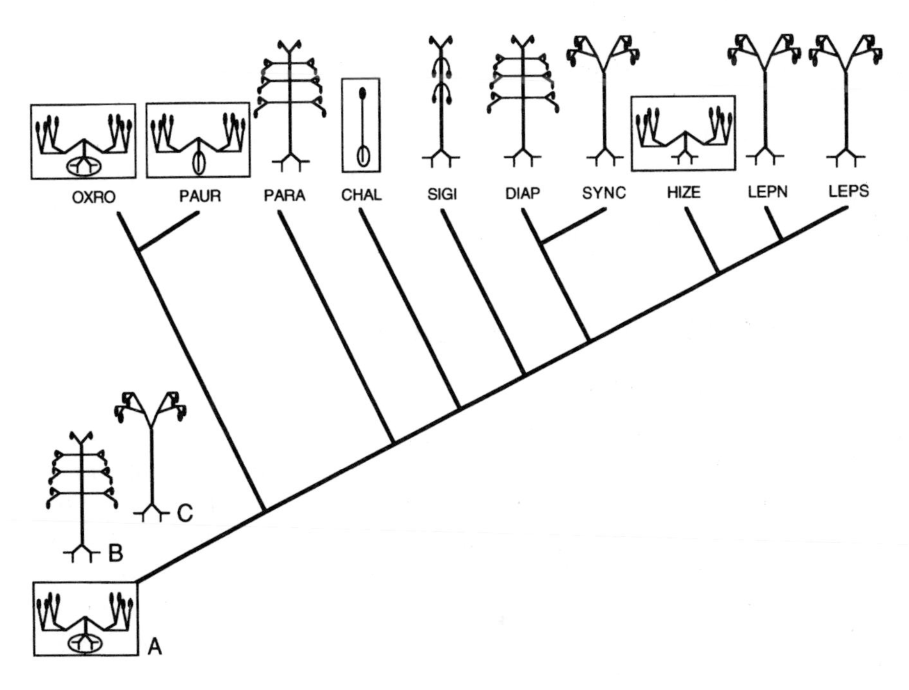

Figure 5 Fully resolved preferred most-parsimonious cladogram of 16 extinct species of rhizomorphic lycopsid, subsequently collapsed to ten monophyletic genera for ease of interpretation. Note the radical differences among mature growth architectures, which require hypotheses of convergence, reversible architectural transitions, loss of entire growth modules, and repeated origins of shrubs from trees (small-bodied, non-tree genera are boxed; ellipses denote compact, 'cormose' rootstocks). These radical architectural transitions are consistent with mutation-driven dichotomous saltation. The cladogram was rooted using an *Oxroadia*-like (A) hypothetical ancestor, though credible alternative hypotheses detailed by Bateman (in press) would allow rooting using *Paralycopodites*-like (B) or *Synchysidendron*-like (C) architectures. Genera as follows: OXRO, *Oxroadia*; PAUR, *Paurodendron*; PARA, *Paralycopodites*; CHAL, *Chaloneria;* SIGI, *Sigillaria;* DIAP, *Diaphorodendron;* SYNC, *Synchysidendron;* HIZE, *Hizemodendron;* LEPN, *Lepidodendron;* LEPS, *Lepidophloios*. Data from Bateman *et al.*, 1992; figure adapted from Bateman, 1992, fig. 12.

plausible example of repeated, profound (and presumably dichotomous) saltational evolution.

Summary

Overall architecture is the largest-scale phenotypic expression of a plant, largely reflecting the relative sizes and spatial relationships of various types of meristem and thereby encapsulating much of its ontogenetic history. There is strong positive feedback between changes in architecture and changes in smaller-scale features of the plant, which may render architectural characters particularly susceptible to non-adaptive (and often ultimately non-directional) saltational evolution.

Their flexibility and scale should not routinely excuse architectural characters from inclusion in cladistic analyses; indeed, we find them particularly illuminating. They are a useful reminder that, despite the large degree of physiological and evolutionary independence enjoyed by particular growth modules, plants nonetheless act primarily as holistic individuals. This point was emphasized in a saltational context by Hay & Mabberley (1991), though we fundamentally disagree with their conclusion that plants should be analysed as holistic units rather than as sets of coded characters. By definition, scientific analysis requires reciprocal illumination between different hierarchical levels of evidence – in this case, of physical scale and degree of phenotypic expression – as well as between pattern and process. The paradigm shift (Kuhn, 1962) advocated by Hay & Mabberley (1991) cannot occur in the absence of a viable alternative paradigm free of characters. We believe that progress requires interaction between character-based analysis and evolutionary theory.

THE NEOGOLDSCHMIDTIAN PARADIGM

Generating hopeful monsters

As noted by several commentators (e.g. Arthur, 1984; Dietrich, 1992; Gould, 1982), Goldschmidt (1940) was mistaken in assuming that the large-scale phenotypic changes inherent in hopeful monsters could only be explained by large-scale genetic modifications. Subsequent knowledge of gene structure and expression has switched the emphasis from the number of base pairs affected by a mutation to the question of whether those base pairs are expressed and, if so, when and how they are expressed (e.g. Britten & Davidson, 1969). The problem was most effectively placed in a developmental genetic context by Arthur (1984), who focused on enzyme-producing structural genes that are in turn controlled by regulatory genes. If the enzyme in question affects only biochemical maintenance of cells, the relevant structural genes are termed S-genes and the regulatory genes R-genes. If, on the other hand, the enzyme is a morphogen affecting ontogeny, both the structural and regulatory genes are termed D-genes. These were in turn divided by McKinney & McNamara (1991) into type I D-genes, which control intercellular development (i.e. rearrange cells), and type II D-genes, which control intracellular development (i.e. determine cell type). As already noted, following their generation by meristems, plant cells rapidly coalesce into more-or-less fixed spatial arrangements; thus, the saltational spotlight becomes firmly fixed on type II D-genes. Although biologically simplistic, this dynamic classification of genes is heuristically useful.

Certainly, D-genes are no mere hypothetical construct. The best-known studies of developmental gene expression focused on the fruit-fly *Drosophila* (e.g. Ashburner, 1989; Duncan, 1986; Gehring, 1987; Lawrence, 1992). A typical arthropod, *Drosophila* consists of repeated segments bearing serially homologous appendages that show differing degrees of functional specialization. Although this modular construction is analogous to vascular plant architecture, *Drosophila* has far fewer degrees of developmental freedom, being confined to three pairs of legs and the inevitable bilateral body symmetry of vagile terrestrial animals. Also, the stepwise metamorphic ontogeny of *Drosophila* contrasts starkly with the gradual ontogeny of plants. Many teratological mutant types have been documented in *Drosophila* that involve either (1) a change in the number of compartmentalized segments or appendages (heterochrony *sensu lato*) or (2) a spatial change in the segment on which a particular appendage is expressed (heterotopy). Most such mutations have been tied to specific type II D-genes, which can be suppressed by inserting short lengths of transposable DNA (transposons). Sequencing of several homeotic genes – notably the homeobox family – has shown that all consist of relatively short lengths of DNA and for much of their lengths have similar base pair sequences, suggesting a common origin (cf. Coen, 1991; Coen & Carpenter, 1992; Marx, 1992; Slack *et al.*, 1993; Smith *et al.*, 1992). A relatively small number of D-genes operate either individually or interactively to control all of the major developmental switches in *Drosophila* (Gehring, 1987; Gehring & Hiromi, 1986; Slack *et al.*, 1993).

Subsequent zoological investigations revealed similar genes in groups that together span most of the organic world, including organisms most favoured for genetic research such as the nematode *Caenorhabditis elegans*, the yeast *Saccharomyces cerevisiae*, and the bacterium *Escherichia coli* (interestingly, these are all species with dominantly molecular rather than morphological expression of the genome; this morphological simplicitly weakens their appointed roles as evolutionary archetypes). The lesser prominence of investigations of the plant kingdom has recently been alleviated by a succession of high-profile papers elucidating D-gene expression in plants via terata and transposons (e.g. Bureau & Wessler, 1992; Wessler, 1988). Thus far, such studies have focused on the floral morphogenesis of dicotyledonous angiosperm herbs such as *Arabidopsis thaliana* (Bowman *et al.*, 1989, 1991; Coen & Meyerowitz, 1991; Hill & Lord, 1989; Mandel *et al.*, 1992; Schultz *et al.*, 1991; Yanofsky *et al.*, 1990), *Antirrhinum majus* (Coen & Meyerowitz, 1991; Schwartz-Sommer *et al.*, 1990), *Primula vulgaris* (Webster & Grant, 1990), *Clarkia concinna* (Ford & Gottlieb, 1992) and *Solanum lycopersicum* (Rasmussen & Green, 1993), though terata also commonly occur among monocots (e.g. orchids: Bateman, 1985; McCook & Bateman, 1990), gymnosperms (e.g. Long, 1977; T. L. Phillips, personal communication, 1993) and pteridophytes (e.g. Leavitt, 1909; Page, 1972). (See also Meyer, 1966; Worsdell, 1916.)

Although environmental modifications can generate teratological flowers (Crozier & Thomas, 1993), most studies have revealed that the expression and differentiation of each whorl of floral appendages (sepals, petals, stamen, carpels) can be explained in terms of control by one or two factors that probably reflect particular D-genes; some are specific to a single whorl, whereas others are more widely expressed (Coen, 1991; Coen & Carpenter, 1992; see also Endress, 1992; Hilu, 1983). Moreover, the floral genes of plants share substantial sequences of base pairs with genes found in species belonging to other kingdoms, suggesting that this family of key developmental genes originated during the earliest stages of life on Earth. They have been progressively modified and coopted for different tasks (not always developmentally related) during major phylogenetic radiations; most homoeotic genes are members of ancient families of D-genes, although their present roles in particular species are often relatively recent. Studies of the genetic control of exclusively vegetative morphogenesis in plants are less advanced (e.g. Aeschbacher *et al.*, 1992; Poethig, 1990; Smith *et al.*, 1992), but similar control of the all-important meristems by very small numbers of D-genes is highly likely.

To summarize, recent studies have demonstrated conclusively that plant development parallels that of animals in being controlled by a small number of key D-genes. Their importance lies not in their size but in the ontogenetic timing and magnitude of their phenotypic expression. Accumulating molecular data increasingly reveal species radiations involving significant morphological differentiation but little molecular change – a much debated example of East African cichlid fishes (cf. Avise, 1990; Greenwood, 1981; Kenleyside, 1991; Meyer *et al.*, 1990) overshadowing equally informative studies of the Hawaiian asteraceans *Tetramolopium* (Lowrey, 1993; Lowrey & Crawford, 1985); and silverswords (Baldwin, 1992; Baldwin et al., 1990, 1991; see also Doyle, 1993). Moreover, the relative simplicity of developmental control reduces the potential of the genome to buffer the effects of D-gene mutations, and increases the likelihood of genetic parallelism – different mutation events having the same effect on the same gene within the same species. Thus, many similar hopeful monsters can occur sporadically throughout the range of their 'ancestral' species. Although exciting, reports of breakthroughs in the understanding of plant developmental control have lacked discussion of the profound evolutionary significance of D-gene mutations. In order to complete the saltational paradigm it is necessary to switch emphasis from the evolutionary play to the ecological theatre.

Establishing hopeful monsters

The architectural development of higher organisms is probably controlled by few D-genes, and like all genes these are susceptible to mutation. Thus, each species constantly generates hopeful monsters throughout its life-span. The limiting factor is not the availability of hopeful monsters but their establishment. This second phase of saltational evolution depends on the interaction between phenotype and environment.

The second fundamental flaw of Goldschmidt's paradigm noted by Gould (1982) was the requirement that successful hopeful monsters should emerge from the saltation event already possessing a level of fitness sufficiently high to compete with other sympatric organisms, both interspecific and intraspecific and including the parent(s) of the monster. This hypothesis appears untenable; organisms arising by saltation will not be competitive with organisms already honed to local optima of fitness by natural selection. Rather, they are faced with 'unwanted' changes that significantly decrease their fitness, and must survive their handicaps as best they can. The solution to their dilemma lies in ecology; specifically, the need to temporarily release the mutant lineage from selection (Arthur, 1984, 1987, 1988; Carson, 1985; DiMichele *et al.*, 1987; Erwin & Valentine, 1984; Levinton, 1988; Valentine, 1980; van Steenis, 1976; Waddington, 1957).

We recognize four establishment thresholds, listed in order of increasing levels of fitness: (1) the lowest threshold is the minimum fitness level necessary for economic establishment of the hopeful monster in its habitat – in the case of a seed plant, germination and growth to maturity. (2) Once the plant reaches maturity it must be capable of generating viable progeny, however inefficient the reproductive process. (3) Further generations are required to achieve continuity of the mutant lineage and to expand its biogeographical and ecological influence. This requires an even higher level of fitness, to compete successfully with sympatric organisms actually occupying the contested niche. (4) There is a theoretical threshold of even greater fitness that would have to be exceeded to outcompete the fittest potential occupant of the niche (because of the non-motility of plants, there is a strong *ad hoc* element to the species that actually occupies a specific niche in a specific habitat at a specific moment in time).

Factors determining establishment are summarized conceptually in Fig. 6. This scenario involves a landscape of five habitats (A–E), each offering one niche for a tree and one for a shrub. Habitats A, B and C–E differ considerably in relative levels of intrinsic stress (S1–S3), which determines fitness thresholds for the successful establishment of plant lineages in the absence of competition (essentially threshold (2) above). Initially, the landscape is occupied by three species of rhizomorphic lycopsid, each already honed by natural selection to a local optimum of fitness in its preferred habitat: a laterally branched tree and an unbranched shrub in habitats B and C, and a laterally branched tree only in habitat D. The parental tree in habitat D continuously produces by mutationally driven saltation two types of teratological propagule, of equal fitness among themselves but of lesser fitness than the pre-existing species in habitats B and C and the parent in habitat D. The first type of hopeful monster, crown-branched trees, are produced by suppression of lateral branching and are potential occupants of the tree niche. The second type of hopeful monster, crown-branched pseudoherbs, are produced by suppression of lateral branching plus profound paedomorphosis, and are potential occupants of the shrub niche. All are of an equal fitness ($\omega 1$) that is appreciably lower than the shared fitness of the adaptively honed pre-existing species ($\omega 2$), and all are dispersed evenly across the landscape.

Figure 6 documents the fates of identical monsters in habitats that differ in intrinsic stress levels and niches that differ in whether they are occupied or vacant (cf. Arthur, 1984; Stanley, 1979). Habitat A offers the highest intrinsic stress (it could be a desert); it exceeds the abilities of both the hopeful monsters and their adaptively honed predecessors to grow and/or reproduce in either niche; thus, the organisms suffer economic or reproductive failure. In contrast, habitat C offers an attractively low level of intrinsic stress; consequently, both niches are already occupied by adaptively honed species. The hopeful monsters may germinate and even reproduce in the habitat, but their long-term prognosis is poor; they will eventually be excluded from both niches by interspecific competition. In habitat D, the hopeful monster is excluded from the tree niche by its own parental stock, but the shrub niche is vacant and the hopeful monster has sufficient fitness to transcend successfully the establishment threshold determined by intrinsic stress. Habitat E also offers low intrinsic stress but nonetheless both niches are vacant due to chance factors (for example, the habitat is a recently formed island or a volcanically sterilized landscape). Thus, any hopeful monster can become established here, irrespective of niche. Lastly, habitat B illustrates the potential for incorrect interpretations of causation. The situation appears identical to that in habitat C,

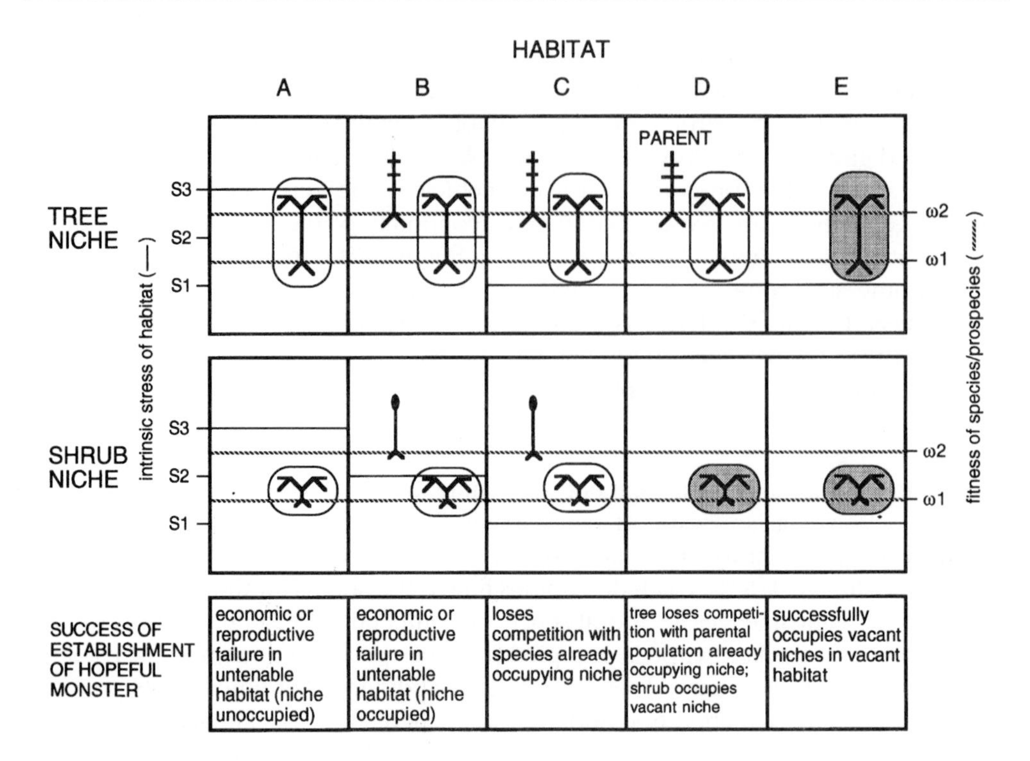

Figure 6 The key role of ecology in the neoGoldschmidtian paradigm. The simple ecological system depicted is a landscape of five habitats (A–E), each offering single niches for a tree and a shrub. These differ in relative intrinsic stress (S: solid line); low-stress habitats are more easily colonized by individuals of low fitness. Initially, the landscape is occupied by three species of rhizomorphic lycopsid: a laterally branched tree and an unbranched shrub in habitats B and C, and a laterally branched tree only in habitat D. These species have been honed by natural selection in their preferred habitats to local optima of fitness (coincidentally the same level for all three species: upper dashed line, ω2). In this scenario, the parental tree in habitat D continuously produces two types of hopeful monster by mutationally driven saltation: crown-branched trees by suppression of lateral branching, and crown-branched pseudoherbs by suppression of lateral branching plus profound paedomorphosis. Again, all are of the same fitness (lower dashed line, ω1), which is appreciably less than that of the pre-existing species. The fates of identical monsters in different habitats are documented. Note that succcessful establishment (stippled) requires vacant niches of relatively low intrinsic stress; any interspecific competition eliminates the monsters due to their suboptimal fitness, which reflects lack of selective honing (see text).

prompting a similar interpretation of competitive exclusion of the hopeful monsters by the incumbents. However, the intermediate stress level of the habitat (S2) means that it is tenable for the adaptively honed species (ω2) but untenable for the hopeful monsters (ω1); the true cause of their failure is intrinsic stress rather than competitive exclusion.

Obviously, the model outlined in Fig. 6 is simplistic. Firstly, many more than two plant niches can be recognized in any one habitat. Secondly, contrasting niches within a particular habitat differ in intrinsic stress; for example, stress was often higher in the shrub niches than the tree niches of many Carboniferous lycopsid forests due to periodic flooding and consequent inundation of smaller-bodied species. Thirdly, the uniformity of fitnesses among species implied here is not mirrored in nature, where fitness is a variable, dynamic, and arguably unmeasurable parameter. Fourthly, the relationship between vegetation and environment is more intimate than the model suggests; plants can profoundly modify their habitats (Lewontin, 1993), thereby often reducing intrinsic stress levels that they encounter.

Nonetheless, we are confident that the basic assumptions of the model are valid. Successful establishment of hopeful monsters requires vacant niches of relatively low intrinsic stress; any interspecific competition eliminates the monsters due to their suboptimal fitness, which reflects lack of selective honing. Thus, the environment acts initially as a passive filter of hopeful monsters rather than the active agent of their evolution. Subsequent imposition of selection stabilizes phenotypic expression, thereby increasingly canalizing the new developmental programme (Carson, 1985; Carson & Templeton, 1984; Iltis, 1983; Valentine, 1980).

Canalization, character loss and complexity

It has long been recognized that major developmental events become increasingly canalized within lineages as they evolve (Levinton, 1988; Rachootin & Thomson, 1981; Waddington, 1942). In particular, developmental control becomes increasingly integrated and interdependent. Also, structures formed later in ontogeny become dependent on earlier events, increasing the burden on the earlier-formed (and generally physically larger) structures. Theoretically, canalization is of adaptive benefit to the long-term survival of the species (Kerszberg, 1989), but it also reduces the chances that the species in question will generate further distinct species by neodarwinian processes, given its declining ability to accommodate developmental variation.

Saltation breaks that canalization, toppling the hopeful monster from the adaptive optimum of its parent(s) but also freeing the potential lineage for radical reorganization of form (Arthur, 1984, 1988; Bateman, in press; Carson, 1985; Erwin & Valentine, 1984; Goldschmidt, 1940; Levinton, 1988; van Steenis, 1976). Rather than creating new D-genes, most mutationally driven saltational events involve the suppression of one or more pre-existing D-genes, resulting in the loss of features coded for by those genes (e.g. Tucker, 1988). Other features that develop later in ontogeny and are attached to the suppressed feature are also liable to be lost in consequence (the 'domino effect' of Bateman, in press), leading to a decrease in overall morphological complexity (cf. Atchley & Hall, 1991; McShea, 1991; Wimsatt & Schank, 1988). In some cases, these secondary losses are avoided by epigenetic readjustment; gene interactions may allow the dependent features to be expressed elsewhere on the body plan (heterotopy). Pleiotropy has the converse effect; suppression of a gene expressed in several parts of the body is likely to have particularly profound effects on morphogenesis, and should generally lead to considerable simplification of form.

Superficially, such character losses appear improbable agents for innovative evolutionary change. However, by breaking canalization and simplifying development, they clear the evolutionary palette for future adaptive innovation. The 'development ratchet' (Levinton, 1988; Vermeij, 1987) is reset at a lower level, leaving a combination of adaptation and contingency to define a new evolutionary trajectory for the lineage should it survive the establishment bottleneck. Indeed, simplification proved to be the most powerful of several forces driving the evolution of architecture among the rhizomorphic lycopsids (Bateman, in press; Bateman *et al.*, 1992). It is difficult to envisage centralized determinate growth and secondary thickening as adaptive in smaller-bodied rhizomorphic lycopsids such as *Isoetes*. Rather, these characters represent a significant waste of resources, and make biological sense only if they are perceived as unbreakable developmental constraints that refelect evolution from large-bodied trees – trees that benefited greatly from the increased fitness conferred by these characters (Bateman, 1992, in press). Similarly, morphological (Mayo, 1993) and cpDNA (French *et al.*, 1993) cladograms concur that the minute aquatics of the Lemnaceae are nested well within the Araceae; this can only be interpreted as an example of extreme paedomorphosis and ecological specialization.

One last point should be made about D-gene suppression. Clearly, a particular gene can be suppressed in many different ways (for example, transposons could be inserted at different points within the active region of the same gene in different genomes), but the net result may be identical in terms of phenotypic expression. Such 'polyphyly' of hopeful monsters is suspected in the case of widespread occurrences of pseudopeloria in the insect-mimicking orchid *Ophrys apifera* (Bateman, 1985), and is probably a more general phenomenon. It constitutes parallelism as restrictively defined by Kellogg (1990). Thus, two or more saltational prospecies may appear morphologically indistinguishable.

Summary

Figure 7 summarizes the neoGoldschmidtian paradigm advocated in this paper, integrating genomic changes, organismal demographics and ecological niches. In this example, a key D-gene promoting the development of a major growth module is suppressed by mutation in a germ-cell lineage (animals) or reproductively potentiated meristem (plants) of a single ancestral individual. The resulting hopeful monster is dispersed as a mutant propagule, and by good fortune germinates in a vacant niche that presents no competition and a tolerably low degree of intrinsic stress. During the ontogeny of the monster, it becomes apparent that it has lost the ability to produce a particular ancestral growth module; this in turn means that smaller-scale features usually occurring on that module can no longer be expressed. Loss of the module also prompts epigenetic and/or pleiotropic readjustment to accommodate the remaining modules.

The reproductively isolated hopeful monster (prospecies) then successfully reproduces. Initially, the population is small. In effect, saltation constitutes an extreme genetic bottleneck that will be reflected in greatly reduced genetic diversity relative to the parental population – a potentially sympatric parallel of the allopatric founder effect. The small size of the population will also render it susceptible to genetic drift (e.g. Ayala & Valentine, 1979). Nonetheless, successful reproduction marks a switch from a non-adaptive saltational mode to a potentially adaptive gradual mode (Carson, 1985; Davis & Gilmartin, 1985; Iltis, 1983). Expansion of the radical new species and/or invasion of its niche by other species gradually introduce the population to competitive selection, thereby reforming coadapted gene complexes. This prompts equally gradual population-level changes in the gene pool; some features initially retained by the hopeful monsters have lost any function and can be eliminated, whereas others have changed their function and can be appropriately modified. Eventually, a large, viable population (taxonomic species) is established. Thus, classical neodarwinism has a significant role to play in the saltational paradigm, by honing

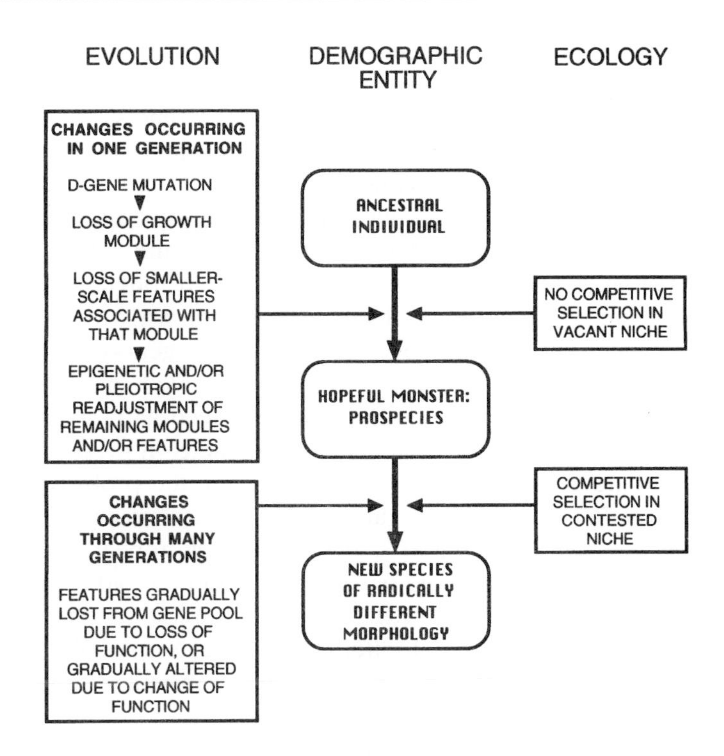

Figure 7 Summary of the neoGoldschmidtian paradigm, integrating genomic changes, organismal demographics and ecological niches. In this example, a key gene controlling early development (D-gene) is suppressed by mutation in a germ-cell lineage of the ancestral individual. The resulting hopeful monster is dispersed as a mutant propagule, which by good fortune germinates in a vacant niche that presents no competition and a tolerably low degree of intrinsic stress. During the ontogeny of the monster it becomes apparent that it has lost the ability to produce a particular ancestral growth module; this in turn means that smaller-scale features usually occurring on that module can no longer be expressed. Loss of the module also prompts epigenetic and/or pleiotropic readjustment to accommodate the remaining modules. The hopeful monster (prospecies) then successfully reproduces, eventually establishing a large, viable and reproductively isolated population (taxonomic species). Expansion of the radical new species and/or invasion of its niche by other species gradually introduce the population to competitive selection, altering or removing features by classic neodarwinian, population-level changes in the gene pool and thereby honing the new lineage to a local fitness optimum.

the saltationally generated lineage to a local fitness optimum. Saltation is supradarwinian rather than antidarwinian.

LIVING VERSUS FOSSIL PLANT SALTATION: LIMITS TO UNIFORMITARIANISM IN EVOLUTIONARY THEORY

Uniformitarianism – the constancy of processes through time – has long been a fundamental principle of geology (Gould, 1990; Hallam, 1983; Lyell, 1830–33). Most physical contraventions of the principle, such as long-term changes in atmospheric composition and in terrestrial weathering rates, can be attributed largely to non-uniformitarianism in the Earth's biota – the former prompted by the evolution of photosynthesis, the latter by the advent of soil-binding roots (e.g. Behrensmeyer *et al.*, 1992). Many neontologists apply uniformitarianism to evolutionary theory, often unconsciously and without serious consideration of the time-span at their disposal; periods of time that can be measured in generations are more tangible than the estimated 420 million years of vascular plant evolution.

Indeed, uniformitarianism is generally applicable to evolution at the molecular level – the generation of variation through mutation and the subsequent history of that genetic variation. However, if one could trace plant phylogeny back through geological time, there can be little doubt that the pool from which that variation is drawn would change, showing an overall decrease in genotypic diversity and a decrease in the phenotypic complexity of the most derived clades present. Reproduction would become increasingly simple – first the seed habit would be lost, then heterospory, then heterothally (e.g. Bateman & DiMichele, in press). Self-fertilization would become progressively more common, increasing the probability of historical continuity for non-lethal mutations. Eventually, one would encounter the earliest vascular plants, such as *Aglaophyton* and *Cooksonia*, which appear to have possessed near-isomorphic, independent haploid gametophytic and diploid sporophytic generations (Kenrick & Crane, 1991; Remy *et al.*, 1993). The haploid phase of the life history may have been as long as the diploid, offering a far greater probability of mutations unbuffered by the continued presence of a second, non-mutant copy of the allele (Knoll *et al.*, 1986).

This reproductive simplification is paralleled by developmental simplification; although the number of D-genes is modest in the more derived extant clades, it was probably still fewer in primitive fossil groups such as the 'rhyniophytes'. This would increase the average phenotypic effect of D-gene mutations, by reducing the probability of alternative developmental pathways and of epigenetic readjustment to compensate for the genomic change.

However, the most profound contraventions of biological uniformitarianism are ecological. Neobotanical ecology focuses heavily on angiosperms, acknowledging their remarkable species-level diversity. We believe that this largely reflects unusually fine niche partitioning. We further believe that this in turn reflects unusually high frequencies of intimate coevolutionary relationships (for example, with pollinators, herbivores and mycorrhizae) rather than the greater genomic 'rigidity' of non-angiosperms invoked by van Steenis (1976). However, passing back through time, the small-scale niche partitioning of the angiosperms would gradually give way to less diverse communities dominated by non-angiosperms (Bateman, 1991; Behrensmeyer *et al.*, 1992; Niklas *et al.*, 1985). The strengths of interspecific links may remain strong, but the potential total number is greatly reduced. By the time one reached the mid- to late Devonian, the ecological scenario presented in Fig. 6 of two-niche ecosystems would be far less of a parody. There was more opportunity for partitioning existing niches and for increasing connectivity among the species occupying those niches. Moreover, many more niches, and even entire habitats, were vacant at any one moment in time. Opportunities for temporary release from selection, allowing the establishment of hopeful monsters, were far more common. In short, a higher frequency of generation of hopeful monsters in the past is possible, but a far higher frequency of their establishment is certain. The further back in time one travelled, the greater would be the significance of non-adaptive saltation relative to adaptive selection.

The observation allows an additional prediction – one of fractal evolutionary patterns through time. Once plants had invaded the land and a basic tool-kit of terrestrial adaptations had evolved in the division Tracheophyta, taxa that we classify as classes and orders on the basis of their greater overall phenotypic dissimilarity should appear earlier in the fossil record than families and genera. Admittedly, estimation of high-level architectural diversity is a highly subjective occupation – witness the recent controversy over the middle Cambrian Burgess Shale arthropods (*Auct. Mult.*, 1992; Briggs *et al.*, 1992; Gould, 1989), which are a plausible marine animal analogue to the problems presented by the terrestrial vascular plant radiation. Nonetheless, if the controversial decision is taken to treat the angiosperms as an order of the class Gymnospermopsida (e.g. Bateman, 1991, fig. 2.5), it can be argued that all eutracheophyte classes originated over a remarkably short period of approximately 70 million years during the late Silurian and Devonian – probably the heyday of botanical saltation.

PLANT VERSUS ANIMAL SALTATION: LIMITS TO ZOOCENTRIC HOLISM IN EVOLUTIONARY THEORY

This chapter began with a mild critique of evolutionary zoocentrism. We now conclude the narrative by returning to a comparison of evolutionary modes in higher plants and higher animals of the terrestrial realm. Given that the fundamental genomic controls of development in animals and plants are similar, any contrasts must reflect differences in (1) frequencies of mutation and polyploidy, (2) modes of gene expression (both determining the generative phase), and/or (3) different ecological roles (determining the establishment phase). Linking the generative and establishment phases is the sessile life-style of plants, and their consequent reliance on numerous localized meristems for open, additive growth and differentiation. This contrasts starkly with the vagile life-style of higher animals; their generalized, closed and often replacive growth (typically involving a greater number of tissues), and their complex, neurally mediated behaviours.

Beginning with the relative frequency of saltational genetic changes, most plants possess numerous meristems that offer greater scope for both polyploidy and mutation; different branches of the same individual often prove to be genetically dissimilar (e.g. Thomson *et al.*, 1991). This is particularly important with respect to germ lineages. In contrast with vertebrates, the gametes or gamete-producing organs of plants are not sequestered early in development. Any one of the many meristems in any one individual is capable of differentiating into reproductive structures that can pass on mutations acquired during normal somatic growth (e.g. Buss, 1987; Darley, 1990). Indeed, plants appear to have developed specific mechanisms to eliminate, or at least buffer the accumulation of, detrimental mutations (Klekowski *et al.*, 1985).

Differences in gene expression are more problematic. Firstly, meristematic growth offers few commands for apical behaviour and these are generally of the binary, 'on-off' variety (Bateman, in press; Borchert, 1983; Borchert & Honda, 1984). The meristem is either indeterminate or determinate, either it divides or it does not, and either it divides isotomously or anisotomously. There is some scope in plants for intermediate responses that would allow gradual, directional evolution of form, notably by positional interplay between appendicular organs and the axial apical meristems via hormonal gradients and/or glycoprotein receptors. Nonetheless, such responses are far more compatible with the generalized growth and cell mobility evident in the more holistic ontogenies of animals. In this context, it is noteworthy that among animals saltation is most easily recognized in arthropods, whose stepwise metamorphic ontogenies and modular segmented growth most closely resemble plant development. Because saltation causes a greater average decrease in fitness among animals than among plants (see later), selection pressures are presumably greater for post-transcriptional and epigenetic flexibility, albeit within heavily prescribed architectural constraints (cf. Gallie, 1993; Goodwin, 1984, 1988, 1993; Goodwin & Saunders, 1992; Løvtrup, 1973; Stebbins, 1992). Furthermore, physiological abortion of hopeful monsters during ontogeny is probably less common in plants than animals; for example, seed plants are presumably more tolerant of developing mutant embryos (including polyploids) than are placental mammals.

The greatest differences between plants and animals concern ecologically mediated establishment; specifically, the theoretical level of fitness required for successful establishment in a particular habitat if vacant, and the actual (higher) level of fitness required for successful establishment in that habitat under specific conditions of competition.

Firstly, intrinsic functional constraints require consideration. In vagile terrestrial animals there is clearly an enormous adaptive advantage to a bilaterally symmetrical body that consists of a core framework subtending paired appendages. Other architectures are frequently generated as hopeful monsters and many transcend the lower, non-competitive fitness threshold for establishment. Unfortunately, the non-competitive threshold is less often relevant for vagile animals than sessile plants. Inevitably, vagile hopeful monsters are immediately assailed by the full force of active and direct competition, for resources such as food and mates, from far fitter organisms.

Plants too have a fundamental architectural constraint, imposed by the facts that (1) their development is centrifugal from a point of origin (prothallus in pteridophytes, seed in spermatophytes) that generates at least one primary axis (stem), and (2) they require distinct negatively geotropic organs for photosynthesis and reproduction, and positively geotropic organs for absorption of water and nutrients; the latter are attached to the former as repeated modular units in unipolar plants, but as a single unit (rootstock) in bipolar plants. Nonetheless, within this constraint lies an enormous range of potential architectures that reflect great flexibility in the meristematic expression of particular organs.

Ecologically, this argument can be placed in a framework of four progressively more stringent thresholds of establishment. Many of these architectures are capable of surpassing the non-competitive establishment threshold for germination and growth (threshold 1). More importantly, the indirect nature of competition among plants gives the hopeful monster a far greater probability of reaching reproductive maturity (threshold 2), decreasing the differential between the non-competitive and actual competitive thresholds (3). Moreover, the potential range of competitors is more restricted due to the slow pace and *ad hoc* unconscious migration of plants relative to animals within particular habitats, as they attempt to accommodate to local changes in resource availability (threshold 4). Thus, competitive displacement is far less common among plants that animals, requiring an environmental perturbation to upset the local ecological equilibrium (DiMichele *et al.*, 1987, 1992). Yet more significantly, all four establishment thresholds for the average plant are lower than for the average animal, and the differentials among those thresholds are also less. Thus, the degrees of freedom for non-selective experimentation in form (the '*patio ludens*' of van Steenis, 1976) are far greater in plants than animals; at any one moment in time, far more of their characters are likely to be adaptively neutral. This is turn leaves much greater opportunity for canalization into a particular suite of developmental constraints that owe their origin to historical accident rather than vectorial adaptation (contingency *sensu* Gould, 1989). Comparison of plant and animal cladograms suggests that architectures are much less likely to be evolutionarily conserved in plants than in animals.

The modular growth of plants and differentiation into permanent core organs and transient appendages renders them far less integrated and holistic than animals (cf. Hay & Mabberley, 1991). Many plant organs, notably appendicular organs such as rootlets, leaves, and reproductive organs but also non-appendicular stems in seed plants possessing axillary branches, terminate ontogenetic cascades. Reiterated and to a large degree independent, they can be evolutionarily modified without a resulting domino effect on other organs formed later in development. Such organs have high degrees of physiological and developmental independence; they can become individually canalized. This in turn permits evolutionary independence in the form of mosaic evolution; differential rates of evolution and degrees of character conservation among different organs (Knoll *et al.*, 1984; Meyen, 1987; Thomas & Spicer, 1987). Bateman (in press) used the rhizomorphic lycopsid phylogeny (Fig. 5) to demonstrate that cladograms offer the opportunity to quantify such mosaicism. He noted that intermediate-scale characters such as vascular anatomy and cone morphology were more highly conserved than large-scale architectural features under pleiotropic control (a few D-genes control the development of many characters) and small-scale features such as spore ornamentation that are unlikely to be subject to strong selection pressures.

Whether these conclusions can be generalized to other clades remains a moot point. Certainly, arguments have been made for the non-adaptive evolution of vascular morphology of extinct progymnosperms (Wight, 1987), and successful saltational changes in angiosperm floral morphology have been well documented among extant Orchidaceae (Bateman, 1985; Leavitt, 1909; McCook & Bateman, 1990). Some occur iteratively, as mutant individuals within widely distributed populations of the 'wild type' morph (e.g. pseudopeloric individuals of *Ophrys apifera*: Bateman, 1985). Others are more abundant, form homogeneous interbreeding populations, and become recognized as species (e.g. the pseudopeloric species *Phragmipedium lindenii*). Yet others originate by saltation but subsequently experience repeated speciation events, potentially by neodarwinian processes (e.g. the type B peloric genus *Thelymitra*).

Vagility and cerebralization also allow mate selection in higher animals, whereas sedentary plants rely on *ad hoc* dispersal of gametes; any incompatibility is purely physiological rather than behavioural. Moreover, most plant species consist of hermaphroditic individuals, and many of these are self-compatible, particularly among the more primitive clades. Thus, many teratological plants have the potential to establish their own mutant lineages without the aid of a partner, nullifying the improbable requirement for two compatible and mutually attracted hopeful monsters that led to the ridicule of Goldschmidt's (1940) original zoological formulation of saltation and continues to preoccupy zoological saltationists (e.g. Arthur, 1984).

Thus, the most important difference between evolutionary modes in higher animals and higher plants is the presence and absence respectively of direct competition, not only for resources but also for reproductive partners. This greatly diminishes the relevance of Malthusian concepts to plant evolution, and thereby of the scenarios of competitive selection that drive modern evolutionary zoology (cf. Eldredge, 1989; Knoll, 1984; Levinton, 1988; Vermeij, 1987). Nonetheless, saltation has undoubtedly also played a major role in the evolution of many animal lineages (e.g. Ahlberg, 1992; Mooi, 1990).

THE SALTATIONAL PARADIGM IN A BROADER EVOLUTIONARY CONTEXT

Figure 8 summarizes our view of a typical pattern of vascular plant evolution through time. Neodarwinian vicariance emphasizes population-level divergences, often of small allopatric or parapatric groups. NeoGoldschmidtian saltation, both dichotomous (mutational) and reticulate (allopolyploid), emphasizes individuals that are often sym-

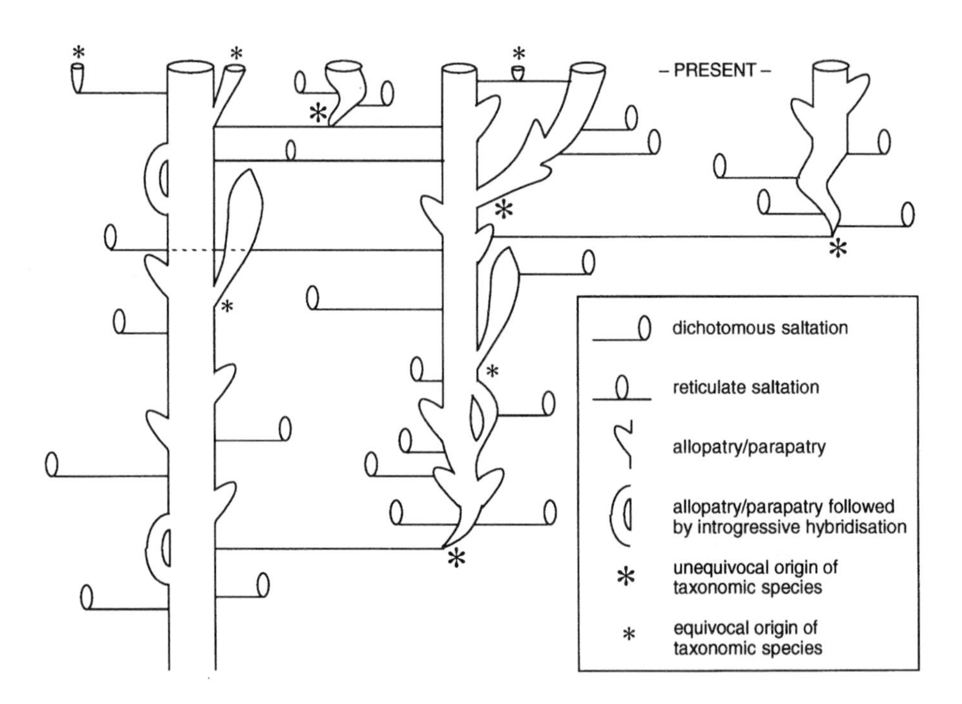

Figure 8 A typical pattern of evolution in vascular plants, viewed as phenotypic changes through time. Prospecies are produced frequently by neodarwinian allopatric vicariance and by both dichotomous (mutational) and reticulate (allopolyploid) neoGoldschmidtian saltation. Most prospecies become extinct rapidly, often by introgressive hybridization in the case of vicariant gradualist lineages. Very few prospecies pass the dual tests of increased abundance and long-term continuity necessary for recognition as taxonomic species (large asterisks). Note that some lineages are difficult to categorize (small asterisks), either because they achieved intermediate population maxima and longevities or because they evolved too recently to assess their longevities. We believe that the ratio of successful to equivocal plus unsuccessful taxonomic speciation events shown in this diagram (4:44) is unrealistically high.

patric with their parent(s). All three mechanisms generate vast numbers of prospecies, though most terminate after a few generations. Termination usually occurs by extinction but can also be caused by introgressive hybridization, particularly in the case of vicariant gradualist lineages. Very few prospecies pass the tests of increased abundance and long-term continuity necessary for recognition as historically significant taxonomic species, and even fewer generate further taxonomic species. Thus we support the emphasis placed on specification and reproduction in the species selection evolutionary model of Futuyma (1987) and Gould & Eldridge (1993), but argue that a new species can be born instantaneously and *in extremis* consist only of a single organism.

Dichotomous saltation is more likely to produce radical new lineages than are reticulate saltation or vicariance, though the magnitude of its maximum effect has diminished through time due to increased developmental canalization and ecological saturation. Nonetheless, saltation may remain the most frequent mode of speciation among vascular plants (van Steenis, 1976). This implies a strong element of chance in evolutionary patterns, and reinforces Gould's (1989) emphasis on contingency; replaying the history of life on Earth would presumably generate a very different biota.

These evolutionary hypotheses require testing by interspecific comparison in a rigorous conceptual framework. At present, comparative biology is pursued on two very different scales – morphological and molecular – that are in practice treated as mutually independent. Morphological phylogenies have the potential to describe the phenotype of the whole organism, but they tend to be static (describing only reproductively mature individuals) and cannot accurately represent underlying genotypic changes. Comparisons of base pair sequences, such as the approximately 1428 base pairs of the phylogenetically popular *rbc*L gene, consider only a tiny proportion of the total genome (i.e. of the available evidence), and only a small proportion of any set of coded sites vary among species and thus are phylogenetically informative. Moreover, many such studies fail to distinguish phenotypically expressed base pairs from the inactive bulk of the genome.

It has long been accepted that the fundamental unit of evolutionary change is the gene but that such changes are mediated via the phenotype of the host organism (the replicators and interactors respectively of Dawkins, 1982, 1986, 1989). Recent studies of D-gene expression in plants (e.g. Coen, 1991; Coen & Carpenter, 1992; Meyerowitz *et al.*, 1992) provide a vital causal link between genotype and phenotype – replicator and interactor – that allows reciprocal illumination between these two contrasting manifestations of the evolutionary process. D-genes can be coded cladistically in order to make the crucial distinction between primitive and derived states and, as we have shown, the resulting cladograms can be used to test competing hypotheses of underlying evolutionary mechanisms. Despite recent advances (Chasan, 1993), biologists have been surprisingly slow to combine relevant concepts of gene expression, developmental control, phylogeny reconstruction, ecological filtering of phenotypes, and evolutionary theory into a truly integrated evolutionary synthesis. This problem has been exacerbated by over-enthusiastic generalization from parochial studies of a few 'flagship' species to all-embracing evolutionary theories. Nonetheless, we are confident that future syntheses will confirm our opinion that plants have their own distinct approach to the evolution of shape and form.

ACKNOWLEDGEMENTS

We are grateful to D.S. Ingram for recognizing the relevance of evolutionary cladistics to this volume, and to V.A. Funk, M. Herschkowitz, J. Pahnke and W.L. Wagner for helpful discussion. We thank P.E. Ahlberg, V.A. Albert, D.H. Erwin, O.S. Farrington, C.J. Humphries, R.G. Olmstead, R. Sattler, R.A. Spicer and J.T. Temple for critically reading the manuscript, while absolving them of its contents. We also thank J. Pahnke for drawing out attention to the (fortuitous) similarity between our opinions on plant evolution and those expressed in an obscure but well-argued paper by C. van Steenis (1976). R.M.B. acknowledges the support of NERC Senior Research Fellowship GT5/F/91/GS/1.

REFERENCES

ABBOTT, R.J., 1992. Plant invasions, interspecific hybridisation, and the evolution of new plant taxa. *Trends in Ecology and Evolution, 7:* 401–405.

AESCHBACHER, R.A., LINSTEAD, P., ROBERTS, K., SCHIEFELBEIN, J.W. & BENFEY, P.N., 1992. A genetic and molecular analysis of root morphogenesis in *Arabidopsis thaliana. American Journal of Botany* (Suppl.), *79:* 25 (abstract).

AHLBERG, P.E., 1992. Coelocanth fins and evolution. *Nature, 358:* 459.

ALBERCH, P., GOULD, S.J., OSTER, G.F. & WAKE, D.B., 1979. Size and shape in ontogeny and phylogeny. *Paleobiology, 5:* 296–317.

ARTHUR, W., 1984. *Mechanisms of Morphological Evolution.* New York, U.S.A.: Wiley.

ARTHUR, W., 1987. *The Niche in Competition and Evolution.* New York, U.S.A.: Wiley.

ARTHUR, W., 1988. *A Theory of the Evolution of Development.* New York, U.S.A.: Wiley.

ASHBURNER, M., 1989. *Drosophila: A Laboratory Handbook.* M.A., U.S.A.: Cold Spring Harbor Laboratory Press.

ATCHLEY, W.R. & HALL, B.K., 1991. A model for development and evolution of complex morphological structures. *Biological Reviews, 66:* 101–157.

AUCT. MULT., 1992. Cambrian and recent morphological disparity. *Science, 258:* 1816–1818.

AVISE, J.C., 1990. Flocks of African fishes. *Nature, 357:* 512–513.

AYALA, F.J. & VALENTINE, J.W., 1979. *Evolving: The Theory and Processes of Organic Evolution.* Menlo Park, C.A., U.S.A.: Benjamin–Cummings.

BALDWIN, B.G., 1992. Phylogenetic utility of the internal transcribed spacers of nuclear ribosomal DNA in plants: an example from the Compositae. *Molecular Phylogenetics and Evolution, 1:* 3–16.

BALDWIN, B.G., KYHOS, D.W. & DVORAK, J., 1990. Chloroplast DNA evolution and adaptive radiation in the Hawaiian silversword alliance (Asteraceae–Madiinae). *Annals of the Missouri Botanic Garden, 77:* 96–109.

BALDWIN, B.G., KYHOS, D.W., DVORAK, J. & CARR, G.D., 1991. Chloroplast DNA evidence for a North American origin of the Hawaiian silversword alliance (Asteraceae). *Proceedings of the National Academy of Sciences, USA, 88:* 1840–1843.

BATEMAN, R.M., 1985. Peloria and pseudopeloria in British orchids. *Watsonia, 15:* 357–359.

BATEMAN, R.M., 1991. Palaeoecology. In C.J. Cleal (ed.), *Plant Fossils in Geological Investigation: The Palaeozoic,* pp. 34–116. Chichester, U.K.: Horwood.

BATEMAN, R.M., 1992. Morphometric reconstruction, palaeobiology and phylogeny of *Oxroadia gracilis* Alvin and *O. conferta* sp. nov., anatomically preserved lycopsids from the Dinantian of Oxroad Bay, SE Scotland. *Palaeontographica, B, 228:* 29–103 + 15 pls.

BATEMAN, R.M. Evolutionary-developmental change in the growth architecture of fossil rhizomorphic lycopsids: scenarios constructed on cladistic foundations. *Biological Reviews* (in press).

BATEMAN, R.M. & DENHOLM, I., 1989. The complementary roles of organisms, populations and species in 'demographic' phytosystematics. *American Journal of Botany* (Suppl.), *76:* 226 (abstract).

BATEMAN, R.M. & DIMICHELE, W.A., 1991. *Hizemodendron,* gen. nov., a pseudoherbaceous segregate of *Lepidodendron* (Pennsylvanian): phylogenetic context for evolutionary changes in lycopsid growth architecture. *Systematic Botany, 16:* 195–205.

BATEMAN, R.M. & DIMICHELE, W.A. Heterospory: the most iterative key innovation in the evolutionary history of the plant kingdom. *Biological Reviews* (in press).

BATEMAN, R.M., DIMICHELE, W.A. & WILLARD, D.A., 1992. Experimental cladistic analyses of anatomically preserved arborescent lycopsids from the Carboniferous of Euramerica: an essay in paleobotanical phylogenetics. *Annals of the Missouri Botanical Garden, 79:* 500–559.

BATESON, W., 1894. *Materials for the Study of Variation Treated with Especial Research to Discontinuity in the Origin of Species.* London, U.K.: Macmillan.

BAZZAZZ, F.A. & CARLSON, R.W., 1979. Photosynthetic contribution of flowers and seeds to reproductive effort of an annual coloniser. *New Phytologist, 82:* 223–232.

BEHRENSMEYER, A.K., DAMUTH, J.D., DIMICHELE, W.A., POTTS, R., SUES, H.-D. & WING, S.L., 1992. *Terrestrial Ecosystems Through Time.* Chicago, U.S.A.: Chicago University Press.

BELL, P.R., 1992. *Green Plants: Their Origins and Diversity* (4th edn). Cambridge, U.K.: Cambridge University Press.

BORCHERT, R., 1983. Phenology and control of flowering in tropical trees. *Biotropica, 15:* 81–89.

BORCHERT, R. & HONDA, H., 1984. Control of development in the bifurcating branch system of *Tabebuia rosea*: a computer simulation. *Botanical Gazette, 145:* 184–195.

BOULTER, M.C., SPICER, R.A. & THOMAS, B.A., 1988. Patterns of plant extinction from some palaeobotanical evidence. In G.P. Larwood (ed.), *Extinctions and Survival in the Fossil Record,* pp. 1–36. Oxford, U.K.: Oxford University Press.

BOWMAN, J.L., SMYTH, D.R. & MEYEROWITZ, D.M., 1989. Genes directing flower development in *Arabidopsis. Plant Cell, 1:* 37–52.

BOWMAN, J.L., SMYTH, D.R. & MEYEROWITZ, D.M., 1991. Genetic interactions among floral homeotic genes of *Arabidopsis. Development, 112:* 1–20.

BRIGGS, D.E.G., FORTEY, R.A. & WILLS, M.A., 1992. Morphological disparity in the Cambrian. *Science, 256:* 1670–1673.

BRITTEN, R.J. & DAVIDSON, E.H., 1969. Gene regulation for higher cells: a theory. *Science, 165:* 349–357.

BUREAU, T.E. & WESSLER, S.R., 1992. *Tourist*; a large family of small inverted repeat elements frequently associated with maize genes. *Plant Cell, 4:* 1283–1294.

BUSS, L.W., 1987. *The Evolution of Individuality.* N.J., U.S.A.: Princeton University Press.

CARSON, H.L., 1985. Unification of speciation theory in plants and animals. *Systematic Botany, 10:* 380–390.

CARSON, H.L. & TEMPLETON, A.R., 1984. Genetic revolutions in relation to speciation phenomena: the founding of new populations. *Annual Review of Ecology and Systematics, 15:* 97–131.

CHASAN, R., 1993. Meeting report: evolving developments [Taos]. *Plant Cell, 5:* 363–369.

CHEPLICK, G.P., 1989. Plant growth regulators as tools for reconstructing the evolution of plant form. *Evolutionary Theory, 9:* 75–83.

CODDINGTON, J.A., 1988. Cladistic tests of adaptational hypotheses. *Cladistics, 4:* 3–22.

COEN, E., 1991. The role of homeotic genes in flower development and evolution. *Annual Review of Plant Physiology and Plant Molecular Biology, 42:* 241–279.

COEN, E. & CARPENTER, R., 1992. The power behind the flower. *New Scientist, 134:* 24–27.

COEN, E. & MEYEROWITZ, E.M., 1991. The war of the whorls: genetic interactions controlling flower development. *Nature, 353:* 31–37.

CRANE, P.R., 1990. The phylogenetic context of microsporogenesis. In S. Blackmore & S.B. Knox (eds), *Microspores: Evolution and Ontogeny,* pp. 11–41. London, U.K.: Academic Press.

CROIZAT, L., 1962. *Space, Time, Form: The Biological Synthesis.* Caracas: author's publication.

CRONQUIST, A., 1968. *Evolution and Classification of Flowering Plants.* Boston, M.A., U.S.A.: Houghton.

CROZIER, T.S. & THOMAS J.F., 1993. Normal floral ontogeny and cool temperature-induced aberrant floral development in *Glycine max* (Fabaceae). *American Journal of Botany, 80*: 429–448.

DAMUTH, J., 1985. Selection among 'species': a formulation in terms of natural functional units. *Evolution, 39*: 1132–1146.

DARLEY, W.M., 1990. The essence of 'plantness'. *American Biology Teacher, 52*: 354–357.

DARWIN, C., 1859. *The Origin of Species by Means of Natural Selection*. London, U.K.: Murray.

DAVIS, J.I. & GILMARTIN, A.J., 1985. Morphological variation and speciation. *Systematic Botany, 10*: 417–425.

DAWKINS, R., 1982. *The Extended Phenotype: The Gene as a Unit of Selection*. San Francisco, U.S.A.: Freeman.

DAVIS, J.I. & GILMARTIN, A.J., 1985. Morphological variation and speciation. *Systematic Botany, 10*: 417–425.

DAWKINS, R., 1986. *The Blind Watchmaker*. Harlow, U.K.: Longman.

DAWKINS, R., 1989. *The Selfish Gene*, 2nd edn. Oxford, U.K.: Oxford University Press.

DE BEER, G.R., 1940. *Embryos and Ancestors*. Oxford, U.K.: Oxford University Press.

DESMOND, A. & MOORE, J., 1991. *Darwin*. London, U.K.: Joseph.

DIETRICH, M.R., 1992. Macromutation. In E.F. Keller & E.A. Lloyd (eds), *Keywords in Evolutionary Biology*, pp. 194–201. Harvard, U.S.A.: Belknap Press.

DIMICHELE, W.A., PHILLIPS, T.L. & OLMSTEAD, R.G., 1987. Opportunistic evolution: abiotic environmental stress and the fossil record of plants. *Review of Palaeobotany and Palynology, 50*: 151–178.

DIMICHELE, W.A., DAVIS, J.I. & OLMSTEAD, R.G., 1989. Origins of heterospory and the seed habit. *Taxon, 38*: 1–11.

DIMICHELE, W.A., HOOK, R.W., BEERBOWER, R., BOY, J.A., GASTALDO, R.A., HOTTON, N. III, PHILLIPS, T.L., SCHECKLER, S.E., SHEAR, W.A. & SUES, H.-D., 1992. Paleozoic terrestrial ecosystems: review and overview. In A.K. Behrensmeyer, J.D. Damuth, W.A. DiMichele, R. Potts, H.-D. Sues & S.L. Wing (eds), *Terrestrial Ecosystems Through Time*, pp. 204–325. Chicago, U.S.A.: Chicago University Press.

DOBZHANSKY, T., 1951. *Genetics and the Origin of Species*, 3rd edn. New York, U.S.A.: Columbia University Press.

DONOGHUE, M.J., 1989. Phylogenies and the analysis of evolutionary sequences, with examples from seed plants. *Evolution, 43*: 1137–1156.

DOYLE, J.A., 1978. Origin of angiosperms. *Annual Review of Ecology and Systematics, 9*: 365–392.

DOYLE, J.J., 1993. DNA, phylogeny, and the flowering of plant systematics. *Bioscience, 43*: 380–389.

DUNCAN, I.M., 1986. The *Bithorax* complex. *Annual Review of Genetics, 21*: 285–319.

ELDREDGE, N., 1989. *Macroevolutionary Dynamics: Species, Niches and Adaptive Peaks*. New York, U.S.A.: MacGraw-Hill.

ELDREDGE, N. & GOULD, S.J., 1972. Punctuated equilibria: an alternative to phyletic gradualism. In T.J.M. Schopf (ed.), *Models in Paleobiology*, pp. 82–115. San Francisco, U.S.A.: Freeman.

ENDRESS, P.K., 1992. Evolution and floral diversity: the phylogenetic surroundings of *Arabidopsis* and *Antirrhinum*. *International Journal of Plant Science, 153*.

ERWIN, D.H. & VALENTINE, J.W., 1984. 'Hopeful monsters', transposons, and Metazoan radiation. *Proceedings of the National Academy of Sciences, USA, 81*: 5482–5483.

FARRIS, J.S., 1983. The logical basis of phylogenetic analysis. In N.I. Platnick & V.A. Funk (eds), *Advances in Cladistics 2*, pp. 7–36. New York, U.S.A.: Columbia University Press.

FERNHOLM, B., BREMER, K. & JÖRNVALL, H. (eds), 1989. *The Hierarchy of Life*. Nobel Symposium 70. Amsterdam, The Netherlands: Excerpta Medica.

FIELD, K.G., OLSEN, G.J., LANE, D.J., GIOVANNONI, S.J., GHISELIN, M.T., RAFF, E.C., PACE, N.R. & RAFF, R.A., 1988. Molecular phylogeny of the animal kingdom. *Science, 239*: 748–753.

FINK, W.L., 1982. The conceptual relationship between ontogeny and phylogeny. *Paleobiology, 8*: 254–264.

FORD, V.S. & GOTTLIEB, L.D., 1992. *Bicalyx* is a natural homeotic floral variant. *Nature, 358*: 671–673.

FOREY, P.L., HUMPHRIES, C.J., KITCHING, I.J., SCOTLAND, R.W., SIEBERT, D.J. & WILLIAMS, D.M., 1992. *Cladistics: A Practical Course in Systematics*. Oxford, U.K.: Oxford University Press.

FRENCH, J.C., HUR, Y. & CHUNG, M., 1993. *cp*DNA phylogeny of the Ariflorae. In P. Wilkinson, S. Mayo & P. Rudall (eds), *Monocotyledons: An International Symposium* (abstract) p. 5. Kew, U.K.: Royal Botanic Garden.

FUNK, V.A., 1982. Systematics of *Montanoa* (Asteraceae: Heliantheae). *Memoirs of the New York Botanical Garden, 36:* 1–135.

FUNK, V.A. & BROOKS, D.R., 1990. *Phylogenetic Systematics as the Basis of Comparative Biology*. Smithsonian Contributions to Botany 73. Washington, D.C.: Smithsonian Institution.

FUTUYMA, D.J., 1986. *Evolutionary Biology,* 2nd edn. Sunderland, M.A., U.S.A.: Sinauer.

FUTUYMA, D.J., 1987. On the role of species in anagenesis. *American Naturalist, 130:* 465–473.

GALLIE, D.R., 1993. Post-transcriptional regulation of gene expression in plants. *Annual Review of Plant Physiology and Plant Molecular Biology, 44:* 77–105.

GEHRING, W.J., 1987. Homeo boxes in the study of development. *Science, 236:* 1245–1252.

GEHRING, W.J. & HIROMI, Y., 1986. Homeotic genes and the homeobox. *Annual Review of Genetics, 20:* 147–173.

GILMOUR, J.S.L. & HESLOP-HARRISON, J., 1954. The deme terminology and the units of microevolutionary change. *Genetica, 27:* 147–161.

GOLDSCHMIDT, R., 1940. *The Material Basis of Evolution*. New Haven, U.S.A.: Yale University Press.

GOODWIN, B.C., 1984. Changing from an evolutionary to a generative paradigm in biology. In J.W. Pollard (ed.), *Evolutionary Theory: Paths into the Future,* pp. 99–120. London, U.K.: Wiley.

GOODWIN, B.C., 1988. Morphogenesis and heredity. In M.-W. Ho & S.W. Fox (eds), *Evolutionary Processes and Metaphors,* pp. 145–162. New York, U.S.A.: Wiley.

GOODWIN, B.C., 1993. Homology, development and hierarchies. In B.K. Hall (ed.), *The Hierarchical Basis of Comparative Biology*. London, U.K.: Academic Press (in press).

GOODWIN, B.C. & SAUNDERS, P. 1992 (eds), *Theoretical Biology: Epigenetic and Evolutionary Order from Complex Systems*. Baltimore, U.S.A.: Johns Hopkins Press.

GOTTLIEB, L.D., 1984. Genetics and morphological evolution in plants. *American Naturalist, 123:* 681–709.

GOTTLIEB, L.D., 1986. The genetic basis of plant form. *Philosophical Transactions of the Royal Society of London, B, 313:* 197–208.

GOTTLIEB, L.D. & JAIN, S.K. (eds), 1988. *Plant Evolutionary Biology*. London, U.K.: Chapman & Hall.

GOULD, S.J., 1977a. *Ontogeny and Phylogeny*. Cambridge, M.A., U.S.A.: Belknap.

GOULD, S.J., 1977b. The return of hopeful monsters. *Natural History, 86:* 22–30.

GOULD, S.J., 1980. *The Panda's Thumb*. New York, U.S.A.: Norton.

GOULD, S.J., 1982. The uses of heresy. Introduction to reprint of R. Goldschmidt, 1940. *The Material Basis of Evolution,* pp. xiii–xlii. New Haven, U.S.A.: Yale University Press.

GOULD, S.J., 1986. Punctuated equilibrium at the third stage. *Systematic Zoology, 35:* 143–148.

GOULD, S.J., 1989. *Wonderful Life: The Burgess Shale and the Nature of History*. New York, U.S.A.: Norton.

GOULD, S.J. & ELDREDGE, N., 1977. Punctuated equilibria: the tempo and mode of evolution reconsidered. *Transactions of the Royal Society of London, B, 3:* 115–151.

GOULD, S.J. & ELDREDGE, N., 1993. Punctuated equilibrium comes of age. *Nature, 366:* 223–227.

GOULD, S.J. & LEWONTIN, R.C., 1979. The spandrels of San Marco and the Panglossian paradigm: a critique of the adaptationist program. *Proceedings of the Royal Society of London, B, 205:* 581–598.

GOULD, S.J. & VRBA, E., 1982. Exaptation – a missing term in the science of form. *Paleobiology, 8:* 4–15.

GOUY, M. & LI, W.-H., 1989. Molecular phylogeny of the Kingdoms Animalia, Plantae and Fungi. *Molecular Biology and Evolution, 6:* 109–122.

GRANT, P.R. & HORN, H.S. (eds), 1992. *Molds, Molecules and Metazoa: Growing Points in Evolutionary Biology*. Princeton, N.J., U.S.A.: Princeton University Press.

GRANT, V., 1971. *Plant Speciation*. New York, U.S.A.: Columbia University Press.

GREENE, H.W., 1986. Diet and arboreality in the emerald monitor, *Varanus prasinus,* with comments on the study of adaptation. *Fieldiana Zoologica, 31:* 1–12.

GREENWOOD, P.H., 1981. *The Haplochromine Fishes of East African Lakes*. Ithaca, U.S.A.: Cornell University Press.

GUERRANT, E.O., 1982. Neotenic evolution of *Delphinium nudicaule* (Ranunculaceae): a hummingbird-pollinated larkspur. *Evolution, 36:* 699–712.

HAECKEL, E., 1868. *Natürlische Scopfüngsgeschichte*. Berlin, Germany: Reimer.

HALDANE, J.B.S., 1932. *The Forces of Evolution*. London, U.K.: Longman.

HALLAM, A., 1983. *Great Geological Controversies*. Oxford, U.K.: Oxford University Press.

HALLÉ, F., 1986. Modular growth in seed plants. *Philospohical Transactions of the Royal Society of London, B, 313:* 77–87.

HALLÉ, F., OLDEMAN, R.A.A. & TOMLINSON, P.B., 1978. *Tropical Trees and Forests: An Architectural Analysis*. Berlin, Germany: Springer.

HARDWICK, R.C., 1986. Physiological consequences of modular growth in plants. *Philosophical Transactions of the Royal Society of London, B, 313:* 161–173.

HARPER, J.L., ROSEN, B.R. & WHITE, J. (eds), 1986. *The Growth and Form of Modular Organisms*. London, U.K.: Royal Society.

HARVEY, P.H. & PAGEL, M.D., 1991. *The Comparative Method of Evolutionary Biology*. Oxford, U.K.: Oxford University Press.

HAY, A. & MABBERLEY, D.J., 1991. 'Transference of Function' and the origin of aroids: their significance in early angiosperm evolution. *Botanische Jahrbücher fur Systematik Pflanzengeschichte, 113:* 339–428.

HENDY, M.D. & PENNY, D., 1989. A framework for the quantitative study of evolutionary trees. *Systematic Zology, 36:* 297–309.

HILL, C.R. & CAMUS, J.M., 1986. Evolutionary cladistics of marattialean ferns. *Bulletin of the British Museum (Natural History), Botany, 14:* 219–300.

HILL, J.P. & LORD, E.M., 1989. Floral development in *Arabidopsis thaliana:* a comparison of the wild type and the homeotic *pistillata* mutant. *Canadian Journal of Botany, 67:* 2922–2936.

HILU, K.W., 1983. The role of single-gene mutations in the evolution of flowering plants. *Evolutionary Biology, 16:* 97–128.

ILTIS, H.H., 1983. From teosinte to maize: the catastrophic sexual transmutation. *Science, 222:* 886–894.

KAUFFMAN, S.A., 1993. *The Origins of Order: Self-Organization and Selection in Evolution*. Oxford, U.K.: Oxford University Press.

KELLOGG, E.A., 1990. Ontogenetic studies of florets in *Poa* (Gramineae): allometry and heterochrony. *Evolution, 44:* 1978–1989.

KENLEYSIDE, M.H.A. (ed.), 1991. *Cichlid Fishes: Behaviour, Ecology and Evolution*. London, U.K.: Chapman & Hall.

KENRICK, P. & CRANE, P.R., 1991. Water-conducting cells in early land-plants: implications for the early evolution of tracheophytes. *Botanical Gazette, 152:* 335–356.

KERSZBERG, M., 1989. Developmental canalization can enhance species survival. *Journal of Theoretical Biology, 139:* 287–309.

KIMURA, M., 1983. *The Neutral Theory of Molecular Evolution*. Cambridge, U.K.: Cambridge University Press.

KLEKOWSKI, E.J. Jr, KAZARINOVA-FUKSNENSKY, N. & MOHR, H., 1985. Shoot apical meristems and mutation: stratified meristems and angiosperm evolution. *American Journal of Botany, 72:* 1788–1800.

KLUGE, A.G. & STRAUSS, R.E., 1985. Ontogeny and systematics. *Annual Review of Ecology and Systematics, 16:* 247–268.

KNOLL, A.H., 1984. Patterns of extinction in the fossil record of vascular plants. In M. Nitecki (ed.), *Extinctions*, pp. 21–68. Chicago, U.S.A.: Chicago University Press.

KNOLL, A.H., 1992. Early evolution of prokaryotes: a geological perspective. *Science, 256:* 622–627.

KNOLL, A.H. & NIKLAS, K.J., 1987. Adaptation, plant evolution, and the fossil record. *Review of Palaeobotany and Palynology, 50:* 127–149.

KNOLL, A.H., NIKLAS, K.J., GENSEL, P.G. & TIFFNEY, B.H., 1984. Character diversification and patterns of evolution in early vascular plants. *Paleobiology, 10:* 34–47.

KNOLL, A.H., GRANT, S.W.F. & TSAO, J.W., 1986. The early evolution of land plants. In Gastaldo, R.A. & Broadhead, T.W. (eds), *Land Plants: Notes for a Short Course,* pp. 45–63. University of Tennessee Geological Sciences Studies in Geology 15.

KNOX, E.B., DOWNIE, S.R. & PALMER, J.D., 1993. Chloroplast genome rearrangements and the evolution of giant lobelias from herbaceous ancestors. *Molecular Biology and Evolution, 10:* 414–430.

KUHN, T.S., 1962. *The Structure of Scientific Revolutions.* Chicago, U.S.A.: Chicago University Press.

LANDE, R., 1986. The dynamics of peak shifts and the pattern of morphological evolution. *Paleobiology, 12:* 343–354.

LAUDER, G.V., 1990. Functional morphology and systematics: studying functional patterns in a historical context. *Annual Review of Ecology and Systematics, 21:* 317–340.

LAWRENCE, P.A., 1992. *The Making of a Fly: The Genetics of Animal Design.* Oxford, U.K.: Blackwell.

LEAVITT, R.G., 1909. A vegetative mutant, and the principle of homeosis in plants. *Botanical Gazette, 47:* 30–68.

LEMEN, C.A. & FREEMAN, P.W., 1989. Testing macroevolutionary hypotheses with cladistic analysis: evidence against rectangular evolution. *Evolution, 43:* 1538–1554.

LEVIN, D.A., 1970. Developmental instability and evolution in peripheral isolates. *Evolution, 104:* 343–353.

LEVIN, D.A., 1993. Local speciation in plants: the rule not the exception. *Systematic Botany, 18:* 197–208.

LEVINTON, J., 1988. *Genetics, Paleontology and Macroevolution.* Cambridge, U.K.: Cambridge University Press.

LEWIS, H., 1962. Catastrophic selection as a factor in speciation. *Evolution, 16:* 257–271.

LEWIS, H., 1966. Speciation in flowering plants. *Science, 152:* 167–172.

LEWIS, H., 1969. Speciation. *Taxon, 18:* 21–25.

LEWONTIN, R.C., 1993. *The Doctrine of DNA: Biology as Ideology.* London, U.K.: Penguin.

LONG, A.G., 1977. Some Lower Carboniferous pteridosperm cupules bearing ovules and microsporangia. *Transactions of the Royal Society of Edinburgh, B, 70:* 1–11.

LØVTRUP, S., 1973. *Epigenetics: A Treatise on Theoretical Biology.* New York, U.S.A.: Wiley.

LOWREY, T.K., 1993. *Phylogeny, Adaptive Radiation, and Biogeography of Hawaiian* Tetramolopium (*Compositae; Astereae*). Smithsonian Contributions to Botany. Washington, D.C., U.S.A.: Smithsonian Institution (in press).

LOWREY, T.K. & CRAWFORD, D.J., 1985. Allozyme divergence and evolution in *Tetramolopium* (Compositae: Astereae) on the Hawaiian Islands. *Systematic Botany, 10:* 64–72.

LYELL, C., 1830–33. *Principles of Geology* (3 vols). London, U.K.: Murray.

MADDISON, W.P., 1990. A method for testing the correlated evolution of two binary characters: are gains or losses concentrated on certain branches of a phylogenetic tree? *Evolution, 44:* 539–557.

MADDISON, W.P., DONOGHUE, M.J. & MADDISON, D.R., 1984. Outgroup analysis and parsimony. *Systematic Zoology, 43:* 83–103.

MANDEL, M.A., GUSCHAFSON-BROWN, C., SAVIDGE, B. & YANOFSKY, M.F., 1992. Molecular characterization of the *Arabidopsis* floral homeotic gene APETALA1. *Nature, 360:* 273–277.

MARGULIS, L., 1993. *Symbiosis in Cell Evolution: Microbial Communities in the Archean and Proterozoic Eons,* 2nd edn. New York, U.S.A.: Freeman.

MARGULIS, L. & SCHWARTZ, K.V., 1988. *Five Kingdoms: An Illustrated Guide to the Phyla of Life on Earth.* San Francisco, U.S.A.: Freeman.

MARX, J., 1992. Homeobox genes go evolutionary. *Science, 255:* 399–401.

MAYNARD SMITH, J., 1972. *On Evolution.* Edinburgh, U.K.: Edinburgh University Press.

MAYO, S.J., 1993. What is the Ariflorae? In P. Wilkins, S. Mayo & P. Rudall (eds), *Monocotyledons: An International Symposium* (abstract), p. 11. Kew, U.K.: Royal Botanic Garden.

MAYR, E., 1963. *Animal Species and Evolution.* Cambridge, M.A., U.S.A.: Harvard University Press.

McCOOK, L.M. & BATEMAN, R.M., 1990. Homeosis in orchid flowers: a potential mechanism for saltational evolution. *American Journal of Botany* (Suppl.), *77:* 145 (abstract).

McKINNEY, M.L. (ed.), 1988. *Heterochrony in Evolution: A Multidisciplinary Approach.* New York, U.S.A.: Plenum.

McKINNEY, M.L. & McNAMARA, K.J., 1991. *Heterochrony: The Evolution of Ontogeny*. New York, U.S.A.: Plenum.

McSHEA, D.W., 1991. Complexity and evolution: what everybody knows. *Biology and Philosophy, 6:* 303–324.

MEEUSE, A.D.J., 1986. *Anatomy of Morphology*. Leiden: Brill.

MEYEN, S.V., 1987. *Fundamentals of Palaeobotany*. London, U.K.: Chapman & Hall.

MEYEN, S.V., 1988. Origin of the angiosperm gynoecium by gamoheterotopy. *Botanical Journal of the Linnean Society, 97:* 171–178.

MEYER, V.G., 1966. Floral abnormalities. *Botanical Review, 32:* 165–195.

MEYER, A., KOCHER, T.D., BASASIBWAKI, P. & WILSON, A.C., 1990. Monophyletic origin of Lake Victoria cichlid fishes suggested by mitochondrial DNA sequences. *Nature, 347:* 550–553.

MOOI, R., 1990. Paedomorphosis, Aristotle's lantern, and the origin of the sand dollars (Echinodermata: Clypeasteroidea). *Paleobiology, 16:* 25–48.

MOSBRUGGER, V., 1990. *The Tree Habit in Land Plants*. Lecture Notes in Earth Sciences 28. Berlin, Germany: Springer.

NIKLAS, K.J., TIFFNEY, B.H. & KNOLL, A.H., 1985. Patterns in vascular land plant diversification: an analysis at the species level. In J.W. Valentine (ed.), *Phanerozoic Diversity Patterns: Profiles in Macroevolution*, pp. 97–128. Princeton, N.J., U.S.A.: Princeton University Press.

O'GRADY, R.T., 1985. Ontogenetic sequences and the phylogenetics of flatworms. *Cladistics, 1:* 159–170.

PAGE, C.N., 1972. An interpretation of the morphology and evolution of the cone shoot of *Equisetum*. *Botanical Journal of the Linnean Society, 65:* 359–397.

PATTERSON, C., 1982. Morphological characters and homology. In K.A. Joysey & A.E. Friday (eds), *Problems of Phylogenetic Reconstruction*, pp. 21–74. Systematics Association Special Volume 21. London, U.K.: Academic Press.

PHILLIPS, T.L. & DIMICHELE, W.A., 1992. Comparative ecology and life-history biology of arborescent lycopsids in Late Carboniferous swamps of Euramerica. *Annals of the Missouri Botanical Garden, 79:* 560–588.

POETHIG, R.S., 1990. Phase change and the regulation of shoot morphologies in plants. *Science, 250:* 923–930.

PRICE, T., TURELLI, M. & SLATKIN, M., 1993. Peak shifts produced by correlated response to selection. *Evolution, 47:* 280–290.

RACHOOTIN, S.P. & THOMSON, K.S., 1981. Epigenetics, paleontology and evolution. In G.G.E. Scudder & J.L. Reveal (eds), *Evolution Today,* pp. 181–193. Proceedings of the Second International Congress of Systematic and Evolutionary Biology.

RAFF, R.A. & KAUFMAN, T.C., 1983. *Embryos, Genes, and Evolution*. New York, U.S.A.: Macmillan.

RASMUSSEN, N. & GREEN, P.B., 1993. Organogenesis in flowers of the homeotic green pistillate mutant of tomato *(Lycopersicon esculentum)*. *American Journal of Botany, 80:* 805–813.

RAVEN, P.H., EVERT, R.F. & EICHHORN, S.E., 1986. *Biology of Plants,* 4th edn. New York, U.S.A.: Worth.

RAYNER, A.D.M., BRASIER, C.M. & MOORE, D. (eds), 1987. *Evolutionary Biology of the Fungi*. Cambridge, U.K.: Cambridge University Press.

REMY, W., GENSEL. P.G. & HASS, H., 1993. The gametophyte generation of some Early Devonian land plants. *International Journal of Plant Science, 154:* 35–58.

RIDLEY, M., 1992. *Evolution*. Oxford, U.K.: Blackwell.

RIEDL, R., 1979. *Order in Living Organisms*. New York, U.S.A.: Wiley.

ROTHWELL, G.W., 1987. The role of development in plant phylogeny: a paleobotanical perspective. *Review of Palaeobotany and Palynology, 50:* 97–114.

SATTLER, R., 1986. *Biophilosophy: Analytic and Holistic Perspectives*. Berlin: Springer.

SATTLER, R., 1988. Homeosis in plants. *American Journal of Botany, 75:* 1606–1617.

SATTLER, R., 1993. Homology, homeosis, and process morphology in plants. In B.K. Hall (ed.), *The Hierarchical Basis of Comparative Biology*. London, U.K.: Academic Press (in press).

SCHINDEWOLF, O.H., 1950. *Grundfragen der Paläontologie*. Stuttgart, Germany: Schweizerbart.

SCHULTZ, E.A., PICKETT, F.B. & HAUHN, G.W., 1991. The FLO10 gene product regulates the

expression domain of homeotic genes AP3 and P1 in *Arabidopsis* flowers. *Plant Cell, 3:* 1221–1237.
SCHWARTZ-SOMMER, Z., HUIJSER, P., NACKEN, W., SAEDLER, H. & SOMMER, H., 1990. Genetic control of flower development by homeotic genes in *Antirrhinum majus. Science, 250:* 931–936.
SELDEN, P.A. & EDWARDS, D., 1989. Colonisation of the land. In K.C. Allen & D.E.G. Briggs (eds), *Evolution and the Fossil Record,* pp. 122–152. London, U.K.: Belhaven.
SIMPSON, G.G., 1944. *Tempo and Mode in Evolution.* New York, U.S.A.: Columbia University Press.
SLACK, J.M.W., HOLLAND, P.W.H. & GRAHAM, C.F., 1993. The zootype and the phylotypic stage. *Nature, 361:* 490-492.
SMITH, L.G., GREENE, B., VEIT, B. & HAKE, S., 1992. A dominant mutation in the maize homebox gene, *Knotted-1,* causes its ectopic expression in leaf cells with altered fates. *Development, 116:* 21–30.
STACE, C.A., 1989. *Plant Taxonomy and Biosystematics,* 2nd edn. London, U.K.: Arnold.
STACE, C.A., 1993. The importance of rare events in polyploid evolution. In D.R. Lees & D. Edwards (eds), *Evolutionary Patterns and Processes,* pp. 159–169. London, U.K.: Academic Press.
STANLEY, S.M., 1975. A theory of evolution above the species level. *Proceedings of the National Academy of Sciences, USA, 72:* 646–650.
STANLEY, S.M., 1979. *Macroevolution: Pattern and Process.* San Francisco, U.S.A.: Freeman.
STEBBINS, G.L., 1950. *Variation and Evolution in Plants.* New York, U.S.A.: Columbia University Press.
STEBBINS, G.L., 1971. *Chromosomal Evolution in Higher Plants.* London, U.K.: Arnold.
STEBBINS, G.L., 1974. Adaptive shifts and evolutionary novelty: a compositionist approach. In F.J. Ayala & T. Dobzhansky (eds), *Studies in the Philosophy of Biology,* pp. 285–338. Berkeley, U.S.A.: University of California Press.
STEBBINS, G.L., 1992. Comparative aspects of plant morphogenesis: a cellular, molecular, and evolutionary approach. *American Journal of Botany, 79:* 589–598.
STEWART, W.N. & ROTHWELL, G.W., 1993. *Paleobotany and the Evolution of Plants,* 2nd edn. Cambridge, U.K.: Cambridge University Press.
STIDD, B.M., 1980. The neotenous origin of the pollen organ of the gymnosperm *Cycadeoidea* and implications for the origin of the higher taxa. *Paleobiology, 6:* 161–167.
STIDD, B.M., 1987. Telomes, theory change and the evolution of vascular plants. *Review of Palaeobotany and Palynology, 50:* 115–126.
SWOFFORD, D.L., 1993. *PAUP: Phylogenetic Analysis Using Parsimony. Version 3.1.1.* Champaign, U.S.A.: Illinois Natural History Survey.
TAKHTAJAN, A.L., 1969. *Flowering Plants: Origin and Dispersal.* Edinburgh, U.K.: Oliver & Boyd.
TAKHTAJAN, A.L., 1972. Patterns of ontogenetic alterations in the evolution of higher plants. *Phytomorphology, 22:* 164–171.
TAKHTAJAN, A.L., 1980. Outline classification of the flowering plants (Magnoliophyta). *Botanical Review, 46:* 225–359.
TEMPLETON, A.R., 1982. Genetic architecture of speciation. In C. Barigozzi (ed.), *Mechanisms of Speciation,* pp. 105–121. New York, U.S.A.: Liss.
THOMAS, B.A. & SPICER, R.A., 1987. *The Evolution and Palaeobiology of Land Plants.* London, U.K.: Croom Helm.
THOMPSON, J.D. & LUMARET, R., 1992. The evolutionary dynamics of polyploid plants: origins, establishment and persistence. *Trends in Ecology and Evolution, 7:* 302–307.
THOMSON, J.D., HERRE, E.A., HAMRICK, J.L. & STONE, J.L., 1991. Genetic mosaics in strangler figs: implications for tropical conservation. *Science, 254:* 1214–1216.
TOMLINSON, P.B., 1982. Chance and design in the construction of plants. In R. Sattler (ed.), *Axioms and Principles of Plant Construction,* pp. 162–183. The Hague: Nijhoff-Junk.
TOMLINSON, P.B., 1987. Architecture of tropical plants. *Annual Review of Ecology and Systematics, 18:* 1–21.
TRAVERSE, A., 1988. Plant evolution dances to a different beat. *Historical Biology, 1:* 277–301.
TUCKER, S.C., 1988. Loss versus suppression of floral organs. In P. Leins, S.C. Tucker & P.K. Endress (eds), *Aspects of Floral Development,* pp. 69–82. Berlin, Germany: Cramer.

VALENTINE, J.W., 1980. Determinants of diversity in higher taxonomic categories. *Paleobiology, 6:* 444–450.

VAN STEENIS, C.G.G.J., 1969. Plant speciation in Malesia with special reference to the theory of non-adaptive, saltatory evolution. *Biological Journal of the Linnean Society, 1:* 97–133.

VAN STEENIS, C.G.G.J., 1976. Autonomous evolution in plants: differences in plant and animal evolution. *Gardens' Bulletin, Singapore, 29:* 103–126.

VERMEIJ, G.J., 1987. *Evolution and Escalation*. Princeton, N.J., U.S.A.: Princeton University Press.

VON BAER, K.E., 1828. *Entwixklungsgeschichte der Thiere: Beobachtung unde Reflexion*. Königsberg: Bornträger.

WADDINGTON, C.H., 1942. Canalisation of development and the inheritance of acquired characters. *Nature, 150:* 563–565.

WADDINGTON, C.H., 1957. *The Strategy of the Genes*. London, U.K.: Allen & Unwin.

WAINWRIGHT, P.O., HINKE, G., SOGIN, M.L. & STICKEL, S.K., 1993. Monophyletic origin of the metazoa: an evolutionary link with fungi. *Science, 260:* 340–342.

WAKE, D.B., 1989. Phylogenetic implications of ontogenetic data. *Geobios Memoire Speciale, 12:* 369–378.

WEBSTER, M.A. & GRANT, C.J., 1990. The inheritance of calyx morph variants in *Primula vulgaris* (Huds.). *Heredity, 64:* 121–124.

WESSLER, S.R., 1988. Phenotypic diversity mediated by the maize transposable elements *Ac* and *Spm*. *Science, 242:* 399–405.

WESTON, P., 1988. Indirect and direct methods in systematics. In C.H. Humphries (ed.), *Ontogeny and Systematics*, pp. 27–56. New York, U.S.A.: Columbia University Press.

WHITE, J., 1979. The plant as a metapopulation. *Annual Review of Ecology and Systematics, 10:* 109–145.

WHITTAKER, R.H., 1969. New concepts of kingdoms of organisms. *Science, 163:* 150–160.

WIEBE, H.H., 1978. The significance of plant vacuoles. *Bioscience, 28:* 327–331.

WIGHT, D.C., 1987. Non-adaptive change in early land plant evolution. *Paleobiology, 13:* 208–214.

WILEY, E.O., 1981. *Phylogenetics: The Theory and Practice of Phylogenetic Systematics*. New York, U.S.A.: Wiley.

WILEY, E.O., SIEGEL-CAUSEY, D.J., BROOKS, D.R. & FUNK, V.A., 1991. *The Compleat Cladist: A Primer of Phylogenetic Procedures*. Lawrence, Kansas: Museum of Natural History, University of Kansas.

WILLIAMS, D.M., SCOTLAND, R.W. & BLACKMORE, S., 1990. Is there a direct ontogenetic criterion in systematics? *Biological Journal of the Linnean Society, 39:* 99–108.

WILLIAMS, G.C., 1992. *Natural Selection: Domains, Levels and Challenges*. Oxford, U.K.: Oxford University Press.

WIMSATT, W.C. & SCHANK, J.C., 1988. Two constraints on the evolution of complex adaptations, and the means for their avoidance: In M.H. Nitecki (ed.), *Evolutionary Progress*, pp. 231–273. Chicago, U.S.A.: Chicago University Press.

WOESE, C.R., KANDLER, O. & WHEELIS, M.L., 1990. Towards a natural system of organisms: proposal for the domains Archaea, Bacteria, and Eucarya. *Proceedings of the National Academy of Sciences, USA, 87:* 4576–4579.

WORSDELL, W.C., 1916. *Principles of Plant Teratology*. London: Ray Society.

WRIGHT, S., 1932. The roles of mutation, inbreeding, crossbreeding and selection in evolution. *Proceedings of the Sixth International Congress of Genetics, 1:* 356–366.

WRIGHT, S., 1968. *Evolution and the Genetics of Populations*. Chicago, U.S.A.: Chicago University Press.

WYNDHAM, J., 1951. *The Day of the Triffids*. London, U.K.: Joseph.

YANOFSKY, M.F., MA, H., BOWMAN, J.C., DREWS, G.N., FELDMAN, K.A. & MEYEROWITZ, E.M., 1990. The protein encoded by the *Arabidopsis* homeotic gene *agamous* resembles transcription factors. *Nature, 346:* 35–39.

ZIMMERMANN, W., 1959. *Die Phylogenie der Pflanzen*. Stuttgart, Germany: Fischer.

CHAPTER

5

On perception of plant morphology: some implications for phylogeny

A. HAY & D.J. MABBERLEY

CONTENTS

We carry with us the wonders we seek without us: there's all Africa and her prodigies in us; we are that bold and adventurous piece of nature, which he that studies wisely learns in a *compendium* what others labour at in a divided piece and endless volume.

Sir Thomas Browne (1605–1682)
Religio Medici I.15 (1642)

Abstract

It is commented that organisms are generally perceived, especially in taxonomy, as fragmentable objects. It is suggested that organisms are not able even to be measured with complete objectivity and that they are more appropriately viewed as complex processes of flow. Attention is drawn to the problem of applying fragment-language to dynamic wholes. It is shown that even the simplest empirical morphological statement is theory-dependent and therefore subjective. The substitution of the term objective by the term intersubjective is proposed. It is argued that taxonomic characters are symbolic, not biological, that it is anachronistic to consider evolution in their terms

Shape and Form in Plants and Fungi
ISBN 0–12–371035–9

and that phylogenetic systematics demands the incorporation of evolutionary and morphological theory. Process morphology is able to be expanded, via a nested hierarchy of parameters, from developmental anatomy to ecology, and is therefore a more appropriate vehicle than fragment morphology for the study of evolution and phylogeny.

INTRODUCTION

Factors that may influence shape and form in plants (and other organisms) can be current (e.g. geneticophysiological or environmental factors) or historical (factors of ancestry, descent and heredity). While discussion will be directed to the latter, specific cases will not be dealt with, nor will hypotheses concerning evolutionary history be developed. Instead, an area of discussion (see Sattler, 1990, 1992), largely unaired in the plant systematic context, will be opened. It concerns the initial approach to problems of ancestry and evolutionary change, which involves acts of perception. The shape and form whose possible determining factors this volume aims to address, are, of course, those which are perceived, and how form is perceived is certainly a factor determining how questions about it may be posed with the aim of understanding its nature. Discoveries about determining factors are thus to no little extent determined by the way morphology is perceived.

In the monumental *Hortus indicus malabaricus*, one of the greatest of works on tropical plants, van Reede published illustrated accounts of the plants known to him on the Malabar coast in the 17th century (Heniger, 1986); most of his plates have been successfully interpreted (Nicolson *et al.*, 1988) and amongst them he depicts three with galls (Manilal *et al.*, 1980). Of these, the one of greatest interest here is 'Ponga' of which van Reede described spiny fruits complete with seeds. These 'fruits' resemble little jak-fruits (*Artocarpus heterophyllus* Lam., Moraceae) and subsequently the plant was named *Artocarpus ponga* Dennst. It was about three centuries before it was realized that the plate represents the galled form of a dipterocarp, now known as *Hopea ponga* (Dennst.) Mabb. (Jenkins & Mabberley, 1994). The 'fruits' are the galls of an undescribed coccid gall-former (Jenkins, 1992) and the 'seeds' are its pupae. Van Reede had seen the resemblance of the galls to the *jak* and was therefore predisposed to consider their contents as seeds. This is an extreme example of a predisposition to pigeonhole morphological entities according to a framework based on experience. Conditioning of this kind is to some degree inevitable and, as Takhtajan (1991) has implied, no observation can be made with total objectivity, or as Bateson (1979) had put it more strongly: 'there can be no objective experience' (see also Inglis (1989), and later).

According to the *Oxford English Dictionary*, morphology is 'that branch of biology which is concerned with the form of animals and plants, and of the structures, homologies and metamorphoses which govern or influence that form'. It is not only a discipline in itself, but also a daily activity of biologists who must apprehend form in a comparative way in order to appreciate similarity, difference and change, and to formulate descriptions and classifications. Morphology can be studied purely empirically, but is given greater meaning within the general theory of biological evolution, a theory embodying change in morphology over time, and its relationship to its surroundings. It is largely this broader, deeper morphology that is addressed here. Nonetheless, 'morphology' is also what is apparent to the layperson, unencumbered with the advantages of electron microscope or DNA-sequencer, so that morphological perception has fundamental importance in the day-to-day ordering of biological surroundings – the old basis of classification of the living world from which modern botanical science is derived.

In a recent book review, Humphries (1992) has stated that the underlying method of classification, taxonomy, 'has undergone many changes over the last 30 years. One consequence is that cladograms depicting the relationships of organisms are now commonplace' and that 'the most fruitful debate on phylogenetic inference of the last 20 years [is] that pattern should clearly be studied before any inference about process'. Further, 'all the evolutionary baggage that plagued the early Hennigians ..., such as the role of ancestors, the concern for "true trees" and functional characters' have been set aside 'as irrelevant to systematics'. As with the pheneticists before them, the (untransformed) cladists' agnostic evolutionary stance is rooted in the perception of characters and their states, but in this case linked to an estimate of their polarity, i.e. which of any pair of states is 'advanced' (apomorphic) judged either by the pattern of character-state distribution in outgroups, or from other generally less favoured evidence. The nub of both approaches is ontological-reductionist. This chapter looks at how morphology in terms of characters is perceived and possible difficulties are discussed. Implications for evolutionary studies are then considered.

Systematists, it must be noted, vary very much in their approach, and even within a particular school, such as the cladists, there are differing opinions and varying levels of commitment. The problems to be highlighted here are to varying degree manifest across contemporary systematic work (including that of the authors), but seem to impact most heavily on aspects of cladistic methodology.

PERCEPTION

Perception is a combination of sensorimotor cognition involving the sense organs and processes of neurotransmission to the brain, together with a selective process of thought: attention is not directed at all the information received. The addition of memory generates experience, and language is added to generate the possibility of consensus about experience and hence a growing body of knowledge about 'reality'. It is the selective thought process which, combined with intelligence, generates insight into the 'nature' of, in this instance, the plant world. The thought processes associated with perception thus influence insights, so it is the way in which organisms are thought about in the act of perception that is discussed.

Several authors within science (see Bohm, 1983), and many outside, acknowledge the unbroken nature of the universe across space and time – unbroken but not even. It is the observer who makes breaks corresponding to features in a topographic continuum. For biology the primary division of the universe, after the division into 'things', is into the living and the non-living, a split which results in the abstraction of organisms from the non-living background (though less dualistic outlooks are possible, directed at the ecosystem or 'Gaia'). (Strictly speaking the split is secondary, as the primary is that between the observer's mind (ego) and all else.) Biologists of course, like other humans, abstract all manner of 'things' from the whole, but it is simply noticed here that biologists abstract organisms from the rest of the universe as a means of establishing the material on which they find the motivation to act.

The morphologist, concerned with shape and form, must also be concerned with a most basic property of an organism: its dimensions, perhaps the primary 'objective' parameter attended to by the observer. Now in order to transmit findings about size and shape, it becomes immediately necessary to use a language (whether mathematical or otherwise as has been the case historically) allowing one to indicate what topographic area of the object is being measured. (Incidentally, 'measurement' includes not only quantification of length, width and depth and the variation of each along another, but also the imposition of shape-qualifying adjectives which approximate and simplify mathematical description of curvature etc. – leaflets 2 cm long × 5 mm wide, ovate-lanceolate.) Arising presumably out of repeated occurrence in the dominant groups of the plant kingdom of similar structure, life cycles, articulations and abscission layers, together with the habit of dismembering plants, words came into being (long before science) such as (or meaning much the same as) leaf, stem, root, flower, fruit and seed. It was entirely natural for morphology, paying greater attention to detail, to develop a greater vocabulary on the same general principles of fragmentation, the words corresponding to zones, organs, tissue types and so on – all parts of which could be grasped where the whole could merely be appreciated. This process of fragmentation and measurement is a form of empirical analysis, the results of which are held to be knowable objectively.

Objective knowledge of this kind is in turn held to be the most 'sound' – somehow safe and reliable – within the domain of science by a great many, if not the vast majority, of present-day natural scientists. An enormous investment is made in trying to exclude subjective (non-empirical) knowledge (see, for example, the extensive and often heated literature on cladism). Unfortunately, much of the basis of objectivity has more to do with common sense than reality – it is more cultural than actual (see Bateson, 1979).

SUBJECTIVITY, MEASUREMENT AND FLOW

In order to know how to measure an organism, it is necessary to know where it begins and ends – surely obvious? Yet in order to get the truly objective measure, its boundaries must be definable with absolute accuracy at least in principle. This is actually impossible in principle at both of two levels. The first is the well-known physical one affected by Heisenberg's uncertainty principle, whereby even if the nature of the boundary of the object is known, it is impossible to determine the position of any of its atoms without affecting that position. It has often been said that the scale of Heisenberg uncertainty is so infinitesimal that it is irrelevant morphologically. That is possibly true to the extent that such accuracy is usually unnecessary, but it affects the type of knowledge one has (objective or subjective) because a decision (conscious or otherwise) has to be made as to the acceptable level of accuracy. Even with the most sensitive instrumentation, ultimately one says 'well, this will do' and the results are instantaneously rendered subjective. Likewise, if it is argued that the atomic and subatomic level is subbiological and therefore again irrelevant morphologically, a similarly subjective decision has to be made about when exactly, as 'magnification' is increased, the 'field of view' ceases to contain biology.

Albeit subjective results of measurement carried out to very high levels of accuracy may be deemed objective for practical purposes. However, there is in principle a further uncertainty at the larger-scale biological level involving the demarcation of the organism from its non-(or other-) organism surroundings. If an organism is defined as a living *thing*, in the present case a plant, the question must be asked whether its boundary is marked by its outer layer of living cells, or the collective plasmalemmas on the outer sides of the outer layer of living cells, or the outer layer of dead cells, or the outer surface of the cuticle, or the outermost bit of secreted mucilage still in contact with the plant, or a vapour zone of secreted volatile oil around the shoots and so on. In fact, as one moves over the surface of the organism, decision upon decision has to be made as to where the surface exactly is, most of the decisions quite arbitrary (what they are is irrelevant here). The organism as a self-defining thing turns out to be fuzzy, and yet it is quite unproblematic to measure for practical purposes trees in metres of height and centimetres of diameter and diatoms in micrometres. The measurements are subjective and reliable (i.e. intersubjective), the observer having, either innately or inculcated, the intuitive ability to default to an appropriate scale. Consequently it is not possible to perform detached monologues about how things really are, even on such simple topics as size. Rather, subject and object participate in a 'dialogue' from which the subject's experience is compiled.

The difficulty associated with regarding an organism as an objectively definable thing lies also in the fact that the material that makes up its living substance is in a more or less rapid and continuous state of flow into and out of it. An organism set against its inanimate background finds a better analogy with a vortex in a stream (an analogy used by Bohm (1983) for all 'things', but particularly pertinent in the context of biology). Firstly, a vortex is clearly recognizable and it comprises matter flowing into and out of it, behaving differently when in. Secondly, it can be measured at an appropriate scale and its shape described, but its position/boundary cannot be defined exactly. The organism then represents a vortical region at a higher level of consciousness than its inanimate surroundings, having highly intricate and characteristic yet fluid and dynamic material structure for which time as well as the usual three dimensions are parametric (see also Hickman, 1991). Consciousness should not be fully equated with self-consciousness nor with acts of thought which represent types of consciousness. Rather consciousness can be understood as a broad spectrum from lowest levels associated with inanimate matter through increasing levels associated with increasing grades of organism to the highest in the biological world – human, and beyond (if one's belief system and experience permit); see Bohm (1983), Wilber (1977). The structure is in a continuous flowing process of unfolding which encompasses the entire unbroken movement from fertilization to death, a movement which initiates similar movements without break (reproduction). Thus viewed, the organism has no static qualities or properties. It is a complex process of flow, not a thing.

An aside

At this point it is necessary briefly to discuss use of 'objective', particularly as it has highly emotive and undesirably value-laden connotations in scientism (objective = good and scientific; subjective = bad and unscientific) and is used in two differing senses (neither of which is usually defined). It is used in a 'hard' sense pointing to 'absolutely real' – a meaning which implies the ability to perceive or disclose something completely in every aspect of its being. We reject the use of the word objective in this sense of naive realism as being absurd (see Sattler, 1986). The second, 'soft', sense in which the word is used is more acceptable but, we argue, misleading and unnecessary. In that sense it is used to convey shared or common or public subjective (i.e. intersubjective) experience or knowledge and is equated with 'scientific' as opposed to private or unique subjective experience equated with 'non-scientific' or 'artistic', 'mystical', etc. We question the utility of the word objective in this sense, for the following reasons: (a) there is common misidentification of intersubjectivity with objectivity in the hard sense; (b) 'objective' often carries the undesirably value-laden connotation of 'real' whether used in the hard or soft sense; (c) since intersubjectivity is a kind of subjectivity it is misleading to use for it a term which can mean something held to lie outside subjectivity; (d) the development of scientific knowledge includes (vitally important) phases of private or unique subjective experience (flashes of insight; brilliant leaps of imagination) which are excluded from science if scientific = objective; (e) the use of the word objective encourages 'either or' thinking (knowledge is either objective or subjective, either scientific or unscientific) where the notion of intersubjectivity allows for degree of intersubjectivity and hence a continuum from unisubjectivity (art) to multisubjectivity (science); and (f) use of the term objective, associated with the notion of the detached observer, is inimical to the incorporation of the participating observer essential in the development of holistic science.

SUBJECTIVITY, LANGUAGE AND FRAGMENTATION

Species are recognizable because complex processes of flow are repeated and multiplied (almost exactly) generation after generation. Reinforced by repetition within the fluid process of ontogeny, zones of the organism are recognizable (e.g. as 'leaves'), as (abstracted) subprocesses of the whole process. The ontogeny concept ought to be applied to the entire organism right through its timespan from the earliest stages of embryogenesis to the latest stages of senescence. Ontogeny is usually held to mean the transition process through early stages of development and differentiation, a meaning which implies that there is some mature living 'endstate'. This is incompatible with the view of organism as an entire unbroken movement from fertilization to death. However, while the form of organisms may be acknowledged as being epigenetic in principle, this is not the way in which morphology is usually perceived in practice, especially by taxonomists. Even when ontogenetic data are used, developmental studies generally *follow* fragmentation (stipule ontogeny, carpel ontogeny, etc.). The *organism* is still not acknowledged as process.

The pointer to perception of morphology is language. Subunits of organisms are organs: petals, carpels, roots and so on. They relate to integrated life processes, yet each is a reification or a part. The problem of inappropriate reification has been raised by Stevens (1991) but there restricted in context to the creation of artificial discontinuities in continuous variation in character-states. The recognition of these subunits derives from empirical analysis of the whole organism carried out with an eye that sees fragments, not flowing process. The idea of a plant having parts has been reinforced by our history of Cartesian dualism and Newtonian mechanism, which have made it acceptable to view organisms as assembled machines rather than epigenetically coming-into-being wholes. It is of course extremely convenient to think in terms of parts and doing so has provided many insights into the structure and function of organisms. However, the existence of parts is now written so deeply into the language of botany that they are seen everywhere.

'Plant A differs from plant B in the relative positions of its leaves.' In order to make this informative statement, the observer has to perceive that plants do have leaves. This seems painfully obvious, but it is not right; nor is it wrong. It is inadequate, however, unless the observer makes it clear that saying the plants *have* leaves is expressing a way of looking as well as the content of the observation itself. Arber (1950) made this point at the morphological level. Although she argued that it was particularly convincing an illusion that leaves were things, especially when they all fell off in winter; she stressed their continuity with the stem, and this led her to the theory that leaves were partial shoots. Rutishauser and Sattler (1985, 1987, 1989) have drawn attention to the advantages of seeing apparently opposing structural morphological interpretations as (potentially) complementary – a helpful notion able to put an end to fruitless debate as to whether, for example, the unusual ever-growing pinnate 'leaves' of *Guarea* and *Chisocheton* (Meliaceae) are really leaves or 'something else' (Fisher, 1986; Fisher & Rutishauser, 1990; Mabberley, 1979), and one leading directly on to consideration of process morphology (Rutishauser & Sattler, 1989).

At this point, however, a morphological argument is not being made as such, rather a psychological one addressing the same sort of problem at an earlier stage in its manifestation. In order to perceive that a plant *has* leaves it is necessary to make a *ceteris paribus* assumption that it would somehow be the same plant without them, but plainly it would be a different plant. The *ceteris paribus* assumption is a very useful and legitimate analytical mental device enabling the observer to think and communicate about a complex phenomenon. But the empirical fact is that the plant does not have leaves, rather it *is* leaves and everything else (except that there is not every*thing* else nor leaves, only the entire unbroken movement of organism). The perception involves the subjective abstraction and reification of a *part* which actually does not exist as such. That which these abstractions intend to represent can of course be cut off and passed to a fellow observer in an act of physical rather than mental excision – a process of empirical extraction rendering nothing but broken pieces quite meaningless in themselves in the context of organism.

Herein there lies a paradox: there is a field or discipline – morphology – dealing with subject matter (living organisms) having the nature of a complex unbroken process of flow coming into being epigenetically. But the discipline uses a language which predominantly conveys analytic reduction to static structure and fragmentation and in doing so largely addresses the animate (organism) as the inanimate (machine). Indeed because the very means of expression is caught up in this distortion, it is difficult either to see or to avoid it. What will be there after the cerebral cuts of analysis impinges heavily on one's view of what was there before (i.e. the nature of the whole). It is now seen that the simplest morphological statement is filled with theory (itself generated in the context of culture and world view (see Sattler, 1986), e.g. the prevalent materialistic scientism of the Western world). 'Plant A differs from plant B in the arrangement of its leaves' seems to be a legitimate and 'objective' sort of remark. Nevertheless it rests on (at least) two tiers of hypothesis: first the covert hypothesis that it is realistic to fragment organisms, or that organisms have parts; second (the rather less covert hypothesis) that there is some sort of sameness or meaningful comparability (homology) about leaves. The statement about plants A and B conveys an empirical observation, but is subjective because it is dependent on chosen theories. It could be made on the basis of process-morphological language – (crudely) plant A 'infoliesces' decussately, whereas plant B 'infoliesces' spirally. Here there is the (by now overt) hypothesis that plants are processes, and the hypothesis that there is some sort of sameness (homology) about the processes of 'infoliescence' in each. Again the statement conveys an empirical observation and is subjective. So the simplest empirical observations are made within the framework of chosen theories even if theory (and the choice) is (has become, or has always been) covert or unconscious. Theory determines even the initial approach (and as has been noted, world view, belief system and personality may influence theory and indeed the motivation to theorize). The elder Bateson pointed this out in the zoological context a century ago (1894), and Sattler, in the botanical, more recently (1986). While the theory may be covert, it is exposed by the language of the approach in which it lurks.

In taxonomy verbal description of morphology is formalized. Here verbs, words which, as the conveyors of movement and process, would allow better description of flowing process of unfolding, are excluded or relegated to participles (usually of a rather simplistic nature – emitting/ed, producing/ed, bearing, etc.) occupying secondary (connecting or qualifying) positions in clauses, while the words which reify static fragmentary structure are elevated as the subjects and objects being connected or qualified. Even the word inflorescence, originally a physiological and morphological process descriptor, is used to denote a thing. This, of course, is at least in part the legacy of the herbarium in which virtually all trace of living character of flowing process is lost in the stiff dead remains that form the main stuff of taxonomy (though some inferences about process can be made, provided there has been detailed prior reference to the living organism). Herbarium specimens are much closer in nature to fossils than they are to living organisms. This simple fact has important consequences for attempts at unifying classification with the general theory of evolution into phylogeny, because morphological perception in classification, which can proceed fairly effectively on the basis of structural fragmentation, gives little insight into the biological (dynamic, flowing, process) morphology necessary to the study of evolution.

MORPHOLOGICAL PERCEPTION AND OBJECT-DOMAIN IN TAXONOMY

It has been shown how perception of form is generally (though not necessarily) fragmentary. In taxonomy, the abstraction of fragments is taken further into the formal recognition of taxonomic characters. How this proceeds epistemologically needs to be understood in order to clarify what is the object-domain to which taxonomic characters belong. In turn there will be a discussion on how recognition of the actual object-domain of characters invalidates a way in which data sets of taxonomic characters are sometimes treated. Wilber (1983) has outlined the traditional psychological and philosophical assertion that there are three major realms of being, and that there are three corresponding modes of attaining knowledge.

To quote from and to paraphrase Wilber (1983): the three realms are the *gross* (the realm of matter and biological life), the *subtle* (the realm of the mental and animic) and the *causal* (the realm of the transcendent and contemplative). In the present discussion only the first two (and in particular confusion between them – category error) are considered. The corresponding means of attaining knowledge are the 'eye of flesh' – the biological eye, the empirical eye, the eye of sensory experience (symbolizing the whole physiologically sensitive human organism) and the 'eye of reason' – the eye of mind (symbolizing the whole thinking human organism). The eye of flesh participates in a world of sensory experience shared by all with a similar eye of flesh, including humans and many higher animals. With technological development, the human eye of flesh has been greatly extended and empowered – the hand lens, the electron microscope, the spectrophotometer and so on. The eye of reason participates in a world of ideas, images, logic, symbols and concepts. Wilber points out (p. 4) 'it is important to remember that the mental eye *cannot* be reduced to the fleshy eye. The mental field includes but transcends the sensory field. While not excluding it, the mind's eye rises far above the eye of flesh: in imagination it can *picture* sensory objects not immediately present, and thus transcend the flesh's imprisonment in the simply present world; in logic it can internally operate upon sensorimotor objects, and so transcend actual motor sequences....' While the mental eye may rely on the eye of flesh for much of its information, it sees much that is invisible to the eye of flesh. The truth of an idea, for example, is not open to sensory perception. (Note that since the eye of mind transcends and includes the eye of flesh, this model is not fragmenting.)

The two means of attaining knowledge being considered here operate in two object-domains. The eye of flesh operates in the domain of sensorimotor objects – 'things' (or events) which from the point of view of the observer are 'out there' and which may be called *sensibilia*. The eye of the mind, however, can operate in both the domain of sensibilia and in the domain of ideas, images, logic, symbols and concepts, or *intelligibilia* (Fig. 1), which exist not 'out there' but within the observer's mind.

The eye of flesh is presymbolic. It directly apprehends visual objects (or objects apprehended through sounds, smells, etc.) – sensibilia – without thought or application of language. It operates at all moments of sensory awareness; it operates *alone* in the milliseconds before a thought arises, when attention is turned.

The eye of reason operates on the apprehensions of the eye of flesh, primarily

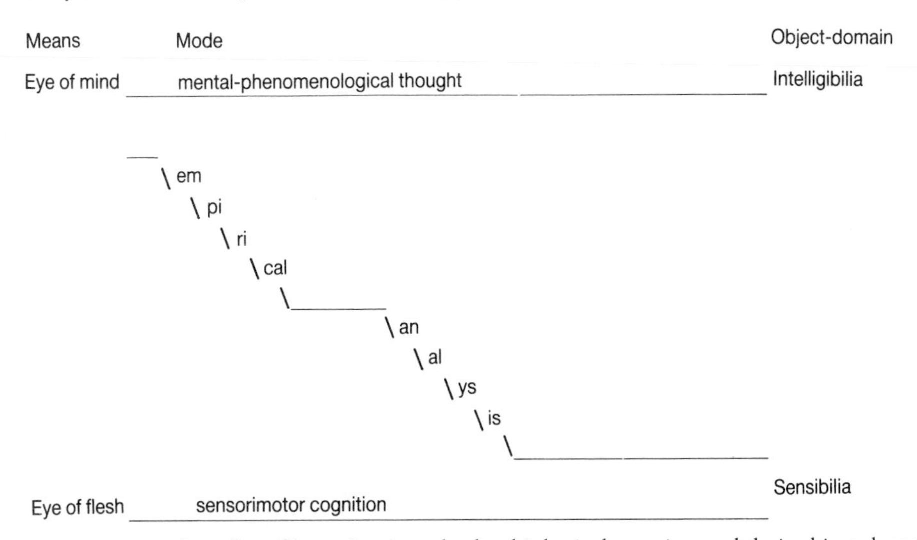

Figure 1 Means and modes of knowing in orthodox biological enquiry, and their object-domains.

labelling, measuring and, in empirical science, organizing and verifying sensibilia. This is empirical analysis. Empirical analysis is thus mental activity grounded in the realm of sensibilia or sensorimotor experience. It generates symbolic expressions – words and numbers, facts, ideas and concepts. The eye of reason also operates directly upon symbolic expressions, facts, ideas and concepts, and when it does so, the mode is mental-phenomenological thought. The objects of mental-phenomenological thought exist in the mind, and this mode is therefore mind on mind, or mind operating on the realm of intelligibilia. Intelligibilia may or may not be traceable to the realm of sensibilia. A logical or mathematical proposition, for example, may have no relation to the realm of sensibilia, yet may be entirely valid within the realm of intelligibilia. While intelligibilia generated by empirical analysis may be traceable to the realm of sensibilia, they are not reducible to that realm. An idea may arise from an empirical apprehension of sensorimotor objects, but it is not reducible to such apprehension. Intelligibilia thus transcend the realm of sensibilia, and are said to be transempirical. As objects of mental phenomenological thought, intelligibilia exist in the observer's (subject's) mind, and knowledge in that realm is therefore subjective.

Pared back to its most basic, taxonomy consists of making a series of binary choices. If a completely inexperienced taxonomist were confronted by a mixed bag of specimens or a forest of plants, he or she might embark upon a series of examinations of the specimens concluding that each is more or less similar to or different from a previously examined one and that there were kinds of plants represented. The basis of choice is similarity/dissimilarity. This, it is presumed, is different from an insect's recognition of a host plant, where some property of the right plant perhaps chemically triggers a particular behavioural response. Nevertheless, it is not an activity demanding sophisticated intellectual skills. The recognition of affinity is a different process, however, from publicizing that recognition, and still more different from the explanation of that affinity, both of which are of course more complex.

Initially the taxonomist directly apprehends, through the eye of flesh via the mode of sensorimotor cognition, the whole plant/specimen in entire aspects of form. The eye of mind is capable then of comparing these direct apprehensions of form (either simultaneously or with form held in memory) and of judging relative similarity. The process is barely intellectual, and even barely conscious – subliminal, instinctive and very powerful. Inglis (1991) has also pointed out that species (and genera) are able to be recognized without knowing the characters (but in his scheme the starting point is the recognition and comparison of parts of the organisms under study, though he later acknowledges that most taxonomy is subliminal). It is the subsequent communicative process that demands characterization and language. The eye of mind, operating in the mode of empirical analysis, abstracts details (in orthodox taxonomy almost invariably on the principle of fragmentation of static structure), compares them on the basis of similarity or difference, names them, and finally discards those facets that neither define nor distinguish the taxon in question, for those are useless in communicating the classification. (Note here in passing that learning fragmentation and labelling can of course lead both to more intensive and complex systematic analysis and to predisposition to abstract details of a particular sort – traditional morphological categories of 'taxonomic value').

Formal taxonomic characters are therefore conceptual or symbolic in nature, not biological. They belong in the mental-subjective object realm of intelligibilia, not the biological 'objective' realm of sensibilia. The character set is not equivalent to nor reducible to (though it is traceable to) the organism. It derives from a reductive process of empirical analysis and is an incomplete metaphorical overlay of the whole organism (or taxon) – each member of the set a badge of identity projected by the taxonomist on to the taxon and its representative organisms. Characters thus represent a higher level of abstraction than do parts or components (see Inglis, 1991). Characters themselves, and the pattern in which their states occur, cannot be examined or treated empirically, since they *result* from empirical analysis, symbolizing in very specific contexts, for the purposes of expression and communication, the concepts of similarity and difference, i.e. relationship.

It has been shown that organisms can be reduced to fragments of static structure, or viewed more legitimately in dynamic terms, and that traditionally the fragmentary approach is used in taxonomy. That it has been is a combination of historical limitations of language, the use of herbaria and simple facility. It is easier to reduce an object to fragments than to grasp its processes. Taxonomy of organisms proceeds along much the same lines as classification of inanimate objects, and the metaphors that it generates about form and relationship are very useful for classification, though biologically unapt. The generation of botanical characters is not the equivalent of generation of insight into or disclosure of the biological nature of plants. It is also known that at fairly low levels in the taxonomic hierarchy, classifications by and large work very well – it is felt instinctively (as well as concluded analytically) that a very large majority of taxa are naturally coherent. In general then, it may be said that it does not matter very much (pragmatically speaking) for phenetic (*sensu lato*) classification that character sets are inferior metaphors biologically.

EVOLUTION, CATEGORY ERROR AND PHYLOGENY

The study of evolution, by contrast, must, in order to succeed, concern itself wholly with generation of creative insight into the living, conscious nature of organisms and the genealogical transformation from one sort of unbroken flowing process into (an)other(s). Few would deny that, in order to study evolution, a classification is necessary serving as a guideline to the more appropriate comparisons needed to propose both evolutionary tracks (directions of transformations from one organic process to others) and evolutionary mechanisms (metaprocesses of transformation of one organic process into another). The unification of classification with the general theory of evolution into phylogeny – arranging taxa according to descent – confronts conflict between the type of morphological perception used in taxonomy and that needed in the study of evolution (see also Corner, 1981). Yet the concept of evolutionary transformation of characters from one state to others (e.g. Donoghue & Cantino, 1988; Humphries & Chappill, 1988; Mickevich & Weller, 1990; Pimentel & Riggins, 1987; Stuessy & Crisci, 1984) is central in cladistics – the most formal, allegedly most rigorous and even 'objective' (e.g. Donoghue & Cantino, 1988; Linder, 1986; Stuessy & Crisci, 1984; Ridley, 1986) method of phylogenetic classification. This concept is ostensibly absent from pattern or 'transformed' cladistics (e.g. Patterson, 1988). That mode of operation of cladistic methodology, divorced entirely, it would seem, from ideas on evolution, is variously regarded as meaningless (Johnson, 1989), unjustified (Ridley, 1986) and, by implication, superficial (Brooks & Wiley, 1985). Curiously it is the persistent claim of cladists that ideas on the nature of evolutionary processes can (and indeed must) be excluded from phylogeny – suffice that there is descent, never mind how. There is then, a claim that the relationship between phylogenetic classification and the study of evolution is one of unidirectional dependence rather than interdependence, a claim which has developed into the 'pattern before process' rubric (Humphries, 1988).

Rather than its logical justification having the historical basis that Darwin deduced evolutionary process from systematic pattern (Brady, 1985) (which simply shows that study of systematic pattern can be, and indeed was, an entry into the problem of phylogeny, no more), 'pattern before process', that is to say pattern *exhaustively* before process, is a red herring feeding on two unfortunate contemporary assumptions. The first is that characters are capable of evolution and the second that characters (in particular the pattern in which they occur across taxa) can be studied empirically. For if they have empirical reality, knowledge of their pattern of occurrence can be interpreted according to the basic scientific principle that the simplest arrangement (parsimony) is the best, as any (and an uncontrollable) number of more complicated arrangements of pattern (cladograms) can be dreamt up involving cases or hypotheses of homoplasy *ad hoc* (e.g. Farris & Kluge, 1985; Kluge, 1984).

It has been shown that characters are symbolic in nature, projected on to the organism not disclosed in it, and it has been seen that by choice they are fragmentary in practice (a character set could be generated wholly from the study of organism as process; this is just not done). Hence the concept of evolutionary transformation of characters (and therefore evolutionary polarity of their states) is metaphorical (as, incidentally, is that of characters' adaptiveness, since characters – as intelligibilia – have no physical or biological contact with the environment – the domain of sensibilia, though there is a connection of reference with a variable level of adequacy; cf. Baum & Larson (1991)). It is implied that, over time, characters can have changed from one state into another. And because many taxonomists are so deeply inured to see fragments of morphology, they become convinced that the metaphor is real, and in turn accept the notion of character transformation as realistic. This is what Whitehead (1967) called the 'fallacy of misplaced concreteness', and what Wilber (1977) and Bateson (1979), after Korzybski (1948), called 'mistaking the map for the territory'. Bateson (1973) also used the more picturesque '... eating the menu card instead of the dinner – an error of logical typing'.

One must, however, remind oneself that while characters may appear to evolve, they exist only in contemporary minds and neither have nor have had biological existence (Hay, 1986). Characters have changed with advancing classification, not with biological evolution. They do not live in little worlds of their own (Ghiselin, 1984), though an assumption that they do lies behind what Humphries & Chappill (1988) call 'one of the main tenets of cladism' – 'that all hypotheses, whether they concern characters or groups must be based on empirical evidence. Hypotheses of parallelism that are not based on criteria such as the congruence test are not based on empirical evidence and as a consequence are untestable and unscientific' – a clear case of scientism (Hay, 1992), for an exposure of the formal self-contradiction of which see Wilber (1983). Once the organism as process is grasped, it becomes obvious that characters and their states are symbolic in nature, and in turn it becomes clearer that the parsimony principle is not of overriding relevance in their treatment. Because character sets and patterns are not able to be disclosed as biological reality, and could be established on more than one basis (e.g. fragmentation vs process descriptors (Sattler & Rutishauser, 1990); words vs mathematical formulae and so on), and could be infinite in number (potentially limitless fragmentation, restrained in practice by traditional fragment categorization), identification of *ad hoc* hypotheses of homoplasy formulated in a sense that could refer adequately to real parallelisms or convergence happenings, is impossible through the medium of taxonomic characters. Therefore the methodological justification for parsimony of character pattern would appear to evaporate.

Rieppel (1991) notes that 'confusion' regarding the application of methodological parsimony, i.e. the failure to distinguish between methodological and evolutionary parsimony, derives from failure to distinguish between the cladogram as a statement of logical relations (concept, intelligibilia; Rieppel uses Popperian 'World 3') as distinct from a more direct statement about historically contingent process (content, sensibilia, 'World 1'). But since the rationale behind method employed in the generation of the cladogram itself rests squarely on a confusion (category error) of precisely the same nature (characters as content, sensibilia, World 1 as opposed to concept, intelligibilia, World 3), the insulation of (currently conceived cladistic) methodological from evolutionary parsimony would seem to fail after all. So inasmuch as a character-state pattern is arranged parsimoniously, so it is being said (metaphorically) that descent is parsimonious – an unambivalent statement (though fashioned in terms of characters) about the nature of evolutionary process (see also Ridley, 1986) and precisely what many cladists wish to avoid. Phylogenetic classification therefore demands the incorporation of evolutionary *and* morphological theory (see also Baum, 1989), and so to become an overtly transempirical activity.

Inasmuch as a character set is a metaphor for (not disclosure of) organism with a greater or lesser degree of adequacy of reference to reality, arrangement of character-state pattern leads to metaphors for (not disclosure of) phylogeny with a greater or lesser degree of adequacy of reference to 'what might really have happened'. The taxonomist's species, (and higher taxa (Nelson, 1989)), because it is defined by characters, is a symbol of relationship between individuals (just as higher taxa are symbols of relationship between lower taxa), rather than the group of organisms itself – the on-the-ground species of ecologists and biogeographers and the 'unit' for many evolutionists. Hence, phylogenetic classification 'discovers' descent with modification of relationship (couched in terms of characters as entities), rather than shedding more direct light onto descent with modification of organic process. The yardsticks for such metaphors, by which some estimate of the degree of adequacy of reference may be made, are the plausibility and explanatory power of morphological *and* evolutionary models argued to be pertinent to the case, though how plausibility is rigorously to be assessed remains a problem yet to be solved.

The development of evolutionary theory, coming as it must from systematic comparison, would be enormously assisted if classification was based on and used a language relating to the study of organisms as living process, since process morphology, as the elucidation of 'grand vivacity' (Corner, 1946), generates superior metaphors for organism. Moreover, organisms as living process do not exist in isolation: the domain for evolution is the ecological milieu, and process morphology can be expanded into ecology. Parameters for process morphology outlined by Jeune and Sattler (1992) are examples pertaining to process at or near the level of meristem. Parameters outlined earlier by Hallé and Oldeman (1970) of tropism, rhythm and podiality, are examples pertaining to process at the level of plant architecture (though their 'module' is a fragment concept), while parameters at a yet higher level, of model conformity, reiteration and senescence (Hallé *et al.*, 1978) are examples pertaining to the development and dynamics of vegetation. Parameters at each level include, but transcend, parameters at each lower level. Each level can be partly elucidated by, but not fully understood in terms of, each lower level. What is emerging is a nested hierarchy of process morphological parameters which address organism from developmental anatomy to ecology in an holistic manner. Fragment morphology, by contrast, when expanded up to the level of forest, results in fictitious 'clouds' of peripheral parts (Hallé *et al.*, 1978). Hence systematics based on fragmentation remains distanced from the evolutionary ecology of flowering forest (the 'real world' which systematic botany aims to address and for which the Durian Theory of Corner (1949), involving elements of both fragment and process morphology, is still the only comprehensive angiosperm evolutionary hypothesis).

The transformation of systematic morphology into a more dynamic mode seems unlikely to be achieved so long as the taxonomic edifice is geared mainly to the herbarium. The importance of living collections in systematic research needs to be reassessed and shifted from its present peripheral position to the central one alongside field studies. With that, a concomitant shift is needed in the way organism is perceived (which in turn demands further exposure of all the tacit assumptions that would otherwise continue to influence our thoughts from the shadows). Initial steps in that direction have been taken by several authors recently. For example, the importance of homoeotic metamorphosis has been expounded in a consideration of the reproductive phase of Fagaceae (Jenkins, 1992, 1993) and Zingiberales (Kirchoff, 1991), as it has in an interpretation of shoot construction in *Azara* (Flacourtiaceae) by Charlton (1991), and a preliminary attempt at a theory of origin and evolution has been made for aroids based mainly on consideration of behaviour and process in living plants (Hay & Mabberley, 1991) – all to some degree in the context of organism rather than atomism (see also Sattler, 1988). The development of computer programs for modelling the dynamics of growth at levels from the anatomical to the ecological, may bring labour intensive process morphology into systematic practicability.

CONCLUSION

Herein criticism has been directed at morphological perception in and aspects of the methodology of cladistics – an explicit and strongly crystallized reductionist method of phylogenetic reconstruction, and one which has arisen from so-called classical taxonomy (a somewhat implicit 'methodology' containing elements of both reductionism and holism) (Hay, 1992). While pointing out what are thought to be serious flaws in certain cladistic methodology, we acknowledge that many insights have been generated by the reductionist approach. From here it is intended to embark on contribution to the development of a complementary explicit holistic methodology for proposing reconstructions of phylogeny. In such, pattern – perceived within systematic comparison of process morphology – and evolutionary process may be considered together in an evolving transempirical theoretical framework in which phylogenetic classification and evolutionary theory are illuminated reciprocally (Ghiselin, 1984), rather than in the vicious circle that some cladists perceive (e.g. Farris & Kluge, 1985).

ACKNOWLEDGEMENTS

The following generously gave their time to reading and providing constructive criticism of earlier drafts of this paper: Dr B.G. Briggs, Professor E.J.H. Corner (to whom we are also indebted for the epigraph), Dr C.J. Humphries, Dr L.A.S. Johnson, Professor C.B.A. McCusker, Professor R. Sattler and Dr R.W. Scotland. Responsibility for the views expressed here is, however, that of the authors.

REFERENCES

ARBER, A., 1950. *The Natural Philosophy of Plant Form.* Cambridge, U.K.: Cambridge University Press.

BATESON, G., 1973. *Steps to an Ecology of Mind.* St Albans, U.K.: Granada.

BATESON, G., 1979. *Mind and Nature – A Necessary Unity.* Sydney, Australia: Bookwise.

BATESON, W., 1894. *Materials for the Study of Variation.* London, U.K.: Macmillan.

BAUM, B.R., 1989. Theory and practice of botanical classification: cladistic, phenetic and classical approaches – critical summary: botanical systematics in 1987. *Plant Systematics and Evolution, 166:* 197–210.

BAUM, D.A. & LARSON, A., 1991. Adaptation reviewed: a phylogenetic methodology for studying character macroevolution. *Systematic Zoology 40:* 1–18.

BOHM, D., 1983. *Wholeness and the Implicate Order.* London, U.K.: Ark.

BRADY, R.H., 1985. On the independence of systematics. *Cladistics, 1:* 113–126.

BROOKS, D.R. & WILEY, E.O., 1985. Theories and methods in different approaches to phylogenetic systematics. *Cladistics, 1:* 1–11.

CHARLTON, W.A., 1991. Homoeosis and shoot construction in *Azara microphylla* Hook. (Flacourtiaceae). *Acta Botanica Neerlandica, 40:* 329–337.

CORNER, E.J.H., 1946. Suggestions for botanical progress. *New Phytologist, 45:* 185–192.

CORNER, E.J.H., 1949. The durian theory or the origin of the modern tree. *Annals of Botany (New Series) 13:* 367–414.

CORNER, E.J.H., 1981. Angiosperm classification and phylogeny: a criticism. *Botanical Journal of the Linnean Society, 82:* 81–87.

DONOGHUE, M.J. & CANTINO, P.D., 1988. Paraphyly, ancestors and the goals of taxonomy: a botanical defense of cladism. *Botanical Review, 54:* 107–128.

FARRIS, J.S. & KLUGE, A.G., 1985. Parsimony, synapomorphy and explanatory power. A reply to Duncan. *Taxon, 34:* 130–135.

FISHER, J.B., 1986. Sun and shade effects on the leaf of *Guarea* (Meliaceae): plasticity of a branch analogue. *Botanical Gazette, 147:* 84–89.

FISHER, J.B. & RUTISHAUSER, R., 1990. Leaves and epiphyllous shoots in *Chisocheton* (Meliaceae) a continuum of woody leaf and stem axes. *Canadian Journal of Botany, 68:* 2316–2328.

GHISELIN, M.T., 1984. Narrow approaches to phylogeny: a review of nine books of cladism. In R. Dawkins & M. Ridley (eds), *Oxford Surveys in Evolutionary Biology,* Vol. 1, pp. 209–222. Oxford, U.K.: Oxford University Press.

HALLÉ, F. & OLDEMAN, R.A.A., 1970. *Essai sur l'architecture et la dynamique de croissance des arbres tropicaux.* Paris, France: Masson.

HALLÉ, F., OLDEMAN, R.A.A. & TOMLINSON, P.B., 1978. *Tropical Trees and Forests. An Architectural Analysis.* Berlin, Germany: Springer.

HAY, A., 1986. *Cyrtosperma* Griff. and the origin of the aroids. Unpublished D.Phil. thesis, University of Oxford.

HAY, A., 1992. Systematic scientism. *Australian Systematic Botany Society Newsletter, 71:* 7–8.

HAY, A. & MABBERLEY, D.J., 1991. 'Transference and Function' and the origin of aroids: their significance in early angiosperm evolution. *Botanische Jahrbücher für Systematik, 113:* 339–428.

HENIGER, J., 1986. *Hendrik Adriaan van Reede tot Drakenstein (1636–1691) and* Hortus Malabaricus. Rotterdam: Balkema.

HICKMAN, C.S., 1991. Functional analysis and the power of the fourth dimension in comparative evolutionary studies. In E.C. Dudley (ed.), *The Unity of Evolutionary Biology. Proc. 4th Int. Cong. Syst. Evol. Biol.,* pp. 548–554. Portland, U.S.A.: Dioscorides.

HUMPHRIES, C.J., 1988. Pattern before process. *Nature, 333:* 300–301.

HUMPHRIES, C.J., 1992. Review of P.H. Harvey & M.D. Pagel, The comparative method in evolutionary biology. *Biological Journal of the Linnean Society, 46:* 312.

HUMPHRIES, C.J. & CHAPPILL, J.A., 1988. Systematics as science: a response to Cronquist. *Botanical Review, 54:* 129–144.

INGLIS, W.G., 1989. Concepts in science, the universe, and everything. *Search, 30 (3):* 71.

INGLIS, W.G., 1991. Characters: the central mystery in taxonomy and systematics. *Biological Journal of the Linnean Society, 44:* 121–139.

JENKINS, R.M., 1992. Taxonomic character change and the significance of galls. Unpublished D.Phil. thesis, University of Oxford.

JENKINS, R.M., 1993. The origin of the Fagaceous cupule. *Botanical Review 59:* 81–111.

JENKINS, R.M. & MABBERLEY, D.J., 1994. A breadfruit among the dipterocarps: galls and atavism. In M.A.J. Williams (ed.), *Plant Galls: Organisms, Interactions, Populations.* Oxford, U.K.: Oxford University Press (in press).

JEUNE, B. & SATTLER, R., 1992. Multivariate analysis in process morphology of plants. *Journal of Theoretical Biology, 156:* 147–167.

JOHNSON, L.A.S., 1989. Models and reality: doctrine and practicality in classification. *Plant Systematics and Evolution, 168:* 95–108.

KIRCHOFF, B.K., 1991. Homeosis in the flowers of the Zingiberales. *American Journal of Botany, 78:* 833–837.

KLUGE, A.G., 1984. The relevance of parsimony to phylogenetic inference. In T. Duncan & T.F. Stuessy (eds), *Cladistics: Perspectives on the Reconstruction of Evolutionary History,* pp. 24–38. New York, U.S.A.: Columbia University Press.

KORZYBSKI, A., 1948. *Science and Sanity.* Lakeville: International Non-Aristotelian Library Publishing Co.

LINDER, H.P., 1986. A review of cladistics for botanists. *South African Journal of Botany, 54:* 208–220.

MABBERLEY, D.J., 1979. The species of *Chisocheton* (Meliaceae). *Bulletin of the British Museum (Natural History), Botany, 6:* 301–386.

MANILAL, K.S., SURESH, C.R. & SIVARAJAN, V.V., 1980. Galled plants under distinct names in *Hortus Malabaricus.* In K.S. Manilal (ed.), *Botany and History of* Hortus Malabaricus, pp. 177–180. New Delhi, India: Oxford & IBH Publishing.

MICKEVICH, M.F. & WELLER, S.J., 1990. Evolutionary character analysis: tracing character change on a cladogram. *Cladistics, 6:* 137–170.

NELSON, G., 1989. Cladistics and evolutionary models. *Cladistics, 5:* 275–289.

NICOLSON, D.H., SURESH, C.R. & MANILAL, K.S., 1988. *An Interpretation of van Rheede's* Hortus Malabaricus. Königstein: Koeltz.

PATTERSON, C., 1988. Homology in classical and molecular biology. *Molecular Biology and Evolution, 5:* 603–625.

PIMENTEL, R.A. & RIGGINS, R., 1987. The nature of cladistic data. *Cladistics, 3:* 201–209.

RIDLEY, M., 1986. *Evolution and Classification: The Reformation of Cladism.* Harlow, U.K.: Longman.

RIEPPEL, O., 1991. Things, taxa and relationships. *Cladistics, 7:* 93–100.

RUTISHAUSER, R. & SATTLER, R., 1985. Complementarity and heuristic value of contrasting models in structural botany. I. General considerations. *Botanische Jahrbücher für Systematik, 107:* 415–455.

RUTISHAUSER, R. & SATTLER, R., 1987. Complementarity and heuristic value of contrasting models in structural botany. II. Case study of leaf whorls: *Equisetum* and *Ceratophyllum. Botanische Jahrbücher für Systematik, 109:* 227–255.

RUTISHAUSER, R. & SATTLER, R., 1989. Complementarity and heuristic value of contrasting models in structural botany. III. Case study on shoot-like 'leaves' and leaf-like 'shoots' in *Utricularia macrorhiza* and *U. purpurea* (Lentibulariaceae). *Botanische Jahrbücher für Systematik, 111:* 121–137.

SATTLER, R., 1986. *Biophilosophy — Analytic and Holistic Perspectives.* Berlin, Germany: Springer.

SATTLER, R., 1988. Homeosis in plants. *American Journal of Botany, 75:* 1606–1617.

SATTLER, R., 1990. Towards a more dynamic plant morphology. *Acta Biotheoretica, 38:* 303–315.

SATTLER, R., 1992. Process morphology: structural dynamics in development and evolution. *Canadian Journal of Botany, 70:* 708–716.

SATTLER, R. & RUTISHAUSER, R., 1990. Structural and dynamic descriptions of the development of *Utricularia foliosa* and *U. australis. Canadian Journal of Botany, 68:* 1989–2003.

STEVENS, P.F., 1991. Character states, morphological variation and phylogenetic analysis: a review. *Systematic Botany, 16:* 553–583.

STUESSY, T.F. & CRISCI, J.V., 1984. Problems in the determination of evolutionary directionality of character state change for phylogenetic reconstruction. In T. Duncan & T.F. Stuessy (eds), *Cladistics: Perspectives in the Reconstruction of Evolutionary History,* pp. 71–87. New York, U.S.A.: Columbia University Press.

TAKHTAJAN, A., 1991. *Evolutionary Trends in Flowering Plants.* New York, U.S.A.: Columbia University Press.

WHITEHEAD, A.N., 1967. *Science and the Modern World.* New York, U.S.A.: Macmillan.

WILBER, K., 1977. *The Spectrum of Consciousness.* Wheaton: Quest.

WILBER, K., 1983. *Eye to Eye.* New York, U.S.A.: Anchor.

CHAPTER

6

A summary of the branching process in plants

ADRIAN D. BELL

CONTENTS

Abstract

A synopsis of current perceptions of plant construction in terms of branching is presented in the form of a single flow chart. A plant consists of a set of shoot units each derived from one apical meristem (bud). The flow chart demonstrates the roles played by shoot units in the dynamic morphology of the developing plant. Shoot units either singly form monopodial axes or in concert form sympodial axes. A given species presents a limited number of axes types termed collectively the architectural unit for that species.

Shape and Form in Plants and Fungi
ISBN 0–12–371035–9

INTRODUCTION

How is the shape or form of a plant to be comprehended? Each subdivision of plant science, whether it be physiology or phylogeny, has more than one conception of the basis and relevance of plant form. Interwoven perceptions may be based on the overall shape of the individual plant or, more frequently of late, may seek to identify constructional subunits from which a plant's form is built. The basic unit may be taken to be the cell, a concept currently under attack (Hagemman, 1992; Kaplan, 1992; Kaplan & Hagemman, 1991) or may implicate organized populations of cells (e.g. Barlow, 1989). Furthermore, it is now customary to recognize that plant form is not static but dynamic (Bell, 1991; Sattler, 1990). The dynamism is genetically controlled (e.g. Coen & Meyerowitz, 1991), operating in terms of ontogeny, and flexible in terms of environmental influence (Fisher, 1992). Plant form is a process (Sattler, 1992). It is a progression that changes with time both in relation to the life of the individual and in relation to evolution. It would seem that even the appreciation of plant morphology is in a state of flux. Nevertheless, there is one tangible constructional unit of the plant that has repeatedly proved to be of great value in elucidating the process of form, the apical meristem ('bud') and the structure developed from it ('shoot unit') (Bell, 1991; Wilson, 1990).

The concept is simple, self-evident and quantifiable. The above-ground framework of a seed plant commences growth by the expansion of one shoot unit developed by one apical meristem. This shoot unit bears additional apical meristems, typically in the axils of leaves, and these develop into additional shoot units. The sequence continues and the plant's form in terms of a framework of shoot units is the result. A newly developed shoot may constitute a single monopodial axis or may represent one of the series of units comprising a sympodial axis. Alternatively, it may form a structure that does not contribute to the main framework and in this discussion will then be relegated to the role of 'ancillary shoot'. The shape and form of the plant clearly depend upon the number, size and position of framework axes as is frequently described but they also depend in many instances on the active reorientation of axes as they develop.

The majority of seed plants (and may be others, Svensson & Callaghan, 1988) are built up in this way although it is just one progression that can be identified in the global changing scheme of process morphology envisaged by Sattler, 1992, in which structure *is* process.

The meristem–meristem cycle (Fagerstrom, 1992) at the heart of the form of most plants has been studied in great depth in recent years and this allows an attempt to be made here to encapsulate the total array of plant form in one single flow chart (Fig. 1).

In essence this chart indicates the options open to each new apical meristem generated in the existing framework of a plant. Such flow charts can appear complex at first sight and for this reason Fig. 1 is broken down into its constituent parts in order for each to be detailed and illustrated. This general flow chart demonstrates alternative potentials for bud/shoot unit development. The reader is directed to specific examples of developmental processes such as: Maillette (1982) and Groenendael (1985), where the precision is at the node/internode level within a shoot unit; Barlow (1989), where hypothetical schemes of development are presented in the form of Petri nets; and Ford (1992), where photosynthesis drives the cycle.

FATE OF AN APICAL MERISTEM

Figure 1 indicates the possible fate of an apical meristem developing within an existing framework. The apical meristem (bud) at the left of the diagram may eventually become an old shoot unit on the right and thus be incorporated into the main structure of the plant. In the most simple instance the apical meristem present in the seed will develop into a single shoot unit and will constitute the entire form of the plant, as in many palm trees. Such a plant can then be represented by just one sequence in terms of branch construction as indicated in Fig. 2.

The meristem cycle at the left is generating new buds located on the single expanded shoot unit, but these all remain dormant. Only one apical meristem has developed into a shoot unit – the monopodial trunk of the palm tree. Eventually, additional meristems will grow out when the plant produces flowering branches but these do not contribute to the main framework. In architectural terms this form corresponds to the model of Holttum or Corner (Halle & Oldeman, 1975) depending upon whether or not the onset of flowering triggers the death of the plant. Some palm trees produce suckers from the base (e.g. *Cocos nucifera* L., the date palm). Reinstatement of the arrow from dormancy to growth in Fig. 2 would reflect this difference; each new shoot unit thus formed will repeat the flow chart sequence of the original seedlings axis and represent a reiteration (Halle *et al.*, 1978 and Fig. 8).

For the majority of plants the branch framework consists of more than one shoot unit and each shoot unit will have a specific developmental route through the flow chart. The parameters used here are based on those now established in the studies of tree architecture (Halle *et al.*, 1978; Barthelemy *et al.*, 1989 and outlined in Barthelemy, 1991, and Bell, 1991). All the principal components of tree architecture, which apply equally to virtually all flowering plants, are contained within Fig. 1 and are discussed here in sequence.

THE MERISTEM–MERISTEM CYCLE

The simple meristem–meristem cycle (Fig. 3) marks the major difference between plants and animals (except the colonal animals such as corals and bryozoans). Each shoot unit as it develops creates new shoot units and the population of units, maybe of differing potential, create the form and shape of the organism. A typical example is illustrated in Fig. 14 which depicts *Cercidiphyllum* Siebold & Zucc., as described in detail by Titman and Wetmore, 1955. Recognition of this dynamism forms a powerful tool (Bilbrough & Richards, 1991; Briand *et al.*, 1991; Ginocchio & Montenegro, 1992; Harper *et al.*, 1986; Haukioja, 1991; Herms & Mattson, 1992; Porter 1983a, b, 1989; de Reffye *et al.*, 1991; Tuomi & Vuorisola, 1989) and allows a quantitative appraisal of plant form in terms of the birth and death of shoot units within the population that constitutes the plant's framework (e.g. Jones & Harper, 1987a, b; Maillette, 1992a; McGraw & Antonovics, 1983; McGraw & Garbutt, 1990; Wilson, 1989).

Thus, the turnover of meristems is established and some proceed to develop further into shoot units. Many will not, entering instead the subloop, Fig. 4.

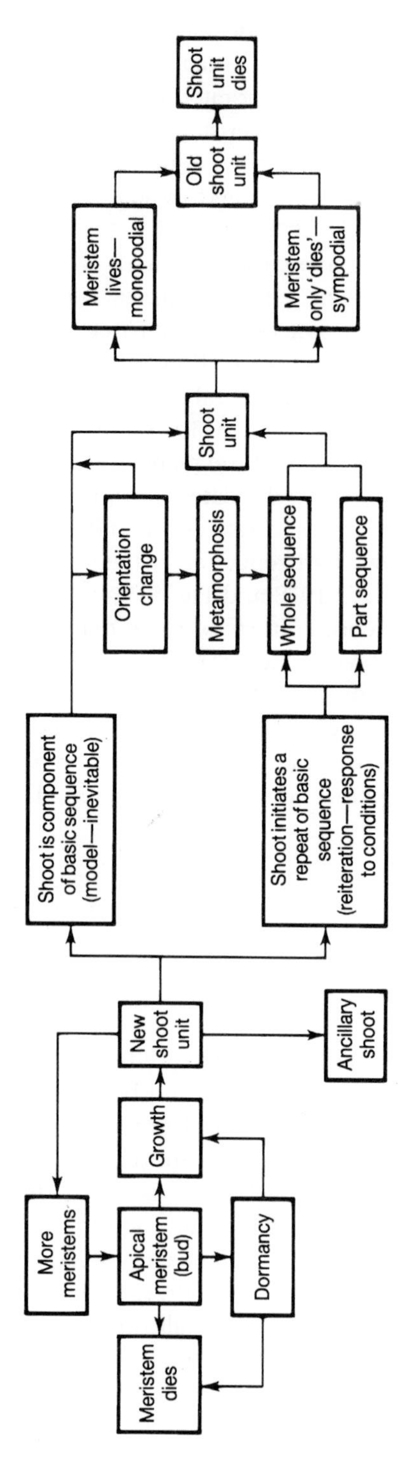

Figure 1 The possible roles of an apical meristem (bud) and its subsequent shoot unit.

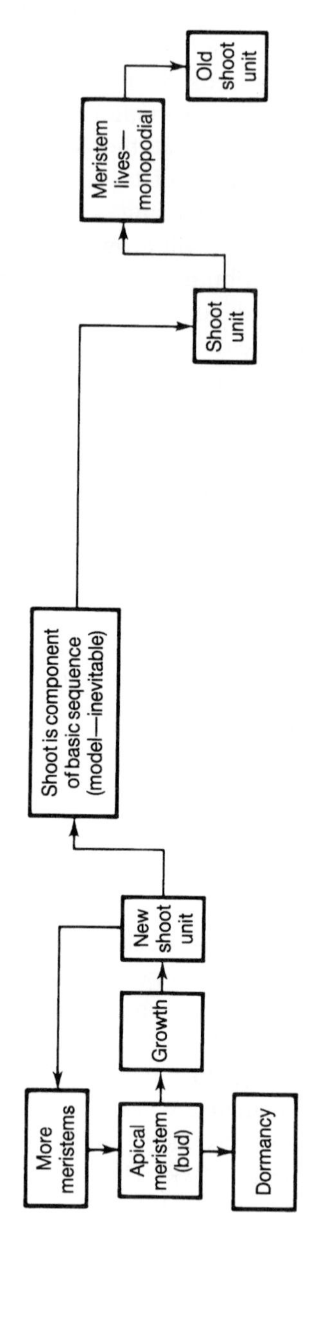

Figure 2 A subset of Fig. 1 that represents the monopodial trunk of a palm tree or a cabbage plant.

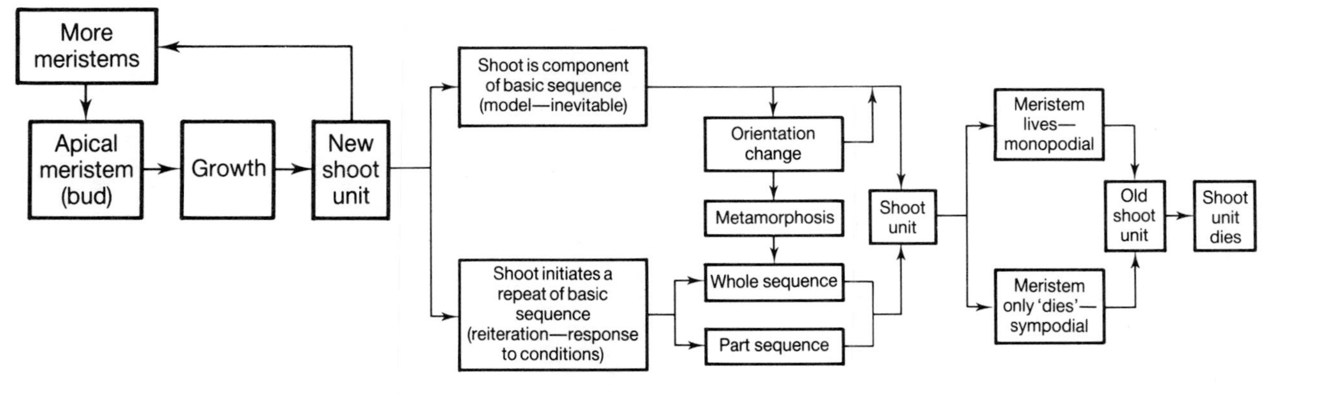

Figure 3 The meristem–meristem cycle.

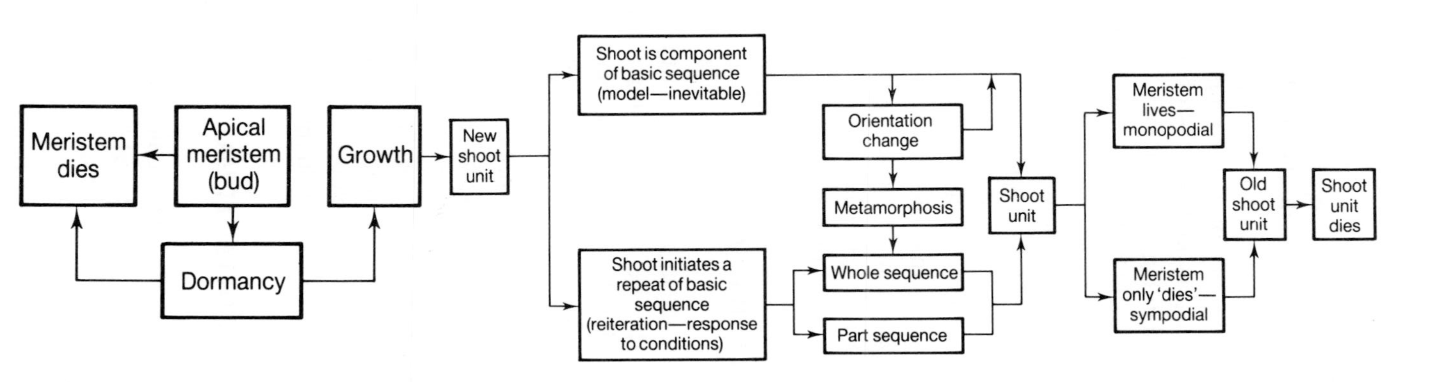

Figure 4 Death of meristems and time delay.

DORMANCY AND MERISTEM DEATH

Hidden in the framework of a plant, particularly that of a long-lived perennial plant such as a tree, there is often an enormous population of dormant buds. In the early years of tree architectural analysis, Corner (1951) was at pains to point out that *most* of the buds in a tree die or remain dormant just as *most* of the shoot units that do develop subsequently die and literally drop out of the system (Fig. 13). This dormant bud population originates morphologically speaking in one of two ways. It represents the majority of apical meristems formed in the conventional positions on existing shoot units, as in the axils of leaves, and in the example here of the palm tree embodied in Fig. 2. Frequently there will be a series of dormant buds associated with each leaf axil (Bell *et al.*, 1992; Burrows, 1990; Tourn *et al.*, 1992). Secondly, many so-called adventitious buds (i.e. buds not associated with leaf axils) may be present, located for example within the bark of an old shoot unit (Fink, 1983, 1984). Whatever the origin, such a dormant bud may eventually develop as a shoot unit (Fig. 3), and then either contribute to the plant framework or represent an ancillary feature (such as a spine or epicaulic flower).

A meristem may die. The consequence of this depends very much on the potential of that particular meristem, could it have contributed to the main framework and can it be replaced (see Fig. 8)? Death may be due to internal factors (Herms & Mattson, 1992) or environmental factors (Haukioja, 1991).

ANCILLARY SHOOTS

The task set in this account is to present a summary of the branching process in plants. This being so, any shoot unit that differentiates into a structure having no constructional function in the plant can be assigned to the category of ancillary shoot (Fig, 1). Flowers, stem tendrils, stem spines, lateral inflorescences (Fig. 15) and even short-shoots (Fig. 14) are thus relegated in this way. However, careful observation is necessary. The short-shoot of *Cercidiphyllum* Siebold & Zucc. (Fig. 14) can occasionally bear a meristem that develops into a long shoot (Titman & Wetmore, 1955), thereby conferring a structural role on the shoot that supports it. Similarly, the lateral inflorescence of *Gongora* Ruiz & Pavon (an orchid) bears a bud that develops into the next vegetative constructional shoot (Barthelemy, 1987). Constructional shoot units themselves can be categorized as two types: the initial units of the young plant (Fig. 5) or subsequent repetitions of the initial sequences (Fig. 8).

THE ARCHITECTURAL MODEL

The concept of the architectural model, first established by Hallé and Oldeman (version in English, 1975) has become both widely applied and refined in detail. In essence, plant species in general and trees in particular are deemed to conform in their branching pattern to one of a limited number of distinctive formats termed architectural models. Outlines are to be found in Hallé *et al.*, 1978, and Bell, 1991. For many plants the model is unequivocal (Fig. 16) and each shoot unit is symptomatically positioned within the basic model sequence (Fig. 5). The fact that every shoot unit of every plant of this type always plays it's correct role in the construction leads to the reasonable supposition that the meristem activity leading to this basic architectural model is genetically controlled. However, this is not proven, as cautioned by Cook (1992). Nevertheless, genetic information of this nature is slowly being unravelled; the single gene switch, for example, that triggers the formation of a floral meristem or a vegetative shoot unit in *Antirrhinum* L. (Coen & Meyerowitz, 1991 and Coen, Ch. 13 this volume). Certainly the precise position of leaves is under genetic control (Hardwick, 1986) and the position of leaves fixes the position of potential shoot units.

Advances in the interpretation of the model concept have led to the introduction of the architectural unit (Barthelemy *et al.*, 1991; Edelin, 1991). The architectural unit is represented by the suite of branch types, and therefore shoot unit types that go to make up the form of a particular plant species. *Araucaria heterophylla* (Salisb.) Franco (Fig. 17), for example, conforms to the general architectural model of Massart (Hallé *et al.*, 1978), a monopodial trunk bearing whorls of monopodial branches. Specifically, however, the architectural unit for the tree is comprised of just three types of branch, each represented by a single type of shoot unit. These would be designated as: axis type 1 (the monopodial trunk); axis type 2 (a branch in a whorl); and axis type 3 (temporary shoot units that die, are shed and replaced (Fig. 17). The fate of these axis types and the potential of their apical meristems are maintained even if detached from the plant framework (Schaffalitzky de Muckadell, 1959). In other species the architectural unit may include an axis type constructed from more than one shoot unit, i.e. each shoot unit forms a constituent part of a sympodial sequence (Fig. 12). Thus a shoot unit may form part of the basic model sequence for the plant. Once in place, however, the contribution of the shoot unit may be modified by a reorientation of the unit in space.

ORIENTATION CHANGE WITHIN THE MODEL

The three-dimensional format of a plant depends upon the juxtaposition of axes. This frame is dynamic in the sense that individual components are enlarging with time and components are constantly being added or lost. In addition, some axes change their orientation with time, thus affecting overall shape and form (Fisher & Stevenson, 1981). There are two principal mechanisms behind such reorientation (Fig. 6). The response to gravity of the shoot apex may change as it develops (Fig. 7a–d) such that a shoot unit initially developing horizontally changes growth direction and proceeds to develop vertically. This particular growth format is typical of a great many rhizomatous plants such as the *Heliconia* L. shown in Fig. 18 and represented by Fig. 7b. In terms of the architectural models of Hallé and Oldeman (1975), this growth form typifies the model of Tomlinson. The converse reorientation, vertical growth direction changing to the horizontal direction, typifies the model of Troll (Fig. 7c) or occurs once for the initial shoot unit of many herbaceous plants such as *Trifolium repens* L. (as Fig. 7a). The form of reorientation indicated by Fig. 7d is a special case which will be discussed under metamorphosis (Fig. 10).

The second mechanism of orientation change involves bending of an existing axis (Fig. 7e–i). This may be interpreted as a bending due to the weight of the axis such as in the continual forward movement of a *Lycopodium* L. stem (Fig. 7h) (Niklas, 1992). For other plant axes the reorientation is due to reaction wood activity forcing the bending (Loup *et al.*, 1991; Mueller, 1988; Wilson & Archer, 1979) (Fig. 7e–g, i).

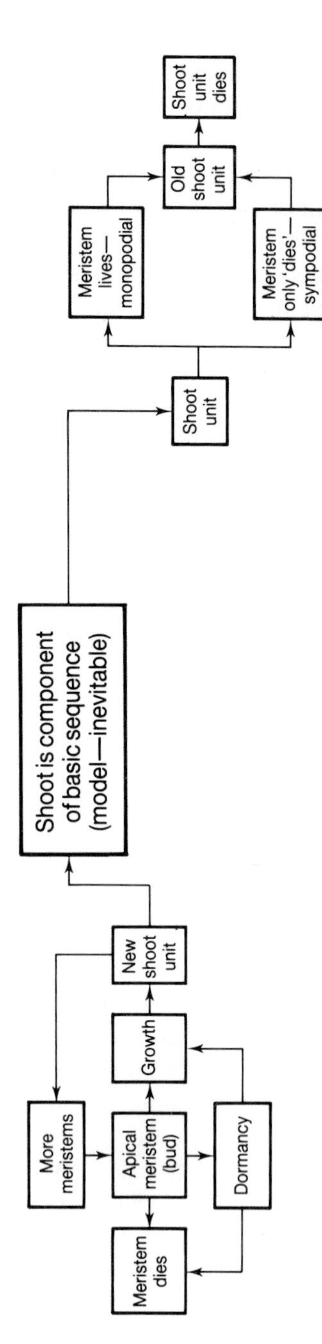

Figure 5 The shoot unit is an elemental component of a plant's form.

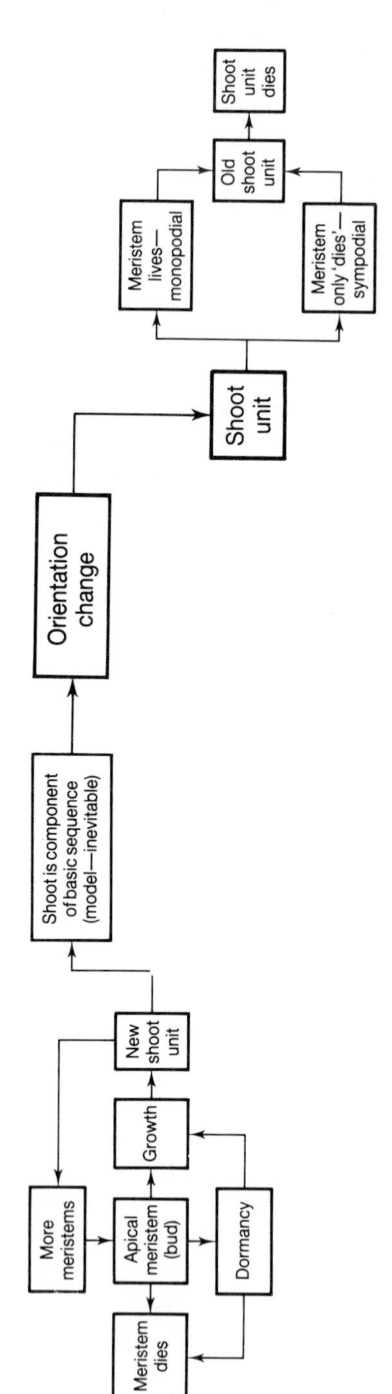

Figure 6 Orientation change within the basic model.

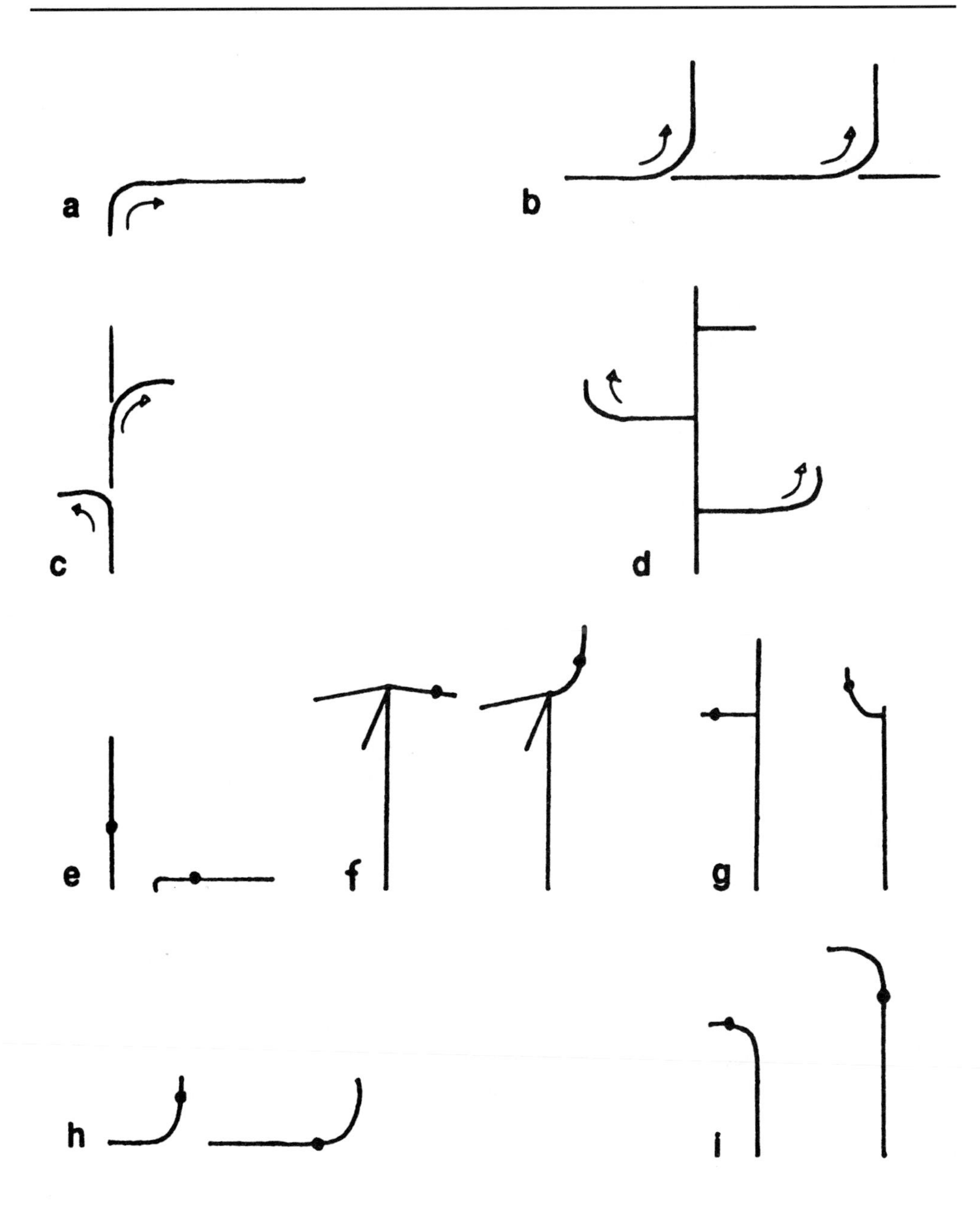

Figure 7 Reorientation. The change of growth direction in a single shoot unit: (a–d) by change of growth direction of the apical meristem; (e–i) by a bending of the existing shoot unit. (a) The first shoot unit of many procumbent herbs; (b) typical development of rhizomatous herbs; (c) development of sympodial trunk in some trees and lianès; (d) metamorphosis, the end of a side branch becomes vertical and will repeat (reiteration) the model sequence of the original trunk (see Fig. 21); (e) the first shoot unit of some woody procumbent shrubs e.g. *Salix repens* L.; (f) formation of a sympodial trunk in some trees (see Fig. 19); (g) reorientation of a side branch reiterating lost trunk (metamorphosis); (h) the continuous prostration of some herbaceous plants e.g. *Lycopodium* L.; (i) the continuous erection of some tree trunks, e.g. *Tamarindus indica* L.

The 'repair' mechanism in Fig. 7g illustrates again the phenomenon of metamorphosis (Fig. 10) but for the most part all these orientation changes take place as predictable features of the development of the plant form. They represent one factor governing the architectural model of the given species. Thus the trunk of *Alstonia macrophylla* Wall. (Fig. 19) is composed of a series of shoot units, the first developing from the epicotyl of the seedling and fitting exactly the sequence of events for the palm (Fig. 2), except that the apical meristem dies at the last step (Fig. 13). The meristems borne on this first shoot unit are dormant except for four located just below the dead apex. These grow to produce new shoot units, one of which subsequently bends due to the formation of reaction wood (Fisher & Stevenson, 1981) to bring it to the vertical and thus to constitute the next unit of the trunk (model of Koriba as described by Hallé & Oldeman, 1975). The growth model of Troll (Fig. 7d) can also develop by the reorientation of components of the trunk from the horizontal to the vertical (Fig. 7i) (Mueller, 1988).

Thus, with or without reorientation, a shoot unit may form an integral component of the plant's basic framework, the basic model. However, other meristems, particularly if released from dormancy, can augment this initial framework.

REITERATION

Activation of the dormant meristems in Fig. 2 allowed shoot units to develop that repeat the format of the original seedling axis. Such additional branching is referred to as total reiteration (Fig. 8) (Hallé *et al.*, 1978) and may be a rare or frequent phenomenon for a given species. It is most unambiguously observed as a response to damage – the classic release from apical control of dormant buds. However, it also occurs habitually in many trees, depending upon growth conditions and is responsible to some extent for the irregular appearance of older plant frameworks, the other erratic feature being the loss of parts. Figure 20 shows a mature tree with two pronounced recent reiteration events. The population of quiescent buds that have the potential to follow the reiteration loop of Fig. 1 and thus initiate additional expressions of the basic sequence is of paramount importance for the form of the plant. This is always true in clonal spreading plants where mechanical constraints are absent (Maillette, 1992b). In a developing tree crown, early reiteration will represent main branch complexes in the construction of the tree which to some extent can be regarded as autonomous entities (Edelin, 1991; Sprugel *et al.*, 1991), but reiteration shoot units located progressively toward the outer surface of the tree will initiate progressively smaller and smaller manifestations of the model sequence (Hallé *et al.*, 1978).

In some circumstances only part of the model sequence is activated – partial reiteration (Fig. 9). Figure 17 shows a horizontal branch of *Araucaria heterophylla* (Salisb.) Franco. Dormant meristems are developing to form short determinate shoot units. This species has a very simple architectural unit, a limited repertoire of axis types as detailed here under the heading architectural model. The short determinate shoots in Fig. 17 are axes type 3 and are temporary structures being shed by the formation of a proximal abscission zone. However, the integrity of the model is maintained by the growth of additional axis 3 shoot units thus replacing part of the original model.

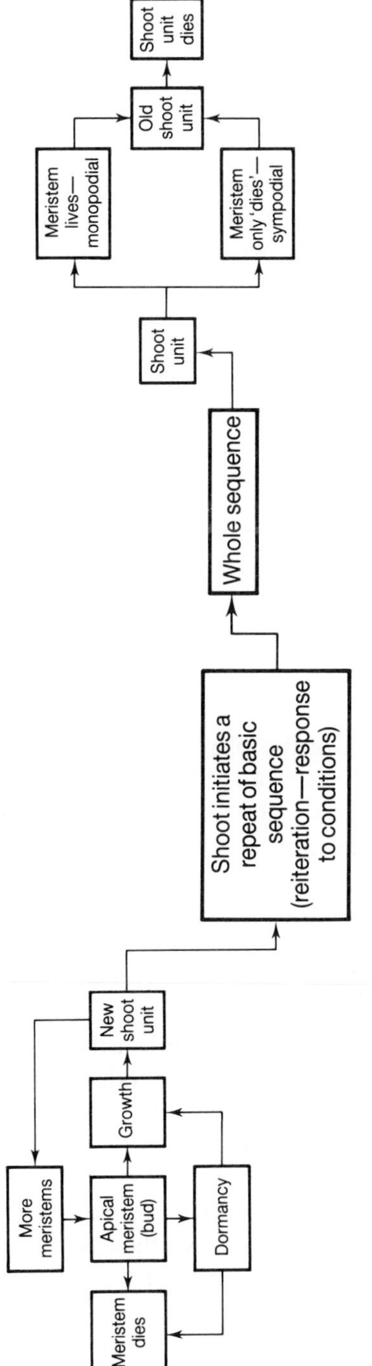

Figure 8 The shoot unit initiates a total repeat of the plant's basic form.

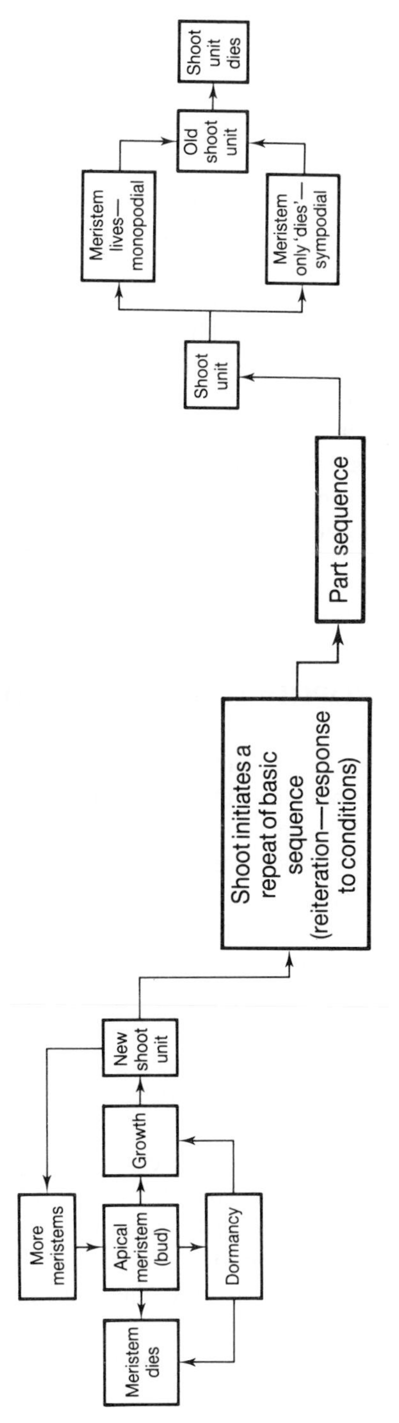

Figure 9 The shoot unit initiates a partial repeat of the plant's basic form.

Partial reiteration involves the development of any axes other than the principal trunk axis, axis type 1 (Barthelemy *et al.*, 1991).

METAMORPHOSIS

In the context of tree architecture the concept of metamorphosis applies to the transformation of a shoot unit that is an integral part of the basic model sequence into a shoot unit that has the new potential to initiate a repeat of the whole model sequence (Fig. 10) Barthelemy *et al.*, 1989. Thus a lateral axis that would have been committed to a subsidiary and horizontal role in the model is reorientated either by apical growth (Fig. 7d) or by reaction wood (Fig. 7g) into the vertical position and then initiates a series of meristem activities forming a total reiteration of the plant's model (Fig. 21). It is not yet clear if a comparable phenomenon occurs in herbaceous plants. Metamorphosis thus represents an important switch in the activity of a shoot unit and is a key feature in the establishment of many tree crowns. It is reiteration based on an already developing axis rather than reiteration from a dormant meristem.

MONOPODIAL AND SYMPODIAL GROWTH

Once a shoot unit is established within the framework of the plant, whether it is a component of the basic model (Fig. 5) or initiates a reiteration (Fig. 8), its apical meristem may continue growth (Fig. 11) or cease to grow (Fig. 12).

If the shoot unit does retain a viable apical meristem and is capable of more or less continued extension, then the axis it forms is termed monopodial (e.g. Fig. 15). If the apical meristem ceases to function (by aborting, forming a terminal flower, etc.) and if an axillary meristem borne on this shoot unit is activated, the axis as a whole is extended by development of this additional shoot unit (Remphrey & Davidson, 1992; Sakai, 1990a, b). Ultimately the axis, which is termed sympodial, will be formed from a row of shoot units (Fig. 18). In a tree, the trunk (axis type 1) and/or the main side branches (axes type 2) may be either monopodial or sympodial, depending upon the architectural model of the species. Often the branch component is built up from a forking of axes where each shoot unit bears two daughter shoot units at its distal end (Fig. 22 and many rhizomatous plants).

Thus, a clear distinction exists between a monopodial axis formed from one shoot unit and a sympodial axis formed from a sequence of shoot units. Nevertheless, these two types of axis may play the same roles in the unit of architecture of the species (Edelin, 1991).

The apical meristem of a monopodial shoot unit may cease to function eventually and therefore the axis will cease to extend. This situation is most apparent in plants in which the apices are transformed into terminal stem spines before meristem death (Fig. 23); the plant framework is built up of monopodial axes and not of series of sympodial shoot units.

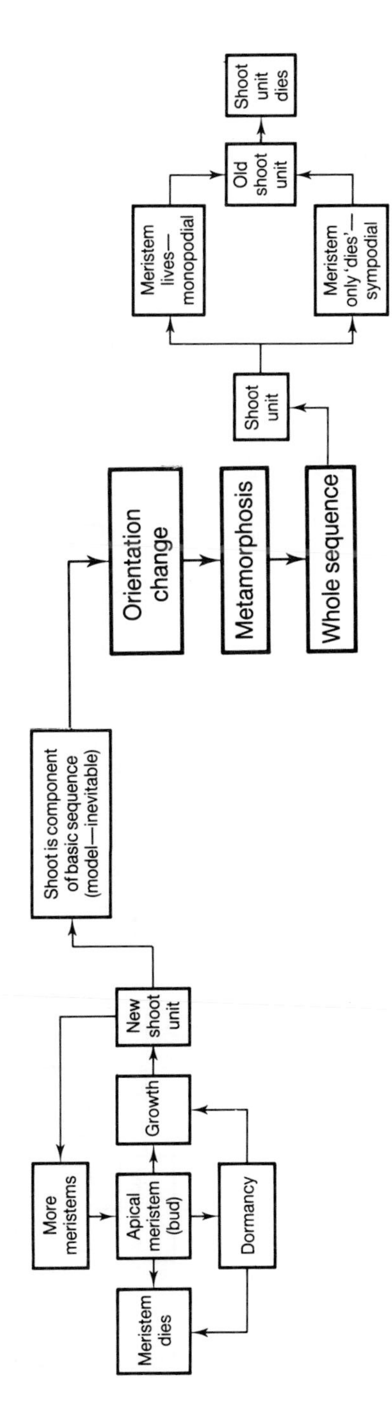

Figure 10 Having been an elemental component a shoot unit assumes a new role and initiates a total reiteration.

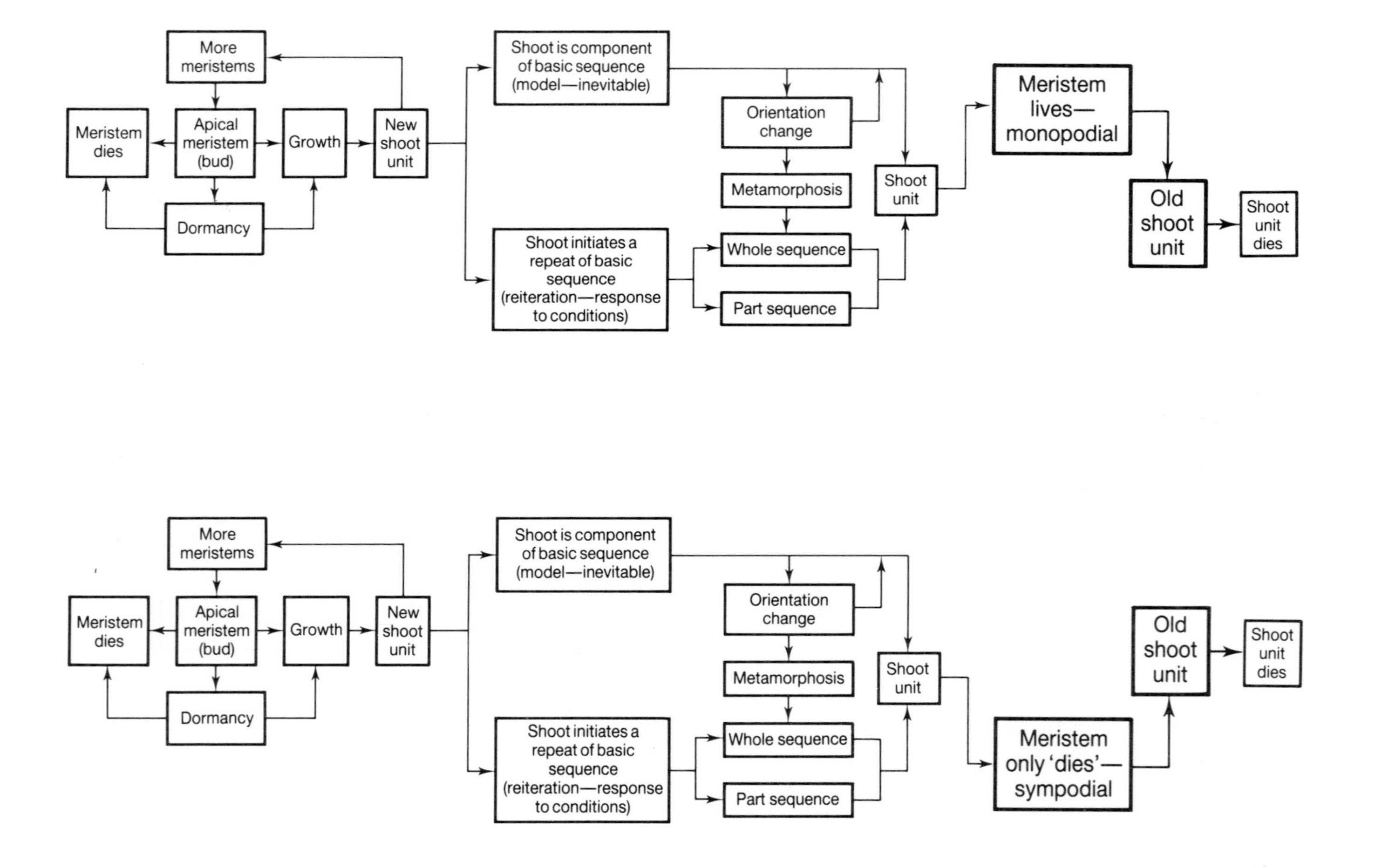

Figures 11 & 12 If the shoot unit's meristem remains active a monopodial axis develops. If the shoot unit ceases to extend it may become one unit of a sympodical axis.

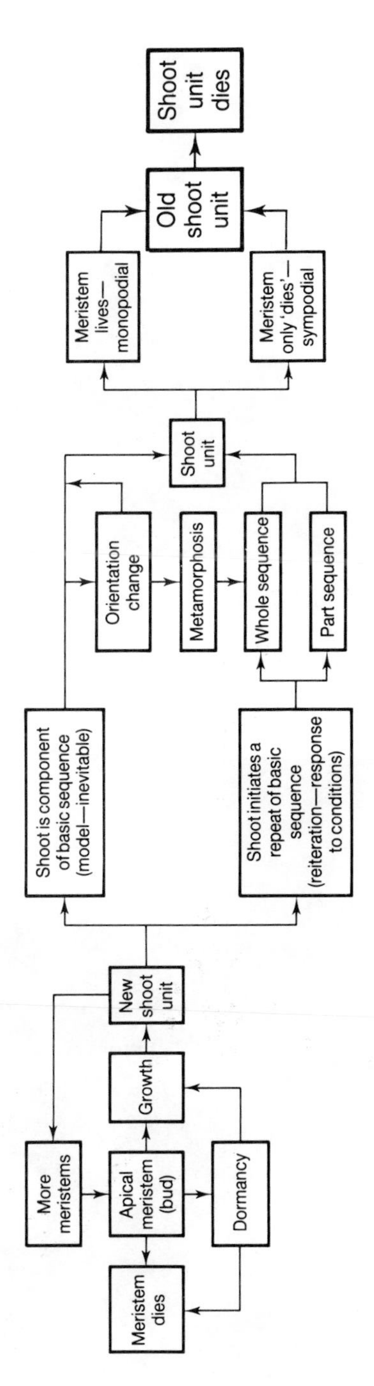

Figure 13 Eventual shoot unit death. This may be predictable.

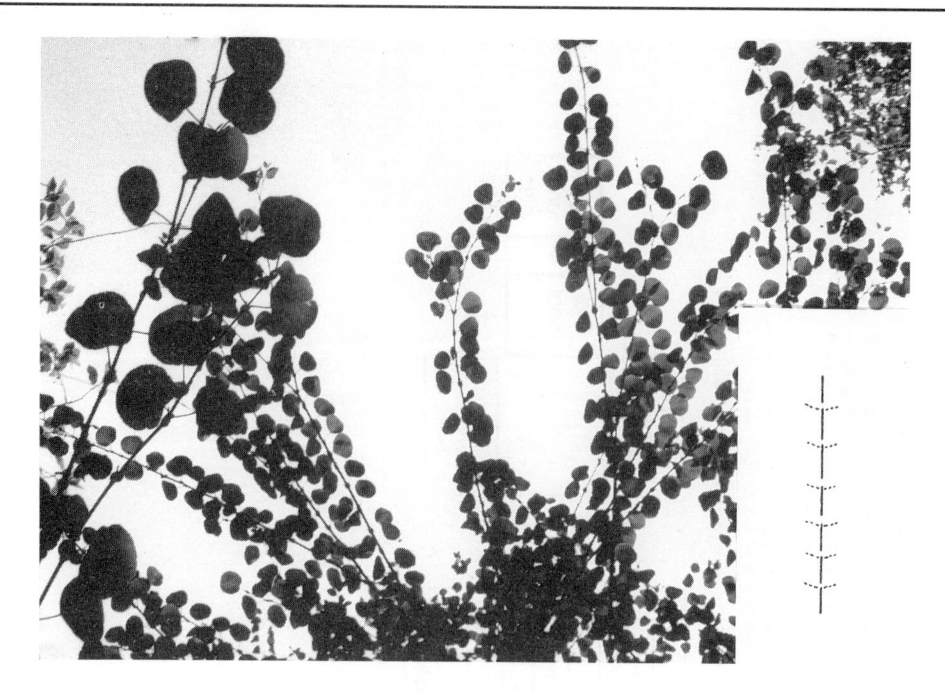

Figure 14 *Cercidiphyllum japonicum* Siebold & Zucc. Each branch is composed of a sympodial series of shoot units. Leaves are borne on very short sympodial side branches. Inset gives a synopsis of the shoot unit disposition; one line equals one shoot unit,

Figure 15 *Viburnum plicatum* Thunb. Each axis consists of a single monopodial shoot unit. Side shoots are represented by inflorescences – ancillary shoots. Inset gives a synopsis of the shoot unit disposition; one line equals one shoot unit.

Figure 16 *Pandanus copelandii* Merrill. Model of Massart. Monopodial trunk (axis type 1) and monopodial side branches (axes type 2). Inset gives a synopsis of the shoot unit disposition; one line equals one shoot unit.

SHOOT UNIT DEATH

The principal events represented in Fig. 1 that have a bearing on plant form are the activation of the apical meristem, the consequent development of a shoot unit, and its incorporation into the plant framework as all (monopodial) or part (sympodial) of an axis. The final phase will be the death of the shoot unit (Fig. 13). Death of the proximal shoot units of a sympodial axis in clonal plants is a common phenomenon (Fig. 18). Less well documented but equally pervasive is the death and shedding of shoot units, often as whole branch complexes, in woody plants (Millington & Chaney, 1973). Death of shoot units can be passive, due to damage, but frequently is due to the active formation of an abscission zone across the base of the unit. Loss of framework components is just as important as their incorporation into the system in the first place (Corner, 1951; Hardwick, 1986).

Figure 17 *Araucaria heterophylla* (Salisb) Franco. Monopodial side branch (axis type 2) bearing temporary axis type 3s – new ones are being formed – partial reiteration. The inset includes the axis type 1 for reference. Inset gives a synopsis of the shoot unit disposition; one line equals one shoot unit.

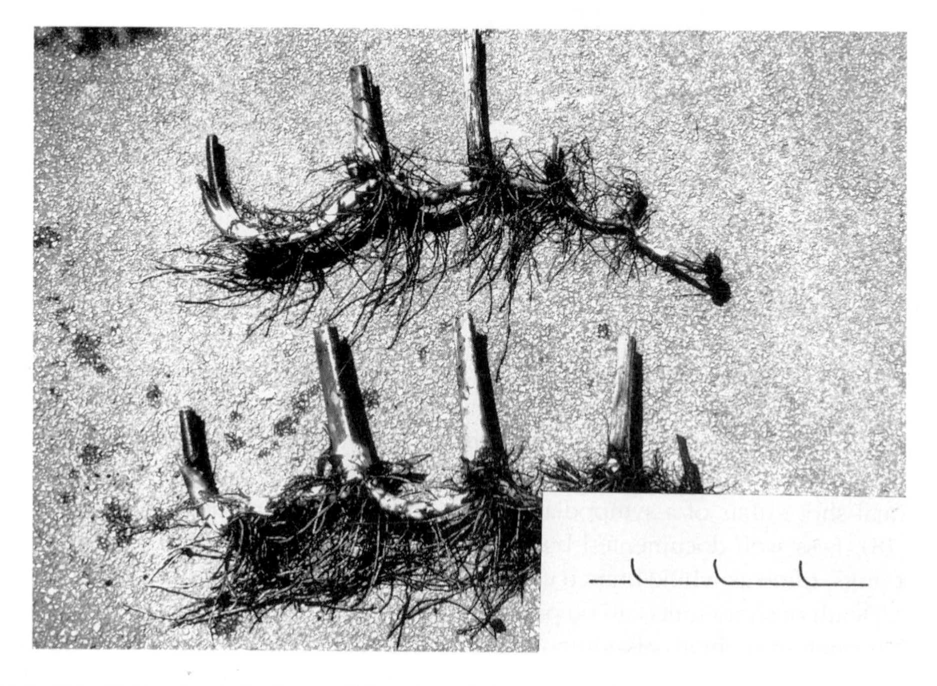

Figure 18 *Heliconia sp.* L. Sympodial series of shoot units (excavated rhizome). Inset gives a synopsis of the shoot unit disposition; one line equals one shoot unit.

SUMMARY

A palm tree consists of one axis structurally composed of one shoot unit developed from one apical meristem. This process is encapsulated in Fig. 2. Figure 2 also precisely represents a barrel cactus or a cabbage plant. Surplus apical meristems (i.e. those located in the axils of leaves) either die, remain dormant or are involved in the production of non-structural components such as flowers and stem spines. For more elaborate plants, additional meristems generated by the meristem cycle (Fig. 3), are incorporated into the branching process and each new shoot unit thus formed will follow one pathway through the alternatives in Fig. 1.

The branching process can be summarized as follows:

1. A population of shoot units characterizes the individual plant. Singly or cumulatively shoot units form axis types. Each shoot unit is born into the existing framework and follows a singular fate. Numbers vary enormously. The vulnerable palm has one shoot unit, the aspen clone many thousands.
2. A suite of axis types characterizes the species. An axis type may be composed of one shoot unit or of a series of shoot units. The suite is termed by Edelin (1991) the architectural unit. The palm tree has one axis type only, *Araucaria* Juss. (Fig. 17) has three. Generally numbers are small, probably less than ten.

In essence the dynamic form of a plant depends upon its branching. To begin to understand this form for a given plant the following need to be known: (a) How many types of branch (i.e. axis types) are involved? (b) How are axis types composed in terms of shoot units as discussed here? (c) What is the demography of shoot units? (d) What is the orientation of axes in space?

Figure 19 *Alstonia macrophylla* Wall. The sympodial trunk (axis type 1) is formed from initially horizontal shoot units that reorientate to the vertical (Fig. 7f). Inset gives a synopsis of the shoot unit disposition; one line equals one shoot unit.

Figure 20 *Acer pseudoplatanus* L. The basic model in this young tree is that of Rauh with oblique side branches (axes type 2). Two dormant meristems have begun to develop, reiterating the vertical growth of the trunk (axes type 1). Inset gives a synopsis of the shoot unit disposition; one line equals one shoot unit.

Figure 21 *Prunus avium* (L) L. Same basic model as Fig. 20 (Rauh). The ends of oblique side branches (axes type 2) are reorientating to the vertical (metamorphosis) and thus beccome transformed into axes type 1 reiterating the basic model. Inset gives a synopsis of the shoot unit disposition; one line equals one shoot unit.

Figure 22 *Phellodendron lavallei* Dode. Each shoot unit bears two daughter shoot units at its distal end. Inset gives a synopsis of the shoot unit disposition; one line equals one shoot unit.

Figure 23 *Colletia paradoxa* (Sprengel) Escal. All shoot units terminate sooner or later in a spine (many are flattened). Inset gives a synopsis of the shoot unit dispositions; one line equals one shoot unit.

REFERENCES

BARLOW, P.W., 1989. Meristems, metamers and modules and the development of shoot and root systems. *Botanical Journal of the Linnean Society, 100:* 255–279.

BARTHELEMY, D., 1987. Une mode de developpement remarquable chez une orchidee tropicale: *Gongora quinquenervis* Ruizet Pavon. *Comptes Rendus de l'Academie des Sciences, 304:* III No. 10, 279–284.

BARTHELEMY, D., 1991. Levels of organization and repetition phenomena in seed plants. *Acta Biotheretica, 39:* 309–323.

BARTHELEMY, D., EDELIN, C. & HALLÉ, F., 1989. Architectural concepts for tropical trees. In L.B. Holm-Nielsen, I.C. Nielsen & H. Balser (eds), *Tropical Forests, Botanical Dynamics, Speciation and Diversity,* pp. 89–100. London, U.K.: Academic Press.

BARTHELEMY, D., EDELIN, C. & HALLÉ, F., 1991. Canopy architecture. In A.S. Raghavendra (ed.), *Physiology of Trees,* John Wiley.

BELL, A.D., 1991. *Plant Form: An Illustrated Guide to Flowering Plant Morphology.* Oxford, U.K.: Oxford University Press.

BELL, A.D., BELL, A. & DINES, T.D., 1992. Bud structure and branch morphology at the canopy surface of a tropical rainforest. In F. Hallé & O. Pascal (eds), *Biologie d'une canopée de foret equatoriale II,* pp. 149–154. Operation Canopée (ISBN 2-9506703-1-8).

BILBROUGH, C.J. & RICHARDS, J.H., 1991. Branch architecture of sagebrush and bitterbrush: use of a branch complex to describe and compare growth patterns. *Canadian Journal of Botany, 69:* 1288–1295.

BRIAND, C.H., POSLUSZNYU & LARSON, D.W., 1991. Differential axis architecture in *Thuja occidentalis* (eastern white cedar). *Canadian Journal of Botany, 70:* 340–348.

BURROWS, G.E., 1990. The role of axillary meristems in coppice and epicormic bud initiation in *Araucaria cunninghamii. Botanical Gazette, 151:* 293–301.

COEN, E.S. & MEYEROWITZ, E.M., 1991. The war of the whorls: genetic interactions controlling flower development. *Nature, 353:* 31-37.

COOK, R.E., 1992. Review of: clonal growth in plants, regulation and function. *Plant Science Bulletin, 38:* 24–25.

CORNER, E.J.H., 1951. *Wayside Trees of Malaya,* Vol. 1. Singapore: The Government Printer.

EDELIN, C., 1991. Nouvelles donnees sur l'architecture des arbres sympodiaux: le concept de plan d'organization. In C. Edelin (ed.) *L'arbre. Biologie et Developpement,* pp. 127–154. Naturalia Monspeliensia Hors de series 1991.

FAGERSTROM, T., 1992. The meristem-meristem cycle as a basis for defining fitness in clonal plants. *Oikos, 63:* 449–453.

FINK, S., 1983. The occurrence of adventitious and preventitious buds within the bark of some temperate and tropical trees. *American Journal of Botany, 70:* 532–542.

FINK, S., 1984. Some cases of delayed or induced development of axillary buds from persisting detached meristems in conifers. *American Journal of Botany, 71:* 44–51.

FISHER, J.B., 1992. How predictive are computer simulations of tree architecture? *International Journal of Plant Sciences, 153:* 5137–5146.

FISHER, J.B. & STEVENSON, J.W., 1981. Occurrence of reaction wood in branches of dicotyledons and its role in tree architecture. *Botanical Gazette, 142:* 82–95.

FORD, E.D., 1992. The control of tree structure and productivity through the interaction of morphological development and physiological process. *International Journal of Plant Science, 153:* 5147–5162.

GINOCCHIO, R. & MONTENEGRO, G., 1992. Interpretation of metameric architecture in dominant shrubs of the Chilean Matorral. *Oecologia, 90:* 451–456.

GROENENDAEL, J.M. van, 1985. Teratology and metameric plant construction. *New Phytologist, 99:* 171–178.

HAGEMANN, W., 1992. The relationships of anatomy to morphology in plants: a new theoretical perspective. *International Journal of Plant Sciences, 153:* 538–548.

HALLÉ, F. & OLDEMAN, R.A.A., 1975. An essay on the architecture and dynamics of growth of tropical trees. Translated by B.C. Stone. Penerbit Universiti, Malaya, Kuala Lumpur, 156 pp.

HALLÉ, F., OLDEMAN, R.A.A. & TOMLINSON, P.B., 1978. *Tropical Trees and Forests.* Berlin, Germany: Springer-Verlag.

HARDWICK, R.C., 1986. Physiological consequences of modular growth in plants. *Philosophical Transactions of the Royal Society London, B, 313:* 161–173.

HARPER, J.L., ROZEN, B.R. & WHITE, J., 1986. The growth and form of modular organisms. *Philosophical Transactions of the Royal Society London, B, 313:* 1–250.

HAUKIOJA, E., 1991. The influence of grazing on the evolution morphology and physiology of plants as modular organisms. *Philosophical Transactions of the Royal Society London, B, 333:* 241–247.

HERMS, D.A. & MATTSON, W.J., 1992. The dilemma of plants: to grow or defend. *Quarterly Review of Biology, 67:* 283–335.

JONES, M. & HARPER, J.L., 1987a. The influence of neighbours on the growth of trees. I. The demography of buds in *Betula pendula. Proceedings of the Royal Society London, B, 232:* 1–18.

JONES, M. & HARPER, J.L., 1987b. The influence of neighbours on the growth of trees. II. The fate of buds in *Betula pendula. Proceedings of the Royal Society London, B, 232:* 19–33.

KAPLAN, D.R., 1992. The relationship of cells to organisms in plants: problem and implications of an organised perspective. *International Journal of Plant Science, 153:* 528–537.

KAPLAN, D.R. & HAGEMMAN, W., 1991. The relationship of cell and organism in vascular plants. *BioScience, 41:* 693–703.

LOUP, C., FOURNIER, M., & CHANSON, B., 1991. Relations entre architecture, mecanique et anatomy d l'arbre. Cas d'un pin maritime (*Pinus pinaster* Soland). In C. Edelin (ed.), *L'Arbre, Biologie et Developpement,* pp. 181–195. Naturalia Monspeliensia No H.S. Actes du 2eme Colloque international sur l'Arbre, Montpellier, 10–15 septembre, 1990.

MAILLETTE, L., 1982. Structural dynamics of silver birch. I. The fate of buds. *Journal of Applied Ecology, 19:* 203–218.

MAILLETTE, L., 1992a. Seasonal model of modular growth in plants. *Journal of Ecology, 80:* 123–130.
MAILLETTE, L., 1992b. Plasticity of modular reiteration in *Potentilla anserina. Journal of Ecology, 80:* 231–239.
MCGRAW, J.B. & ANTONOVICS, J., 1983. Experimental ecology of *Dryas octopetala* ecotypes. II. A demographic model of growth branching and fecundity. *Journal of Ecology, 71:* 899–912.
MCGRAW, J.B. & GARBUTT, K., 1990. Demographic growth analaysis. *Ecology, 71:* 1199–2004.
MILLINGTON, W.F., & CHANEY, W.R., 1973. Shedding of shoots and branches. In T.T. Kozlowski (ed.), *Shedding of Plant Parts,* pp. 49–204. New York, U.S.A.: Academic Press.
MUELLER, R.J., 1988. Shoot tip abortion and sympodial branch reorientation in *Brownea ariza* (Leguminosae). *American Journal of Botany, 75:* 391–400.
NIKLAS, K.J., 1992. *Plant Biomechanics: An Engineering Approach to Plant Form and Function.* Chicago, U.S.A.: The University of Chicago.
PORTER, J.R., 1983a. A modular approach to analysis of plant growth. I. Theory and principles. *New Phytologist, 94:* 182–190.
PORTER, J.R., 1983b. A modular approach to analysis of plant growth. II. Methods and results. *New Phytologist, 94:* 191–200.
PORTER, J.R., 1989. Modules, models and meristems in plant architecture. In G. Russell, B. Marshall & P.G. Jarvis (eds), *Plant Canopies, Their Growth Form and Function.* Society for Experimental Biology Seminar Series 32. Cambridge, U.K.: Cambridge University Press.
de REFFYE, Ph., ELGUERO, E., & COSTES, E., 1991. Growth units construction in trees: a stochastic approach. *Acta Biotheoretica, 39:* 325–342.
REMPHREY, W.R. & DAVIDSON, C.G., 1992. Branch architecture and its relation to shoot tip abortion in mature *Fraxinus pennsylvanica. Canadian Journal of Botany, 70:* 1147–1153.
SAKAI, S., 1990a. Sympodial and monopodial branching in Acer: implications for tree architecture and adaptive significance. *Canadian Journal of Botany, 68:* 1549–1553.
SAKAI, S., 1990b. Sympodial and monopodial branching in Acer (Aceraceae) – evolutionary trend and ecological implications. *Plant Systematics and Evolution, 171:* 187–197.
SATTLER, R., 1990. Towards a more dynamic plant morphology. *Acta Biotheoretica, 38:* 303–315.
SATTLER, R., 1992. Process morphology: structural dynamics in development and evolution. *Canadian Journal of Botany, 70:* 708–714.
SCHAFFALITZKY DE MUCKADELL, M., 1959. Investigations on ageing of apical meristems in woody plants and its importance in silviculture. *Report of the Danish Forest Experimental Station, Copenhagen Denmark, 25:* 307–455.
SPRUGEL, D.G., HINCKLEY, T.M. & SCHAAP, W., 1991. The theory and practice of branch autonomy. *Annual Review of Ecology and Systematics, 22:* 309–334.
SVENSSON, B.M. & CALLAGHAN, T.V., 1988. Apical dominance and the simulation of metapopulation dynamics in *Lycopodium annotinum. Oikos, 51:* 331–342.
TITMAN, P.W. & WETMORE, R.H., 1955. The growth of long and short shoots in *Cercidiphyllum. American Journal of Botany, 42:* 364–372.
TOURN, G.M., TORTOSA, R.D. & MEDAN, D., 1992. Rhamnaceae with multiple lateral buds: an architectural analysis. *Botanical Journal of the Linnean Society, 108:* 275–286.
TUOMI, J. & VUORISOLA, T., 1989. What are the units of selection in modular organisms? *Oikos, 54:* 227–233.
WILSON, B.F., 1989. Tree branches as populations of twigs. *Canadian Journal of Botany, 67:* 434–442.
WILSON. B.F., 1990. The development of tree form. *Horticultural Science, 25:* 52–54.
WILSON, B.F. & ARCHER, R.R., 1979. Tree design: some biological solutions to mechanical problems. *BioScience, 9:* 293–298.

CHAPTER

7

A commentary on some recurrent forms and changes of form in angiosperms

B.L. BURTT

CONTENTS

Abstract
Examples are quoted where morphological change due to a single gene mutation has been demonstrated. Amongst these, distinction is made between those that would be lethal under natural conditions, those that are reversions (involving the reappearance of features that have probably been suppressed in the course of evolution), and those that introduce new features (many of which may be liable to recur in different species – homologous variations). A few morphological changes are then examined that could, possibly, have been brought about by single mutations involving location of growth (for instance in floral symmetry or the development from phanerocotylar to cryptocotylar germination – both recurrent changes in angiosperms), and finally similar changes that act as a springboard for further step-by-step evolution (as in the sequence that leads from simple anisocotyly to a variety of novel growth forms in Gesneriaceae, or that leading to autumnal leafless flowering in *Crocus* and *Colchicum*).

THE EFFECT OF SINGLE GENES

Shape and form are the very stuff of taxonomy. Yet, although experimental taxonomists may study their inheritance, very few delve into their causal development. Here some recurrent patterns, of differing complexity, are selected for comment in a search for wider patterns lying behind their origins.

Shape and Form in Plants and Fungi
ISBN 0–12–371035–9

A taxonomist who knows or suspects that a difference between two plants is due to a single gene, will tend to think that the feature is of little taxonomic value: an old view that drew renewed strength from Vavilov's law of homologous variations presented in 1922 (Vavilov, 1922, 1949). There are hundreds of examples to endorse this view. However, for the study of the way in which genes control form, changes under single gene control may be of great value. They make an appropriate starting point.

I have used, in the first place, two papers that review the effect of single genes: one by Hilu (1983), the other by Gottlieb (1984). A few of the examples thrown up by these reviews, and the apparent readiness to regard them as of high taxonomic and evolutionary importance, deserve comment.

A difference between the two major subfamilies of Caryophyllaceae is that in Alsinoideae (chickweeds, stitchworts, etc.) the calyx consists of five free sepals, whereas in Caryophylloideae (campions, carnations and pinks) it is tubular with five apical teeth. An abnormal *Silene vulgaris* Garcke (bladder campion) was found in which the tubular calyx was split to the base into five segments. A single recessive gene was involved. This was unfortunately cited as showing that a character of subfamilial rank was under monogenic control (Turrill, 1940: 59; and so quoted by Gottlieb, 1984). That was not so. This calyx with a tube split into five segments was just that: it was in no way equivalent to a calyx with five free sepals. The calyx tube is an important functional feature of the *Silene* flower, because it holds the long-clawed petals and stamens in position. The flower with split calyx just flopped open, and did not set seed naturally; it occurred in nature only as an isolated abnormality.

The family Compositae has several conspicuous characters that seem to be under simple genetic controls; for example, the presence or absence of ray flowers, or receptacular paleae, or pappus. Taking ray flowers first: Hilu and Gottlieb both cite the example of *Haplopappus* Cass.: H. aureus A. Gray has rays but *H. venetus* (HBK) Blake lacks them. They were once placed in separate sections, largely because of this difference. But they are interfertile (though one is annual, the other of shrubby habit) and the radiate condition is dominant (Jackson & Dimas, 1981). Hilu commented 'These macromutations are of evolutionary importance . . .'. This reference to macromutations reads as though the author thought that the mutation produced a fully developed ray where there had been only a tubular flower before. Personal experience of taxonomic routine work in Compositae led to the widely accepted view that the absence of rays in genera or species that normally had them was due to a mutation suppressing their development, a mutation little different in value from the sort that puts a block in the biosynthetic pathway of anthocyanin, and results in white flowers. Certainly plants with radiate and discoid heads have been happily accommodated in *Aster tripolium* L. for 150 years or so, and there are many other examples of radiate and discoid heads within a single species.

It may be much the same with presence or absence of receptacular paleae, although they vary less frequently. The genus *Crepis* L. was long thought to be without paleae: then it was found that the neighbouring monotypic genus *Rodigia* Spreng., distinguished by its paleaceous receptacle, was fully interfertile with *Crepis foetida* L. and closely resembled it; therefore Babcock (1947: 697) reduced it to a subspecies. *C. foetida* is, on all counts, an advanced species of *Crepis*: paleae on the receptacle can therefore scarcely be the retention of a primitive character; it is much more likely to represent release from suppression.

J.R. Harlan (1982), writing on agricultural grasses, was emphatic that evolution working through suppression could be reversed. In the grass family much taxonomic differentiation has taken place through contraction and sterilization of the inflorescence and spikelets. Some of this suppression has been undone in cultivated races. One example from many: in wild species of *Hordeum* L. (barley) there are three spikelets at a node, but only the central one is fertile, the laterals are either reduced to males or are neuter: these barleys are described as two-rowed. The commonly cultivated six-rowed barley has the lateral spikelets restored to full fertility. Sometimes this reappearance of ancestral characters can result in the cultivated race belonging technically to another genus or even tribe. Harlan in 1983 called this form of reversion 'genetic recall'. A comparable reversion has just been reported in wild *Tripsacum dactyloides* (Jackson *et al.*, 1992).

In the genus *Valerianella* L. there is a remarkable case of fruit dimorphism controlled by a single gene. The fruit in this genus is three-celled with two cells sterile. In North America for a long time two species were recognized, *V. ozarkana* and *V. bushii.* It was known that they sometimes grew together, but it has now been established that they actually interbreed and are fully fertile, that the remarkable differences in the fruits are controlled by a single gene, and that *bushii* is the double recessive (Ware, 1983). Now the *bushii* fruit (Fig. 1B) is not remarkable in the genus, many other species are similar (Fig. 1C); but the wing-like sterile locules and the bristles of *ozarkana* are remarkable (Fig. 1A). Are these features all controlled by a single gene? Surely it is much more likely that *ozarkana* has evolved these specialized fruits, but that, within the species, a gene mutation has occurred that has blocked the pathway to their development and the fruit has matured in its more or less ancestral form as *V. bushii* Dyal, now called *V. ozarkana* forma *bushii* (Dyal) Eggers. This is akin to Harlan's genetic recall.

The examples cited so far are ones where genetic information is available. Except for the split *Silene* calyx, these mutations are not lethal. There are, of course, many other known mutations, such as those that produce leaves instead of carpels or petals

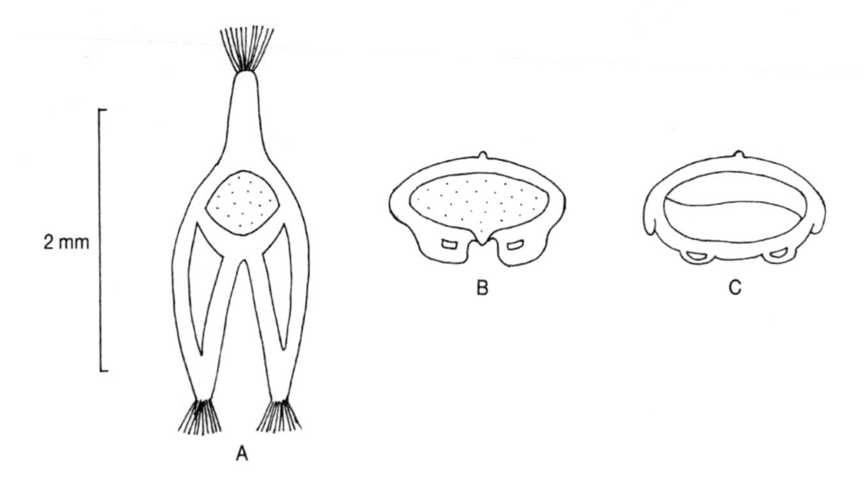

Figure 1 Transverse sections of the fruits of *Valerianella.* **A** *V. ozarkana* Dyal forma *ozarkana*; **B** *V. ozarkana* forma *bushii* (Dyal) D.M. Eggers; **C** *V. dentata* (L.) Poll. **A** & **B** reproduced with permission from Ware (1983).

instead of stamens, which would not be viable in the wild. These can scarcely be of major evolutionary importance, however valuable they may be in the study of gene action.

FLORAL SYMMETRY

Sometimes, despite the lack of genetic evidence, it seems probable or possible that a major change is caused by a single gene, with results that do have an evolutionary significance.

The flowers of Gesneriaceae (the family of gloxinias, African violets, etc.) are usually bilaterally symmetrical (zygomorphic). One or two isolated genera have radially symmetrical (actinomorphic) flowers. These may well be secondarily actinomorphic; in other words, somewhere way back their ancestors had zygomorphic flowers. This view is based largely on their very scattered distribution within the classification of the family (Burtt, 1970). There are perhaps ten such genera altogether. One example is *Cyrtandroidea* Jones from the South Pacific (Fig. 2); a typical species of *Cyrtandra* J.R. & G. Forst. from the same area (Fig. 3) only differs in the symmetry of the flower. *Cyrtandroidea* was properly reduced to *Cyrtandra* by the late G.W. Gillett (1973). A similar example is shown by the long-established *Fieldia* A. Cunn. from New South Wales and the recently described *Lenbrassia* G.W. Gillett from Queensland (Gillett, 1974). *Fieldia* has an actinomorphic corolla and five stamens, *Lenbrassia* a zygomorphic corolla and only two stamens. There are no other major differences and the

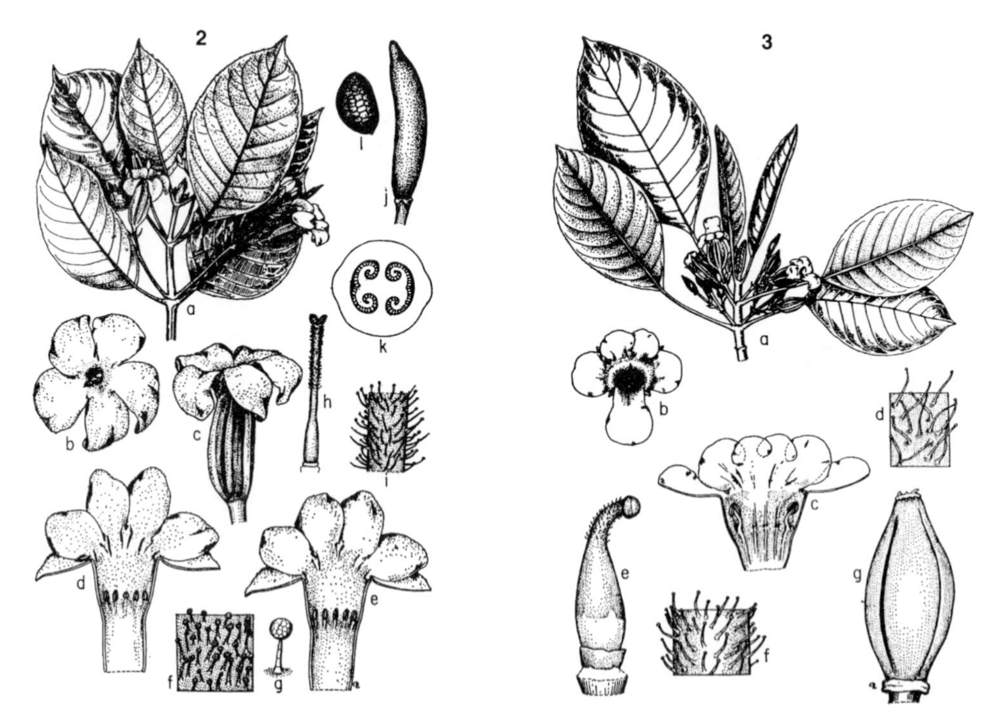

Figure 2 *Cyrtandra jonesii* (F. Brown) G.W. Gillett. **Figure 3** *Cyrtandra nukuhivensis* F. Brown. Reproduced with permission from G.W. Gillett, the genus *Cyrtandra* in the South Pacific. Copyright © 1973, The Regents of the University of California.

genera must be united under *Fieldia*, the older name. These two examples show that conspicuous differences in floral symmetry can occur in plants that are closely related. This is well-known to horticulturists who have developed true-breeding actinomorphic forms of the African violet (*Saintpaulia* H. Wendl.) and a good many other plants. Radial symmetry was undoubtedly normal somewhere back in the evolutionary lineage of Gesneriaceae. But the radially symmetrical flowers of today are 'reversions' only in this one sense of symmetry. What is seen is the floral genome expressing itself under a radial instead of a bilateral discipline. An intermediate state between radial and bilateral symmetry of the corolla is improbable. It therefore seems likely that the change from one to the other is controlled by a single gene.

SEEDLINGS

Dicotyledonous seeds normally germinate in one of two ways: with the cotyledons carried above ground (phanerocotylar or epigeal), either photosynthetic or, much more rarely, ± colourless but with large food reserves; or with the cotyledons remaining hidden within the seed (cryptocotylar and often hypogeal), the cotyledons themselves either food-storing or haustorial, withdrawing reserves from a copious endosperm.

There has been debate in recent years about which mode is primitive. A taxonomic survey of the distribution of hidden cotyledons, which is the less common mode (Burtt, 1991) showed that it occurs in all but one (Caryophyllidae) of the six subclasses, in 36 of the 64 orders and in 84 of the 318 families of dicotyledons in Cronquist's system (Cronquist, 1981), and in 373 genera, of which 45 have both modes. This very scattered distribution of cryptocotyly, and the fact that it frequently occurs in genera that are believed, on other grounds, to be advanced in their families (e.g. *Juglans* L. and *Carya* Nutt. in Juglandaceae, *Corylus* L. in Betulaceae, *Quercus* L. in Fagaceae – full details in Burtt, 1991) gives very strong reasons for believing cryptocotyly to be repeatedly and independently derived from phanerocotyly.

When green cotyledons appear above ground it is largely due to the elongation of the hypocotyl, which pulls them out of the seedcoat and then pushes them (with the plumule between them) up into the air. In cryptocotylar germination the plumule, hypocotyl and radicle are pushed out of the seed by elongation of the cotyledonary petioles and the plumule is carried upwards by growth of the epicotyl. I have suggested that the change from phanerocotyly to cryptocotyly could result from a simple displacement of growth from the hypocotyl to the petioles of the cotyledon, and that the necessary further changes would follow as epigenetic adjustments to the new environment. It is a recurrent change that takes place both within families and genera. It is not known if the pattern of change followed is always the same, as the much-needed detailed comparative studies are lacking.

The work by Stebbins and Yagil (1966) on hooded barley showed that this abnormality is dependent on a gene that induces the development of a meristem on the back of the lemma of an otherwise normal flower. From this meristem develops the 'hooded' structure, which is in effect a rudimentary inflorescence developed in an abnormal position. As Stebbins and Yagil put it 'all the morphological characteristics of the hooded lemma are probably produced by the action of a single gene which alters the course of development at an early primordial stage and initiates an entirely new epigenetic sequence of development'. That hypothesis was put forward after a full developmental study. My suggestion that a similar explanation lies behind the change from phanerocotylar to cryptocotylar germination is merely guesswork.

Cotyledons that become unequal after germination are characteristic of most of the Old World members of the family Gesneriaceae (subfamily Cyrtandroideae: *Aeschynanthus* Jack, *Saintpaulia*, *Streptocarpus* Lindl., etc.). New World genera (subfamily Gesnerioideae: *Achimenes* Pers., *Columnea* L., *Gloxinia* L'Hérit., *Sinningia* Nees, etc.) do not show this feature.

The seeds of these Old World Gesneriaceae are very small (commonly less than 1 mm long), without endosperm. Thus photosynthesis is an urgent requirement as soon as they germinate. A meristem at the base of one cotyledon provides increased photosynthetic tissue, without the delay of the organization of a plumule and further leaves. This may be the selective value of anisocotyly in Gesneriaceae (Burtt, 1970). In the caulescent species of *Streptocarpus* and in some related genera (such as *Chirita*), one cotyledon develops a form similar to that of the subsequent foliage leaves, while the other remains very small and eventually withers. There is usually a delay in the organization of the plumule but once this has taken place development is normal, and the foliage leaves do not show continuing growth. However, many gesneriads are not caulescent. The extreme possibility is that the enlarged cotyledon becomes the only foliage leaf of the plant. This happens conspicuously in some African species of *Streptocarpus* (such as *S. grandis* N.E. Br., reaching 38 × 33 cm) or in species of the Malesian genus *Monophyllaea* R. Br. (such as *M. tetrasepala* B.L. Burtt, 35 × 35 cm): the inflorescences of these plants arise on the midrib at the base of the lamina, or are sometimes displaced up the midrib, or sometimes they arise on the leaf stalk. There are also rosulate species of *Streptocarpus* and a few species that have stems but clearly belong to the acaulescent group (Hilliard & Burtt, 1971: 5–33; Jong & Burtt, 1975; Jong, 1978). In these 'acaulescent' *Streptocarpus* the leaf-organs produced subsequent to the cotyledonary one are not normal foliage leaves but are replicas of the enlarged cotyledon, complete with basal meristem and epiphyllous inflorescence; they also share with the enlarged cotyledon the peculiar attribute of being able to form an abscission layer shortly above the leaf-base when the cold unfavourable season sets in: the terminal part of the lamina is shed and growth is renewed from the basal meristem when favourable conditions return.

A few species of *Monophyllaea* also develop a stem and here again the subsequent leaves resemble the cotyledonary one. However, *Monophyllaea* belongs to the ever-wet tropics of Malesia and has no mechanism for formation of an abscission layer.

Monophyllaea and *Streptocarpus* have independently reached a unifoliate pattern: they are not closely related and are placed in separate tribes. Other genera with a single leaf (e.g. *Acanthonema* Hook. f., *Trachystigma* C.B. Clarke) are more closely related to *Streptocarpus* but nevertheless probably achieved that state independently.

A quite different pattern of growth is also dependent on the accrescent cotyledon; this is found in some species of the genus *Epithema* Blume and was investigated in the Central African *E. tenue* (Hallé & Delmotte, 1973). The enlarged cotyledon is the only leaf during the first growing season (Fig. 4): in the ensuing dry weather it dies off, leaving the swollen base of the plant where the third leaf is already beginning to develop (Fig. 5B). When the rains come this third leaf grows out, looking at first very like the previous season's enlarged cotyledon; but there is an essential difference.

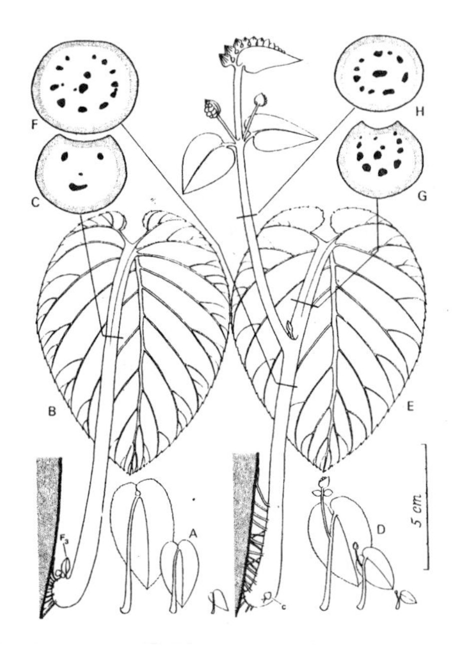

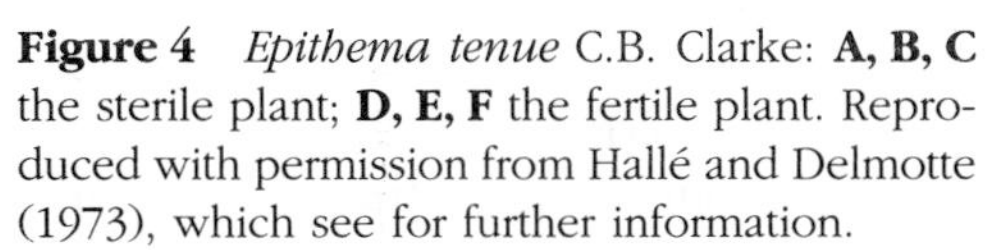

Figure 4 *Epithema tenue* C.B. Clarke: **A, B, C** the sterile plant; **D, E, F** the fertile plant. Reproduced with permission from Hallé and Delmotte (1973), which see for further information.

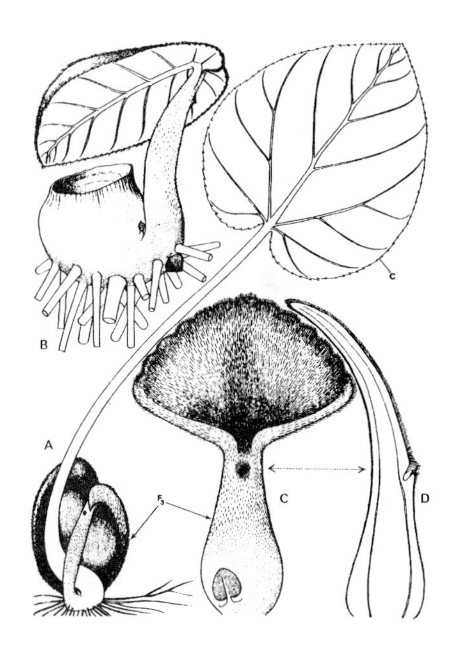

Figure 5 *Epithema tenue* C.B. Clarke. The sterile plant in the dry season: **A** while the mature tissues of the large cotyledon wither in the weeks that follow the last rains, the young tissues of leaf 3 survive without injury the 4 months of the dry season; **B** the young plant after the fall of the large cotyledon; **C** leaf 3 seen from above, the depression at the base of the lamina encloses the apical meristem; **D** longitudinal section of **C**. Reproduced with permission from Hallé and Delmotte (1973).

The third leaf carries just below the lamina a sunken meristem (Fig. 5C, D), from which arises the flowering shoot.

The acquisition of a basal meristem in one cotyledon has given the Old World Gesneriaceae an increased potential that has been expressed in a variety of ways. This anisocotyly may, like cryptocotyly, have originated by a relatively simple genetic mutation, but there is a major difference. Cryptocotyly may be envisaged as recurrent mutations evoking a series of epigenetic events that have roughly the same effect on the different seedlings. Real innovation on this pattern is small. Anisocotyly in Gesneriaceae, on the other hand, seems to have resulted from a simple mutation that not only embodied an immediate gain but gave a potential for considerable adaptive change in growth patterns, within which recurrent forms, such as the unifoliate habit, are again found.

Two genera, *Crocus* L. (Iridaceae) and *Colchicum* (Liliaceae) resemble these anisocotylous Gesneriaceae in exhibiting a series of simple changes each of which provides a platform for a further change (Burtt, 1974). Given an underground storage organ, rapid growth is possible as soon as conditions become favourable; furthermore leaf and flower buds can develop in the protection of the soil, so that only warmth and moisture are needed to give very rapid flowering in spring. Most often (*Tulipa* L.,

Narcissus L., *Scilla* L.) such flowers are borne aloft on their peduncles, but the peduncle may remain underground and the flower may be raised by the growth of the perianth tube, as in *Crocus* and *Colchicum*. Thus the ovary retains the protection of the soil and needs only to be raised (by the delayed elongation of peduncle or stem) when the seed is ready for dispersal in early summer. But where the growing season is short spring is the best time for seedling growth. Therefore, a shift of flowering to autumn, followed by underground ripening of the fruit and seed-shedding in spring, might be advantageous: but spring would still be the best time for leafing. Thus the 'naked' flowers found in autumn crocuses (*C. nudiflorus* J.E. Smith, *C. speciosus* M. Bieb, etc.) and meadow saffron (*Colchicum autumnale* L. etc.) may have evolved. None of these steps seems to need more than a simple genetic change in the timing and location of growth.

CONCLUSIONS

The gene effects considered here can be roughly classified into four main, but not always mutually exclusive, groups.

1. Genes that act directly on a single character without affecting the viability of the plant. It is these that taxonomists reasonably consider to be of little taxonomic value: simple colour changes, albinos, Vavilov's homologous variations.
2. Genes that act directly on a single character but often render the plant sterile. Examples are double flowers with petals replacing stamens, or flowers with leaves replacing carpels.
3. Genes that permit reactivation of processes for which the genetic mechanisms are present in the plant but have been suppressed. Genes responsible for the effect that Harlan called genetic recall belong here, as do those permitting reappearance of ray-flowers in normally discoid genera or species of Compositae.
4. Genes that produce their effect early in a developmental sequence, or initiate a new one. Two subtypes may be recognizable in this putative group; putative because only for hooded barley (a) is the genetic position known.
   a. Those where a simple genetic change initiates a new or altered epigenetic sequence. The suggested mode of change from phanerocotylar to cryptocotylar germination would be an example.
   b. Those where a simple genetic change has a morphogenetic effect that is not only advantageous in itself but provides a platform for new, independent, genetic changes. The accrescent cotyledon of Gesneriaceae, leading to the possibility of new and diverse growth patterns would be one example; the sequence leading to autumn-flowering crocus and colchicum another.

This review underlines the need for taxonomists to have a deeper understanding of the characters they use. Perhaps too much attention has been paid to those within the reach of simple genetic experiments (groups 1 & 3 above) and too little to the morphogenesis of deep-seated features (group 4). The value of studies in this latter field is attested, for instance, in the recent work on floral ontogeny of Leguminosae (Tucker, 1992) or on developmental differences in the fruits of Umbelliferae (Theobald, 1971), to give only two examples. Taxonomy will benefit more from results in this field than it will from highly speculative attempts to discover phylogenetic relationships. It is not that such relationships are unimportant, but they require a great deal of data and a deep understanding of the characters used. Such studies are metataxonomies (Burtt, 1964). They should come after the morphological classifications. At present these should be given priority as they need to be much more soundly based than they are. Only then shall we be able to see where the real tensions between a 'phylogenetic' classification, including, for example, molecular data, and a morphological one lie. Tensions from which we shall learn more about the plants and about evolution.

REFERENCES

BABCOCK, E.B., 1947. *The Genus* Crepis, part 2, University of California Publications in Botany, vol. 22. Berkeley, U.S.A.: University of California Press.

BURTT, B.L., 1964. Angiosperm taxonomy in practice. In V.H. Heywood & J. McNeill (eds), *Phenetic and Phylogenetic Classification* (Systematics Association Publication No. 6), pp. 5–16). London, U.K.: The Systematics Association.

BURTT, B.L., 1970. Studies in the Gesneriaceae of the Old World: XXXI. Some aspects of functional evolution. *Notes from the Royal Botanic Garden, Edinburgh, 30:* 1–10.

BURTT, B.L., 1974. Patterns of structural change in the flowering plants. *Transactions of the Botanical Society of Edinburgh, 42:* 133–142.

BURTT, B.L., 1991. On cryptocotylar germination in dicotyledons. *Botanische Jahrbücher für Systematik Pflanzengeschichte und Pflanzengeographie, 113:* 429–442.

CRONQUIST, A., 1981. *An Integrated System of Classification of Flowering Plants.* New York: Columbia University Press.

GILLETT, G.W., 1973. The genus Cyrtandra (Gesneriaceae) in the South Pacific. *University of California Publications in Botany, 66:* 1–59.

GILLETT, G.W., 1974. Lenbrassia (Gesneriaceae): a new genus endemic to N. Queensland. *Journal of the Arnold Arboretum, 55:* 431–434.

GOTTLIEB, L.D., 1984. Genetics and morphological evolution in plants. *American Naturalist, 123:* 681–709.

HALLÉ, F. & DELMOTTE, A., 1973. Croissance et floraison de la Gesnériacée africaine Epithema tenue C.B. Clarke. *Adansonia sér. 2, 13:* 273–287.

HARLAN, J.R., 1982. Human interference with grass systematics. In J.R. Estes, R.J. Tyrl & J.N. Brunker (eds), *Grasses and Grasslands*, pp. 42–47. Norman: University of Oklahoma Press.

HILLIARD, O.M. & BURTT, B.L., 1971. *Streptocarpus.* Pietermaritzburg: University of Natal Press.

HILU, K.W., 1983. The role of single-gene mutations in the evolution of flowering plants. *Evolutionary Biology, 16:* 97–128.

JACKSON, L.L., DEWALD, C.L. & BOHLEN, C.C., 1992. A macromutation in *Tripsacum dactyloides* (Poaceae): consequences for seed size, germination and seedling establishment. *American Journal of Botany, 79:* 1031–1038.

JACKSON, R.C. & DIMAS, C.F., 1981. Experimental evidence for systematic placement of the *Haplopappus phyllocephalus* complex (Compositae). *Systematic Botany, 6:* 8–14.

JONG, K., 1978. Phyllomorphic organization in rosulate *Streptocarpus. Notes from the Royal Botanic Garden, Edinburgh, 36:* 369–396.

JONG, K. & BURTT, B.L., 1975. The evolution of morphological novelty exemplified in the growth patterns of some Gesneriaceae. *New Phytologist, 75:* 297–311.

STEBBINS, G.L. & YAGIL, E., 1966. The morphogenetic effects of the hooded gene in barley, I. The course of development in hooded and awned genotypes. *Genetics, 54:* 727–741.

THEOBALD, W.L., 1971. Comparative anatomical and developmental studies in the Umbelliferae. In V.H. Heywood (ed.), *The Biology and Chemistry of the Umbelliferae*, pp. 177–197 (Supplement I to the Botanical Journal of the Linnean Society 65). London, U.K.: Academic Press.

TUCKER, S.C., 1992. The role of floral development in studies of legume evolution. *Canadian Journal of Botany/Revue canadienne de botanique, 70 (4):* 692–700.

TURRILL, W.B., 1940. Experimental and synthetic plant taxonomy. In J.S. Huxley (ed.), *The New Systematics*, pp. 47–71. Oxford, U.K.: Clarendon Press.

VAVILOV, N.I., 1922. The law of homologous series in variation. *Journal of Genetics, 12:* 47–89.

VAVILOV, N.I., 1949. The law of homologous series in the inheritance of variability (translated by K. Starr Chester). *Chronica Botanica, 13:* 54–94.

WARE, D.M. EGGERS, 1983. Genetic fruit polymorphism in North American Valerianella (Valerianaceae) and its taxonomic implications. *Systematic Botany, 8:* 33–44.

CHAPTER

8

The representation of shape and form by computer

R.J. PANKHURST

CONTENTS

Abstract
As computerized taxonomy has gained ground, it has been gradually realized that the lack of sufficient organization in taxonomic data is itself a problem, let alone the questions of processing such data. An account is given of progress in data processing of plant morphological descriptions. The fitting of polynomial functions to shapes is then discussed, and then finally the use of Lindenmayer systems for modelling of patterns of plant growth. The relation of these techniques to the biology of plant development and to classification is discussed, and this provides a novel insight into the relation between plant development and computation.

INTRODUCTION

The classification and identification of plants or animals depends very largely on macroscopic characters, such as structure, shape and colour. Characters that require sophisticated equipment, even nothing more than a microscope, are used much less often. This is even more true for characters depending on modern techniques such as electron microscopy or molecular studies. This is probably nothing more than a matter of practical convenience. Measurements and statements of size are also used as characters, but sparingly. This is perhaps because such characters tend to be even more variable than the qualitative ones. With the rise of statistical techniques during

Shape and Form in Plants and Fungi
ISBN 0–12–371035–9

the 20th century, increased use of numerical characters might have been expected, but this does not seem to have generally been the case. There is a sense in which all characters are capable of being expressed numerically, even though it may not be practical to do so. Since the estimation of qualitative characters is often a subjective matter, so that different observers may obtain different results, it is also to be expected that quantitative versions of characters would be more objective. Description of colours, for example, could be expressed in terms of precise physical measurements of hue, chroma and reflectivity. Descriptions of shapes, e.g. of the leaves of plants, normally expressed with terms such as 'ovate' or 'cordate', can also be expressed as mathematical formulae. Nevertheless, it is also possible to give greater precision to purely qualitative descriptions.

Traditionally character data are collected by a human observer, but it is also possible to collect it via computer-controlled measuring devices or computerized images. Numerical data may be collected directly by such means, and it is also possible to recognize qualitative character states via a process of pattern recognition known as feature extraction. These methods will not be discussed in detail, but for more information, see the recent reviews by Rohlf (1993a, b).

Three quite different approaches will now be considered: (1) codification of plant descriptions with characters; (2) approximation of shapes with polynomial functions; and (3) repeated substitution of strings of symbols. In each case, the biological significance of the representation will be considered.

CHARACTERIZATION

Computerized schemes for the description of characters of plants and animals have been in use since the beginning of the 1970s. The characters used have been mainly macroscopic morphological characters. The data format used for this is called DELTA (Dallwitz, 1980, for DEscription Language for TAxonomy) and is now an official international standard. Given that the data are written in this formal manner, programs are available in CONFOR (Dallwitz, 1980) and PANKEY (Pankhurst, 1986) for producing the equivalent as written text, for the construction of keys, for various other identification aids and for numerical analysis. For more information, see the DELTA manual (Dallwitz, 1980) and a textbook on taxonomic computing (Pankhurst, 1991). A brief explanation of DELTA is given below.

Figure 1 shows an example of a file in DELTA format. The data has a TITLE, and then follow definitions of all the characters after the line CHARACTER DESCRIPTIONS. The characters and their states are numbered in a convenient sequence. Every time a string of letters occurs as part of the data, e.g. 'absent', this is terminated with a solidus (/). The first character shown is a qualitative character with two states, the presence of the stem. Comments are permitted and are enclosed with sharp brackets, e.g. <presence>. Character 2 is a real quantitative character and does not need a list of states, but it does have units (cm). Character 14 is an integer quantitative character. The DEPENDENT CHARACTERS are very important because the computer could not otherwise know what are the logical rules that connect characters and control when certain characters may or may not occur; in the example, 1,1:2–4 means that if character 1 (stem presence) is equal to state 1 (absent) then characters 2 through 4, which are various properties of the stem, are impossible. After ITEM DESCRIPTIONS a

```
*HEADING
                    JURINEA TEST/

..
..

*CHARACTER DESCRIPTIONS
#1. Stem ⟨presence⟩/
 1. absent/
 2. present/

#2. Stem ⟨height⟩/ cm/

..
..

#14. Capitula ⟨no.⟩/

..
..

*DEPENDENT CHARACTERS 1,1:2–4:12–13

*ITEM DESCRIPTIONS

#1.J. linearifolia/
 1,2 2,12–40 3,3 4,2 5,1 6,1 7,2/3 8,1 10,2 11,V 12,1 13,1 15,0.5–1.8
 16,1 17,1 18,1 19,5/6/7 20,3.5–4.5 21,2 22,1 23,3 24,2.5–3.5

..
..

*END
```

Figure 1 Brief example of DELTA data file for *Jurinea*.

description of *Jurinea linearifolia* DC. is given. Each group of numbers represents a character (followed by a comma) and the state(s) which go with it. For example, 3,3 means that character 3 (stem leaf distribution) has state 3 (leafy throughout). Similarly, 19,5/6/7 means that character 19 (colour of external bracts) is variable and has states 5, 6 or 7 (various shades of pink through purple), and 20,3.5–4.5 means that the fruit (strictly, the achene) is 3.5 to 4.5 mm long.

An attempt to apply this formalism to describe the diversity of an actual flora was described by Pankhurst (1983). Fifty taxa were chosen, each from a different taxonomic order, from the flora of Britain (about 1700 species) in order to represent as much diversity as possible. This resulted in a DELTA data file with 320 characters and 520 rules for dependent characters, in a hierarchy of nine levels. A selection of these rules is shown in Fig. 2. The characters were selected in order to generalize the description of plant variation at all taxonomic levels, and to express the significant known differences, and therefore necessarily included many of the characters used in angiosperm classifications. To this extent, the characters do have some taxonomic significance. Whether the characters are properly homologous or not is difficult to

Level		Character state	Depends on character no. (at level no.)
0	8 64	stem (any kind) = present and leaves = present	—
1	13	aerial stem (not rhizome or stolon) = present	8(0)
2	14	stems (aerial) = all alike (not different sterile and fertile)	13(1)
3	42	stem branches = present	13(1), 14(2)
4	69	branch leaves = present	64(0), 42(3)
5	93	branch leaves = simple (i.e. not compound)	69(4)
6	106	branch leaves cut to 1st order or cut to 2nd order	93(5)
7	110	branch leaf 1st order segment depth = half to three-quarters or more than three-quarters	106(6)
8	107	(arrangement of) branch leaf 1st order segments = pinnate	110(7), 106(6)

Figure 2 A sequence of conditional characters.

establish, but they are homologous in the sense that they have been made comparable, in as far as this is possible. Some difficulty was experienced in formulating the characters in any way at all which would permit a single description scheme to apply to them all and to find a workable scheme of dependencies. In this sense, the characterization of these taxa is not arbitrary. Evidently other character schemes could be constructed for the same purpose, but it is not yet known how much alternative schemes might differ, or whether the dependency structure has any intrinsic significance.

Another example of an application of the DELTA format is a study on the form of inflorescences (Pankhurst, 1984). The traditional descriptors for inflorescences are not merely a set of technical terms but involve the use of a classification of forms. Various names are used for the recognized forms, such as a raceme. It has long been recognized that the classification does not cover all the forms that actually occur. Similar remarks might also be made about the description of fruit types. Two examples of inflorescences which do not fit the classification are found in two common British plants, *Geranium molle* L. and *Malva sylvestris* L., shown diagrammatically in Fig. 3. The Geranium inflorescence is a simple one, but of a type that has no name. It might be regarded as a raceme below and as a spike above, except for the fact that the flowers occur everywhere in pairs. The *Malva* inflorescence might be regarded as a panicle, except that the flowers usually occur in fascicles rather than singly, and it likewise has no name. In the British plant project, about 5% of species had unclassifiable inflorescences. One way to make it possible to describe all inflorescences would be to replace the classification with a set of more detailed characters, as shown in Table 1, where some of the familiar inflorescence types are coded. This approach results in a long list of characters, but does at least enable the actual forms to be described accurately. A much more thorough study of inflorescences with a typological approach (Weberling, 1989), although excellent in itself, does not help much with

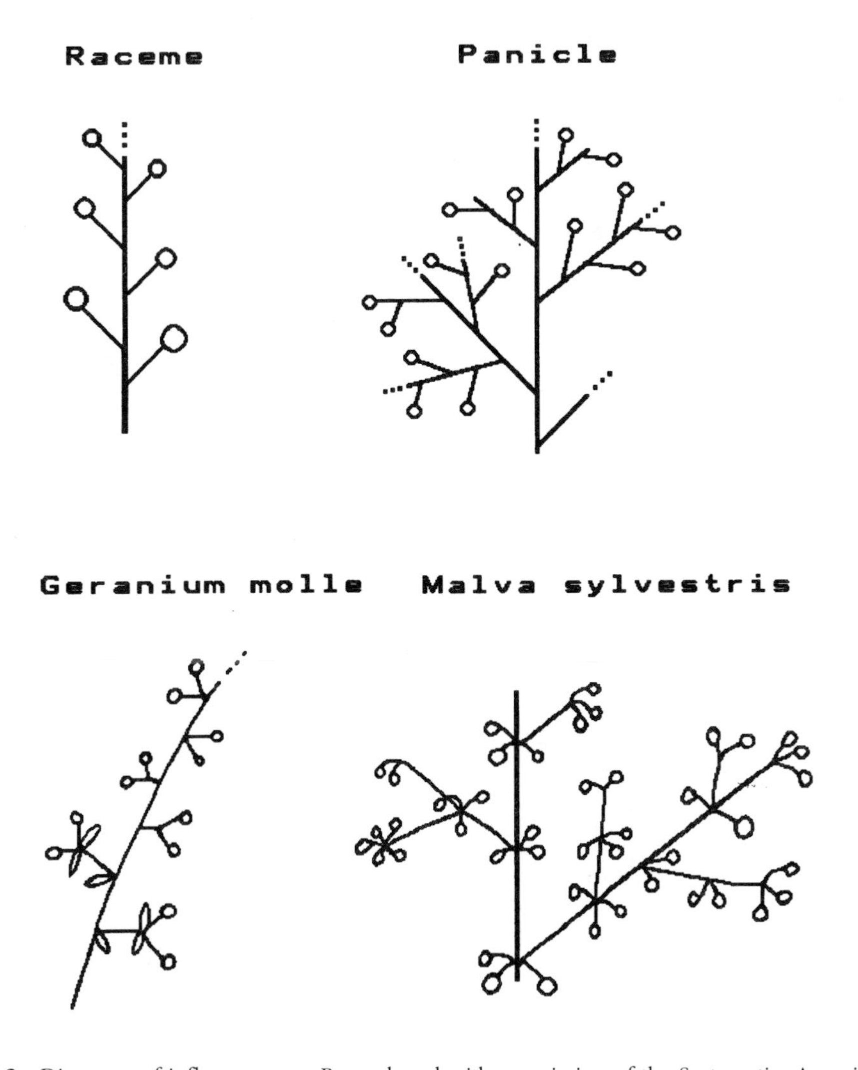

Figure 3 Diagrams of inflorescences. Reproduced with permission of the Systematics Association.

the description of actual inflorescences since it categorizes them not by their actual form, but by the way that they are known or presumed to have developed. Another way to rectify this situation might be to define new types of inflorescence. No extended scheme of inflorescence description is in general use, and were it to be proposed it might well not find favour, on account of the large number of new forms that would have to be created. An example of where this second approach has been helpful is the study on the architecture of tropical trees by Hallé *et al.* (1978). Figure 4 shows a form of tree branching called Rauh's model. This corresponds to a logical scheme of classification, so that models can be identified by a diagnostic key. This particular model is described as being monopodial and rhythmic with lateral inflorescences. Do these character-based descriptive schemes have any relation to the underlying processes of development? Possibly, but this question will be considered later.

Table 1 Inflorescence characters

	General							First-order branches					
	Limited or unlimited	No. of highest definite order	Order no. of first flowers if indefinite	Shape	Flowers stalked	Main axis terminal flower	Axis straight or curved	Branches present	Flowers present	Clusters present	Branching type	Monopodial branching type	Sympodial branching type
Raceme	L	1	X	E	+	–	St	+	+		M		X
Panicle	U	X	1	E	+	–	St	+			M		X
Corymb				FT		–							
Cyme				FT		+							
Umbel				FT	+	–		+		+	S	X	F
Spike or spadix	L	1	X	Cy	–	–	St	+	+	–	M		X
Concinnus + helicoid cyme						+	C	+	+		M	Os	X
Scorpioid cyme						+	C	+	+		M	A	X
Scape	L	0	X		+	+		–					
Dichasium	U	X	1			+		+	+	–	S	X	Di
Capitulum	L	1	X		–	–			+	–	S	X	H

+, present; –, absent; X, inapplicable; A, alternate; C, curved; Cy, cylindrical; Di, dichasial; E, elongated; F, fascicle; FT, flat-topped; H, head; L, limited; M, monopodial; Os, one-sided; S, sympodial; St, straight; U, unlimited.

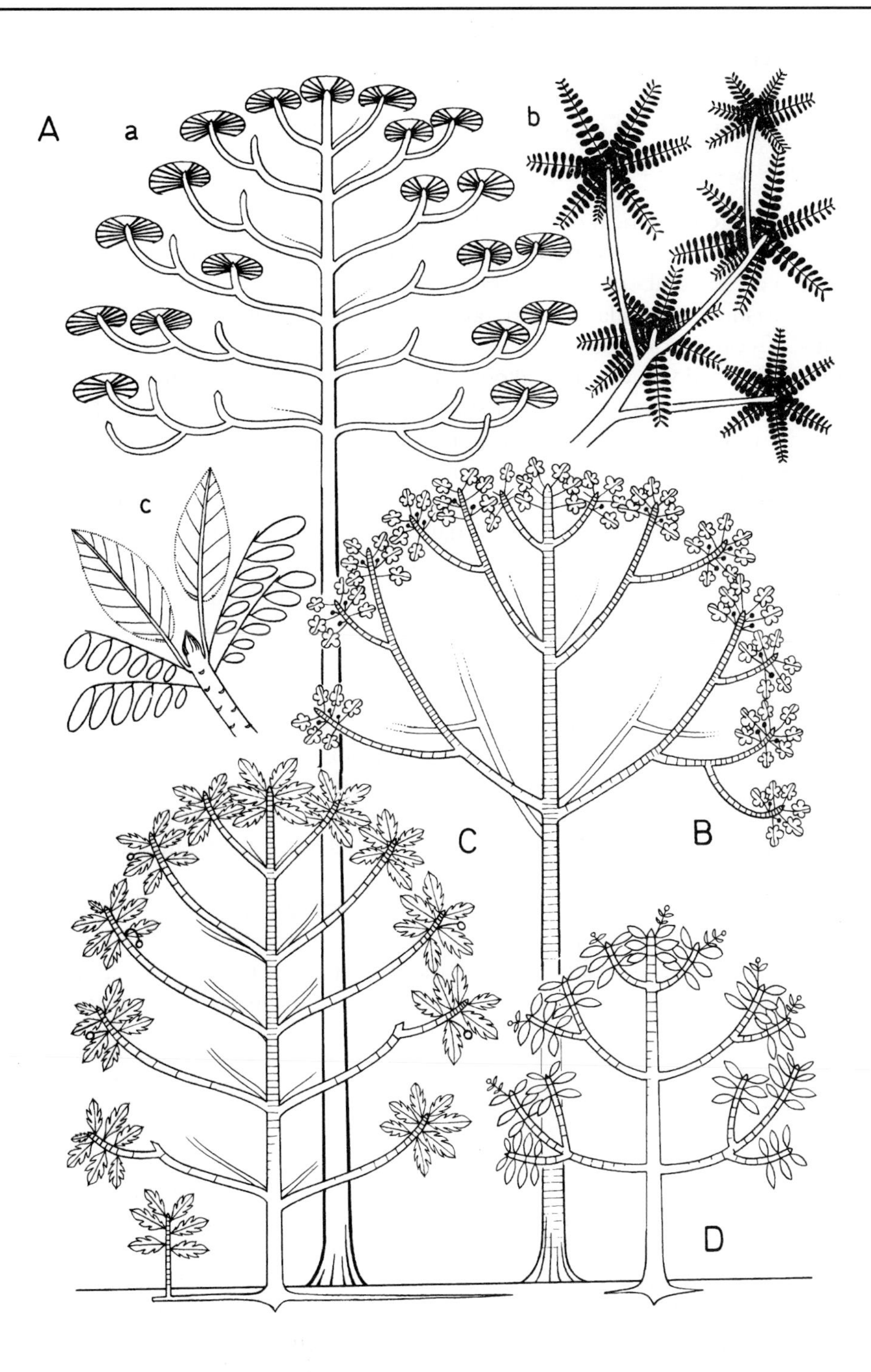

Figure 4 Tree branching pattern, Rauh's model. Reproduced with permission of Springer-Verlag, Prof. Tomlinson and Dr Hallé.

POLYNOMIAL APPROXIMATIONS

A polynomial function of order n of a variable x may be written as:

$$f_n(x) = a_0 + a_1x + \ldots + a_nx^n$$

where the a's are numerical coefficients. If the values of the coefficients can be found to make the function fit to the boundary of some part of a plant, e.g. the shape of a leaf, then these numbers can be used as a basis for some kind of numerical analysis, such as measures of similarity, correlation or principal components analysis. This is a means of converting qualitative shapes into some quantitative measures instead. The functions used will often be alternately even and odd functions of x as the value of n increases. An even function is symmetric about zero, and an odd function is antisymmetric about zero. These two kinds of function then serve to capture, respectively, the symmetric and the antisymmetric features of the actual shape. The functions used are often orthogonal functions, which is a property that enables the coefficients to be calculated. Orthogonality will usually mean that if f_n is multiplied by f_m and integrated over a range of interesting values, e.g. the length of a leaf, then the result is zero unless $n = m$. The Fourier series is a very well-known function of this kind, where

$$f_n(x) \text{ is } \sin nx \text{ or } \cos nx$$

The expansions of sin and cos as functions of x are infinite, and so it is usual to choose arbitrarily some maximum value of n as a stopping point. A biological example of the fitting of a Fourier series is taken from Mou and Stoermer (1992), applied to the shape of the valves of diatoms of the genus *Tabellaria* and is shown in Fig. 5. In these examples the maximum value of n is 20, and it appears that the rate of improvement of fit falls off very rapidly for higher values of n.

Another type of polynomial suitable for curve fitting is the Legendre polynomial. This is again an orthogonal polynomial, but of finite length. Calculation of different orders of polynomials is made easier by the existence of a variety of different recurrence relations between polynomials of one order and the next. For example, some lower order polynomials are:

$$P_0 = 1$$
$$P_1 = x$$
$$P_2 = 1/2\ (3x^2 - 1)$$

Examples of fitting with Legendre polynomials are taken from Stroemer and Ladewski (1982) applied to the shape of the diatom *Didymosphenia geminata.* Figure 6 shows a sequence of the shapes of polynomials of orders 0 to 9, and the even and odd alternation of the successive polynomials is evident. Figure 7 shows successive approximations to the valve of a diatom (from Patrick & Reimer, 1975). Once again, the rate of improvement of fit falls off rapidly for higher n.

Quite a different kind of mathematical function which might be fitted to biological shapes is the fractal. Fractals have the property that the detailed shape of the function is repeated again and again at higher degrees of approximation. The famous example of this is a natural coastline, whose total length seems to increase indefinitely when approximated with higher degrees of accuracy. Fractals also have the property that, although the shape generated may be complex, its numerical representation may be

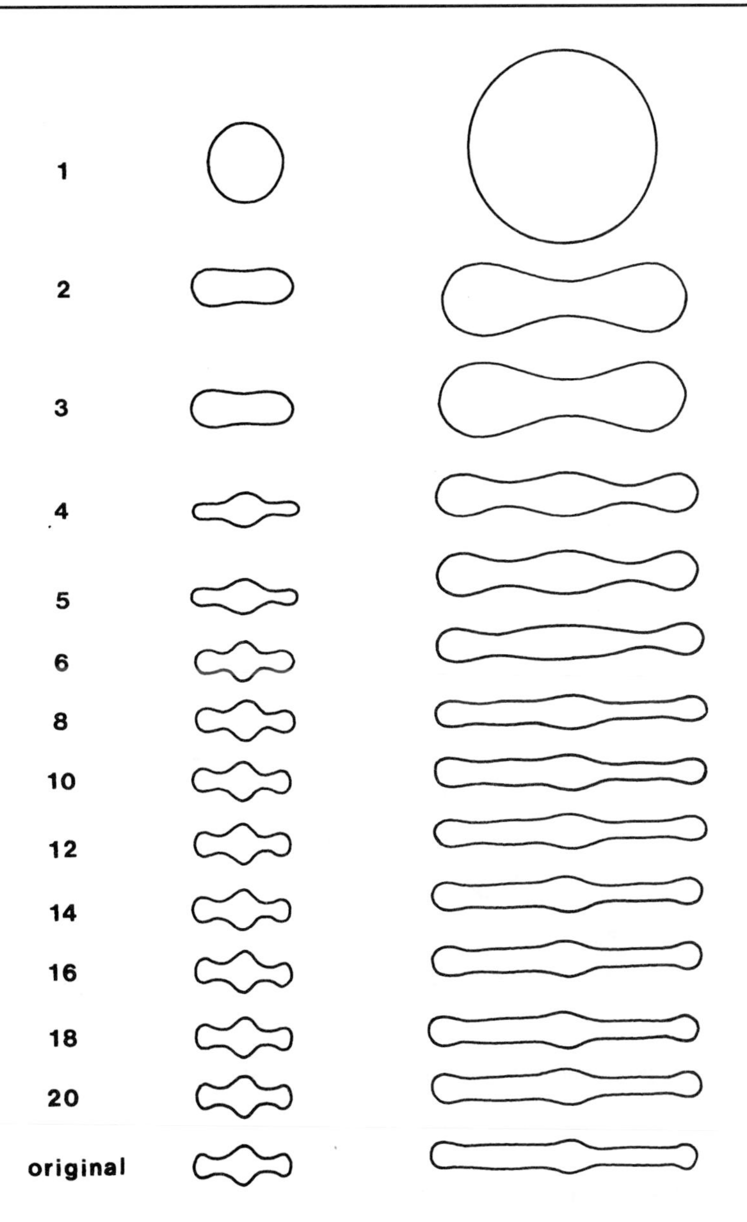

Figure 5 Successive reconstruction of *Tabellaria* valves. Reproduced with permission of Dr Stoermer.

very simple. However, it seems that most simple shapes in biology, such as those just described, are not fractal in nature, so that fractal approximations are inappropriate. This will be discussed again later.

It is now appropriate to ask whether the polynomial methods of approximation have any biological significance. No indication of this has been seen in the literature and it seems inherently unlikely. These numerical methods are useful, for example, for practical problems, such as quantifying populations of organisms so that they may be compared and distinguished.

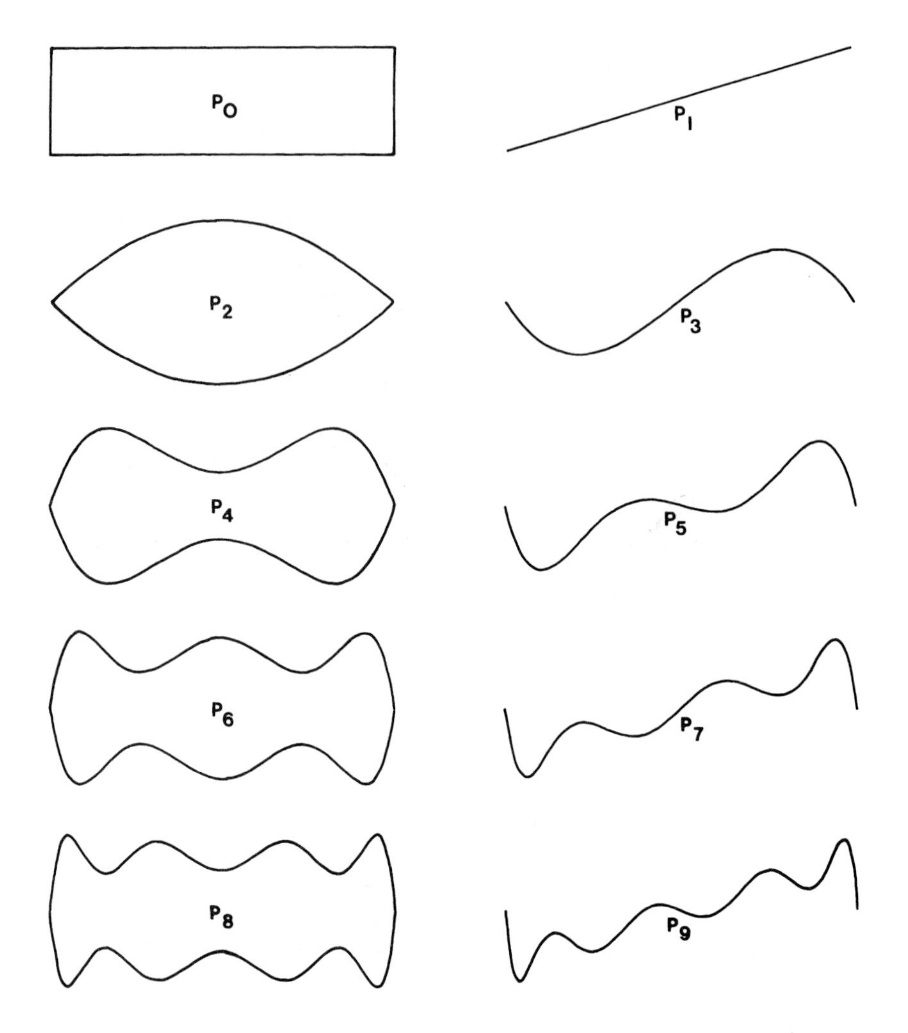

Figure 6 Legendre polynomials of order 0 to 9. Reproduced with permission of Dr Stoermer.

LINDENMAYER SYSTEMS

Lindenmayer systems or L-systems (Lindenmayer, 1968) are string rewriting systems. In computing terms, a string is a sequence of symbols (letters or numbers or punctuation) of any length. A rewriting system consists of a starting string (called an axiom) and a series of rules (production rules) for substituting symbols for one or more other symbols. The symbols do not have to represent anything, but a worked example from Garbary and Corbit (1992) models the growth of linear cells in a plant which branches from a central axis.

Suppose that *F* stands for a cell
f stands for a join between cells

letters *a b c d* stand for (invisible) growth points
\+ means branch to the left

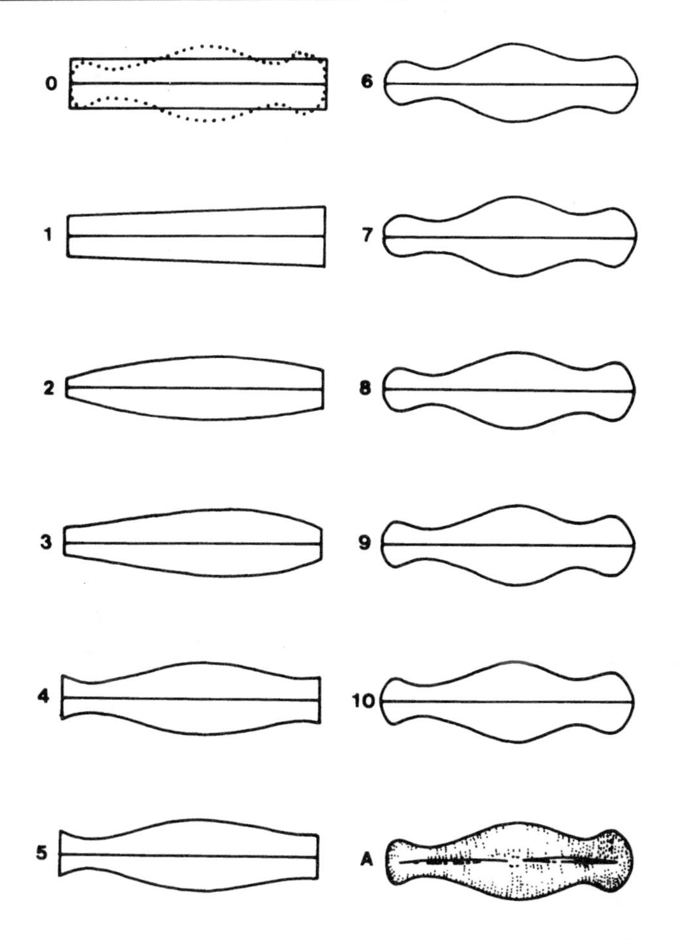

Figure 7 Reconstruction of the shape of *Didymosphenia*. Reproduced with permission of Dr Stoermer.

– means branch to the right
[means start a branch
] means stop a branch

Take as an axiom *aF*
i.e. a single cell with a growth point *a* and four rules:

1	$a \rightarrow bFf$	subapical cell
2	$b \rightarrow Fd[+fFd]fc$	segment with left branch
3	$c \rightarrow Fd[-fFd]fb$	segment with right branch
4	$d \rightarrow Fd$	cell elongation

These rules are applied repeatedly in the order given. If a rule has no effect then it is ignored; just pass on to the next one. The expansion goes as follows:

axiom 0 *aF*
1 *bFfF*
2 *Fd[+fFd]fcFfF*
3 *Fd[+fFd]fFd[–fFd]fbFfF*

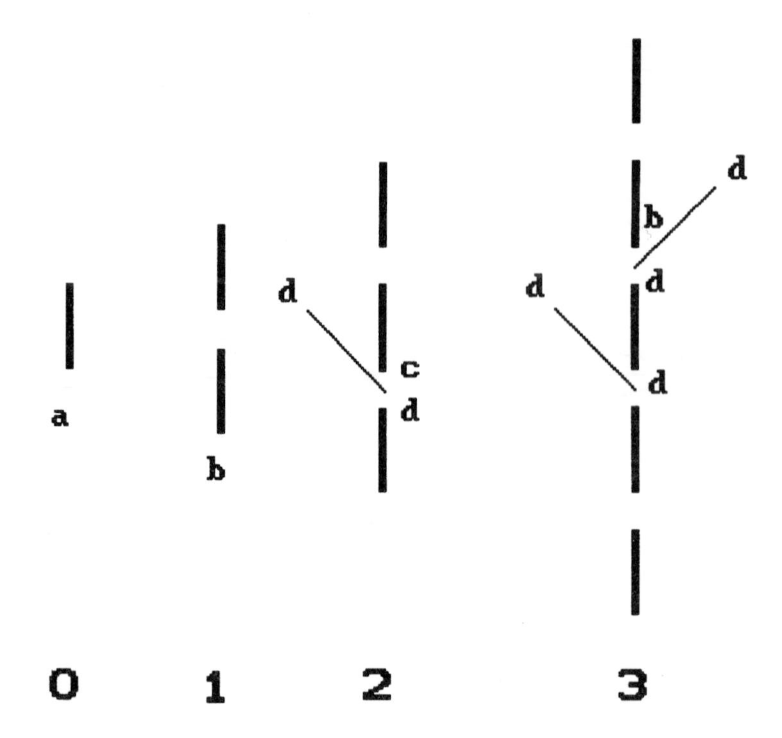

Figure 8 Growth of an alternately branched plant.

The diagrams in Fig. 8 show this same process, with the lines representing the linear cell *F* and the gaps between the lines the intercellular space *f*. A more realistic example is given by Garbary and Corbit (1992) for the red alga *Antithamnion tenuissimum* (Hauck) Schiffner. Figure 9 shows a specimen of this species and Fig. 10 shows a

Figure 9 Specimen of *Antithamnion tenuissimum*. Reproduced with permission of Dr Garbary.

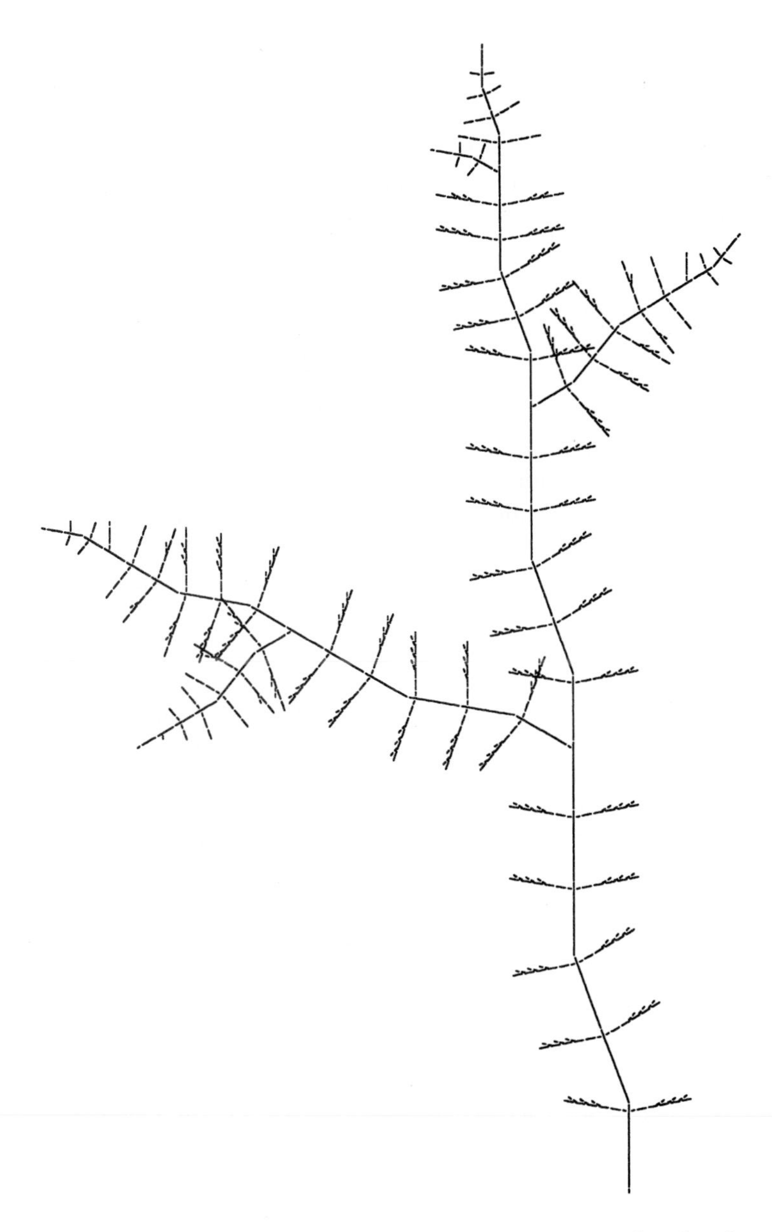

Figure 10 Model of *A. tenuissimum* generated by an L-system. Reproduced with permission of Dr Garbary.

computer-drawn model based on a sequence of 31 production rules. Further examples can be found in Schneider and Walde (1992) and Corbit and Garbary (1993). The operation of sets or production rules such as the above is best carried out with a computer. Any word processing program on a desktop computer could also be used to do this, just by copying and substituting strings by pasting.

The L-system just described is operating linearly in two dimensions. Other more complex systems permit three dimensions and other kinds of growth, such as leaf surfaces and the growth of cells, and not just for axial systems. It is also possible to allow for context dependence, where the operation of rules depends on the symbol(s) found before and/or after the symbol being substituted. For example, a stem would not be able to produce a flower until a certain number of internodes had been reached. Probabilities can be introduced, so that each rule has some probability of being obeyed. This allows for the random variation of plant structures. L-systems can be parametrized, i.e. can have numerical arguments set into them, to allow for structures with different proportions but still using the same overall plan. In short, models for a very wide range of plant structures have been modelled using L-systems. These are clearly described and well illustrated by Prusinkiewicz and Lindenmayer (1990). The modelling of herbaceous plants has been pursued further than that of trees. Trees require more complex models, since the interactions between different branches (context dependence) and between the plant and its immediate environment are more important. Nevertheless, the tree structures of Hallé *et al.* (1978) have been successfully modelled with L-systems.

The basic idea of the L-systems, the substitution of symbols in strings, has fundamental significance in mathematics and computer science. Turing (1936) gave mathematical expression to the idea of an automaton, which can be visualized as a simplified computing machine, Fig. 11. This machine has a single read/write head which passes over a tape. The tape is infinite in length and has sections or cells in which data can be stored, read or written. It is sufficient for the data to be stored in binary, as zeros or ones. The head can read or write the contents of a cell and decide to move the tape in either direction. Turing proved mathematically that this simple model of a computer is capable of all possible computations. In addition, Post (1943) showed that the process of substituting strings with production rules was exactly equivalent to the operation of Turing's automaton. A readable account of the mathematical theorems is provided by Minsky (1967). In fact, there are other possible models of computation as well, such as the neural networks that have recently been given a good deal of publicity. The actual implementation of personal desktop computers with their integrated circuits, memory chips and magnetic discs is also, of course, equivalent to an automaton, except that it is finite. Hence, the L-systems are much more than an ingenious way of modelling plant growth; L-systems are totally equivalent to computer programs or algorithms.

Something rather like L-systems occurs in computer science for the specification of computer languages, such as FORTRAN or C. The starting unit is the program, which is broken into subroutines, which are subdivided into statements, which may be of different kinds, each with their own rules, and so on. Another way of saying the same thing is that these are the rules of syntax for a grammar. There is also an analogy between grammars for computer languages and grammars for human languages, although human language is far more complex.

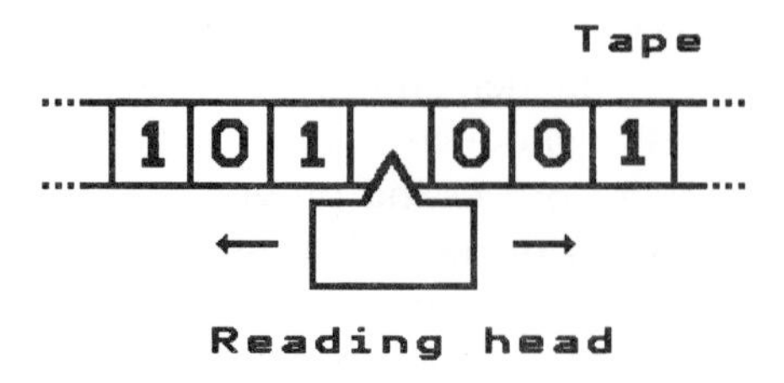

Figure 11 Diagram of a Turing machine.

Biologists will hardly need to be reminded of the analogy with the genetic code, with the letters ACGT and the substitution and replication of these in DNA, and of the enormous potential of this scheme for expressing development and variation. It is worth mentioning at this point that speculation has begun on the subject of nanotechnology, i.e. the possibility that computers could be built at a very small scale from organic molecules (Drexler, 1987). States of memory and the transmission of signals would be represented by the physical movement of molecules. The attraction of this is that, although such physical movement is about a hundred times slower than electronic circuits, the overall scale is about a million times smaller, giving the possibility of great increases in speed of computation.

Given that L-systems can be found that will provide concise models for patterns of growth in plants, it is reasonable to ask, given a pattern, how to find an L-system to model it. Prusinkiewicz and Lindenmayer (1990) discuss this briefly and say that the situation is not very satisfactory. Most of the models published so far seem to have been found by trial and error, and by the elaboration of simpler models. Another question of importance to taxonomists is whether the L-system model for a particular pattern is unique. If so, then the features of the L-system could be used as taxonomic characters. None of the literature reviewed here makes any comment on this, but given the equivalence of L-systems with computer algorithms, uniqueness seems unlikely. It is well known that there are occasions when different algorithms can be written to solve the same problem. It is also well known that organisms with different genetic codes and different ancestries can produce similar responses, at least in their external gross morphology, to the same environment, although this presumably is a weaker analogy.

Mention was made above of the possibility of using fractals as approximations to shapes, and it should be mentioned that some L-systems, although not all, can behave like fractals. Some kinds of plant growth, such as the leaf patterns of ferns, e.g. *Dryopteris* spp., and umbellifers such as *Anthriscus* spp. involve the repetition of the same pattern at smaller scale. Fractals have the property of self-similarity, where the overall pattern is repeated at a smaller size at finer levels of detail. Fractals, however, are infinite, whereas in plants the process of subdivision is, and has to be, finite. There is also an interesting parallel with the mathematics of space-filling, non-intersecting and non-overlapping curves, of which the ice crystal in a snowflake is the best-known example. These properties are natural biological constraints, since patterns such as the leaves just mentioned cannot overlap and cannot grow except into free space.

CONCLUSIONS

The successful application of L-systems to the modelling of many patterns and different kinds of plant growth, and their equivalence to computer algorithms, seems to be of great significance. It is likely that the fundamental mechanisms of plant growth will be found to be equivalent to computer processors, and that the processes of development will be algorithmic. The analogy to models of computation need not be taken too literally, certainly not in terms of the hardware of modern desktop microcomputers. Nevertheless, it seems plausible that, by analogy with L-systems, plant development will be found to be implemented as substitutions of groups of molecules by sets of rules.

ACKNOWLEDGEMENT

Thanks are made to Dr David Mann for bringing to the author's notice various papers on fitting shapes and L-systems to algae.

REFERENCES

DALLWITZ, M.J., 1980. User's guide to the DELTA system. *Report No. 13.* CSIRO Division of Entomology, P.O. Box 1700, Canberra City, ACT 2601, Australia.

DREXLER, K.E., 1987. *Engines of Creation: The Coming Era of Nanotechnology.* Oxford: Fourth Estate.

CORBIT, J.D. & GARBARY, D.J., 1993. Computer simulation of the morphology and development of several species of seaweed using Lindenmayer systems. *Computers & Graphics, 17:* 85–88.

GARBARY, D.J. & CORBIT, J.D., 1992. Lindenmayer-systems as models of red algal morphology and development. In F.E. Round & D.J. Chapman (eds), *Progress in Phycological Research*, vol. 8, pp. 143–77. Bristol: Biopress.

HALLÉ, F., OLDEMAN, R.A.A. & TOMLINSON, P.B., 1978. *Tropical Trees and Forests. An Architectural Analysis.* Berlin, Germany: Springer-Verlag.

LINDENMAYER, A., 1968. Mathematical models for cellular interaction in development. *Journal of Theoretical Biology, 18:* 280–315.

MINSKY, M., 1967. *Computation, Finite and Infinite Machines.* N.J., U.S.A.: Prentice-Hall.

MOU, D. & STOERMER, E.F., 1992. Separating *Tabellaria* (Bacillariophyceae) shape groups based on Fourier descriptors. *Journal of Phycology, 28:* 386–395.

PANKHURST, R.J., 1983. The construction of a floristic database. *Taxon, 32:* 193–202.

PANKHURST, R.J., 1984. On the description of inflorescences. In R. Allkin & F. Bisby (eds), *Databases in Systematics*, Systematics Association Special vol. 26, pp. 309–320. London, U.K.: Academic Press.

PANKHURST, R.J., 1986. A package of computer programs for handling taxonomic databases. *Computer Applications in the Biosciences, 2:* 33–39.

PANKHURST, R.J., 1991. *Practical Taxonomic Computing.* Cambridge, U.K.: Cambridge University Press.

PATRICK, R. & REIMER, C.W., 1975. *The Diatoms of the United States*, vol. 2 (1). Academy of Natural Sciences of Philadelphia, Monograph no. 13.

POST, E., 1943. Formal reductions of the general combinatorial decision problem. *American Journal of Mathematics, 65:* 197–268.

PRUSINKIEWICZ, P. & LINDENMAYER, A., 1990. *The Algorithmic Beauty of Plants.* New York: Springer-Verlag.

ROHLF, F.J., 1993a. Computer vision needs in systematic biology. In R. Fortuner (ed.), *Advances in Computer Methods for Systematic Biology*, pp. 365–373. Baltimore: Johns Hopkins University Press.

ROHLF, F.J., 1993b. Feature extraction in systematic biology. In R. Fortuner (ed.), *Advances in Computer Methods for Systematic Biology*, pp. 375–392. Baltimore: Johns Hopkins University Press.

SCHNEIDER, C.W. & WALDE, R.E., 1992. L-system computer simulations of branching divergence in some dorsiventral members of the tribe Polysiphonieae (Rhodomelaceae, Rhodophyta). *Phycologia, 31:* 581–590.

STOERMER, E.F. & LADEWSKI, T.B., 1982. Quantitative analysis of shape variation in type and modern populations of *Gomphoneis herculeana. Beiheft Nova Hedwigia, 73:* 347–386.

TURING, A., 1936. On computable numbers with an application to the Entscheidungsproblem. *Proceedings of the London Mathematical Society (Series 2), 42:* 230–265, *43:* 544–546 (1937).

WEBERLING, F., 1989. *Morphology of Flowers and Inflorescences.* Cambridge, U.K.: Cambridge University Press.

CHAPTER

9

Cell divisions in meristems and their contribution to organogenesis and plant form

PETER W. BARLOW

CONTENTS

Abstract

The meristems of plant organs are sites of intense anabolism where chromatin replicates, nuclei divide, cytoplasm is synthesized, and phragmoplasts and new walls are elaborated. The last two structures partition the continually increasing mass of protoplasm into cells and establish a three-dimensional network of walls within the growing organ. This is one of the primary functions of meristems. Two main consequences follow: the protoplasm is regionalized into units of biochemical specialization, often recognized as tissues, and a 'supporter' system is developed which resists or dissipates mechanical stresses imposed by both the external environment and the turgor-generated pressure of the internal cellular environment. The elaboration of species-specific cell wall branching patterns in the distal portion of apical meristems of shoots and roots is a means of optimizing these two consequences in the context of organ growth and function. The cell divisions that define the main tissue units are primarily those which take place during early embryogenesis; the subsequent development and

Shape and Form in Plants and Fungi
ISBN 0–12–371035–9

growth of the root and shoot poles of the embryo then perpetuate the earlier-formed differentiated states, as well as establish the future apical meristems with their distinctive patterns of division. Both histogenesis and cell wall patterning are recapitulated in neo-embryonic sites of lateral root and shoot primordia. Another possible consequence of the cellular condition is that cell walls and their associated cytoskeleton together serve as a source of epigenetic information for morphogenesis, this being the process by which the form of an organ is developed. Then, as a result of feedback processes, cell wall patterning may be modified in response to preceding morphogenetic events. On the other hand, morphogenesis need not directly involve cells, as evidenced by the distinct patterns of development in multinucleate organisms such as occur in the Cladophorales.

Morphogenesis is primarily a supracellular phenomenon that relies on directed changes in the local rates of expansion of the outer boundary layer of organs and certain tissues, the epidermis and pericycle, for example. It follows that the patterning, by epigenetic means, of cells and their walls within an organ is a secondary morphogenetic process. New cells, formed by division, fill in the new volume created by the growth of the organ and contribute to the supporter system. The siting of meristems in a growing organ is another consideration in the development of plant form. An apical location of the meristem and growth zone may be a means of securing some positional advantage for the organ within the environment. For example, actively elongating cells near the tip of an organ might provide a more efficient means of initiating a tropism than would occur if these cells were sited in intercalary or basal locations. In addition, the pattern of radial growth at the apices of roots and some shoots maximizes their penetration of resistant media such as soil.

Tropisms (and nutations) should be considered as a class of morphogenetic movements since they also result from a controlled pattern of deformations of boundary layers. Like other morphogenetic events, tropisms contribute to the optimal exploration and transformation of the environment.

SHAPE AND FORM

Shape and form are terms often used interchangeably – for example, two possible definitions from the *Shorter Oxford English Dictionary* are: 'Shape, the external form or contour', and 'Form, a body considered in respect to its outward shape' – yet in biology, perhaps more than in other disciplines, there are circumstances where it is necessary to distinguish between the two. The shape of a plant is certainly contained in its 'external contour'; the silhouette of a tree, for example, can be useful for identifying its species. In this context, shape relies on perception of features in two dimensions. To speak of a plant's form, however, indicates acquaintance with its three-dimensional structure, and perhaps even of how the individual organs are interrelated. In the present essay, which discusses the relationship of cells to plant form, this distinction between two-dimensional shape and three-dimensional form (Table 1) will continue to be made.

The shape of cells can be determined from their outline (usually polygonal) when, for example, they are seen in sections prepared for microscopy. Less easy to discern, and less frequently referred to, is the form of cells, although this topic was painstakingly investigated by a dedicated group of North American plant anatomists in the

Table 1 Elements of shape and form at three levels of plant organization, together with nominees for the supporter and boundary subsystems at each of these levels

Level	Shape (two-dimensional)	Form (three-dimensional)	Supporter	Boundary
Cell	Polygon	Polyhedron	Cytoskeleton, extracellular matrix	Plasma membrane
Organ	Silhouette	Cylinder	Cell wall network	Cuticle/epidermis
Organism	Silhouette	Architectural model	Sclerenchyma, xylem (heartwood)	Epidermis/periderm

1940s. Cell form depends on the mutual arrangement of the edges and facets of the cell walls. In the root apex of tomato, for example, polyhedral cells with 13, 14 and 15 facets are the most frequent, and the commonest arrangements of the facets involve those with six, seven or eight edges (Duffy, 1951).

Likewise, plant organs, and plants themselves, have distinctive shapes and forms. In the case of organs, numerous adjectives exist to describe their shape and are commonly used for taxonomic purposes. For example, leaves are said to be linear, lanceolate, trullate, etc. A sense of external form is conveyed by the epithets cylindrical, spherical, etc., which adequately describe the general form of many plant organs. In the case of whole organisms, the potentially complex problem of describing their shape and form is simplified by a tendency to treat root and shoot systems separately, thereby ignoring the possibility that the forms of the two systems may be related in some way. However, in a number of taxonomically unrelated temperate species no such correlation was discovered by Fournier (1979), one of the few authors to consider the unity, as opposed to the disjunctness, of the root and shoot systems in the context of the morphology of the whole plant.

The forms of shoot systems, particularly those of tropical trees, have been codified by Hallé and Oldeman (1970) in terms of a set of architectural 'models' (a translation of the French term 'modèles' [*pl.*]) which take account of branching pattern and orientation of the axes (see also Hallé *et al.*, 1978). Likewise, an attempt to classify the forms of root systems has been made by Jeník (1978), though he found it impossible to establish a hierarchical classification, as was made for shoot systems, owing to the rather small number of diagnostic features available. Again, the proposed models for roots relate to those displayed by various species of tropical trees. A more general descriptive classification of the forms of root systems was earlier proposed by Cannon (1949) which contrasts with recent topologically based classifications (Fitter, 1987) using quantifiable parameters such as 'altitude' and 'total exterior path length'. However, this latter approach refers more to the shape of a fibrous root system than it does to providing a description of its form.

In addition to the shape and form of an organism is its size. So far, there is little understanding of why different species are of different size, even though they may possess similar amounts of genetic material and may live in similar environments. Plants can sense their size (height) and use this information in the unfolding of their developmental pathway, particularly in relation to flowering (McDaniel, 1992). Presumably, there is some means of titrating, at sensitive points in the plant, the amounts of mobile substances that act as morphogens (see Khait, 1986). An associated property is an ability to regulate the size of each organ in relation to the size of the whole; this is the problem of biological 'scaling' discussed by Prothero (1986). It is also a topic that is rarely discussed in relation to plants, yet instances where the same organ can vary greatly in relation to the size of the organism (e.g. the variation in cotyledonary size mentioned in Chapter 7 of this volume by B.L. Burtt, and also the facultative anisophylly displayed by the Gesneriaceae (e.g. Dengler & Sanchez-Burgos, 1988)) or, on the contrary, where organ sizes vary coordinately, as in bonsai and other miniaturized plants versus their full-sized counterparts, suggest that there are some interesting mechanisms of growth regulation at the level of the whole plant still awaiting discovery.

Unfortunately, space precludes discussion of these important topics. In what follows, the emphasis will be on the relation of meristems and their constituent cells to the development of individual organs within the root and shoot system, the latter being comprised of stems and leaves. Many details of the shape and form of root and shoot systems, and the way in which meristems contribute to them, have been discussed by Barlow (in press), while Barthélémy (1991) presents a useful summary of the progressive changes of shoot form that occur during the lifespan of plants.

THE CELLULAR SUPPORT OF FORM

Central to plant morphogenesis are the means whereby plant form is developed and maintained. Some forms are structurally self-supporting as a result of mutual interactions between their components. Famous examples of this principle are the geodesic domes developed by Buckminster Fuller which, interestingly, bear resemblance to the faceted external forms of unicellular radiolarians (e.g. *Aulonia hexagona* and *Circogonia icosahedra*) and some colonial algae such as *Volvox*. The forms of some pluricellular aggregates of diatoms (e.g. *Tabellaria fenestrata*) may also be determined by the degree of mutual support or adhesion between cells. However, most plants and their organs require special internal support structures to assist their much more elaborate patterns of growth, particularly as they ascend into the aerial environment. These support structures are closely allied to the continued development of form and provide one of the means by which living systems are maintained. Other support structures, such as stilt roots, might be regarded as external supports for the shoot system (Barlow, 1994). Shoots and stilt roots are examples of mutually supporting structures that cooperate in the development of plant form (*vide* the remarks earlier).

Living systems can be regarded as being hierarchically organized into 'levels' that accommodate not only the components of which such systems are made but also the communities of which they are a part. In the present context of plant life, three levels of increasing complexity are relevant, namely cells, organs and organisms. According to the analysis of J.G. Miller (1978), propounded at length in his book *Living Systems*, a set of 20 subsystems is common to, and may even be said to define, each level of organization (Miller & Miller, 1990). These subsystems are concerned with one or more of three vital functions: reproduction and the processing of matter and information. The second and the eleventh subsystems in Miller's classification are the 'boundary' and the 'supporter', respectively; both are involved in matter–information processing. At the level of the cell, the supporter is the cytoskeleton and extracellular matrix; at the level of the organ, it is the network of cell walls; while at the level of the organism, the supporter is the xylem of the vascular cylinder and specialized tissue such as sclerenchyma (Table 1). All these items comprise an internal scaffold for each of the levels and assist in the support of physical loads.

The boundary subsystem (Table 1) is particularly germane to the problem of form, for it is the boundary of organs and organisms (i.e. the cuticle and epidermis) that is perceived in the recognition of plant shape and, more importantly, is modelled and remodelled in the genesis and development of form at both these levels. Boundaries of cells (the extracellular matrix, or wall), too, are remodelled in the development of cellular form as, for example, is demonstrated by the crenulated form of leaf mesophyll cells. In this case, the modelling and remodelling of cell shape is assisted by the cytoskeletal system (Wernicke & Jung, 1992). The supporter and boundary subsystems are of especial interest, both in the context of how these subsystems are related to morphogenesis of multicellular plants (Barlow, 1993b) and also as part of an ongoing analysis (Barlow, 1987, 1993a) of how Miller's ideas apply to plant life.

Although the 20 subsystems confer a measure of autonomy on each of the three levels mentioned, the levels are nevertheless interconnected. Each level has both a structural and a functional role that are fully expressed in a triplet of successive levels such as those considered here (Barlow, 1993a). The structural role is evident from the fact that each level is composed of elements that constitute the level below. Thus, cells compose organs, and organs collectively comprise the organism. The functional role is evidenced in processes that result from the passage of information down through the levels – organism to organ, and organ to cell. Because of this dual role, each level is able not only to sustain its own growth (thus increasing the span of the level in question) but also to respond to damage by regeneration and to adapt to environmental variability through morphogenetic plasticity (e.g. sun-leaf and shade-leaf morphologies (e.g. Dengler, 1980) in response to high or low light intensities)) (Barlow, 1993a).

The significance of cells to organogenesis is manifold. The contribution of their walls to the supporter subsystems of organs and organisms has already been mentioned. The specialization of cellular products within the confines of the organ enables the elaboration of other subsystems, including that of boundary, at the organ level. Thus, the spatial differentiation of cellular properties within the organ results in the recognition of various types of tissues (e.g. epidermis, cortex – terms which, for historical reasons, often have spatial rather than functional connotations). A specialized zone that is usually differentiated towards one end of young organs is dedicated to the synthesis of new protoplasmic material consisting of nucleic acids, cytoplasm and polysaccharides. Such zones are the meristems (the 'reproducer' subsystem of the organ in Miller's (1978) terminology), the products of which are new cells. Taking place at even more restricted sites within the meristem (at least in roots) is the replication of symbiotic organelles, the mitochondria and plastids (Kuroiwa *et al.*, 1992), whose activities confer particular patterns of metabolism on the protoplasm in their vicinity.

THE CONTRIBUTION OF CELLS TO PLANT FORM

If cells, by virtue of the rigidity of their walls, contribute to the supporter subsystem of an organ, do they and their divisions also have a role in developing the form of the organ they support? It is often implicitly accepted that cells are the elements deployed in, or even responsible for, morphogenesis. However, this is to fall into the often-repeated error of supposing that cell division is the cause of growth when in actuality it is an optional accompaniment of growth. 'The cellular basis of morphogenesis', a ringing phrase, which has even been used as the title of a book (Evered & Marsh, 1989), suggests a concept that, for all its allure, should be viewed with caution by plant scientists.

For some animal scientists concerned with morphogenesis this phrase may present no particular problem; but if that is so, it might point up an important difference between the way in which form is acquired in animals and plants. In plants, as argued here, morphogenesis is principally a process that is regulated at the boundaries of organs, whereas in animals a good deal of cellular behaviour is regulated by the properties inherent to the surfaces of individual cells. Although these properties may be particularly important in regulating tissue differentiation and the cellular movements that occur during this process, particularly during embryogenesis, do they actually contribute to form of the organ or embryo in which these processes occur? It is difficult to believe that properties at the boundary (e.g. ectodermal and epidermal cell layers) of developing animal organs do not also affect their final form (e.g. feather and scale patterning on the skin of avian embryos (Dhouailly, 1984)), as well as that of the organism, as they do in plants.

Cell walls, by their prominence in histological preparations, attract attention. This has caused especial importance to be attached to cells, as perusal of nearly any textbook will show, and it would be easy to claim for them a role in plant morphogenesis that may not be truly theirs. As already indicated, and as argued later, one of the major attributes of cells is that their walls are involved in structural support for the organism whose growth is driven by the increasing mass of the cytoplasm. (This was perceived over a century ago by Herbert Spencer (1867, p. 4) who wrote that 'Increase in mass is primary . . . Increase of structure is secondary, accompanying or following increase in mass with more or less regularity'.) Morphogenesis is a phenomenon related to the surface of the organism and involves a continual modelling and remodelling of the boundary layer at the epidermis and, in the case of roots, at the pericycle boundary also.

The independence or otherwise of morphogenesis from the cellular condition has for a long while been one of the key questions in developmental biology. In the last century, Anton de Bary was scathing of the 'hegemony of the cell', a phrase he coined to lament the grip which Schleiden's cell theory had at that time on the framing of concepts of plant development (Barlow, 1982a). Even today, morphogenesis is still perceived as a supracellular phenomenon that is independent of the properties or the activities of individual cells (Kaplan & Hagemann, 1991). This view is appealing because, if combined with the concept of a hierarchical organization of the plant, it means that morphogenesis should properly be regarded as a process pertaining to the level of the organ, rather than to that of the cell. Morphogenesis thus occurs independently of cells, not because of them, even though large numbers of cells are accommodated in the organ being formed. The true significance of cells is that their walls provide support for the developing organ and compartments for its physiological functions.

Organ growth is primarily a function of an increase of protoplasmic mass in space and time, coupled with the continued elaboration of a boundary to contain that mass.

The protoplasm enclosed by this boundary is structured into metabolic domains. This is elegantly demonstrated by the hexagonal spacing of nuclei in the multinucleate algal cells of *Griffithsia pacifica* where each nucleus resides within, and apparently controls, a certain volume of cytoplasm (Goff & Coleman, 1987). The hexagonal arrangement of nuclei may be analogous to the generally hexagonal shape of cells in meristems of multicellular plants; the necessity for a regular spacing of the nuclei within the protoplasm is reinforced by a corresponding patterning of the walls. In *G. pacifica*, and evidently in multicellular plants also, nuclear chromatin replication occurs (with or without nuclear division) as a result of the growth of the cytoplasm in which the nuclei reside, and hence restores to the nucleus the ability to govern its surrounding cytoplasmic domain.

The form of an organ has nothing to do with the way in which its protoplasmic mass, or its cells, is subdivided. This is an interpretation of the somewhat enigmatic aphorism of de Bary that 'the plant forms the cells, not cells the plant'. De Bary took what might be considered to be a 'top downward' view of development, where the plant, or the organ, is the fundamental unit of growth, and that cells and their properties are simply the outcome of the propensity of the growing unit to produce phragmoplasts and then to partition and differentiate the newly created volume of cytoplasm in certain ways. Thus, in addition to any structural role that they might have, another significance of cells is that their walls subdivide the growing protoplasm into metabolic units, or tissues, whose various functions need to be separated as required by the overall developmental stratagem of the organism (Barlow, 1982b).

This partitioning into metabolic units occurs very early in postzygotic development and coincides with the differentiation of the spherical protoplasmic mass of the embryo into epidermis, plerome and periblem. The earliest divisions may also establish root and shoot domains in the more distal, proembryonic zone, and a suspensor in the proximal zone. Figure 1 shows a minimal embryonic cell lineage (for *Sphenoclea zeylanica*) upon which is superimposed the differentiation of the various histogenetic regions. The creation of histogens at this early stage is significant since all development that follows perpetuates this newly established pattern of differentiation, perhaps doing so by a process of homoeogenetic induction. This latter process, however, may be no more than the inheritance by the dividing nuclei of a particular configuration of their chromatin so that they continue to function in the same way as their progenitors.

An apparent irrelevance of cells to the acquisition of form has been argued from two lines of evidence. First, some plants are not constructed of individual cells but are one giant, multinucleate 'cell'; yet such organisms have a definite form that is reproducible from generation to generation and can be recreated after damage or inversion within a gravity field (Matilsky & Jacobs, 1983; Sinnott, 1960). Examples are found within the green algae (Chlorophyta), *Bryopsis* and *Caulerpa* being two such genera. The supporter subsystem of these giant cell organisms is comprised of a network of trabeculae connecting opposite sides of the organism. Moreover, some of these algae consist of regions (organoids) whose forms resemble those of higher plants – for example, cauloids (stems), phylloids (leaves) and rhizoids (roots) (Kaplan & Hagemann, 1991).

A boundary structure (or subsystem) is common to both multinucleate (algal) and multicellular (angiospermous) plants, and the form of each class of organism is likely to be the result of directed changes or deformations of the boundary which then

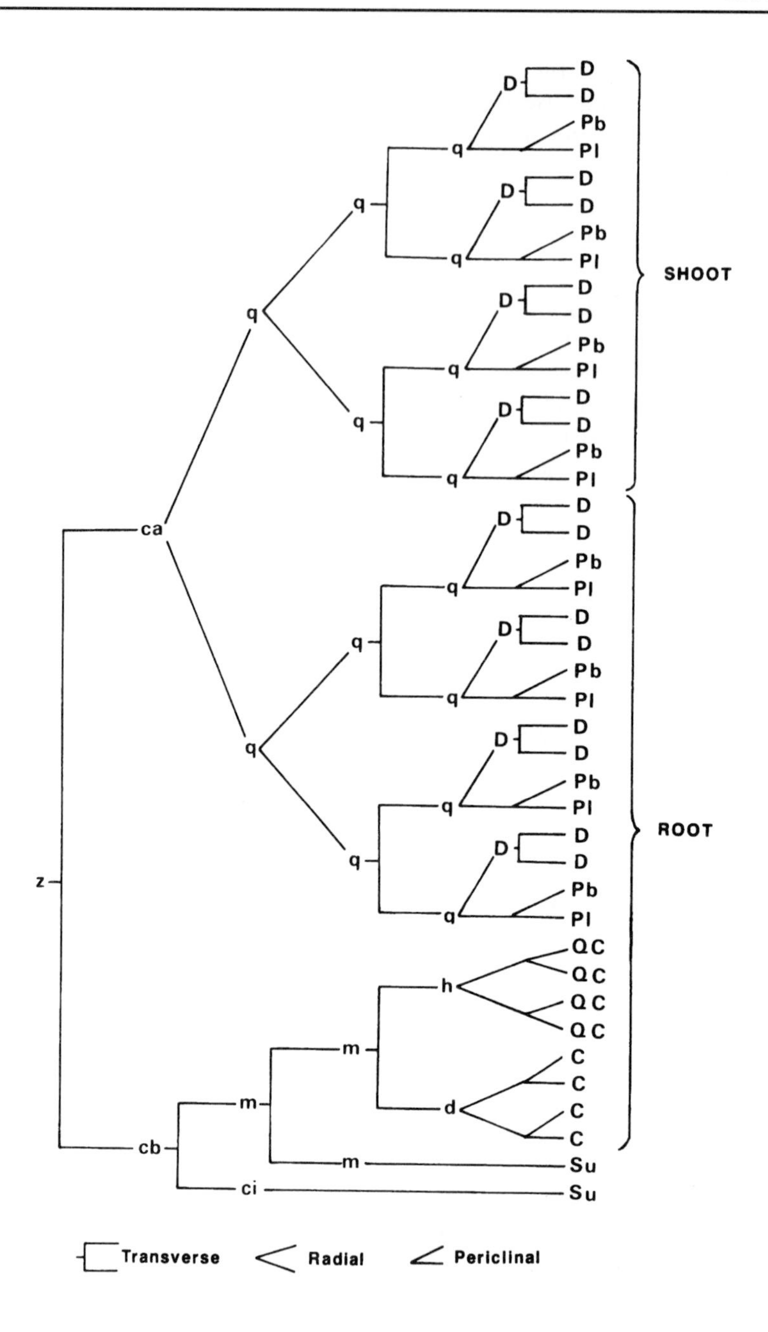

Figure 1 A cell division lineage in an angiosperm embryo that generates cell layers from which will arise the different histogens (C, D, Pb, Pl) and cell types (QC, Su), and from them, the embryonic root and shoot regions. C, calyptrogen; D, dermatogen; Pb, periblem; Pl, plerome; QC, quiescent centre; Su, suspensor. Early types of cell are designated z (zygote), ca, cb, ci, d, q, h (hypophysis) and m, letters that conform to embryological terminology (see Mestre, 1967). The scheme is based on the lineage for embryos of *Sphenoclea zeylanica* (Sphenocleaceae) described by Swamy and Padmanabhan (1961). The three types of division, transverse, radial (anticlinal) and periclinal, are indicated by the three types of branched lines, the explanations for which are shown at the foot of the diagram.

enable the out-pushing of enations that subsequently continue to develop as new organs, e.g. leaves, buds. Attention to changes in the boundary at these points of out-pushing would possibly reveal processes that are primary to morphogenesis. In this context, the earliest stages of rhizoid formation by the zygote of the brown alga, *Fucus*, represents a fruitful model system (Quatrano *et al.*, 1985).

The homology of structural forms such as phylloids and leaves suggests that all organ forms, regardless of whether they are associated with a unicellular or a multicellular state, belong to one of a number of alternative potential (essential) forms available to plants. The reduplication of similar forms in both algae and higher plants is an example of so-called mimicry, or convergent evolution, a condition familiar in the animal kingdom, though it, too, is known in plants (Went, 1971). However, although organs such as the roots and shoots of ferns, gymnosperms and angiosperms each possess similar general forms (cylindrical, with or without leafy appendages), the shape, form and size of their cells and, more importantly, the positioning, or patterning of the cell walls within the growing apices, can differ markedly from taxon to taxon. In fact, within roots, different cellular patterns are so characteristic of different taxa that it is possible to place them into some sort of evolutionary sequence (Fig. 2) (Voronin, 1969). Similar categorizations of cellular patterns in shoot apex structure have been discussed by Popham (1960). Moreover, within an ovule, embryos develop similar shapes and tissues in spite of differences in the sequence

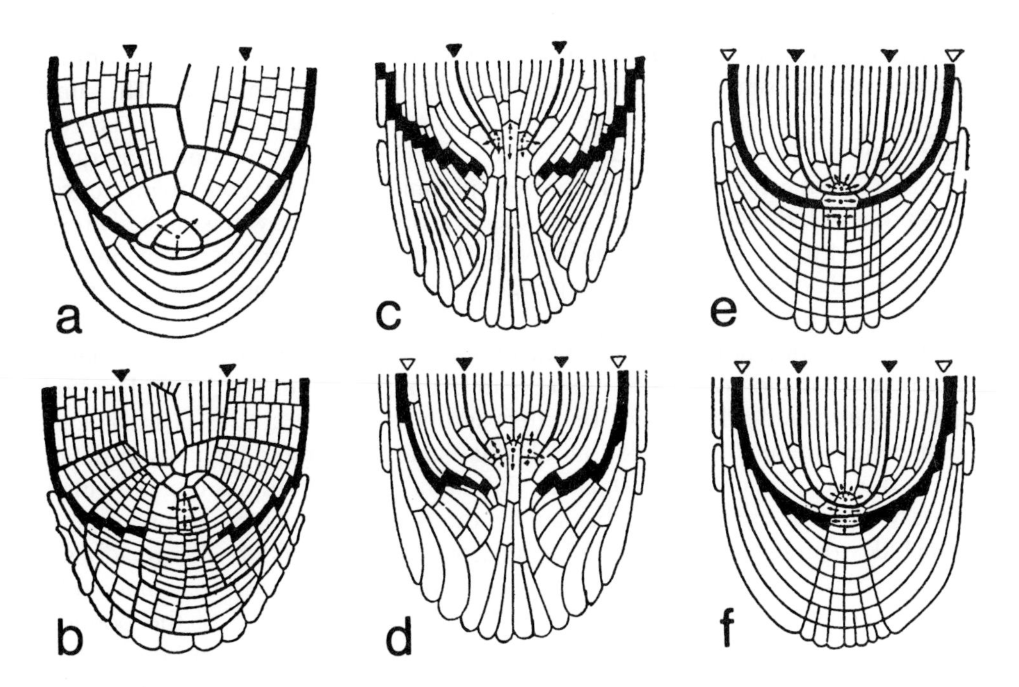

Figure 2 Schematic drawings of various arrangements of cell files that can construct root apices of different groups of plants. **a**, leptosporangiate fern; **b**, marratiaceous fern; **c**, gymnosperm; **d**, monocotyledon (open type of apex); **e**, monocotyledon (closed type); **f**, dicotyledon (closed type) Filled darts (▶) indicate the boundary between plerome and periblem; open darts (▷) indicate the boundary between plerome and calyptrogen. Heavily inked walls represent the outer boundary wall of the epidermis. In the text, these boundaries are considered to be important for morphogenesis through variation in their rate of extension. After Voronin, 1969.

with which the cell walls are inserted into the growing mass of protoplasm (Mestre, 1967). Hence the walls and the cells that comprise the embryo appear to be relatively incidental features with respect to the development of its form. The same is evidently true of the roots shown in Fig. 2.

The important events of morphogenesis are those that affect the pattern of growth at the embryo or organ surface. Accordingly, the development of the epidermis and its cuticle, which are both formed at early stages of embryogenesis (Bruck & Walker, 1985), may be important for shaping both the embryo and later stages. Moreover, cell division in roots and shoots can be abolished by treating them with colchicine (Foard *et al.*, 1965) or radiation (Foard, 1971), yet the capacity for further, though limited, organogenesis is retained. In roots of colchicine-treated wheat (and similar observations have been made on other species), certain cells of the pericycle are still able to grow and form 'primordiomorphs' (Foard *et al.*, 1965). These structures resemble lateral root primordia in size and form, yet they lack the normal complement of cells; new walls do not form and the primordiomorphs consist of a few cells each enclosing a large polyploid nucleus. Whereas, in shoots, the deformable boundary layer associated with leaf primordia is that of the outer epidermal boundary layer, the development of lateral root primordia is an example of morphogenesis being controlled by an internal boundary, in this case the boundary between the plerome and periblem.

In both shoots and roots, some process of wall loosening, associated with the presence of islands of particularly actively growing protoplasm, causes the relevant boundaries to become deformed. This in turn leads to the out-pushing of a primordium due to the osmotic pressure generated in the underlying cells. The limited amount of growth that occurs when division is experimentally inhibited, as shown by the results of Foard and colleagues (1965, 1971), suggests that one consequence of division is that genes responsible for protoplasmic growth are continually activated by this process. If division is suppressed, then the period for growth is similarly restricted because of the lack of this gene activity. Hence, mitosis and meristems may provide the conditions that permit morphogenesis since this is necessarily a growth-driven process.

An important question in an acellular or supracellular view of morphogenesis, is how a new orientation of growth at an organ boundary is controlled. But, here, what might be cause and what might be effect is unclear, as it always is in biological systems where there is feedback control (quite apart from the difficulties in defining cause and effect in such systems). Does, for example, the protoplasm underlying the boundary have an organization that guides the overall orientation of organ growth that includes particular temporal and spatial patterns of boundary deformations? Or is the converse true, and that indications of protoplasmic order (e.g. microtubule orientations) are a response to the preceding pattern of growth of the boundary, much in the way that actin filaments (stress fibres) in fibroblasts can be oriented in a way that reflects externally imposed deformations of the outer membrane of the cell (Burridge *et al.*, 1987)?

At the whole plant level, it has been established for trees that part of the supporter subsystem comprised of reaction wood (in angiosperms) and compression wood (in gymnosperms) is a response to the configuration of the branches in the shoot system (Zimmermann & Brown, 1971). Likewise, at the organ level, the pattern of cell walls can be influenced by extracellular forces (Lintilhac, 1984). At each of these levels, the supporter subsystem probably has a particular type of organization that is genetically determined, but can also be modified by forces created by the growth (to which the subsystem also contributes) of the level in question. In this context, there are many similarities with the supporter subsystem of mammalian organisms (i.e. the skeleton) and its responses to external physical forces that actually help to shape and maintain the form of bones and skeletal structures (Lanyon, 1992).

Counterarguments to the acellular view can be presented. In the first place, multinucleate organisms, or even multinucleate stages of development within an otherwise multicellular organism, are rare or transient (e.g. endosperm). This may be because the multinucleate condition does not permit sufficient morphological or biochemical specialization, and that the cellular condition has some particular selective advantage, even in the earliest stages of embryonic development. It is noteworthy that multinucleate stages in embryogenesis, as for example in gymnosperms and in *Paeonia* (Cave *et al.*, 1961; Yakovlev & Yoffe, 1957), or in the Dipteran, *Drosophila* (Sonnenblick, 1950), are of relatively short duration and are superseded by cellular stages. The multicellular condition may, therefore, fulfil a requirement for specialized protoplasmic, histogenetic domains that is better served by the compartmentation of protoplasm by plasma membranes and walls (cf. Fig. 1) than by any other means. The multicellular condition may also enable larger and more elaborate forms to be constructed through the turgor pressure generated simultaneously by many cells whose growth is multidirectional. However, all this may not bear directly on the acquisition of form.

The question of convergence of forms (mimicry) at different levels of organization (cf. examples from the cellular and organ levels, mentioned above in connection with algae), and even within the same level, is intriguing, but may be explicable on the basis that all extant organismal forms represent minimal energy (entropy) states associated with the bonding of macromolecules and their topological configurations. That is, certain configurations of cell wall networks (and perhaps of biochemical compartments and their networks also) may be favoured over others. These favoured networks, being at the cellular level, may in turn associate with the acquisition of certain organ forms which themselves represent particular minimal entropy states at this next higher level in the living system hierarchy.

In the second place, a multiplicity of cellular patterns associated with a common form (cf. Fig. 2) could simply mean that the attendant patterns of morphogenesis are all achieved in a similar way, there being redundancy in the routes involved. Thus, even though the fine details of cell division may be relatively unimportant in these morphogenetic processes, the cells themselves may not be indispensable because their walls and, perhaps more importantly, their cytoskeleton, transmit patterns of strain that constitute a type of epigenetic information for the final emergence of structure (Lintilhac, 1974, 1984). Moreover, while the forms of roots and shoots may be approximately similar, perhaps because each fulfils one of the minimal entropy states, the internal cellular patterns of roots and shoots definitely differ, and probably do so because the biophysical forces in the meristematic growing points are deployed in different ways. These forces could themselves be transduced into morphogenetic processes at the meristem boundaries. In other words, the walls of cells may contribute additional information for morphogenesis that cannot be supplied by the cytoplasm alone. On the other hand, as already mentioned, cell wall patterns may be modified in response to forces transmitted into the organ as a result of the contour of the organ boundary (Lintilhac, 1974) or its rate of growth. In this way, the diverse cellular patterns of roots and shoot apex may be generated (Hejnowicz, 1989; Nakielski, 1987). Although cells generate the internal pressure necessary for growth, and through which transcellular forces might be created, the overall form is determined independently of this via effects on the boundary. Thus, boundary deformations should be regarded as primary morphogenetic events, whereas other epigenetic effects promoted by cell wall patterning are secondary.

The problem of form in relation to cells thus seems far from settled. Total abandonment of cellular concepts in the development of form would amount to a Kuhnian paradigm shift, even though those concepts are not particularly well defined, but at present a new paradigm to replace it is not sufficiently well formulated. It is easy to recognize the insistent 'pull' exerted by the cell concept in relation to algal development, for instance, when Goff and Coleman (1987) stated that the giant coenocytic units of the alga *Griffithsia pacifica* are 'in effect multicellular'; in reality these units are multinucleate and manifestly have no requirement to be otherwise. Maybe processes at both the cellular and the supracellular level contribute to morphogenesis, and that to try to marshall evidence in favour of one at the expense of the other obscures the development of a satisfying concept (Rutishauser & Sattler, 1985).

THE POSSIBLE SIGNIFICANCE OF CELL DIVISION PATTERNS TO MORPHOGENESIS

One way in which the relationship between cell division and morphogenesis has been explored is through modelling using L-systems. This approach takes as a premise that the pattern of cell formation and the associated ordered pattern of growth of each cell wall leads to the acquisition of form and is the result of some type of epigenetic information. The information can be notated by an algorithm for development; it is assumed that a certain context is provided for dealing with that information. Thus, an initial cellular state (state 0) has embedded within it information for the production of a subsequent state (state 1). State 1, in turn, then becomes a source of information for the attainment of state 2, and so on. The sequence of state transitions proceeds until the informational potential inherent in the cells and in their interactions with neighbouring cells, is exhausted or cancelled. In this way, cell and organ form are the outcome of the potentialities of the initial condition or state, coupled with an inherent capacity of the protoplasm for growth. The determinism of this sequence resides in the state transitions and in the context in which they take place.

It was the use of such a concept in relation to meristem structure in ferns, formalized in the language of L-systems, that enabled Lindenmayer (1984) to extract from an analysis of cell patterns in roots of *Azolla pinnata* a set of state transition rules that successfully formalized the sequence of cell division and cell growth; by which Lück *et al.* (1988) were able to simulate the cellular patterns of the shoot epidermis of *Angiopteris lygodifolia*; and by which Lück and Lück (1991) have been able to model the production of cells that form the shoot apex of *Ceratopteris thalictroides*. Also, the branched pattern of leaves (some of which resemble the shapes of the green algal species mentioned earlier) have been shown to be interpretable by state transitions (Lindenmayer, 1975). This list of achievements is not exhaustive, as perusal of the book *The Algorithmic Beauty of Plants* by Prusinkiewicz & Lindenmayer (1990) will show. Generally, these systems reproduce or replicate already existing structures, be they cell walls or organ boundary elements. They also contain information to specify form by means of rules that determine details of cell wall growth, including that of the wall which serves as the organ boundary.

The general principle behind work with L-systems and cell division patterns may be illustrated with a relatively simple example from *Angiopteris lygodifolia* (Barlow, 1991; Lück *et al.*, 1988) whose shoot apex was described by Imaichi (1986). In common with the other examples of ferns mentioned earlier, anatomical observation of the *Angiopteris* apex shows that all its cells appear to derive from a tetrahedral apical cell, which has a triangular shape when seen in surface view (Fig. 3). Division of the apical cell causes each of its three internal facets (represented by the sides of a triangle) to be incorporated into the respective daughter cells, or merophytes. Division proceeds in either a clockwise or counterclockwise manner, ejecting each of the three facets in turn into a daughter cell and in the process creating new replacement facets. The fourth, superficial facet, never enters a daughter cell, though by being part of the surface it does nevertheless expand, thereby compensating for the erosion of its area as a result of its contributions to a succession of daughter cells. Concurrently, each new merophyte itself divides to yield two daughters which later also divide. Thus, each merophyte grows and becomes filled with a network of cell walls of different age, the walls having been inserted with specific orientations during earlier rounds of division (Fig. 3). The same pattern of division is repeated in each merophyte so that three identical sectors, each occupying a 120° arc of the surface, are formed.

Although it is easy to observe the pattern of walls on the shoot surface, it should be remembered that these walls are records of only the anticlinal divisions in the superficial, epidermal layer of cells. Other periclinal, anticlinal and transverse div-

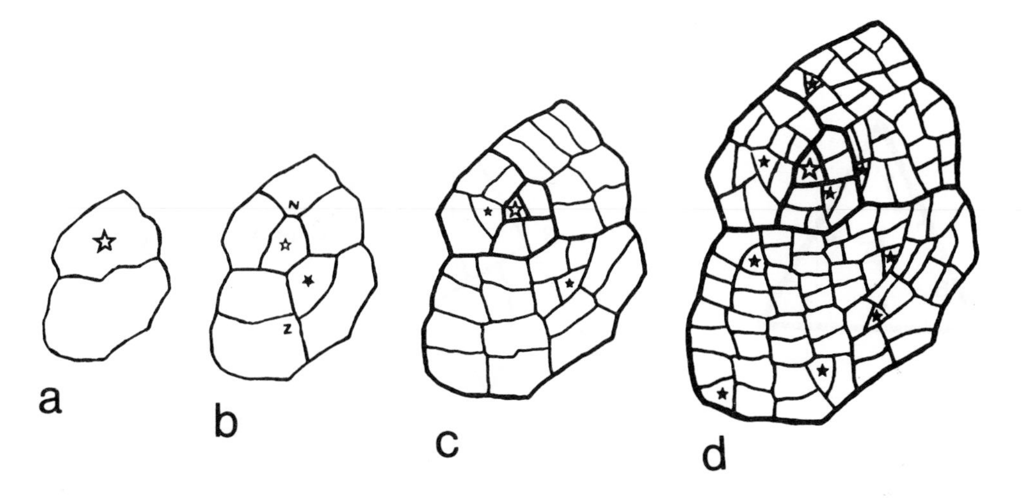

Figure 3 **a–d**, Evolution of cell complexes in the epidermal layer of the shoot apex of *Angiopteris lygodifolia*. The apex was in the condition illustrated in **d** at the time of collection, but earlier stages of cellular development can be traced from the sequence in which the walls have been inserted, so that **a** shows an early state with an apical cell and one daughter merophyte. Open star (☆), the apical cell; filled stars (★), small triangular cells. In (b) the letter z indicates the type of tetrad; it signifies the type of wall junction between related pairs of cells. Modified from Barlow, 1991; original observation in Imaichi, 1986.

isions also occur in deeper parts of the merophyte which extend into the bulk of the shoot. Lack of information about the division patterns in these internal cells can frustrate attempts to find regularities in the division pattern in the epidermis since division events that occur unseen in the deeper regions may influence those that occur more superficially. Despite this, however, the division pattern in the epidermal layer conforms to a simple L-system algorithm that can be notated as follows: $1 \rightarrow 5/_1 23$, $2 \rightarrow 4$, $3 \rightarrow 5/_1 2$, $4 \rightarrow 3$, $5 \rightarrow 4$.

The algorithm is a set of state production rules where each number corresponds to the state of a particular cell wall. State 1 is that of the newest wall. At the first step – and here a step corresponds with the act of cell division – this wall is renumbered $5/_1 23$. The wall at state 3 is also simultaneously renumbered. The slash, /, represents the site at which a new wall attaches when the cell divides; the subscript 1 indicates that the new wall is to be numbered 1. The new wall spans older walls that were previously at states 1 and 3, and does so at a specific site on each. Other walls, numbered 2, 4 and 5, also proceed to their designated state. At the next step the production rules are followed again. The initial shape of the cell is triangular (like the apical cell itself) and its walls, like the walls of all other cells, are designated as being in states 1, 2, 3, 4, 5 (Fig. 4a). These last-mentioned features (triangular shape and wall states) are, in this case, taken as the starting points for development. The amount of wall growth at each step is not specified in this particular algorithm and the initial cell is progressively subdivided at each step. However, this need not be the case if wall growth rules are also specified.

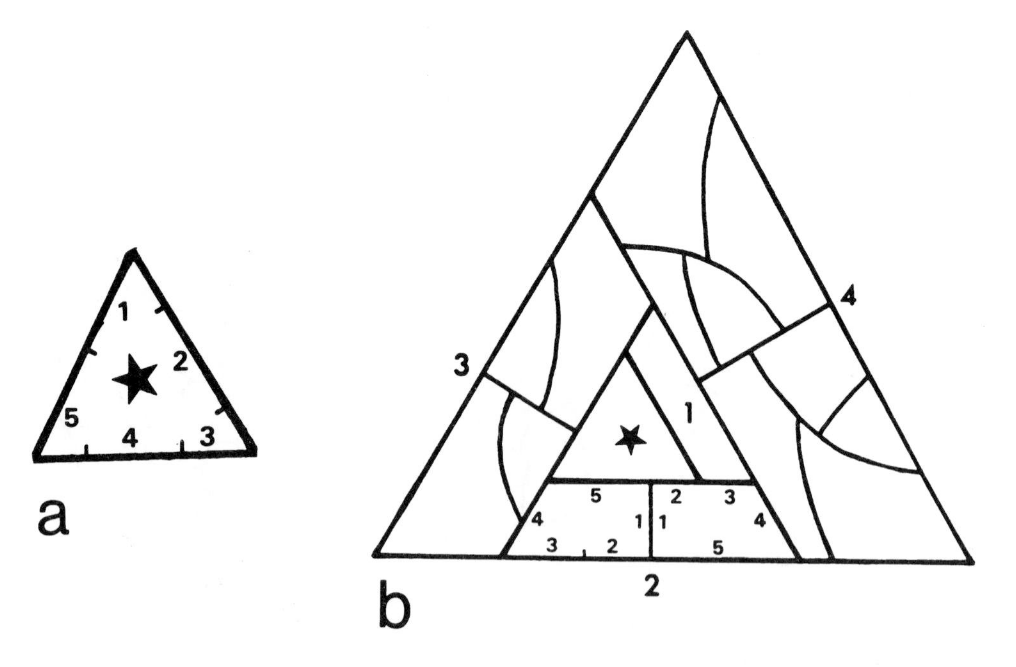

Figure 4 Evolution of a group of 16 cells from a triangular apical cell (filled star, ★) such as found in the shoot of *Angiopteris* (see Fig. 3). The pattern of cells is derived from the operation of the D0L-system algorithm mentioned in the text. The walls of the apical cell in **a** are numbered (1–5); these represent different states. In **b** a pair of derivative cells are numbered similarly to show that the states persist. Large numbers (**1**, **2**, **3**, **4**) against four of the cell complexes indicate the successive 'merophytes' derived by the operation of the algorithm. Modified from Barlow, 1991.

Operation of the algorithm through four steps results in a group of 15 cells arranged in four merophytes surrounding an initial cell which maintains its triangular shape and wall states (Fig. 4b). The derivative cells within the merophytes have various shapes dictated by the division scheme. They also correspond closely with those actually seen in the epidermis of the *Angiopteris* shoot apex (Fig. 3). An interesting feature of the simulated development is the emergence of small triangular cells amongst the descendants (Fig. 3c, d). Because of their shape, these cells could begin to operate as new, secondary, apical cells. In a real apex they might take on the function of a growth centre for a new apex, causing it to bifurcate; or they might be the precursors of a leaf primordium. Another feature to note is that when a group of four cells of common descent, or tetrad, arises, the junction where the two youngest walls meet the next oldest wall resembles either an S or a Z. The tetrads are thus known as either S or Z tetrads. The above algorithm generates Z tetrads only, which coincides with the predominant type of tetrad seen in the surface cell layer of the *Angiopteris* apex (Fig. 3b, c).

The process described here represents only a partial, and perhaps intrinsically insignificant, aspect of apical morphogenesis. No actual shaping of the tissue (either in reality or as generated by the algorithm) has taken place since the surface of the shoot to which these anticlinal walls attach grows approximately uniformly in all directions. The shaping of the apex would only come about if it were thrown into folds or humps as a consequence of certain areas of the surface expanding at a faster rate than others. Such areas may coincide with cells delimited by a particular pattern of anticlinal walls. These patches of more rapid growth could continue development as leaf primordia. In the example of the *Angiopteris* apex, differential wall expansion may commence at sites coinciding with the small triangular derivative cell (Fig. 3c, d). As yet, this aspect of differential surface growth rate has been little explored through L-systems, though an attempt has been made by Lück and Lück (1979) to model in this way the pattern of epidermal cell growth found on the rhizophore of *Selaginella kraussiana.*

Descriptions of development in terms only of abstract algorithms are rather unsatisfying since the contribution of biophysical processes known to be associated with growth is ignored. That is, because the algorithm used is a deterministic, interactionless, D0L-system, there are none of the feedbacks from the developing form itself that might influence later patterns of growth. A further criticism of this formal approach might be that, when all is said and done, it is only another way of describing the pattern of division and that there is a lack of any general predictive power since every algorithm is necessarily unique to the system for which it was devised. These remarks are only partially true, however, since it is now theoretically possible to increase the number of inputs to a given cell by including some type of interaction from neighbouring cells. But, as Lindenmayer commented, 'it seems best to start with simple control systems that are adequate for some known division patterns . . .', and '. . . tractable formulation may give a better insight into the phenomenon than a more realistic one because in the former case one gains more easily access to the consequences of one's assumptions' (Lindenmayer, 1984). The fact that developmental algorithms can be devised at all suggests that there may be a propagation state in morphogenesis, and that they also intimate an ordered cellular structure containing epigenetic information by which cellular patterning is determined. The algorithms are simply a way of codifying that information.

Patterns of cell wall expansion are often linked with the orientation of cortical microtubules in the underlying cytoplasm of the cell, and it is assumed (on the basis of rather limited experimental evidence) that the observed pattern of microtubule orientation is a factor contributing to the orientation of cell growth rather than a consequence of it. Therefore, any alterations to the wall growth pattern, particularly those relating to the inception of an area that assumes a new growth direction, such as must occur in the development of a primordium, require shifts in the alignments of microtubules underlying such an area. These shifts might be possible if, for example, cortical microtubule arrays align themselves along the shortest peripheral path between opposite walls by means of a 'self-cinching' mechanism (Lloyd & Barlow, 1982). Then, the formation of small cells with length/breadth ratios appreciably different from those of their neighbouring cells may offer the opportunity to alter the prevailing orientation of wall expansion. Figure 5 shows how this might operate in the development of a leaf primordium on the *Angiopteris* shoot apex. The small triangular cells that arise in development (both in reality and through the operation of the aforementioned algorithm) may provide new starting conditions for a recapitulation of apical development. However, on this apex there are obviously more triangular cells than there are future primordia (Fig. 3d), so there must be a means of selecting only those cells that will actually function as initials for the leaf.

It is possible that the continuing differentiation of the vascular system, whose pattern is governed by diffusion-reaction mechanisms, has a decisive influence on primordium selection (cf. Larson, 1983), perhaps by inducing changes in the external boundary layer of the apex necessary for its deformation and out-pushing. In this connection, it is interesting to find that, in shoot apices of *Pisum sativum* and *Vicia faba*, the first stages of xylem differentiation can occur only a few cell lengths away from the surface and that they precede any overt signs of leaf primordium formation

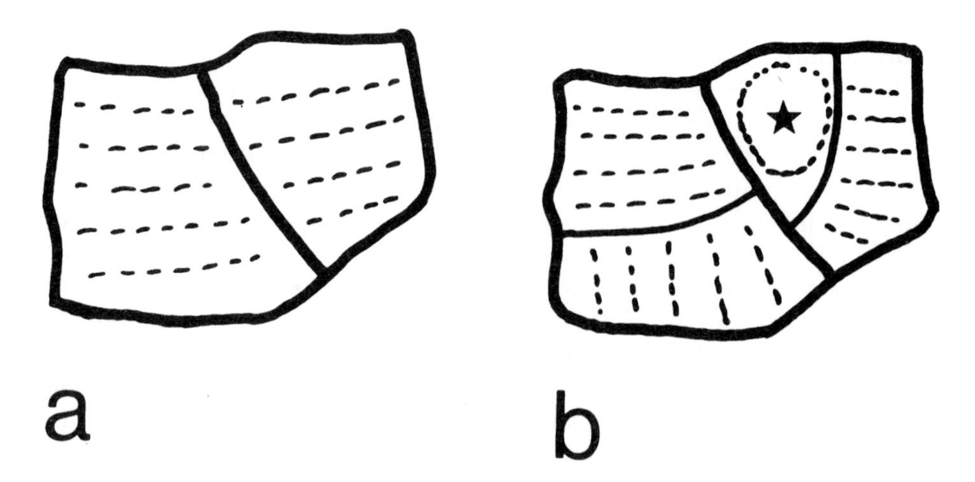

Figure 5 Hypothetical way in which a change in microtubule (dotted lines) orientation can come about as the result of the changed shape of a cell following division. The cells are taken from an area shown in Fig. 3. One of the cells in **a** divides so creating a small triangular daughter cell, indicated by a filled star (★) in **b**. Here, microtubules form a hoop around the anticlinal walls favouring the growth of this cell in a plane perpendicular to that of the page; microtubules in the other cells hoop around the periclinal walls and favour growth of the cells in a plane parallel with that of the page.

(Gahan & Bellani, 1984). Although the generality of this finding is not clear, such a coupling between a differentiation event and a morphogenetic event within an angiosperm shoot apices might indicate that the primary morphogenetic process (boundary deformation) has a physiological basis, perhaps fostered by diffusion-reaction mechanisms, that override regulatory schemes based on lineage patterns which appear to be more evident in mosses and liverworts (Barlow, 1990). However, this may be a feature of the scale over which morphogenetic controls of leaf primordium development operate in the two groups of plants: primordia are comprised of many cells in the case of angiosperms, but of single cells in the case of mosses; in the latter case, physiological controls operating over a distance comparable to the width of a single cell would be indistinguishable from a cellular, or lineage, type of control.

A coupling of internal physiology and morphogenesis would also draw the emphasis away from purely cytoskeletal and cellular schemes which dominate some proposals for morphogenetic change (cf. Green & Selker, 1991) and integrate them with other regulatory factors of shoot development. Similar considerations may also apply in the positioning of lateral root primordia. In roots of some ferns, for example, primordia occur at specific sites in the endodermis and indicate a relationship with a specific pattern of division in the merophytes with which they are associated (Lin & Raghavan, 1991). But whether these divisions give rise directly to altered cell shapes which then favour the reorientation of cell growth and the emergence of new organ axes is not known. Relevant, however, is the finding that, at least in young maize roots, pericycle cells, from which lateral root primordia arise, have randomly oriented cortical microtubules against their periclinal walls (Baluška *et al.*, 1992). It can be imagined that such arrays are readily reoriented to bring about a new direction of growth at the periblem boundary leading to the establishment of a primordium. These boundary effects could be influenced by gradients of regulatory factors originating in the the vascular cylinder (Barlow, 1990, in press) and could even have an origin in particular patterns of electrical currents that might lead to new polarities of organ axis growth (see, for example, Gorst *et al.*, 1987).

CELL DIVISION AND INTERNAL STRUCTURE

Most of the cell divisions occurring in the distal portion of shoot and root apices are concerned with elaborating the numerous files of cells that run the length of these organs. Only proximal to the 'formative' zone where this occurs do cell divisions assume a predominantly transverse orientation. These are the proliferative divisions that increase the number of cells within each file. Both proliferative and formative divisions multiply the number of radial and periclinal walls, oriented at right angles to each other, which comprise the supporter subsystem. The production of walls by means of all classes of divisions in the meristem may be under deterministic control (Barlow, 1991, 1993b; Lindenmayer, 1984; Lück *et al.*, 1994).

The regular patterns of walls formed by such divisions may be a means of optimizing the structural support, and of dissipating the stresses, within the root and shoot organ. In an extreme case, as exemplified by roots of the turtle grass, *Thallasia testudinum,* transversely arranged bands of long and short cells alternate along the axis (Tomlinson, 1969). Here, a definite support, and protective, role has been attributed to this cellular pattern since the short cells act as diaphragms which can seal the end of the root, preventing embolisms, should it be broken. This observation recalls the role of walls as sites of organ fission in less complex organisms (fungi, algae, mosses). Here, chains of cells can fracture across transverse walls (or septa) and lead to the dispersal of the cellular fragments. The tmema cells of moss protonemata, developed by an unequal transverse cell division when growth conditions are unfavourable, represents a special type of cell directed to this end (Bopp *et al.*, 1991).

As mentioned earlier, there are major differences in the way the formative cell divisions are organized in roots that characterize the higher-order taxonomic groupings (Voronin, 1969, and Fig. 2). The more closely related the species, the finer the distinctions between cell division patterns. The reason for the more major differences in patterns (compare the pattern in Fig. 2a with that in Fig. 2e, for example) seen between the higher-order taxa can be partly attributed to differences in the local rates of increase of protoplasmic mass within the volume of the apex (Hejnowicz, 1989) and may also be attributable to variations in the degree and spatial extent of the deformability of the outer epidermal boundary layer. Even the relatively minor difference between an open and a closed meristem in roots of the same species (e.g. *Helianthus annuus*) is interpretable in this way (Clowes, 1981).

It is unlikely that the differences in formative division patterns, as seen in Fig. 2, are related to alterations in the patterns of tissue differentiation since similar tissues develop in similar relative positions within the apex whatever its division pattern. It is more likely that these patterns of differentiation, although confined to discrete cellular areas, have their basis in chemical diffusion-reaction patterns and homoeogenetic induction (Barlow, 1984) and are established first in the embryo and later in the neo-embryonic primordia along the developed axes. However, it should be recalled that in roots of *Azolla pinnata* and *A. filiculoides* there is an exact correspondence between the sequence of formative divisions and the ensuing tissue differentiation such that the divisional history of a cell precisely predicts the type of tissue that will form from the various cellular descendants (Gunning *et al.*, 1978). In roots of angiosperms there is less overt evidence of such a high level of correspondence (Barlow, 1984). In stems and leaves the same may be true, leading to the belief that cellular location within the organ rather than cellular ancestry is an important determinant of cell differentiation (Barlow, 1990; Dawe & Freeling, 1991). The small size of the *Azolla* root might mean that physiological controls of differentiation act over distances that correspond to cellular dimensions, making it seem that specifications of differentation based on cell lineage have precedence.

Because formative cell divisions are parallel to the axis of the cell file to which the dividing mother cell belongs, they help create a branched system of cell files. All branching systems, whether they pertain to roads, rivers, or the axes which comprise shoot and root systems, represent the evolution, within the system, of an optimal flow (traffic, water) or means of capture (light, solutes) of a resource or a product. The branching of shoot and root axes also tends to optimize the distribution of mass (weight). This optimization of a physical property may underlie the cell file branching pattern found in organs as occurs, for example, in the root cortex where branching seems to be particularly precisely regulated (Barlow, 1993b). It may have to do with optimizing the pattern of strains imposed by stresses sensed at the external boundary. Even the junction of one wall with another as, for example, when a new wall is inserted at cytokinesis has this property. In this sense, cell-wall and cell-file branching reflect an optimizing of the function of the supporter subsystem. In the case of roots, the paraboloidal form of the tip, and also the presence of a widening growth zone a little behind the meristem, maximizes the ability of the apex to force itself between, and then open up, cracks in the soil to permit its onward passage (Atwell, 1993). Another possibility is that the cellular patterns created by the branching of cortical cell files permit effective growth responses (tropisms) to external stimuli as the consequence of differentiation of a particular file of reactive cells.

Tropic responses are a class of morphogenetic movements brought about by controlled deformations (in the longitudinal direction) of boundary layers: at the epidermis in the case of shoots (Firn & Digby, 1977), and at the endodermis/pericycle boundary in the case of roots (Björkman & Cleland, 1988). In both organs the tropisms appear to be intimately connected with changed electrical patterns around and within them (Iwabuchi *et al.*, 1989; Mulkey *et al.*, 1981). Moreover, the zone where differential elongation is located is often just behind the meristem of roots and shoots, thus enabling the tip to divert readily its direction of growth so that it can penetrate and explore new regions of the environment as well as respond, by tropic growth, to physical displacements. Actually, the sensitivity of the tip to environmental stimuli, coupled with its potentiality to alter the direction of its growth, argues for the functional necessity of locating the meristem at the apex of roots and shoots so that it can closely influence the production of new cells available for immediate elongation.

The contribution of cell-file branching to tissue differentiation at an early stage of embryonic development (Fig. 1) has already been remarked upon. However, later in development, additional divisions, which are usually but not always transversal, contribute to the further differentiation of certain optional cell types, such as hairs and trichomes (Barlow, 1984). One recently studied case of cellular branching (Seago & Marsh, 1989), which relates to an optional path of differentiation, has an especially intimate relationship with the supporter subsystem. Young shoot-borne (or adventitious) roots of the bulrush, *Typha glauca*, develop an aerenchyma in their cortex. This partly comes about by cell death (lysigeny), though some cell splitting (schizogeny) is also involved. As if in compensation for the loss of mechanical support that this might cause, a multilayered hypodermis is developed (Fig. 6a). As the root apex ages, aerenchyma formation ceases and the hypodermis no longer develops (Fig. 6b). The hypodermis is formed by periclinal and radial divisions in the distal meristematic zone of the cortex. Whether or not these divisions occur and, importantly for the development of the supporter subsystem, whether or not the additional walls are formed, is apparently regulated coordinately with the signal for aerenchyma formation.

According to Seago and Marsh (1989), the correlation between periclinal divisions in the outer cortex and aerenchyma formation in the central cortex is evident in a wide range of wetland species. Indeed, the cortex of such species shows many anatomical features that are related to the shape and form of both its cells and its intercellular spaces, and how these bear on both the mechanical stability and the ventilation of the tissue. Interestingly, the intercellular spaces comprise part of another of Miller's proposed subsystems, the 'distributor'. Much of organogenesis and the elaboration of organ form is concerned with the development of this subsystem, but space precludes all except this fleeting mention (though see Barlow, 1994, for some further discussion). The topic is mentioned in order to indicate that form, tissue differentiation, and subsystem development and function are all intimately interconnected. Clearly, an holistic perspective that recognizes hierarchical levels and associated sub-

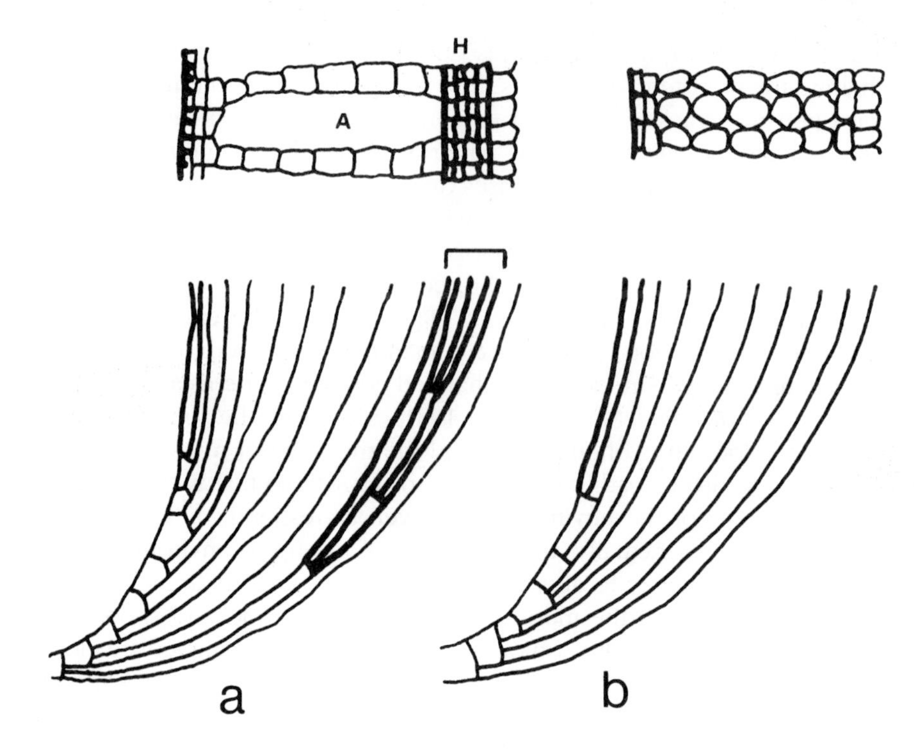

Figure 6 Correlation between additional periclinal and radial divisions in the outer cortex and aerenchyma formation in adventitious roots of *Typha glauca*. In **a**, which are drawings of longitudinal and transverse sections from the tip of a young root, additional divisions (brace and thickened lines) in the outer cortex have correlated with the development of a hypodermis (H) and aerenchyma (A). In older roots **b**, these additional divisions do not occur and no hypodermis or aerenchyma has formed. Redrawn from Seago and Marsh, 1989.

systems is one that can provide a satisfying coherence to the overall development of plants and reveal the significance of their form.

CONCLUSIONS

One of the significant contributions of cell division to organogenesis is that it generates walls that serve as the supporter subsystem for the organ throughout its subsequent life. Furthermore, collections of specialized walls, such as those that characterize sclerenchyma or heartwood, serve as supporter for the whole plant. Coincidental with wall formation, especially in embryos and neo-embryonic sites (primordia), is the establishment of tissues within which occur specialized biochemical processes, vital for the overall function of the plant. The pattern of tissue differentiation, which reflects a spatial segregation of cellular functions, is then carried forward into the new organ axis by a process of homoeogenetic induction. These processes of intraorgan differentiation, facilitated by division, are complemented at the cellular level by the specialization of the cytoplasm through the differentiation of its complement of organelles (vacuoles, Golgi apparatus, mitochondria, etc.) whose activities are important in tissue differentiation and function (Clowes, 1971).

Although the pattern of division in the meristem may be important in creating a supporter subsystem of optimum rigidity, the sequence in which the new walls are formed may be unimportant as long as an adequate frequency and ratio of periclinal and radial walls is maintained within the growing mass. However, the sequence of wall inception could become important for organ morphogenesis if, as a result of specific divisions, certain walls were thereby enabled to expand more rapidly than others. In fact, the central process of morphogenesis in both multicellular and unicellular organisms is for there to be a specific pattern of differential cell wall expansion, especially at the external boundary of an organ or organism. The crucial regulatory factors for this process require identification, but may involve electrical or ionic currents (Nuccitelli, 1988; O'Shea, 1989) whose effects within cells then need to be stabilized by a specific 'axis-stabilizing complex' (Quatrano, 1990) involving linkages between cell wall, plasma membrane and cytoskeleton. The pattern of walls and the underlying cytoskeleton within the growing points of an organ may be an additional source of epigenetic information for morphogenesis and lead to self-sustaining, concatenated sequences of developmental states that can be formalized and simulated by means of developmental algorithms. At the same time, these sequences may feed information back to the system and modify its subsequent development.

ACKNOWLEDGEMENT

The author thanks Prof. Rolf Sattler for helpful comments on an early draft of this paper.

ADDENDUM

After this chapter had been prepared the proceedings of a symposium in honour of Katherine Esau came to hand (published in *International Journal of Plant Sciences, 153,* No. 3, Part 2, 1992). In this volume are papers by P. Sitte, D.R. Kaplan, W. Hagemann, and T.J. Cooke & B. Lu (amongst others) which express views complementary to some of those presented here. Unfortunately, it has not been possible to discuss them in the body of the present work.

REFERENCES

ATWELL, B.J., 1993. Response of roots to mechanical impedance. *Environmental and Experimental Botany, 33:* 27–40.

BALUŠKA, F., PARKER, J.S. & BARLOW, P.W., 1992. Specific patterns of cortical and endoplasmic microtubules associated with cell growth and tissue differentiation in roots of maize (*Zea mays* L.). *Journal of Cell Science, 103:* 191–200.

BARLOW, P.W., 1982a. 'The plant forms cells, not cells the plant': the origin of De Bary's aphorism. *Annals of Botany, 49:* 269–271.

BARLOW, P.W., 1982b. Root development. In H. Smith & D. Grierson (eds), *Botanical Monographs,* vol. 18. *The Molecular Biology of Plant Development,* pp. 185–222. Oxford, U.K.: Blackwell Scientific.

BARLOW, P.W., 1984. Positional controls in root development. In P.W. Barlow & D.J. Carr (eds), *Positional Controls in Plant Development,* pp. 281–318. Cambridge, U.K.: Cambridge University Press.

BARLOW, P.W., 1987. The hierarchical organization of plants and the transfer of information during their development. *Postepy Biologii Komorki, 14:* 63–82.

BARLOW, P.W., 1990. Rhythm and polarity are the bases of morphogenesis in plant apices. Unpublished manuscript, 30 pp. Abstract in *Bulletin du Group d'Etude des Rhythmes Biologiques, 22 (1),* 2.

BARLOW, P.W., 1991. From cell wall networks to algorithms. The simulation and cytology of cell division patterns in plants. *Protoplasma, 162:* 69–85.

BARLOW, P.W., 1993a. The response of roots and root systems to their environment. An interpretation derived from an analysis of the hierarchical organization of plant life. *Environmental and Experimental Botany, 33:* 1–10.

BARLOW, P.W., 1993b. The cell division cycle in relation to root organogenesis. In J.C. Ormrod & D. Francis (eds), *Molecular and Cell Biology of the Plant Cell Cycle,* pp. 179–199. Dordrecht, The Netherlands: Kluwer Academic Publishers.

BARLOW, P.W. 1994. The origin, biology and diversity of shoot-borne and adventitious roots. In T.D. Davis & B.E. Haissig (eds), *Biology and Adventitious Root Formation,* pp. 1–23. New York, U.S.A.: Plenum Press.

BARLOW, P.W. From cell to system: repetitive elements in shoot and root development. In M. Iqbal (ed.), *Growth Patterns in Vascular Plants.* Portland, OR., U.S.A.: Dioscorides Press (in press).

BARTHÉLÉMY, D., 1991. Levels of organization and repetition phenomena in seed plants. *Acta Biotheoretica, 39:* 309–323.

BJÖRKMAN, T. & CLELAND, R.E., 1988. The role of the epidermis and cortex in gravitropic curvature of maize roots. *Planta, 176:* 513–518.

BOPP, M., QUADER, H., THONI, C., SAWIDIS, T. & SCHNEPF, E., 1991. Filament disruption in *Funaria* protonemata: formation and disintegration of tmema cells. *Journal of Plant Physiology, 137:* 273–284.

BRUCK, D.K. & WALKER, D.B., 1985. Cell determination during embryogenesis in *Citrus jambhiri.* I. Ontogeny of the epidermis. *Botanical Gazette, 146:* 188–195.

BURRIDGE, K., MOLONY, L. & KELLY, T., 1987. Adhesion plaques, sites of transmembrane interaction between the extracellular matrix and the actin cytoskeleton. *Journal of Cell Science* (Suppl. 8): 211–229.

CANNON, W.A., 1949. A tentative classification of root systems. *Ecology, 30:* 542–548.

CAVE, M.S., ARNOTT, H.J. & COOK, S.A., 1961. Embryogeny in the California peonies with reference to their taxonomic position. *American Journal of Botany, 48:* 397–404.

CLOWES, F.A.L., 1971. Cell organelles and the differentiation of somatic plant cells. In J. Reinert & H. Ursprung (eds), *Results and Problems in Cell Differentiation,* vol. 2. *Origin and Continuity of Cell Organelles,* pp. 323–342. Berlin, Germany: Springer-Verlag.

CLOWES, F.A.L., 1981. The difference between open and closed meristems. *Annals of Botany, 48:* 761–767.

DAWE, R.K. & FREELING, M., 1991. Cell lineage and its consequences in higher plants. *Plant Journal, 1:* 3–8.

DENGLER, N.G., 1980. Comparative histological basis of sun and shade leaf dimorphism in *Helianthus annuus* L. *Canadian Journal of Botany, 58:* 717–730.

DENGLER, N.G. & SANCHEZ-BURGOS, A.A., 1988. Effect of light level on the expression of anisophylly in *Paradrymonia ciliosa* (Gesneriaceae). *Botanical Gazette, 149:* 158–165.

DHOUAILLY, D., 1984. Specification of feather and scale patterns. In G.M. Malacinski (ed.), *Pattern Formation. A Primer in Developmental Biology,* pp. 581–601. New York, U.S.A.: MacMillan.

DUFFY, R.M., 1951. Comparative cellular configurations in the meristematic and mature cortical cells of the primary root of tomato. *American Journal of Botany, 38:* 393–408.

EVERED, D. & MARSH, J. (eds), 1989. *The Cellular Basis of Morphogenesis.* CIBA Foundation Symposium, 144. Chichester, U.K.: J. Wiley.

FIRN, R. & DIGBY, J., 1977. The role of the peripheral cell layers in the geotropic curvature of sunflower hypocotyls: a new model of shoot geotropism. *Australian Journal of Plant Physiology, 4:* 337–347.

FITTER, A.H., 1987. An architectural approach to the comparative ecology of plant root systems. *New Phytologist, 106* (Suppl.): 61–77.

FOARD, D.E., 1971. The initial protrusion of a leaf primordium can form without concurrent periclinal divisions. *Canadian Journal of Botany, 49:* 1601–1603.

FOARD, D.E., HABER, A.H. & FISHMAN, T.N., 1965. Initiation of lateral root primordia without completion of mitosis and without cytokinesis in uniseriate pericycle. *American Journal of Botany, 52:* 580–590.

FOURNIER, A., 1979. Is architectural radiation adaptive? *Diplome d'Etudes Approfondies d'Ecologie Générale et Appliquée.* Montpellier, France: Université des Sciences et Techniques du Languedoc.

GAHAN, P.B. & BELLANI, L.M., 1984. Identification of shoot apical meristem cells committed to form vascular elements in *Pisum sativum* L. and *Vicia faba* L. *Annals of Botany, 54:* 837–841.

GOFF, L.J. & COLEMAN, A.W., 1987. The solution of the cytological paradox of isomorphy. *Journal of Cell Biology, 104:* 739–748.

GORST, J., OVERALL, R.L. & WERNICKE, W., 1987. Ionic currents traversing cell clusters from carrot suspension cultures reveal perpetuation of morphogenetic potential as distinct from induction of embryogenesis. *Cell Differentiation, 21:* 101–109.

GREEN, P.B. & SELKER, J.M.L., 1991. Mutual alignments of cell walls, cellulose, and cytoskeletons; their role in meristems. In C.W. Lloyd (ed.), *The Cytoskeletal Basis of Plant Growth and Form,* pp. 303–322. London, U.K.: Academic Press.

GUNNING, B.E.S., HUGHES, J.E. & HARDHAM, A.R., 1978. Formative and proliferative divisions, cell differentiation, and developmental changes in the meristem of *Azolla* roots. *Planta, 143:* 121–144.

HALLÉ, F. & OLDEMAN, R.A.A., 1970. *Essai sur l'Architecture et la Dynamique de la Croissance des Arbres Tropicaux.* Paris, France: Masson & Cie.

HALLÉ, F. OLDEMAN, R.A.A. & TOMLINSON, P.B., 1978. *Tropical Trees and Forests. An Architectural Analysis.* Berlin, Germany: Springer-Verlag.

HEJNOWICZ, Z., 1989. Differential growth resulting in the specification of different types of cellular architecture in root meristems. *Environmental and Experimental Botany, 29:* 85–93.

IMAICHI, R., 1986. Surface-viewed shoot apex of *Angiopteris lygodiifolia* Ros. (Marratiaceae). *Botanical Magazine (Tokyo), 99:* 309–317.

IWABUCHI, A., YANO, M. & SHIMIZU, H., 1989. Development of extracellular electric pattern around *Lepidium* roots: its possible role in root growth and gravitropism. *Protoplasma, 148:* 94–100.

JENÍK, J., 1978. Roots and root systems in tropical trees: morphologic and ecologic aspects. In P.B. Tomlinson & M.H. Zimmermann (eds), *Tropical Trees as Living Systems,* pp. 323–349. Cambridge, U.K.: Cambridge University Press.

KAPLAN, D.R. & HAGEMANN, W., 1991. The relationship of cell and organism in vascular plants. *BioScience, 41*: 693–703.

KHAIT, A., 1986. Hormonal mechanisms for size measurement in living organisms in the context of mature juvenile plants. *Journal of Theoretical Biology, 118:* 471–483.

KUROIWA, T., MAKATO, F. & KUROIWA, H., 1992. Studies on the behavior of mitochondrial DNA: synthesis of mitochondrial DNA occurs actively in a specific region just above the quiescent center in the root meristem of *Pelargonium zonale. Journal of Cell Science, 101:* 483–493.

LANYON, L.E., 1992. Control of bone architecture by functional loading. *Journal of Bone and Mineral Research, 7* (Suppl. 2): S369–S375.

LARSON, P.R., 1983. Primary vascularization and the siting of primordia. In J.E. Dale & F.L. Milthorpe (eds), *The Growth and Functioning of Leaves,* pp. 25–51. Cambridge, U.K.: Cambridge University Press.

LIN, B.-L. & RAGHAVAN, V., 1991. Lateral root initiation in *Marsilea quadrifolia.* I. Origin and histogenesis of lateral roots. *Canadian Journal of Botany, 69:* 123–135.

LINDENMAYER, A., 1975. Developmental algorithms for multicellular organisms: a survey of L-systems. *Journal of Theoretical Biology, 54:* 3–22.

LINDENMAYER, A., 1984. Models for plant tissue development with cell division orientation regulated by preprophase bands of microtubules. *Differentiation, 26:* 1–10.

LINTILHAC, P.M., 1974. Differentiation, organogenesis, and the tectonics of cell wall orientation. III. Theoretical considerations of cell wall mechanics. *American Journal of Botany, 61:* 230–237.

LINTILHAC, P.M., 1984. Positional controls in meristem development: a caveat and an alternative.

In P.W. Barlow & D.J. Carr (eds), *Positional Controls in Plant Development*, pp. 83–105. Cambridge, U.K.: Cambridge University Press.

LLOYD, C.W. & BARLOW, P.W., 1982. The co-ordination of cell division and elongation: the role of the cytoskeleton. In C.W. Lloyd (ed.), *The Cytoskeleton in Plant Growth and Development*, pp. 203–228. London, U.K.: Academic Press.

LÜCK, J. & LÜCK, H.B., 1979. Two-dimensional, differential, intercalary plant tissue growth and parallel graph generating and graph recurrence systems. *Lecture Notes in Computer Science, 73:* 284–300.

LÜCK, J. & LÜCK, H.B., 1991. Double-wall cellwork systems for plant meristems. *Lecture Notes in Computer Science, 532:* 564–581.

LÜCK, J., LINDENMAYER, A. & LÜCK, H.B., 1988. Models of cell tetrads and clones in meristematic cell layers. *Botanical Gazette, 149:* 127–141.

LÜCK, J., BARLOW, P.W. & LÜCK, H.B., 1994. Cell genealogies in a plant meristem deduced with the aid of a 'bootstrap' L-system. *Cell Proliferation* **27**.

MATILSKY, M.B. & JACOBS, W.P., 1983. Regeneration in the coenocytic marine alga, *Caulerpa*, with respect to gravity. *American Journal of Botany, 70:* 635–638.

McDANIEL, C.N., 1992. Determination to flower in *Nicotiana*. *Current Topics in Developmental Biology, 27:* 1–37.

MESTRE, J.-C., 1967. La signification phylogénétique de l'embryogénie. *Révue Générale de Botanique, 74:* 273–322.

MILLER, J.G., 1978. *Living Systems.* New York, U.S.A.: McGraw-Hill.

MILLER, J.G. & MILLER, J., 1990. The nature of living systems. *Behavioral Science, 35:* 157–163.

MULKEY, T.J., KUZMANOFF, K.M. & EVANS, M.L., 1981. Correlations between proton-efflux patterns and growth patterns during geotropism and phototropism in maize and sunflower. *Planta, 152:* 239–241.

NAKIELSKI, J., 1987. Spatial variations of growth within domes having different patterns of principal growth direction. *Acta Societatis Botanicorum Poloniae, 56:* 611–623.

NUCCITELLI, R., 1988. Ionic currents in morphogenesis. *Experientia, 44:* 657–665.

O'SHEA, P., 1989. Biophysical mechanisms of development. Signal transduction and self-organization. In J. Krekule & F. Seidlová (eds), *Signals in Plant Development*, pp. 25–44. The Hague, The Netherlands: SPB Academic Publishing bv.

POPHAM, R.A., 1960. Variability among vegetative shoot apices. *Bulletin of the Torrey Botanical Club, 87:* 139–150.

PROTHERO, J., 1986. Methodological aspects of scaling in biology. *Journal of Theoretical Biology, 118:* 259–286.

PRUSINKIEWICZ, P. & LINDENMAYER, A., 1990. *The Algorithmic Beauty of Plants.* New York, U.S.A.: Springer-Verlag.

QUATRANO, R.S., 1990. Polar axis fixation and cytoplasmic localization in *Fucus.* In A. Mahowald (ed.), *Genetics of Pattern Formation and Growth Control*, pp. 31–46. New York, U.S.A.: Wiley-Liss.

QUATRANO, R.S., GRIFFING, L.R., HUBER-WALCHLI, V. & DOUBET, R.S., 1985. Cytological and biochemical requirements for the establishment of a polar cell. *Journal of Cell Science* (Suppl. 2): 129–141.

RUTISHAUSER, R. & SATTLER, R., 1985. Complementarity and heuristic value of contrasting models in structural botany I. General considerations. *Botanische Jahrbucher für Systematik, 107:* 415–455.

SEAGO, J.L. & MARSH, L.C., 1989. Adventitious root development in *Typha glauca*, with emphasis on the cortex. *American Journal of Botany, 76:* 909–923.

SINNOTT, E.W., 1960. *Plant Morphogenesis.* New York, U.S.A.: McGraw-Hill.

SONNENBLICK, B.P., 1950. The early embryogeny of *Drosophila melanogaster.* In M. Demerec (ed.), *The Biology of Drosophila*, pp. 62–167. New York, U.S.A.: Wiley.

SPENCER, H., 1867. *The Principles of Botany.* Vol. II. London, U.K.: Williams and Norgate.

SWAMY, B.G.L. & PADMANABHAN, D., 1961. Embryogenesis in *Sphenoclea zeylanica*. *Proceedings of the Indian Academy of Sciences, 54B:* 169–187.

TOMLINSON, P.B., 1969. On the morphology and anatomy of turtle grass, *Thallasia testudinum* (Hydrocharitaceae). II. Anatomy and development of the root in relation to function. *Bulletin of Marine Science, 19:* 57–71.

VORONIN, N.S., 1969. Apikal'nye meristemy v kornyakh golosemennykh rastennii i printsipy ikh graphicheskoii interpretatsii (Apical meristems of the root in gymnosperms and the principles of their graphical interpretation). *Botanicheskii Zhurnal, 54:* 67–76 (in Russian).

WENT, F.W., 1971. Parallel evolution. *Taxon, 20:* 197–226.

WERNICKE, W. & JUNG, G., 1992. Role of the cytoskeleton in cell shaping of developing mesophyll of wheat (*Triticum aestivum* L.). *European Journal of Cell Biology, 57:* 88–94.

YAKOVLEV, M.S. & YOFFE, M.D., 1957. On some peculiar features in the embryogeny of *Paeonia* L. *Phytomorphology, 7:* 74–87.

ZIMMERMANN, M.H. & BROWN, C.L., 1971. *Trees. Structure and Function.* Berlin, Germany: Springer-Verlag.

CHAPTER

10

The contribution of chimeras to the understanding of root meristem organization

LIAM DOLAN, PAUL LINSTEAD, R. SCOTT POETHIG & KEITH ROBERTS

CONTENTS

Abstract

Chimeric plants have been useful tools for investigating meristem organization and cell fate in shoots. They have revealed that the dicotyledon shoot meristem is composed of three layers that constitute three separate lineages. Chimeras induced at different times in development have provided useful information on the number of founder cells and their fate in organs. Fewer studies using chimeric plants to understand root meristem organization and cell lineage have been reported. These studies suggest that there may be a small number of quiescent initials in the root meristem and indicate that lateral roots are derived from a small number of cells. The results of these studies are reviewed here.

Shape and Form in Plants and Fungi
ISBN 0–12–371035–9

INTRODUCTION

Meristems are self-maintaining, proliferating populations of cells from which the post-embryonic architecture of the plant is derived. They are generally composed of fast-dividing, meristematic cells and a population of relatively quiescent cells (Steeves & Sussex, 1989). Cells that are 'left behind' by the growing meristem elongate and differentiate. The cues that direct these differentiation events are thought to be positionally dependent and not lineage restricted (Irish, 1991; Langdale *et al.*, 1989). This is most clearly illustrated in chimeric plants with a genetically green epidermis overlying an albino subepidermal tissue in which a rare division moves a cell of the epidermal lineage into the mesophyll (Stewart, 1978). This displaced cell abandons its epidermal character and assumes a subepidermal developmental fate becoming visible as a dark-green sector in otherwise albino tissue. While positional information is important in specifying cell fate, studying the cell lineage provides information on the numbers and organization of cells that give rise to organs (Furner & Pumfrey, 1992; Irish & Sussex, 1992; Poethig & Sussex, 1985). It is in this area that chimeric plants have been informative.

Histology provides a static view of the cellular organization in meristems. It does not permit the determination of cell fates but rather allows suggestions to be made as to the probable fates of cells. For example, misinterpretation of histological data led to the suggestion that the cells of the developing lamina of a dicot leaf are produced by the activity of a meristem at the leaf margin, known as the marginal meristem (Avery, 1933). While a number of quantitative studies indicated that the margins of leaves could not function as meristems, it was a clonal analysis that provided unequivocal evidence against the existence of the marginal meristem and showed that dividing cells in the margin of the tobacco leaf contributed relatively little to the growth of the lamina (Dolan & Poethig, 1991; Dubuc-Lebreaux & Sattler, 1980; Poethig & Sussex, 1985).

Clonal analysis allows the fates of single marked cells to be determined (Poethig, 1987). It depends on the ability to mark cells phenotypically which allows clones of descendant cells to be identified. Since these clones are usually marked with recessive mutations, these plants are known as genetic mosaics or chimeras. Sectors can be induced at key times in development to determine the contribution of specific cell populations to the tissue of the mature plant. Stable chimeras are formed if meristematic initials are labelled with a marker gene. The maintenance of a stable population of marked initial cells showed that the shoot apical meristem is composed of three layers each with a characteristic fate (Satina *et al.*, 1940; Stewart & Dermen, 1970).

The scorable markers used in plants are generally mutations in genes involved with the synthesis of chlorophyll, anthocyanin or wax, but those affecting morphological traits, such as hair morphology have also been used (Irish, 1991; Poethig, 1987). They should be scorable at the single-cell level so that clone boundaries are easily identified. They must also be cell-autonomous so that non-clonally related, adjacent cells do not become marked, i.e. the phenotype should reflect the genotype of the cell and not that of neighbouring cells. The marker should not damage the cells expressing it or put them at a competitive disadvantage. The frequency of sectors must be sufficiently low so that each sector is distinct and there is no doubt that each sector arose independently. A number of methods have been used to make genetic mosaics.

Chromosome breakage induced by ionizing radiation has been used to uncover recessive marker genes in many species (reviewed by Poethig, 1987). One method uses B chromosome translocations carrying dominant marker genes that are lost stochastically during development in maize (Dudley & Poethig, 1993). The dominant *Semigamy* mutation of cotton (*Gossypium barbadense*) causes the formation of maternal/paternal/zygotic mosaics in developing embryos resulting in the formation of chimeric plants which have been used to determine the organization of the developing embryonic meristem in cotton (Christianson, 1986; Turcotte & Feaster, 1967). Excision of transposable elements from marker genes 'marks' cells and their derived clones and has been used extensively in studying lineage in the maize flower (Dawe & Freeling, 1990, 1992; Dellaporta *et al.*, 1991; Dudley & Poethig, 1993). Chromosomal rearrangements and changes in ploidy have been used to identify clones in plants (for example Bain & Dermen, 1944; Satina *et al.*, 1940). Cytochimeras are chimeras that contain clones of cells with scorable chromosomal abnormalities and were first used to illustrate the organization of the cells in the shoot apical meristem of *Datura* (Satina *et al.*, 1940). Cytochimeras often arise during grafting and result in the formation of polyploid nuclei. Polyploidy can also be induced in cells by treatment with colchicine, while cells with other scorable chromosomal rearrangements can be induced with irradiation (Brumfield, 1943; Satina *et al.*, 1940).

CHIMERAS AND THE ORGANIZATION OF SHOOT MERISTEMS

Histological studies led to the proposal that there were distinct layers of initials in the shoot meristem (Schmidt, 1924). The epidermal and subepidermal layers constitute the tunica and divide anticlinically expanding the surface area of the growing meristem while the central layer of cells, the corpus, divides anticlinally and periclinally forming the core. Prompted by Baur's (1909) descriptions of chimeras, Satina *et al.* (1940) made cytochimeras to determine the contribution of each of these layers to the mature plant.

In their experiments Satina *et al.* (1940) treated seeds with colchicine and subsequently examined axillary meristems and mature tissue for the presence of polyploid cells. Treatment with colchicine induces a variety of sectors some of which are restricted to a single meristematic layer. Since axillary meristems are derived from a small number of cells in three meristematic layers, some lateral buds have entire layers derived from mutant cells. Plants in which one meristematic layer is genetically different from the other two are known as periclinal chimeras. A number of chimeric meristems with polyploid cells localized in individual layers were obtained and the contribution made by each layer could be determined by the histological examination of structures in the mature plant. The outer layer of cells (L-I) gave rise to the epidermal layer of the plant. The subepidermal layer (L-II) gave rise to large portions of the leaf and peripheral regions of the stem while the core (L-III) formed the central portion of these organs, including most of the vascular tissue. These experiments have been repeated many times using conventional genetic markers (Stewart, 1978).

While these periclinal chimeras clearly illustrate the distinct contribution made by the three layers of the shoot meristem, they give no indication of the number of initials in any one of the layers. In an investigation of the fate map of the formation of the embryonic shoot apex of cotton, the recovery of one-third and two-thirds sectorially chimeric plants revealed that the shoot meristem is derived from three cells in each of the L-I, L-II and L-III layers of the developing embryonic meristem (Christianson, 1986). It has been suggested that the maintenance of these sectorial chimeras during the growth of the plant indicates that three initials remained in each layer during the growth of the shoot, although one of the sectors is usually displaced in similar chimeras of cotton (Christianson, 1986). Such displacement of initials in meristems is not uncommon (Stewart, 1978). Nevertheless, the stability of the small number of apical initials in the shoot of other species is substantiated by the appearance of a one-third and two-thirds sectorial pattern through approximately 100 nodes in cranberry, 50 nodes in poinsettia and 26 nodes in privet (Bain & Dermen, 1944; Stewart & Dermen, 1970; Stewart, 1978). Stewart (1978) interpreted these, and other, chimeras as indicating that the small number (two to three) of initials in each of the three meristematic layers of the growing shoot (making nine in total) can be very stable (Bartels, 1961).

WHAT CHIMERAS HAVE TAUGHT US ABOUT ROOTS

Few experiments using chimeras to determine the organization of the root meristem in dicots have been undertaken. They are described in the following section. Most experiments involved the generation of cytochimeras. Some of the root chimeras described were induced by treatment with ionizing radiation or colchicine while others arose during grafting or occurred spontaneously in interspecific hybrids. The results suggest that there is a small number of root initial cells located in the quiescent centre, surrounded by a large population of actively dividing cells (meristematic cells) whose descendants will eventually leave the meristem.

In one experiment, *Crepis capillaris* seeds were germinated and grown for 2 days to a length of 2 mm and irradiated with 300 rads of X-rays to induce the formation of chromosomal rearrangements and polyploidy (Brumfield, 1943) (see scheme in Fig. 1). The plants were grown for a further 3 weeks, fixed and transverse sections made in the meristematic zone. It was estimated that the period of 3 weeks would allow all marked meristematic cells that were not initials to pass out of the meristematic zone. Thus, any marked cells present in the meristem at this time would be the descendants of permanent initial cells marked at the time of irradiation. Thirteen sectors were obtained in 12 roots. Nine sectors occupied approximately one-fourth of the transverse area of the root (Fig. 2). Sectors encompassed all tissue layers of the root (epidermis, cortex and stele). Three sectors occupied half the cross-sectional area of the root and again the marked cells were present in all cell layers. A single sector was restricted to four cells of the epidermis. These patterns suggest that there is a set of approximately four initials in the meristem of the 2-day-old seedling and each initial contributes to all cell layers of a wedge-shaped portion of the root. These initials are predicted to be organized as a disc of four cells (Fig. 3).

In a similar experiment *Vicia faba* seedlings were grown for 4 days to a length of 3 cm and irradiated (Brumfield, 1943). Plants were again allowed to grow for 3 weeks before they were fixed and sectioned. Three sectors were reported, two of which resembled the sectors obtained in *Crepis*, occupying approximately one-fourth of the cross-sectional area of the root with marked cells in all tissue layers. These sectors are again consistent with the presence of a small number (perhaps four) initials in

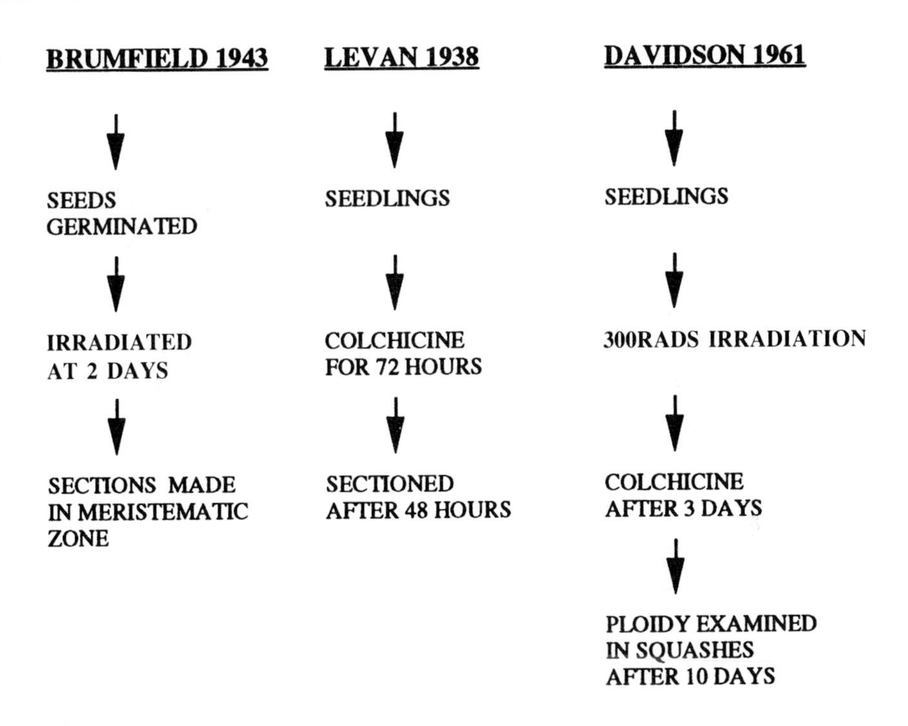

Figure 1 Three experimental schemes designed to determine the organization of initials in root meristems.

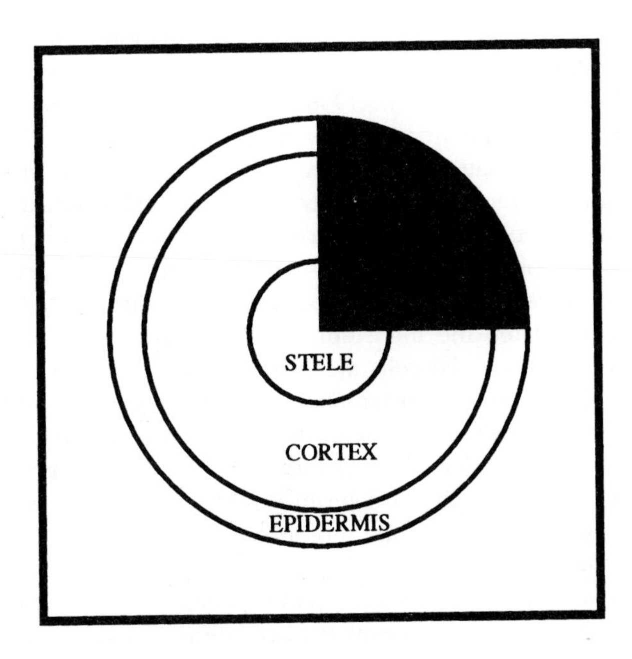

Figure 2 Schematic representation of a sector induced by irradiation in *Crepis capillaris* (Brumfield, 1943). A cross-section in the meristematic zone of an irradiated plant showing the location of a sector (black). The sector occupies approximately one-quarter of the cross-sectional area of the root and encompasses all tissue layers from the epidermis to the stele.

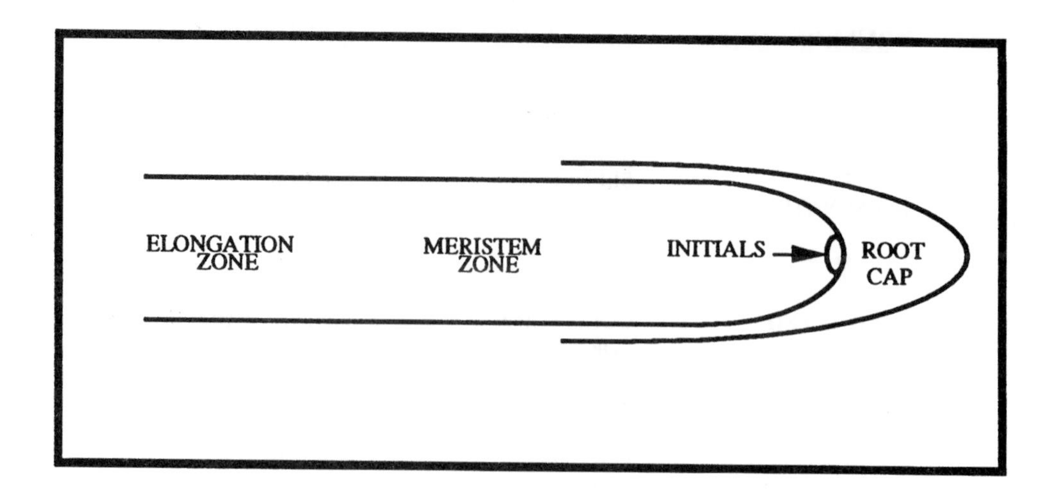

Figure 3 Zonation of the root tip as predicted by the chimeras reported by Brumfield (1943). These experiments suggested that a plate of four initial cells was located in the root tip.

the apex. A diagonal sector running through the centre of the root was also found and proved difficult to explain in terms of the 'four initials in a disc' model but was thought to have arisen as a result of cell death and/or a rare rearrangement among the cells of the initials. Such cell death events are common in irradiated plants. The first two sectors lend support to the model of a plate of four cells comprising the initials of the root.

Further evidence for the existence of a discrete population of initials, from which all cells in the meristem ultimately derive, came from investigations of the effect of colchicine on mitosis in roots (Levan, 1938). Colchicine is a drug that destabilizes microtubules and consequently treated cells fail to undergo mitosis after DNA replication, resulting in the formation of cells with an extra chromosome complement. Obviously, this change in ploidy is seen only in dividing cell populations while non-dividing cells maintain their normal chromosome number. Treatment of roots with colchicine-induced polyploidy in most cells of the meristem, suggesting that many cells were dividing. Nevertheless, a small population of diploid cells located at the apex eventually repopulated the meristem. This illustrates that a small number of slowly dividing (quiescent) initial cells are present at the root apex next to a large population of actively dividing (meristematic) cells which eventually leave the meristem. In one treatment, *V. faba* plants were treated with colchicine for 72 h, grown for a further 48 h in the absence of colchicine, sectioned (transversely) throughout the meristematic zone and the ploidy determined. Polyploid cells were present in all sections in the meristematic zone while diploid cells were only observed in the sections taken in the apical regions of the root. This suggests that there is a pool of colchicine-resistant and therefore slowly dividing cells in the apex of the root that contribute cells to the colchicine-sensitive population (actively dividing or meristematic cells). When roots were grown for longer periods after colchicine treatment, these diploid cells repopulated the entire meristem while the polyploid cells were lost. This demonstrates that the slowly dividing diploid cells near the apex are those cells from which the meristematic cells are derived. The disappearance of polyploid cells during this period of further growth is evidence for the transitory nature of these cells and therefore they cannot be considered as initials. An alternative explanation for these results is that the susceptible cells are no longer able to divide due to the colchicine treatment. This possibility is not considered likely by Davidson (1959) since it was found that polyploid sectors induced in the primary root contributed cells to normal lateral roots.

Another set of experiments using a combination of both X-irradiation and colchicine provided more evidence for the location of root initials in the quiescent centre (Davidson, 1961). If it were possible to induce the quiescent cells to divide by some experimental treatment, a subsequent treatment of these roots with colchicine would then induce polyploidy in these cells. If these cells are the initials of the root, the combination of treatments should lead to the formation of stable cytochimeras, since the initial cells will have become polyploid. It has been shown that treatment of roots with X-rays stimulates the cells of the quiescent centre to undergo DNA replication within 3 days and this is followed soon after by cell division (Clowes, 1956). Therefore, cells were irradiated with 300 rads and 3 days later treated with 0.025% colchicine (Davidson, 1961). After growing for a further 10 days, root squashes were made of the meristematic zone. Of the cells in the meristem 65% were tetraploid compared with 14% in the control (plants which had no irradiation treatment before exposure to colchicine). The results suggest that irradiation rendered the cells of the quiescent centre sensitive to colchicine resulting in the persistence of polyploid cells in the meristem. The persistence of these cells suggest that the cells of the quiescent centre are the initial cells of the root, i.e. the cells which give rise to all other cells of the root. Analysis of the types of chromosome complements induced in these plants estimated the number of initials to be between 2 and 8 cells with a mean of 4.3, a number that is remarkably close to that determined for the roots of *Crepis capillaris* and *Vicia faba* (Brumfield, 1943).

While the above experiments describe the determination of the number of root initials it gives no indication of the number of dividing cells in the meristem. The number of these meristematic cells surrounding the population of initials of the root of *Vicia faba* was determined by irradiating one-day-old seedlings and counting the number of cells with rearrangements at various times after irradiation (Davidson, 1959). Since the arrangement of the chromosomes was observed on root-squash material, the three-dimensional arrangement of the sectors could not be determined. Nevertheless, this study revealed that the root meristem (the dividing population of cells) is composed of 30–50 cells. These cells are not stable initials because they are eventually lost from the meristem. Sectors present in the lateral roots that were formed in the irradiated primary roots indicated that laterals arose from a small number of cells of the pericycle of the primary root. One to five sectors per lateral root were found, indicating that at least six cells contribute to the formation of the lateral root primordium. The persistence of these abnormal cells in the lateral meristems indicates that such cells are capable of forming root structures and their disappearance in the above studies is a consequence of meristem organization and not a result of the inability of cells with rearranged chromosomal complements to grow normally.

These experiments indicate that there is a small population of initials (probably four cells) located in the quiescent centre of the root apex of *V. faba*. This small population of cells is next to a larger population of 30–50 actively dividing cells. While estimates of the size of the actively dividing population have not been made in *C. capillaris* Brumfield's (1943) experiments suggest that there are also approximately four root initials in this species. A remarkable characteristic of these initial cells is that they give rise to sectors containing cells in all tissue layers from the epidermis to the stele, i.e. they are not restricted to any particular tissue type.

EXAMPLES OF STABLE SECTORIAL ROOT CHIMERAS

The experiments of Levan (1938), Brumfield (1943) and Davidson (1961) illustrate the existence of a relatively small number of initials each of which contributes to all tissue layers of the root. Stable sectorial chimeras in a variety of plant species had also indicated that a small number of initials exist at the root apex. These chimeras arose in plants with unstable karyotypes that were prone to somatic chromosomal rearrangements and from plants derived from graft chimeras where cells with abnormal karyotype arise spontaneously during the formation of organs from the callus at the graft-union site. A fraction of roots formed in plants of these types contained cells that could be distinguished on the basis of their chromosomal organization or number, i.e. they were cytochimeras. The distribution of sectors is described later. These chimeras should be interpreted with caution. Since it is not known at what stage this chimerism was induced it is possible that the patterns reflect the number of founder cells of roots rather than the number of stable initials in root meristems. Nevertheless, the arrangement of sectorial patterns is similar to those induced by irradiation of seedlings (Brumfield, 1943).

In one study, trisomic plants of *Crepis tectorum* were open pollinated, producing offspring with not only the expected parental chromosomal abnormality but also a number of novel chromosomal arrangements (Navashin, 1931). It is likely that the triploid chromosomal complement renders these cells susceptible to such somatic rearrangements. Each cell examined from an individual had the same karyotype in all plants but one. In this plant, two different karyotypes were observed. This suggests that at some time after the first embryonic division a cell underwent a somatic rearrangement marking the cell and all its descendants with an abnormal karyotype. Both sectors contributed to the formation of the root tissue of the plant. In one root the sector occupied approximately one-quarter of the cross-sectional area. This sectorial chimerism is identical to that described by Brumfield in *V. faba* (1943). Results of another study of *C. tectorum* plants grown from old seed indicate that the pericycle of the primary root is derived from three cells (Navashin, 1933). Lateral roots from plants displaying more than one karyotype in primary roots were sectioned. In approximately one-third of roots, all cells were of the same aberrant karyotype. The remaining two-thirds were entirely normal.

The somatic instability of abnormal chromosome complements found in species hybrids can result in the formation of spontaneous cytochimeras. In tobacco, roots of one such hybrid have been examined (Kostoff, 1930). One root was found in which half the tissue was haploid and the other half tetraploid. While this observation does not prove that there are four initials in this root it indicates that there may be a small number, perhaps two. Another common form of chromosomal rearrangement occurs during grafting when tetraploid cells are often formed. In roots formed on plants regenerated from callus of the graft union site in tomato, sectorial chimeras were observed and individual sectors encompassed all cell layers of the root from the epidermis to the stele (Lesley, 1925). This again shows that initials can contribute cells to many tissue layers of the mature root.

The observations on stable sectorial chimeras are difficult to interpret clearly. On the one hand, they appear to confirm the conclusions of Brumfield (1943) and Davidson (1961) that there is a relatively small number of stable initials in the meristem, each of which gives rise to cells in all layers of the root. On the other hand, it is possible that the organization of the chimeras is a reflection of the arrangement of founder cells that originally formed the roots and not the organization of initials at their growing apices. Since the stage at which the chimerism was induced is unknown, it is difficult to distinguish between these alternatives. Since the primary root of many species is derived from the basal tier of four cells in the octant stage embryo it is possible that the one-third and one-quarter sectors which are described in the 'stable' mosaics merely reflect the embryonic organization of these cells. If one of these cells undergoes a chromosomal rearrangement it will form a quarter sector in the developing primary root. Consequently, a quarter of the initials will carry the rearrangement. In this case it will appear as though there are four stable initials in the meristem while there may be many more.

Although a number of root chimeras have been described in the literature, no periclinal chimeras have been reported. Their absence suggests that there is no layering of cells into discrete lineages in the root meristem as there is in the shoot meristem. Brumfield's (1943) chimeras are consistent with this interpretation. However, lateral roots are derived from a single-cell layer that simultaneously becomes organized in layers (unlike the shoot meristem that is initiated from cells in three layers). Therefore most root sectors would be expected to be sectorial in organization.

CHIMERAS AND THE ORIGIN OF ADVENTITIOUS ROOTS

Adventitious roots are those that develop from shoot tissue or occasionally root tissue that does not normally form roots (Esau, 1977). Anatomical studies have shown that they arise endogenously (from cells in the vascular system) (Esau, 1977). By making root cuttings from periclinal chimeras of *Bouvardia* cv. Bridesmaid, Bateson recovered only cv. Hogarth regenerants (Bateson, 1916). Since it was known that this cultivar was a periclinal chimera with 'Hogarth' tissue in the L-III, it was concluded that adventitious roots were derived from the L-III. The initiation of adventitious roots from a single tissue layer is in agreement with histological data. It is in striking contrast with the formation of shoot meristems that are derived from cells derived in all three meristematic layers.

ARABIDOPSIS ROOT CHIMERAS

The *Arabidopsis thaliana* root is currently being used by a number of laboratories for the study of root development (Schiefelbein & Somerville, 1990; Schiefelbein & Benfey, 1991). Its small size and simple cellular organization combined with the ability to use *Arabidopsis* in genetic analysis, make it an ideal system for the study of morphogenesis. The histological organization of the *Arabidopsis* root has been described (Dolan *et al.*, 1993) and it is reviewed briefly here.

A scanning electron micrograph (Fig. 4) shows the organization of the cells in the epidermis. At maturity the root epidermis is composed of hair cells and non-hair cells. Cells in the meristematic region are small and hairs form in cells of certain files after elongation has ceased. Figure 5 is a longitudinal section through a primary root and illustrates the organization of the files of cells. A transverse section of a root in the elongation zone is illustrated in Fig. 6. There are eight endodermal files and eight cortical cell files surrounded by 19 epidermal cells. Fourteen pericycle cell files surround the vascular cells. The two protophloem cells have already differentiated in this specimen while the protoxylem cells have yet to form. The stele is therefore said to be diarch in form. There are two cell types in the epidermis, hair cells and non-hair cells. Serial sections through the root reveals that the cells that form hairs are usually those epidermal cells located in the cleft between two underlying cortical cells (Dolan, manuscript in preparation). This cellular organization is maintained until secondary thickening when derivatives of the stele divide and the outer layers of the root are lost.

While this description provides valuable information on the numbers of cells comprising the root, a more dynamic picture of meristem development is needed to describe how many initials there are and which tissues in the mature root they give rise to. A clonal analysis will provide this information and we have resorted to using transgenic plant lines for this purpose. This is necessary because of the paucity of appropriate genetic markers expressed in roots. Sectors induced in this way are free from the complications of cell death that are caused by ionizing radiation and the disruptive effects of colchicine.

The product of the *gus* A (*uid* A) gene of *Escherichia coli*, β-glucuronidase (GUS), is widely used as a reporter gene in plant systems because it catalyses the formation of a blue precipitate from the colourless substrate X-gluc (X-glucuronide). Under appropriate conditions, the blue precipitate is localized to those cells expressing the enzyme, i.e. it is cell a autonomous marker. It is therefore an ideal system for marking cells in clonal analysis. Insertion of the transposable element *Ac* into the untranslated

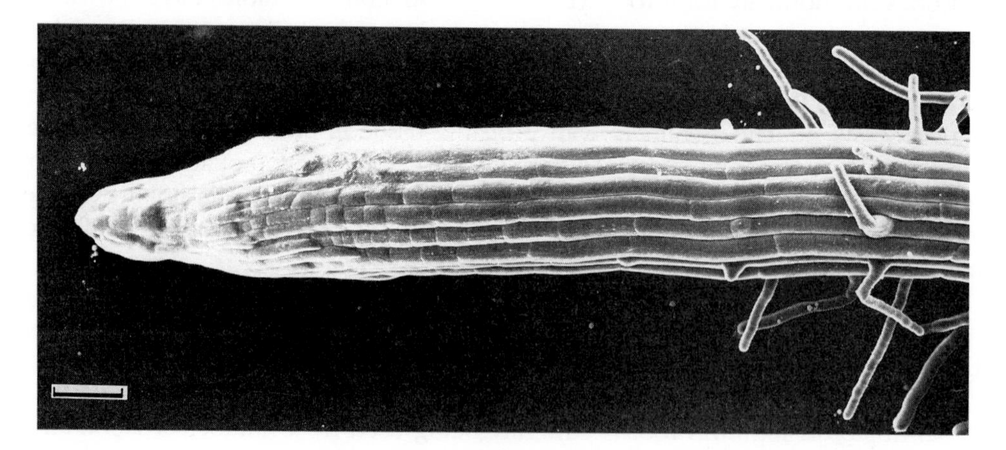

Figure 4 Scanning electron micrograph of a 5-day-old *Arabidopsis* primary root. The mature epidermis is composed of files of root hair cells and files of non-hair cells. The hairs form in the region where cell elongation has ceased. The region where hairs form is considered to be the zone of differentiation. The zone where the cells are expanding before hairs are initiated is known as the elongation zone. Scale bar 30 μm.

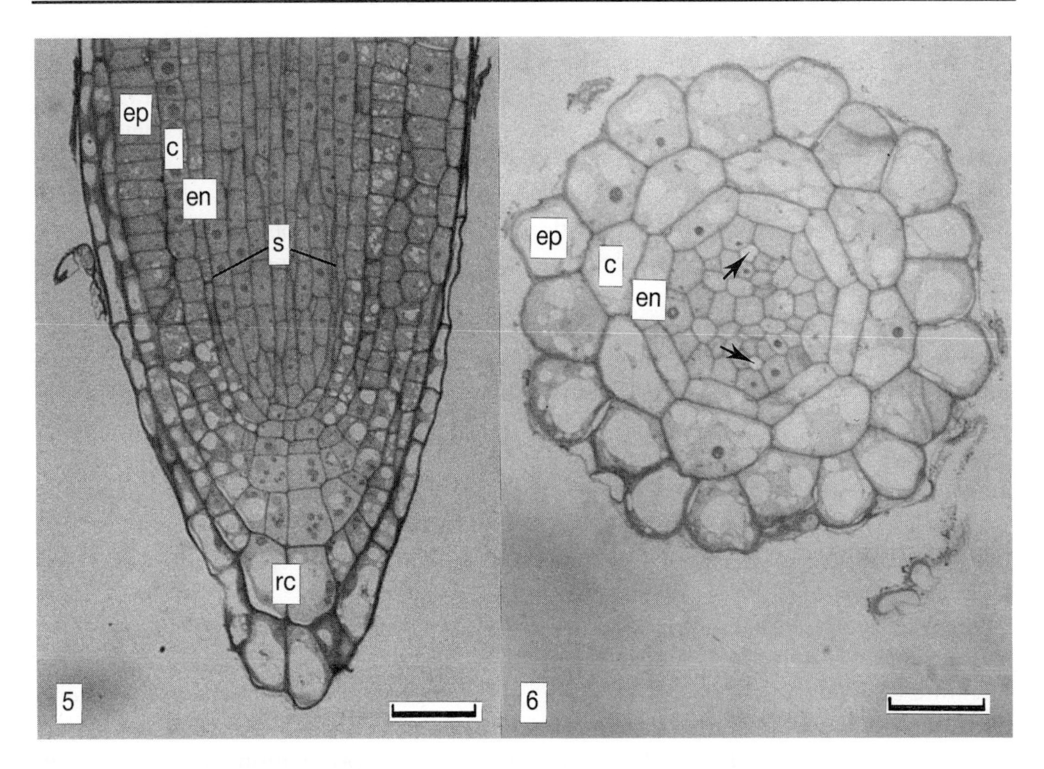

Figure 5 Median longitudinal section through a 10-day-old *Arabidopsis* root. rc, root cap; ep, epidermis; c, cortex; en, endodermis; s, stele. Scale bar 20 μm.

Figure 6 Transverse section in the elongation zone. The numbers of cortical and endodermal cells are invariant; each layer is composed of eight cells in transverse section. There are on average 19 cells in the epidermis. The two protophloem sieve tube elements have differentiated. The protoxylem has not yet formed. ep, epidermis; c, cortex; en, endodermis. Arrows points to protophloem elements. Scale bar 20 μm.

leader of the *uid* A gene blocks its expression in transformed plants (E. Lawson and C. Dean, personal communication; Finnegan *et al.*, 1989). The roots of such plants appear uniformly white. *Ac* element excision in a fraction of the cells restores GUS activity to those cells and their descendants. The random nature of these excision events means that they can be used to mark cells at different stages in development. The size and extent of these sectors provides information as to the number of meristematic cells and their ultimate fate (Poethig, 1987; Finnegan *et al.*, 1989).

Our preliminary data illustrate the usefulness of the transposon-based system for the study of root development. The clonal relationship of the lateral root cap and epidermis has been clearly illustrated. Our histological analysis had suggested that these two tissue types are derived from common initials. This prediction was verified by clonal analysis since a number of individual sectors were found to contain cells of both the lateral root cap and epidermis. No sectors in the lateral root cap were found running into the central root cap (columella) confirming the predicted clonal distinction between these tissues. In addition, sectors induced in the central root cap cells (columella) were restricted to that position and did not run into the lateral root cap. Sectors induced in the primary roots of the plant were generally small suggesting that the meristematic cells and not the initials had been labelled that implies that the initial cells may be located in the quiescent centre.

Many of the sectors in the lateral roots were larger than those observed in the primary root, indicating that either the 'initial' cells in laterals are not in fact quiescent or may be the result of an early excision event in the developing primordia when it was composed of only a few cells. Developmental regulation of the transposon resulting in earlier transposon excisions in lateral root primordia could also account for the difference in sector organization between these roots. These preliminary results attest to the value of the clonal analysis in understanding the cellular aspects of root morphogenesis.

Future studies on the organization of the root meristem by clonal analysis in concert with other methods will allow the determination of the number and organization of initial cells in the root meristem of *Arabidopsis*.

CONCLUSIONS

Chimeras have been useful tools for understanding the cellular organization of both shoot and root meristems. Since so few root chimeras have been reported in the literature there is no general consensus as to the organization of initials in root meristems. Nevertheless the results discussed here indicate that there may be a small number of initial cells that rarely divide surrounded by a population of more actively dividing cells. The root chimeras reported in this chapter suggest that each of these initial cells may form cells in all tissue layers of the root (including the root cap). Similarly, shoot chimeras have indicated that there is a small number (three) of initials in each of the three meristematic layers of dicots. Nevertheless, this interpretation has been dismissed by other workers (e.g. Clowes, 1956, 1959) indicating that more studies are necessary to clarify the situation. One problem with the interpretation of root chimeras is that in most cases the chimerism was induced by irradiation or treatment with drugs that may cause a disruption of meristem structure. Therefore it is possible that the patterns that we observe have more to do with the numbers and fates of cells that are involved with meristem regeneration and not normal meristem maintenance. Future studies using less destructive techniques to determine fate, such as those described here for *Arabidopsis*, will be informative. They will identify the location and number of initials in the root meristem and will reveal the contribution of these cells in the differentiated root. These analyses will lay the basis for other genetic and developmental studies of pattern formation and differentiation in the root.

ACKNOWLEDGEMENTS

The authors thank Kate Duckett, Clare Grierson, Katarina Schneider and Chris Staiger for helpful comments on earlier drafts of this manuscript. They also thank Sue Bunnewell for photography, and acknowledge Emily Lawson and Caroline Dean for providing transgenic lines. The AFRC and USDA (Grant 90-37261-5620) support root research in the Norwich and Philadelphia laboratories.

REFERENCES

AVERY, G.S., 1933. Structure and development of the tobacco leaf. *American Journal of Botany, 20:* 565–592.

BAIN, H.F. & DERMEN, H., 1944. Sectorial polyploidy and phylotaxy in the cranberry (*Vaccinium macrocarpon*). *American Journal of Botany, 31:* 581–587.

BARTELS, F., 1961. Zur Entwicklung der Keimpflanze von *Epilobium hirsutum*. IV der Nachweis eine Scheitelzellenwachstums. *Flora, 150:* 552–572.

BATESON, W., 1916. Root cuttings, chimeras and sports. *Journal of Genetics, 6:* 75–80.

BAUR, E., 1909. Das Wesen und die Erblichkeitverhältnisse der Var. albomarginate hor von *Pelargonium zonale. Z. Indukt. Abstammungs-Vererbungsl. 1:* 330–351.

BRUMFIELD, R.T., 1943. Cell lineage studies in root meristems by means of chromosome rearrangements induced by X-rays. *American Journal of Botany, 30:* 101–110.

CHRISTIANSON, M.L., 1986. Fate map of the organizing shoot apex in *Gossypium. American Journal of Botany, 73:* 947–958.

CLOWES, F.A.L., 1956. Localization of nucleic acid synthesis in root meristems. *Journal of Experimental Botany, 7:* 307–312.

CLOWES, F.A.L., 1959. Reorganization of root apices after irradiation. *Annals of Botany, 23:* 205–210.

DAVIDSON, D., 1959. Changes in chromosome complements of cells in cells of *Vicia faba* roots following irradiation. *Journal of Experimental Botany, 10:* 391–398.

DAVIDSON, D., 1961. Mechanisms of reorganization and cell repopulation in the meristema in roots of *Vicia faba* following irradiation and colchicine. *Chromosoma, 12:* 484–504.

DAWE, R.K. & FREELING, M., 1990. Clonal analysis of the cell lineages in the male flower of maize. *Developmental Biology, 142:* 233–245.

DAWE, R.K. & FREELING, M., 1992. The role of initial cells in maize anther morphogenesis. *Development, 116:* 1077–1085.

DELLAPORTA, S.L., MORENO, M.A. & DELONG, A., 1991. Cell lineage analysis of the gynoecium of maize using the transposable element *Ac. Development* (Suppl. 1): 141–147.

DOLAN, L. & POETHIG, R.S., 1991. A genetic analysis of leaf development in cotton. *Development* (Suppl. 1): 39–46.

DOLAN, L., JANMAAT, K., WILLEMSEN, V., LINSTEAD, P., POETHIG, R.S., ROBERTS, K. & SCHERES, B. 1993. Cellular organisation of the *Arabidopsis thaliana* root. *Development, 119:* 71–84.

DUBUC-LEBREUX, M.A. & SATTLER, R.A., 1980. Development des organes foliacles chez *Nicotiana tabacum* et le problème des meristemes marginaux. *Phytomorphology 30:* 17–32.

DUDLEY, M. & POETHIG, R.S., 1993. The heterochronic *Teopod 1* and *Teopod 2* mutations of maize are expressed non-cell autonomously. *Genetics, 133:* 389–399.

ESAU, K., 1977. *Plant Anatomy.* New York, U.S.A.: John Wiley.

FINNEGAN, E.J., TAYLOR, B.H., CRAIG, S. & DENNIS, E.S., 1989. Transposable elements can be used to study cell lineages in transgenic plants. *Plant Cell, 1:* 757–764.

FURNER, I. & PUMFREY, J.E., 1992. Cell fate in the apical meristem of *Arabidopsis thaliana. Development, 115:* 755–764.

IRISH, V.F., 1991. Cell lineage in plant development. *Current Opinion in Genetics & Development, 1:* 169–173.

IRISH, V.F. & SUSSEX, I.M., 1992. A fate map of the *Arabidopsis* embryonic shoot apical meristem. *Development, 115:* 745–753.

KOSTOFF, D., 1930. A chromosomal chimera of tobacco. *Journal of Heredity, 21:* 445–448.

LANGDALE, J.A., LANE, B., FREELING, M. & NELSON, T., 1989. Cell lineage analysis of the maize bundle sheath and mesophyll cells. *Developmental Biology 133:* 128–139.

LESLEY, M.M., 1925. Chromosomal chimeras in tomato. *American Naturalist, 59:* 570–574.

LEVAN, A., 1938. The effect of colchicine on root mitoses in *Allium. Hereditas, 24:* 471–486.

NAVASHIN, M., 1931. Spontaneous chromosome alterations in tomato. *Univ. California Publ. Agric. Sci., 6:* 201–206.

NAVASHIN, M., 1933. Altern der Samen als Ursache von Chromosomenmutationen *Planta, 20:* 233–243.

POETHIG, R.S., 1987. Clonal analysis of cell lineage patterns in plant development. *American Journal of Botany, 74:* 581–594.

POETHIG, R.S. & SUSSEX, I.M., 1985. The cellular parameters of leaf development in tobacco: a clonal analysis. *Planta, 165:* 170–184.

SATINA, S., BLAKESLEE, A.F. & AVERY, A.G., 1940. Demonstration of the three germ layers in the apex of datura by means of induced polyploidy in periclinal chimeras. *American Journal of Botany, 25:* 895–905.

SCHIEFELBEIN, J.W. & BENFEY, P.N., 1991. The development of plant roots: new approaches to underground problems. *Plant Cell, 3:* 1147–1154.

SCHIEFELBEIN, J.W. & SOMERVILLE, C., 1990. Genetic control of root development in *Arabidopsis. Plant Cell, 2:* 235–243.

SCHMIDT, A., 1924. Histologische Studien an phanerogramen Vegetationspunkten. *Botan. Arch., 8:* 345–404.

STEEVES, T.A. & SUSSEX, I.M., 1989. *Patterns in Plant Development,* 2nd edn. Cambridge, U.K.: Cambridge University Press.

STEWART, R.N., 1978. Ontogeny of the primary body in chimeral forms of higher plants. In S. Subtleney & I.M. Sussex (eds), *The Clonal Basis of Development.* New York, U.S.A.: Academic Press.

STEWART, R.N. & DERMEN, H., 1970. Determination of number and mitotic activity of shoot apical initial cells by analysis of mericlinal chimeras. *American Journal of Botany, 57:* 816–826.

TURCOTTE, E.L. & FEASTER, C.V., 1967. Semigamy in Pima cotton. *Journal of Heredity, 58:* 55–57.

CHAPTER

11

The contribution of auxin and cytokinin to symmetry breaking in plant morphogenesis

D.E. HANKE & S.J. GREEN

CONTENTS

Abstract

New symmetry relations are generated within homogenous and essentially amorphous tissues when a patterned distribution of specialized cells forms *de novo* during the wound regeneration sequence. The origins of the pattern can be traced to the unique physicochemical properties of auxin and cytokinin and the proteins catalysing their metabolism and transport. Positive feedback effects lead to the accumulation of these otherwise cell-permeant growth substances in some cells and their depletion from others, converting an initially homogeneous distribution into a patterned variation in concentration. Supporting evidence is provided from the results of original unpublished experimental work on the patterns of xylogenesis in storage pith of jerusalem artichoke tuber and cytokinin habituation in stem pith of tobacco.

Shape and Form in Plants and Fungi
ISBN 0–12–371035–9

INTRODUCTION

A most ingenious paradox

W.S. Gilbert

In development, new features emerge from groups of cells in which those features did not exist before in any recognizable form. They arise like Melchisedech without apparent antecedents and their 'spontaneous generation' is compelling. The most hypnotically fascinating examples are those in which a group of identical cells, mitotic products of the same parent cell, sort themselves into different types. Two examples of this sort of symmetry breaking will be considered later. That such a group of cells should acquire increased complexity as a completely autonomous process, independent of outside influences, seems to contravene some cybernetic version of the first law of thermodynamics. However, appearances are deceptive. Complexity is not actually being created *sui generis* but only interconverted. One form of complexity, the genetic information, is being translated into another, the physical phenotype that is the expression of that genetic information.

So much is dogma. The paradox lies in the fact that the base sequence is only one-dimensional whereas the organism into which it is translated is three-dimensional. At the level of a single cell the disparity is not that hard to reconcile if we focus on the process of automatic folding of nascent polypeptides, spontaneously converting the one-dimensional sequence into three-dimensional objects. From here, by way of the specific contribution of each of many proteins to the cytoskeleton, it is a surprisingly short step to the control of cell shape. However, if we move up to a large scale and consider groups of cells, the control system has to operate over a commensurately larger scale, coordinating the responses of many cells, which places it some considerable way further down the line of command from the molecule of DNA. This in turn means that the black box the geneticist has to invoke for delivering phenotype from genotype is even bigger and blacker than usual, and so has room inside it for a biochemist. To the biochemist, this black box is like a rathole to a Jack Russell terrier, even to the extent that it may be difficult to persuade him or her to extricate themselves long after it has been conclusively demonstrated that the rat is no longer at home.

When individual cells within a group become different from each other (the unit of differentiation is the cell), they can do so only if they influence each other which, for plant cells, invariably seems to involve diffusion of hormonal signals across the intervening cell walls.

Hormone and receptor are male and female, mobile versus passive, cell-pervasive versus cell-autonomous, locking in everlasting duality. Over the years, from Alan Turing (1954) to Louis Wolpert (1971), the perceived status of the signal has waned as more and more examples have pointed up the overriding importances of sensitivity, of the cell-specific nature of the response, of changing patterns of responsivity in the responding cells. It is our contention that the distribution of control between the signal and the responding cell is not constant, but is different for different responses. While in the majority of cases the responding cell may well contribute an overwhelming proportion of the total control, there are circumstances in which the level of signal is decisive. One such occasion is in symmetry breaking, when the progressive amplification of a tiny, initial inhomogeneity is focused by feedback between cells and culminates in its orderly resolution as a sharp prepattern of morphogen.

THE MODEL

To build up a large difference between cells as a result of amplifying a tiny difference in signal strength requires positive feedback processes such as autocatalysis. The classical example is the reaction-diffusion model of Gierer and Meinhardt. There are four component processes (Fig. 1). Activator A activates its own production (process 1), and stimulates the production of inhibitor I (process 2), which suppresses activator A production (process 3) and diffuses much faster than activator A (process 4). Processes 2 and 3 constitute a negative feedback loop, tending to bring any uniform increase or decrease in the concentration of either A or I back to the original concentration. Processes 1 and 4 will amplify any local increase in the concentration of A because the inhibitor produced in response diffuses away faster and suppresses activator production in the area outside that of local activation. Within the area of local activation, the positive feedback effect of process 1 tends to increase the concentration of A, and a stable, non-uniform distribution of A and I is generated (Meinhardt, 1974).

THE EXAMPLES

Over the years, we have experimented with two plant tissue culture systems in which an initially homogeneous set of cells of the same type sort themselves out into a mixture of two different types under the influence of a plant growth substance. We think that in these examples the mechanism for amplifying differs from the reaction-

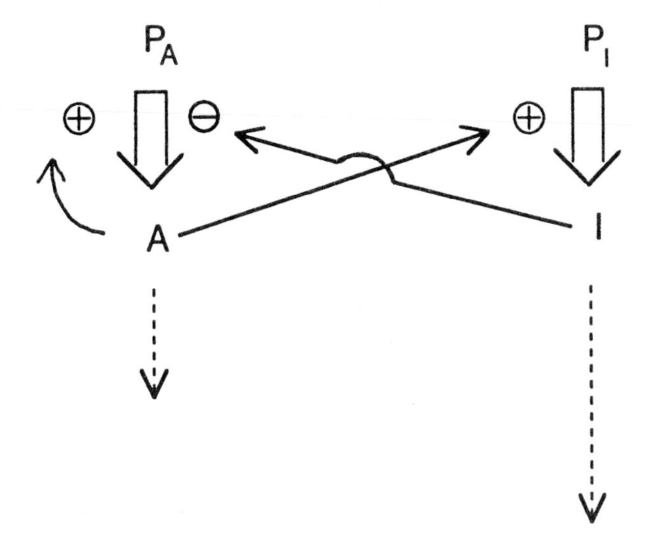

Figure 1 Schematic diagram of the reaction-diffusion model. A is activator, P_A its precursor; I is inhibitor, P_I its precursor. Solid arrows denote stimulation (⊕) or inhibition (⊖). Dashed arrows denote diffusion (after Meinhardt, 1974).

diffusion model in several interesting ways. It seems that control of the intercellular movement of growth substances may play a much more important role than the control of their synthesis, and there is no absolute requirement for the inhibitor, an economy to be expected in view of the constraints of evolution.

Auxin-induced xylogenesis in jerusalem artichoke tuber pith

For the past 15 years we have run an undergraduate practical in which cylindrical tuber pith explants (1 mm high, 4 mm diameter) of the Yeoman strain (Yeoman & Evans, 1967) of *Helianthus tuberosus* L., 'Bunyard's Round', are incubated on nutrient medium containing a xylogenic concentration of indol-3-ylacetic acid (details as in Dalessandro, 1973). The hypertonic medium shrinks the appressed surface of the explant, distorting the explant; and auxin and nutrients diffuse in via its lower perimeter which forms a ring of contact with the medium. Seven days later a network of xylem has differentiated under the surface, within an outer layer of newly formed wound callus (Fig. 2A).

As shown for *Coleus* stem pith (Comer, 1978), this morphogenesis involves two component processes, cell division followed by cell differentiation. Division is initiated in cells situated a few cell diameters below a cut surface, generating a continuous sheet of dividing cells. Auxin is required and, by implication, putative volatile regulators that enable the proximity of the cut surface to be sensed. The pattern of cell divisions is therefore guided by the position of the cut, enabling tissue regeneration to follow the unpredictable topography of a damaged surface. Within the sheet some, but not all, dividing cells differentiate as wound vessel members.

How is it that some of the dividing cells differentiate, but not their neighbours on either side? Xylem differentiates as strands and networks within the continuous sheet of dividing cells. How is this pattern ensured and mass differentiation of all the cells in the sheet avoided?

There is a separate and specific requirement for auxin in xylogenesis, over and above the requirement for auxin in initiating the preceding cell divisions. The clearest evidence is that in some tissues, auxin treatment specifically induces xylogenesis without prior division (Fukuda & Komamine, 1980). We can explain the pattern of xylem if it is prefigured by local accumulation of xylogenic levels of auxin in strands and networks with a complementary pattern of interstrand depletion of auxin, and there are good reasons for predicting that just such a pattern will develop spontaneously as auxin diffuses across a uniform array of otherwise identical cells.

By stimulating H^+-extrusion from cells, auxin promotes its own uptake and transport (evidence reviewed in Goldsmith, 1977), the autocatalytic mechanism for a positive feedback effect. As auxin diffuses across a uniform field of cells, those cells which by random chance contain slightly more auxin than their neighbours at the same position in the gradient will tend to acquire it from them, strengthening their ability to accumulate more. The down-gradient neighbours of such privileged cells may well be disadvantaged relative to this upstream process, but will enjoy an advantage relative to their neighbours at the same position in the gradient, *usw.* The system is analogous to a sand tray demonstration of the origin of water courses and, as for a sand tray, a 'drainage tree' pattern (Fig. 2B) involving progressive capture might be predicted. Similar patterns can be obtained more quickly by spreading oil paint on a piece of glass, covering it with non-absorbent paper and peeling off the paper in

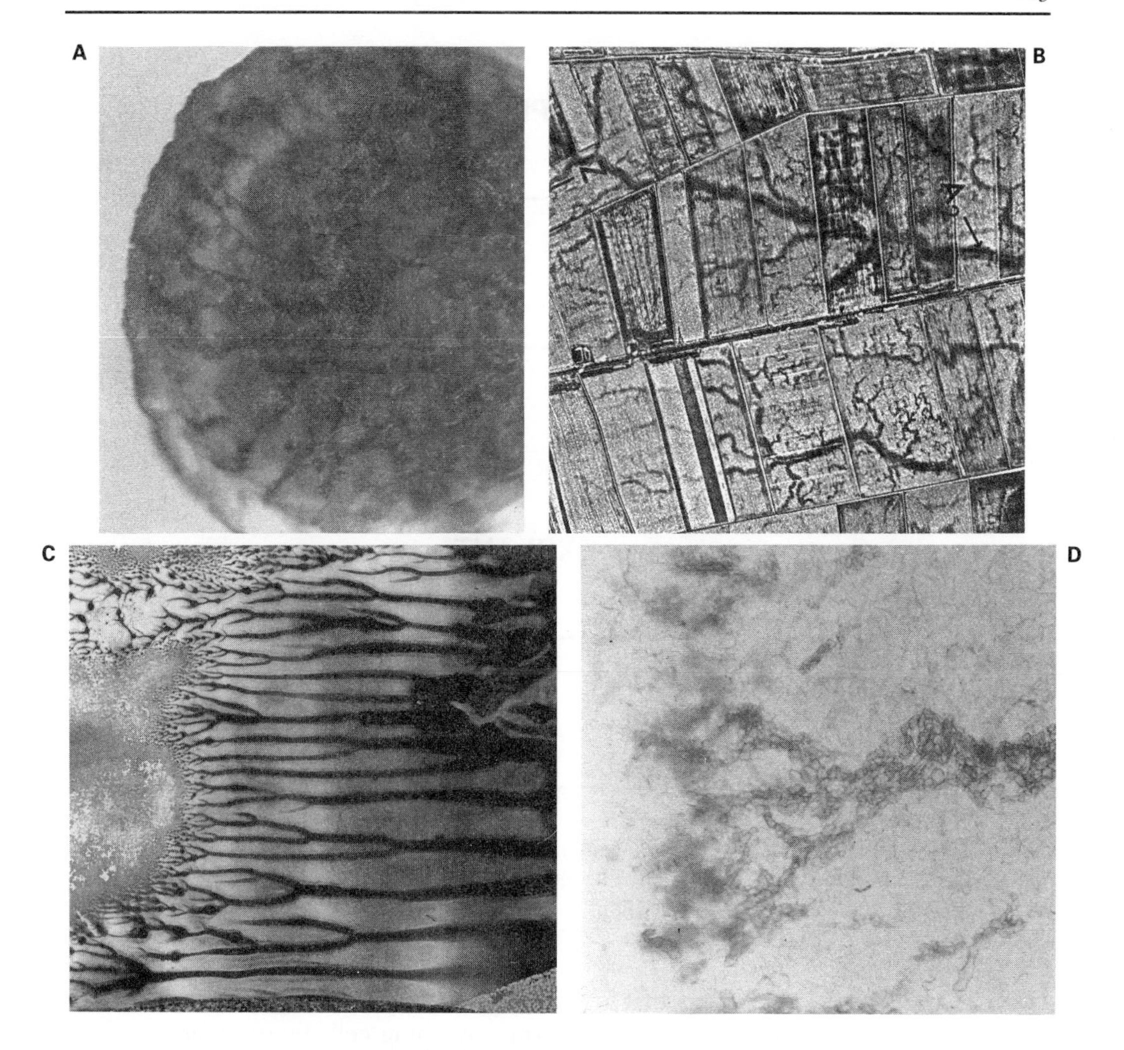

Figure 2 **A** Low-power micrograph of a 4×1 mm jerusalem artichoke tuber pith explant after 7 days of incubation on a xylogenic concentration of auxin, cleared and stained with safranin as in Dalessandro (1973). **B** Negative print of an aerial photograph of a fossilized creek system. Silt in a former marshland drainage system shows up light (dark in this negative) against dark fenland soil (light in this negative). **C** Painting by Pery Burge. **D** Micrograph of the cleared and stained result of an experiment in which an auxin gradient and a cytokinin counter-gradient were set up across a thin sheet of jerusalem artichoke tissue. The source of auxin was at the left, the source of cytokinin far to the right, of the region of the sheet depicted here. The frontispiece shows a coloured version of this figure.

one direction (Fig. 2C). In this case, surface tension (the high chemical affinity of oil for oil relative to glass and paper) is the basis of positive feedback.

As a simple test of this idea, we generated an artificial gradient in auxin concentration across a sheet of jerusalem artichoke tissue. Using the techniques and media of Dalessandro (1973), a sheet of tuber pith 50 mm × 20 mm × 1 mm was laid on growth regulator-free nutrient medium. A 35 mm × 5 mm × 1 mm slip of 10 g l^{-1} agar, 5 μM naphth-1-ylacetic acid (a xylogenic concentration), was laid on top of the tissue across one end (Fig. 3A). After 14 days in the dark at 25°C, the tissue was cleared and stained. Wound vessel members had differentiated as scattered groups and individuals in callus that formed on the explant in the area all around the agar slip. No 'drainage tree'

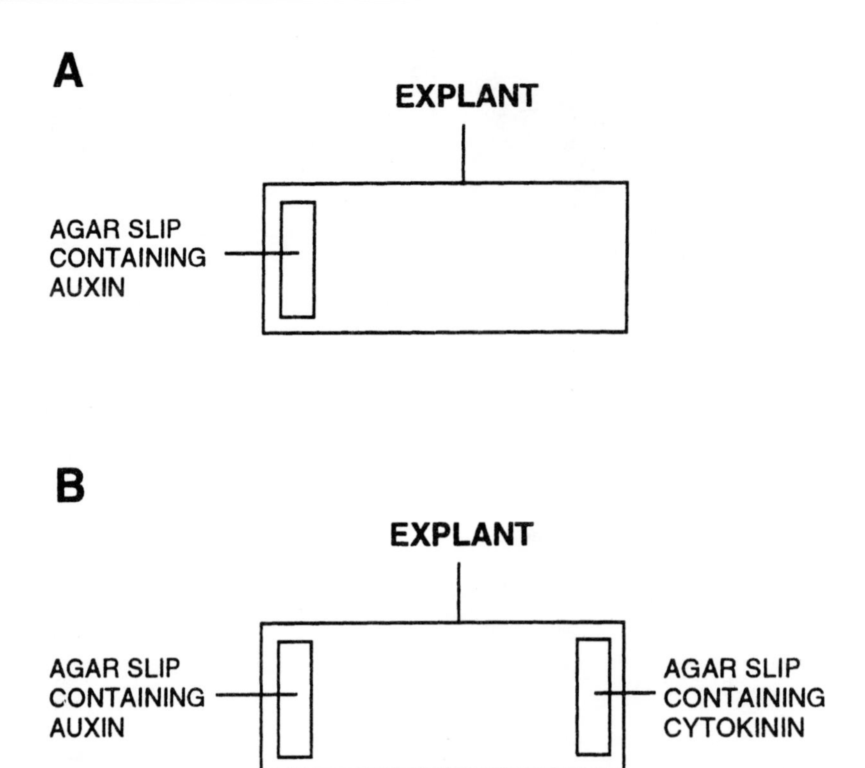

Figure 3 Schematic diagram of the arrangement of agar slips on a sheet of jerusalem artichoke tuber pith tissue in an experiment designed to set up: **A** a diffusion gradient of auxin, **B** a gradient of auxin with a counter-gradient of cytokinin.

patterns were seen. Reasoning that we might be able to enhance the difference between high auxin-containing and low auxin-containing cells by providing a counter-gradient of an auxin-antagonist, we repeated the experiment but added an agar slip of 5 μM kinetin, placed on the upper surface of the explant at the opposite end to the auxin source (Fig. 3B). The result was the differentiation of strands of xylem running from close to the auxin source a short way towards the cytokinin source. Close to the auxin source, the xylem took the form of many fine strands. Further away from it, the strands joined in typical 'drainage tree' patterns as if captured by larger strands (Fig. 2D).

In that this pattern exactly matches the prediction, we have some justification for continuing to test the model proposed here for the formation of patterns of wound vessel members. We need, for example, to establish the existence of the prepattern of channels of high auxin-level, high auxin-transport capacity embedded in the auxin-impoverished mass of tissue. From these results we cannot say that activator alone can generate the pattern because, although the spacing of xylem is well defined for disc-shaped explants on auxin alone, we could only obtain strands within sheets if a cytokinin was applied to the other end of the sheet. However, it does appear that the autocatalytic basis of positive feedback need not involve synthesis, but could be a result solely of the control of auxin movement by auxin. These, and other ideas on the role of auxin in pattern formation, have been discussed by Tsvi Sachs (1981).

Cytokinin-induced habituation in tobacco stem pith

The second example also involves a set of uniform pith cells sorting themselves out into two different types, in this case: cells stably dependent on exogenous cytokinin for division in the presence of auxin (C^-) *versus* cells stably independent of a supply of cytokinin for division (C^+). The cells in the pith of a tobacco stem at the time it is cut are in neither of these states. Instead they are unstably cytokinin-dependent in that after explantation onto nutrient medium they do not divide in response to auxin unless cytokinin is supplied, but a proportion of them spontaneously become C^+. The process is referred to as habituation, the resulting C^+ types as habituated cells. The cells that do not habituate (always the great majority) do not stay the same; they become C^-. The phenomenon has been intensively investigated by Fred Meins Jr whose research group has characterized various features of habituation, reviewed in Meins (1989). We have been using the system in an undergraduate practical class on radioimmunoassay of zeatin-cytokinins for the past 10 years.

Significantly, exogenous cytokinin at very low concentrations will induce a much higher proportion of the pith cells to become C^+, provided the cytokinin is supplied soon after cutting, i.e. before the cells have become C^- or C^+. The concentration of cytokinin required to promote habituation is an incredible 10^{-10} M, only one-thousandth of that required to induce divisions in the same cells. That supplying a minute amount of cytokinin can bring about a transition to a state in which the cell is independent of any exogenous cytokinin suggests an autocatalytic effect of cytokinin on the internal level of cytokinin, i.e. that the freshly cut pith cells are in a state of readiness to amplify dramatically the tiniest traces of this growth substance. In support of this idea, there seems to be a very sharp threshold for cytokinin-promoted habituation between ineffectually low and maximally effective concentrations, as would be expected for an autocatalytic effect (Meins & Lutz, 1980).

What is the biochemical basis for the autcatalytic effect?

Using our radioimmunoassay for zeatin-cytokinins (Turnbull & Hanke, 1985), we found that both C^- and C^+ lines of callus tissue from stem pith of *Nicotiana tabacum* L. cv. 'Havana 425' had the same, very low levels of zeatin-cytokinins: 1 to 3 pmol zeatin riboside equivalents g^{-1} fresh weight. This means that C^+ lines do not have cytokinin degradation blocked or synthesis stimulated, but instead either the cell cycle block that requires cytokinin for alleviation is now non-functional or the cells are cytokinin-hypersensitive and even the very low endogenous levels are now sufficient to sustain cell division.

For experiments in which we measured cytokinin levels during the early stages in habituation we used a 35°C treatment to increase the proportion of pith cells likely to habituate instead of cytokinin treatment, with a view to avoiding the complications due to exogenous cytokinin. The scheme of the experiments is shown in Fig. 4. Using sterile technique, 5 mm × 5 mm × 5 mm cubes of stem pith were cut from internodes 9 to 15 of 3-month-old greenhouse grown plants of 'Havana 425' and explanted onto solidified Linsmaier & Skoog, 11 μM naphth-1-ylacetic acid, following the standard Meins protocol (Meins & Lutz, 1980) as closely as possible. Control tissue was kept at 25°C in the dark for 14 days and did not habituate (25N). Tissue kept for 7 days at 35°C only revealed whether or not each 5 mm cube had produced habituated tissue after a period of 7 days at 25°C to allow growth. Those showing habituated sectors were dissected into clearly habituated (35H) and clearly non-habituated (35N(–H))

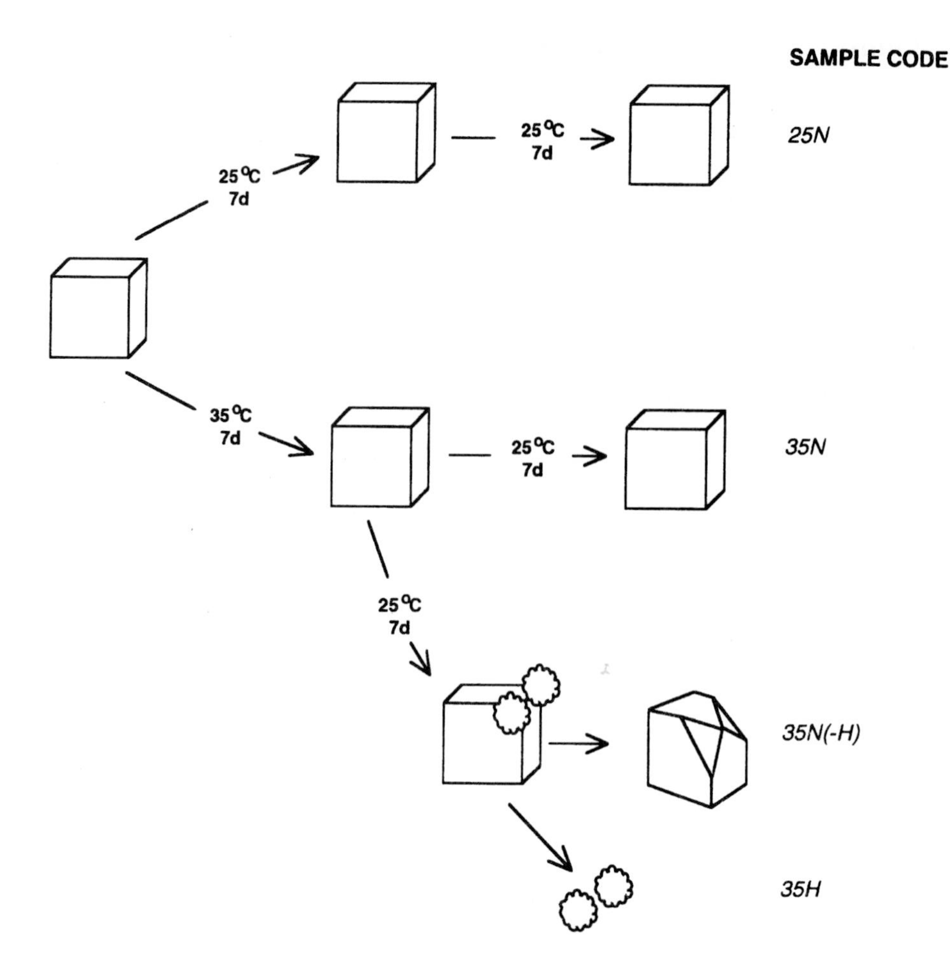

Figure 4 Design of temperature treatments (25°C vs 35°C) and sampling for an experiment to investigate the level of zeatin-cytokinins in cytokinin-independent habituated (H) and cytokinin-dependent non-habituated (N) tissue at 14 days after explantation of pith cubes from tobacco stem on to solidified nutrient medium containing auxin.

samples. Tissue samples were extracted, part purified by SEP-PAK and radioimmunoassayed as described in Turnbull & Hanke (1985). Estimates of the total zeatin-cytokinins in each of these different types of tissue at 14 days after excision are presented in Table 1. Two important points emerge. First, the 35°C treatment does not by itself affect the level of zeatin-cytokinins. If anything zeatin-cytokinin levels fall during culture. Second, C^+ tissue at 14 days contained levels of zeatin-cytokinins up to ten times higher than those in the adjacent C^- tissue of the same pith cube.

It could be argued that the extra cytokinin is a product of the cell division that is restricted to the C^+ tissue in these conditions, but the fact that with time the divisions continue while the excess of cytokinin disappears disproves that idea.

We resolved the zeatin-cytokinins into individual compounds by reversed-phase HPLC (Table 1), and it turns out that almost all of the excess zeatin in the early C^+ tissue is as the ribotide.

Now, cytokinins are very different from most other plant morphogens in their physicochemical properties. The other plant growth substances are hydrophobic but

Table 1 Tissue content of zeatin-cytokinins

Tissue type	Total zeatin-cytokinins before HPLC	Zeatin ribotide	Zeatin riboside	Zeatin free base	Zeatin 9-glucoside
Stem pith at time zero	14	3	0.4	n.d.	0.2
25N	3	2	n.d.	n.d.	0.6
35N	3	5	0.2	4	0.5
35N(–H)	6	8	0.2	n.d.	n.d.
35H	42	25	3	n.d.	4

n.d., none detected.

carry a carboxyl group so that they undergo anion trapping in the cytoplasm of plant cells as follows. Outside the cell they exist predominantly in the neutral hydrophobic protonated form and diffuse across the plasma membrane by passive, non-saturable diffusion – they simply dissolve in and out of the membrane. Once through it, these molecules deprotonate in the mildly alkaline cytosol. Now in the form of a charged anion the growth regulators become lipid-insoluble and must continue to accumulate here so long as the cell continues to pump out the H^+ ions the molecules of growth regulator bring in with them.

Compare this with our picture (Fig. 5) of cytokinin transport at the cellular level as

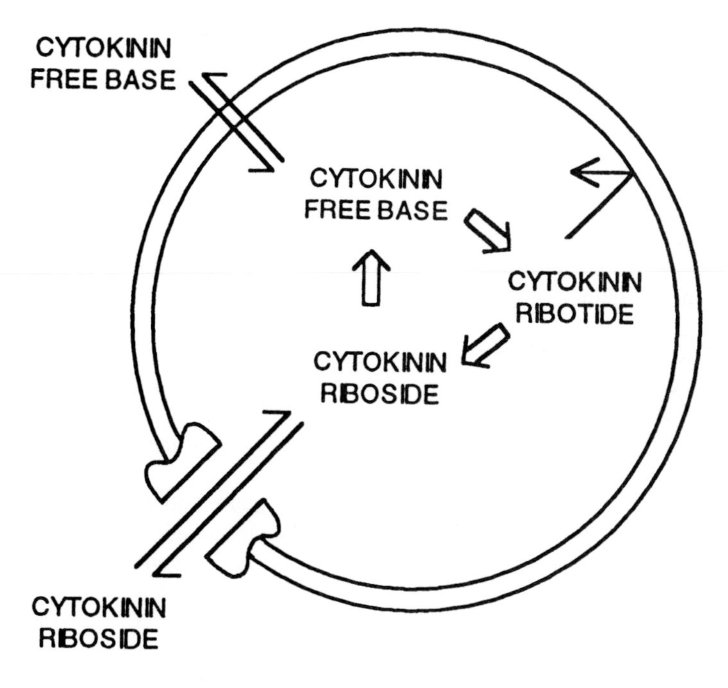

Figure 5 Schematic diagram of the features of cellular transport of the different metabolic forms of cytokinins (after Laloue *et al.*, 1981). The diagram depicts a plasma membrane with one purine nucleoside transporter. Open arrows depict metabolic conversion, solid arrows diffusion.

determined by Laloue *et al.* (1981). In the free base form cytokinins are sufficiently hydrophobic to pass both in and out of the cell by passive, non-saturable diffusion. Once inside the cell they are subject to exactly the same metabolism as the other purines, adenine and guanine, and are efficiently phosphoribosylated to the ribotide and in this charged, anionic form they cannot get back out at all. Ribotide is a source of riboside by dephosphorylation. Cytokinin ribosides are free to move in and out of cells, provided the cells concerned have the purine nucleoside transporter in the plasma membrane. The transport of cytokinin ribosides across this membrane is therefore passive but saturable.

It now seems fairly clear to us that the positive feedback mechanism involved in habituation is based on the accumulation of cytokinin, as the ribotide, at the expense of cytokinin picked up from other cells. We do not yet know how. It could be because cytokinins promote phosphoribosylation of purines, or inhibit dephosphorylation of ribotide, or some other connection.

CONCLUSION

Any positive feedback mechanism for symmetry breaking across a group of cells that is built around a molecule that stimulates its own uptake is simpler than one based on a molecule that stimulates its own synthesis. Using synthesis it is necessary to postulate a second component, an inhibitor, to 'talk' to the opposite cell type, rushing out to dampen them down. Using transport the other cells are damped down by being starved of the activator. It seems to us that when both auxins and cytokinins form morphogenic patterns across fields of cells, in both cases the autocatalytic core device at the heart of the positive feedback necessary to amplify signal up to active levels is one based on transport, not synthesis.

REFERENCES

COMER, A.E., 1978. Pattern of cell division and wound vessel member differentiation in *Coleus* pith explants. *Plant Physiology, 62:* 354–359.

DALESSANDRO, G., 1973. Interaction of auxin, cytokinin and gibberellin on cell division and xylem differentiation in cultured explants of jerusalem artichoke. *Plant and Cell Physiology, 14:* 1167–1176.

FUKUDA, H. & KOMAMINE, A., 1980. Direct evidence for cytodifferentiation to tracheary elements without intervening mitosis in a culture of single cells isolated from the mesophyll of *Zinnia elegans*. *Plant Physiology, 65:* 61–64.

GOLDSMITH, M.H.M., 1977. The polar transport of auxin. *Annual Review of Plant Physiology, 28:* 439–478.

LALOUE, M., PETHE-TERRINE, C. & GUERN, J., 1981. Uptake and metabolism of cytokinins in tobacco cells: studies in relation to the expression of their biological activities. In J. Guern & C. Péaud-Lenoël (eds), *Metabolism and Molecular Activities of Cytokinins*, pp. 80–96. Berlin, Germany: Springer-Verlag.

MEINHARDT, H., 1974. The formation of morphogenetic gradients and fields. *Berichte der Deutschen Botanischen Gesellschaft, 87:* 101–108.

MEINS, F., Jr, 1989. Habituation: heritable variation in the requirement of cultured plant cells for hormones. *Annual Review of Genetics, 23:* 395–408.

MEINS, F., Jr & LUTZ, J., 1980. The induction of cytokinin habituation in primary pith explants of tobacco. *Plant, 149:* 402–407.

SACHS, T., 1981. The control of the patterned differentiation of vascular tissues. *Advances in Botanical Research, 9:* 152–262.

TURING, A.M., 1954. The chemical basis of morphogenesis. *Philosophical Transactions of the Royal Society of London, B, 237:* 37–72.

TURNBULL, C.G.N. & HANKE, D.E., 1985. The control of bud dormancy in potato tubers. Measurement of the seasonal pattern of changing concentrations of zeatin-cytokinins. *Planta, 165:* 366–376.

WOLPERT, L., 1971. Positional information and pattern formation. *Current Topics in Developmental Biology, 6:* 18.

YEOMAN, M.M. & EVANS, P.K. 1967. Growth and differentiation of plant tissue cultures II. Synchronous cell divisions in developing callus cultures. *Annals of Botany, 31:* 323–332.

CHAPTER

12

The regulation of shape and form by cytosolic calcium

A.J. TREWAVAS & M. KNIGHT

CONTENTS

Shape and Form in Plants and Fungi
ISBN 0–12–371035–9

Abstract
The calcium hypothesis which regards calcium homoeostasis as critical to plant cell signal transduction is outlined. The special emphasis this hypothesis places on the measurement and imaging of calcium is discussed. Results obtained with a new method of calcium measurement, involving transformation with aequorin genes are described briefly. Touch and wind are two mechanical signals which induce immediate increases in intracellular calcium, $[Ca^{2+}]_i$. The consequences of these observations for the function of growth substances and for theories that involve mechanical influences in shape and form are considered.

LIVING ORGANISMS TRANSDUCE ENVIRONMENTAL SIGNALS TO SURVIVE

Living organisms must grow and reproduce in a varying, often hostile, environment. In general, animals constrain environmental damage to development by the use of the protected environment of an egg or uterus for embryogenesis. Thus, subsequent growth of the free-living animal takes place in prespecified tissues. In the mature animal elaborate sensing systems are integrated together with a rapid communication network of hormones and nerves. These networks permit rapid responses, often movement, to signals that specify food, predators or other unfavourable environments. Both internal and external signals are perceived by receptors that are usually located on the cell surface. The activated state of the receptor is then transduced so that the stimulated cell can construct a response.

The sessile life-style of plants has necessitated the evolution of different types of behaviour. The environment has to be tolerated rather than avoided. Although plants undergo a superficially similar embryological phase with the formation of the seed they also continue to grow and develop throughout their life-cycle specifying and generating new tissues. Plants have evolved a modular type structure (environmental and predator loss is easily replaced) and development is often plastic (Trewavas, 1986a). Plasticity, an integrated response of development to environmental variation, involves relatively novel control circuitry and plant growth substances may be a fundamental part. However, the essentials of signalling are still present. Although plants have not evolved elaborate and specialized sensory organs (most plant cells seem able to sense most signals), environmental signals are sensed by cellular receptors and the signals transduced. A response is constructed which maximizes the chances of survival and subsequent reproduction.

CALCIUM: A CENTRAL MOLECULE IN SIGNAL TRANSDUCTION

The transduction systems of both plant and animal cells seem to use a limited number of molecules to elaborate a cellular response to signals. One of these molecules is cytosolic calcium, $[Ca^{2+}]_i$. The general $[Ca^{2+}]_i$ relations of cells are now better understood, thanks to the recent development of technologies which have revolutionized the measurement and imaging of $[Ca^{2+}]_i$ in living cells (Read *et al.*, 1992a). These techniques have shown that cells maintain a low $[Ca^{2+}]_i$ (about 100 nM) concentration in the face of high extracellular concentrations of millimolar or higher. Equivalent high concentrations are found in internal stores such as the vacuole. When the cell is signalled, $[Ca^{2+}]_i$ can be increased either by the opening of calcium channels in the plasma membrane or in the vacuole or rough endoplasmic reticulum membranes. Internal membrane calcium channels may be opened by inositol phosphates which are released from the phospholipids of the cell surface membrane by G protein-activated or receptor-activated phospholipase activity. There is little doubt that cellular calcium signalling is complex. Most importantly, signalling may have very local effects on regions of the cytoplasm because the spatial diffusion of calcium in the cytoplasm is severely constrained. $[Ca^{2+}]_i$ is a suitable molecule for differentially activating metabolic activities within a single cytoplasm. $[Ca^{2+}]_i$ signals are sensed by a variety of calcium-binding proteins, of which calmodulin is the most well-known, and further interpreted by a variety of calcium–calmodulin dependent enzymes, most notably protein kinases.

THE MEASUREMENT OF $[Ca^{2+}]_i$

The Molecular Signalling Group in Edinburgh has had nine years of experience in the use of calcium-sensitive fluorescent dyes for measuring and imaging calcium in living plant cells (Read *et al.*, 1992a). Based on experience with 11 different plant and fungal cells it has been concluded that fluorescent dyes do provide suitable methods for measuring calcium, but there are considerable difficulties with their routine use. While solutions to these difficulties can usually be found, in several examples, most notably fungal hyphae, it has not yet been possible to find a solution (Read *et al.*, 1992b). The primary difficulty in all cases is getting the calcium-sensitive fluorescent dye into the cytoplasm. The most reliable method is microinjection and this unfortunately limits $[Ca^{2+}]_i$ measurement to single-cell systems which can tolerate this very perturbing process. Even if loading is successful, many cell types accumulate dye in organelles which have different $[Ca^{2+}]_i$ relations. While there is a useful number of single-cell systems that can be investigated, there are many other signals for which a multicellular system is essential. It is the realization of this fundamental limitation of fluorescent dyes that has been the driving force for the search for an alternative, described later: transformation with aequorin genes.

CAGED CALCIUM AND INOSITOL PHOSPHATES DEMONSTRATE POTENTIAL FOR PHYSIOLOGICAL REGULATION BY $[Ca^{2+}]_i$

A more valuable way of testing the calcium hypothesis is by the use of chemically caged molecules. Caged calcium and caged inositol triphosphates are inactive molecules in which calcium or IP_3 can be released by photolysis with UV radiation. Once the caged molecule is loaded into the cell, the $[Ca^{2+}]_i$ and $[IP_3]_i$ may be manipulated by the extent of UV irradiation. When combined with methods for measuring and imaging $[Ca^{2+}]_i$, an extremely effective method for demonstrating the potential of these molecules to control physiological processes is available. It has been shown that elevating artificially the cytoplasmic concentration of calcium or IP_3 can mimic: (1) the effects of abscisic acid and other closing stimuli on guard cells; (2) the effects of red light on the swelling of etiolated wheat leaf protoplasts; (3) the effects of incompatible stigma extracts; (4) external electrical fields on the growth of pollen tubes (Franklin-Tong *et al.*, 1993; Gilroy *et al.*, 1990; Shacklock *et al.*, 1992). In all these cases evidence has also been obtained to show that the physiological stimulus increases $[Ca^{2+}]_i$. These are powerful technologies and they give a clear-cut result, but experimentation is still confined to single cells or protoplasts.

THE USE OF AEQUORIN-TRANSFORMED PLANTS FOR MEASURING PLANT CELL CYTOSOL CALCIUM

Aequorin is a calcium-sensitive luminescent protein which can be isolated from the luminous jellyfish, *Aequorea*. Aequorin has been used for many years for the measurement of calcium and $[Ca^{2+}]_i$ and has been microinjected into single animal cells, but with some difficulty. Aequorin is composed of a 21 kDa protein, apoaequorin, and a luminophore, coelenterazine, which reconstitute themselves into active aequorin in living cells of *Aequorea*. Reconstitution *in vitro* requires the vigorous and unphysiological conditions of high salt and the presence of a reducing agent. During the reaction with calcium, coelenterazine is oxidized to coelenteramide and apoaequorin has to be recharged with new coelenterazine before it can be used again.

To provide a method of measuring $[Ca^{2+}]_i$ in a multicellular plant, it was decided to transform tobacco plants with apoaequorin cDNA and then to find out whether the synthesized apoaequorin would reform aequorin plant cells by incubation of seedlings in coelenterazine, a strongly hydrophobic compound. The success of these experiments, considered initially to be high-risk, is described in two papers (Knight *et al.*, 1991a, 1992). As a spin off, a simple method was developed for measuring $[Ca^{2+}]_i$ in bacteria transformed with apoaequorin cDNA (Knight *et al.*, 1991b). In short, luminous plants have been produced whose luminosity directly reports $[Ca^{2+}]_i$.

SIGNALS THAT MODIFY PLANT CELL $[Ca^{2+}]_i$

Using transformed seedlings ten signals have been detected which will increase $[Ca^{2+}]_i$ in one-week-old tobacco seedlings. These signals are touch, wind, cold shock, fungal elicitors, wounding, pH, salination, hydrogen peroxide, exogenous calcium and ethylene. All are relevant to the understanding of calcium homoeostasis in plant cells and establish its dynamic nature. The kinetics of $[Ca^{2+}]_i$ change is different for each signal. The effects of touch or wind produce transients which last only a few seconds; those of exogenous calcium, pH, ethylene or salination last many minutes and the others lie somewhere in between. The different signals have an unexpectedly specific effect on the kinetics of $[Ca^{2+}]_i$ change.

THE SPECIFIC EFFECTS OF TOUCH AND WIND STIMULATION UPON PLANT GROWTH AND REPRODUCTION

The relevant signals for the present discussion are those of touch and wind. Both are described as mechanical signals and the physiological effects of both are to alter specifically growth patterns and, in appropriate plants, to reduce yield (Grace, 1977).

Wind

The specific effects of wind upon plant growth are in part the result of changes in tension and compression in the regions of the stem (and petioles) where movement occurs. The cell walls are thickened in these particular tissue areas and the diversion of carbohydrate resources into the cell wall and the corresponding reductions in vegetative growth lead to reductions in crop yield. Some investigations have concentrated on the effects of wind on plant cell water relations to try to explain changes in growth and yield, but these do not seem to have been successful.

The data collected, which demonstrate an immediate impact of wind on $[Ca^{2+}]_i$, open an entirely new chapter in the study of this much-neglected signal. It was possible to demonstrate that the increase in $[Ca^{2+}]_i$ is the result of wind-induced bending in tobacco seedlings and only continues while the tissue is in motion. When motion ceases, $[Ca^{2+}]_i$ immediately returns to basal levels. Figure 1 suggests in a simple diagrammatic way that the altered cell shape during motion causes $[Ca^{2+}]_i$ increase, but it should be emphasized that it is not known whether it is confined solely to cells whose shape is altered. Attempts have been made to image $[Ca^{2+}]_i$ in these transformed seedlings and data do suggest that transmission of signals occurs from the point of impact of the signal on the plant. Bending the plant by means other than wind also increases $[Ca^{2+}]_i$.

A primary effect of tissue movement in wind, with the associated thickening of plant cell walls, is the stiffening of the stem or branch which then resists further movement. This is a clear example of plasticity in plant development (Trewavas, 1986a) and results in stems of different plants having different thicknesses. The factors that contribute to this plasticity are as follows:

1. Environmentally induced variations in the carbohydrate resources devoted to growth in both the root and the shoot, i.e. the root/shoot ratio. In different individuals of the same species the extremes of the root/shoot ratio can vary by up to 20-fold (Trewavas, 1986a). An individual with a high root/shoot ratio should be much more resistant to wind-induced movement and the root/shoot ratio is modified by soil conditions and nutrition.

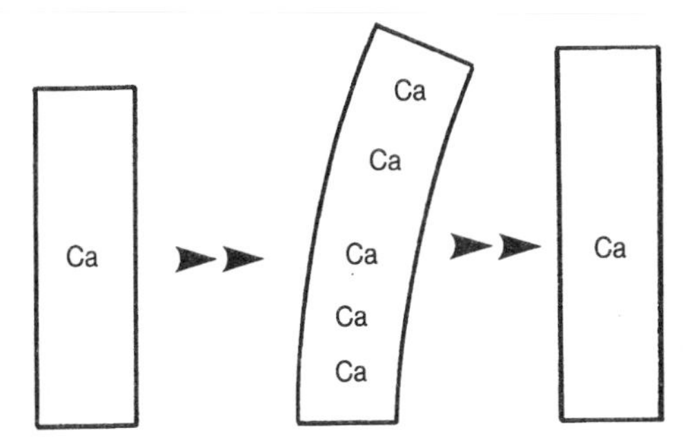

Figure 1 When cell shape is changed mechanically cytosolic calcium is increased. A simple diagram suggesting that the change in cell shape induced by tissue bending is responsible for the increase in $[Ca^{2+}]_i$. It is further suggested by the diagram that, after bending, the cell will increase slightly in length due to plastic extensibility in the cell wall. Tissues and cell bending which increase $[Ca^{2+}]_i$ have obvious relevance to understanding the processes of circadian movement of stems and petioles, the growth of roots around obstacles and tropic bending.

2. The thickness of the stem or branch which is the integrated result of both genetic and environmental influences such as far-red light (thus shading density) or the weight of foliage or prior stimulation by winds.
3. The amount of foliage determined by the extent and growth of individual branches and the extent of branching itself. The degree of apical dominance can be modified by many environmental variants and there is a defined genetic contribution as well.

Plasticity in stem thickness and thus in the response to stimulation by wind is necessary since plants do not determine the sites in which they grow; they must possess the flexibility to optimize resource utilization and adjust their structure to cope with the specific environment in which each individual grows. The response to wind helps to optimize the mechanical characteristics of the structure to further wind stimulation and to ensure that the stem can support the foliage it has produced without terminal wind damage. Very little is understood of the molecular basis of environmentally induced plasticity (Trewavas, 1986a), but on the basis of what has been described, $[Ca^{2+}]_i$ would seem to be a primary candidate for transduction and integration.

The physiology of plant responses to wind stimulation has attracted little attention, but plants can be exquisitely sensitive to movement. Neel and Harris (1971) mimicked the effects of wind on movement by shaking young plants for 30 s day^{-1}. These treatments can reduce growth by up to five-fold and can even induce premature dormancy. In little-known data Bunning (1941) and Bunning and Lempnnau (1954) reported the effects of shaking some 30 years earlier, but the results derived from experiments with etiolated plants and *Mimosa*, are little different from the above. Anatomical analysis again showed the increased diversion of resources into stem thickening with a consequent loss of carbohydrate for other important developmental processes such as reproduction and yield. It is generally accepted that the major difference between the yields of field-grown plants and the equivalent glasshouse-grown plants is the result of wind stimulation.

Touch

Investigations of the sensitivity of plants to touch is usually confined to *Mimosa* and insectivorous plants in which an obvious response can be observed. A summary of this literature, much of it old, can be found in a recent text (Simon, 1992). Bunning (1941) and Bunning and Lempnnau (1954) reported that the growth of etiolated plants can be dramatically reduced by gentle stroking with sheets of paper. The effects of stroking mimicked the effects of white light! The inhibitory effect on growth rates of stroking or rubbing plant stems was again reported by Jaffe (1973) and has been intermittently examined by others since then (references in Braam & Davis, 1990). Jaffe *et al.* (1985) believed that the production or synthesis of ethylene was responsible for the pronounced inhibition of growth and associated stem thickening which he and others had observed.

A primary step forward in understanding touch (and wind) stimulation and transduction was taken by Braam and Davis (1990) who described five touch genes. Expression of these genes was specifically increased by touch stimulation and the increase in mRNA levels can be up to 10–100-fold within 0.5 h of stimulation. Sequence analysis indicated that one of these genes controlled the production of calmodulin and two others controlled production of related proteins. The observations which show touch stimulation to increase $[Ca^{2+}]_i$, immediately therefore have interesting consequences for calmodulin gene expression; this suggests the intriguing possibility that calmodulin can regulate its own synthesis. The $[Ca^{2+}]_i$ pool mobilized by wind stimulation seems likely to be intracellular.

WHAT IS THE ROLE OF ETHYLENE IN MEDIATING GROWTH RESPONSES AFTER WIND OR TOUCH STIMULATION?

The effects of both touch and wind on the levels of $[Ca^{2+}]_i$ and calmodulin gene expression occur well before any increase in the biosynthesis or accumulation of ethylene, which normally takes several hours. These observations have important consequences for understanding the real functional role of ethylene and it is first necessary to describe the important experimental facts.

Inhibition of stem growth by applied ethylene has been demonstrated numerous times, although there are exceptions such as rice in which shoot growth is promoted. In etiolated shoots with an apical hook, ethylene treatment inhibits hook opening. Exogenous ethylene will also inhibit the growth of roots and it is thought that exogenously applied auxin may also inhibit root growth by inducing ethylene synthesis and accumulation. Growth responses to ethylene by etiolated shoots and roots are believed to provide a means for seedlings to deal with soil compaction. It was Darwin (1887) who first reported that the growth of roots was sensitive to touch. More recent reports by Hanson and collaborators (summarized in Hanson & Trewavas, 1982) show that touch, wounding, stroking and cold shock, all treatments which immediately increase $[Ca^{2+}]_i$, also cause inhibition of root growth within seconds. Recovery of growth can take up to 2 h. The implication behind these observations is that increased $[Ca^{2+}]_i$ and increased ethylene effectively initiate the same physiological response; growth inhibition. To confuse the matter further it has also been observed that ethylene treatment of tobacco seedlings can increase $[Ca^{2+}]_i$.

Why then do plants increase ethylene biosynthesis if an increase in $[Ca^{2+}]_i$ is sufficient to produce the response? One possibility is that the accumulation of ethylene simply lengthens an initially transient inhibition of growth by $[Ca^{2+}]_i$; ethylene function is then simply to promote the persistence of the response. But this would hardly explain why a volatile chemical is used for this purpose rather than some other intracellular enzyme events modifying cell wall secretion or other aspects of growth. But the answer may lie in the volatile characteristics of ethylene: ethylene can diffuse between cells and thus help to synchronize effectively the growth responses of large areas of stem or root tissue.

Many single-cell $[Ca^{2+}]_i$ measurements show that cells usually differ from each other in their response to a single uniform stimulus (for example, Shacklock *et al.*, 1992, and additional references in Trewavas, 1991). The variations are in the kinetics of change, the size of the change, the latency (time of initiation) which can often be a continuum. Usually, a proportion of cells do not respond at all. Furthermore, the stem is composed of cells that differ in their mechanical strength (most notably the vascular tissue) and the distribution of air spaces to adjacent cells will be quite variable. In other words, the act of bending will stimulate mechanically the component cells to differing extents and this is in cells whose $[Ca^{2+}]_i$ kinetics are already highly variable. As it stands a uniform tissue response using only $[Ca^{2+}]_i$ is unlikely.

If, on the other hand, ethylene biosynthesis is induced and ethylene diffuses between the different stem cells and induces growth inhibition, the pattern of cellular response will become much more uniform and mechanically more stable. Cellular variability in the induction of ethylene biosynthesis is now not so important because diffusion smooths out the variations and thus the responses between individual cells. More interestingly, prior response to bending by $[Ca^{2+}]_i$ increase may render the cells refractory to ethylene. The function of ethylene may therefore be to help synchronize and make more uniform the response of a heterogeneous tissue. Synchronizing functions for growth substances have been proposed before (Trewavas, 1986a, b, 1991).

MECHANICAL STRESS AND THE SPECIFICATION OF PLANT FORM

Contributions by Sachs, D'Arcy Thompson and Green

Notions that plant form is determined by mechanical influences and stresses between plant cells was enunciated by both Sachs (1887) and D'Arcy Thompson (1942). However, the ideas were largely hypothetical and were often disregarded because there was little evident way in which they could be tested. Both these workers emphasized that plant form is an holistic quantity, 'Die Pflanze bildet Zellen, nicht die Zelle bildet Pflanzen', and that form results from a continuity of forces running through the entire organism, independent of the number, magnitude and form of the individual cells which 'enter into the fabric like a froth'. Self-evidently the contiguous nature of the cell walls in plants provides for a possible mechanical field of force through tissues. Mechanical distortions in just one cell, if instituted, cause them to have influences on the shape and form of the cell wall contiguity over a much wider area. Perhaps most important in the generation of some of these ideas were the early observations which showed that growing tissues were in a state of mechanical stress and tension. Simply slicing a growing stem in half longitudinally and placing it in water leads to tissue bending (Sachs, 1887). Differences in tension between different cell types are a fact of life for plants.

D'Arcy Thompson (1942), like Hooke and Grew before him, was impressed by the apparent similarity between the appearance and shapes of cells in parenchymatous tissue and an aggregate of soap bubbles. He suggested that the application of simple knowledge about surface tension helped explain the shape and form of cells. He also suggested that newly formed cell wall behaves like a fluid film, and the shape of parenchymatous cells follows directly when the surface tension (the surface energy) between cells adopts a minimum value. A corollary of this is that a minimal surface energy should also imply a minimal surface area of interaction. The planes of new cell division will therefore be set to minimize this value.

Some interesting consequences follow from these hypotheses. The plane of new cell division will be set so as to minimize the tension and compression stress within the whole tissue. It has been shown that pressure, externally applied to dividing insect cells, can be used to rotate the mitotic axis (Gurdon, 1974). A good example of the plane of division being set to minimize stress is to be found in reaction wood formation (Sinnott, 1960). The principle of minimal surface area can be used to account for what are the otherwise puzzling characteristics of the planes of early divisions of a number of lower plant embryos and thus to account for Errera's rule (D'Arcy Thompson, 1942). In dividing meristematic tissue the new partition wall will initially meet the old walls at right angles because its tension is small compared with that of the older, more rigid, walls. But with time this new wall increases its rigidity, no doubt as a result of cross-linking, and assumes a tension closer to that of the older walls. An angle of 120° is now predicted in agreement with observation. Epidermal cell cross-walls are often set at right angles to the outer, much more rigid, cell wall. 'The behaviour of cells in the meristem is determined not by any character or properties of their own but by their position and the forces to which they are subject in the system of which they are part. As soon as the tension of the adjacent cell walls becomes unequal then the form alters' (D'Arcy Thompson, 1942).

D'Arcy Thompson (1942) also emphasized the rather special character of the epidermis and drew such structures with a thicker outer cell wall. Measurements which substantiate this notion have only recently been made (Kutschera, 1992). A higher tension exerted by the outer epidemal cell walls on the inner tissues, with a requirement to minimize the stress throughout, should lead to tissues with a circular cross-section. With a specified and very localized area of cell division the form of many shoots and roots can be partly understood. If cell division is not localized but occurs throughout, then the resulting structure will be spherical as in fruits; a sphere being the shape with minimal and uniform mechanical stress.

More recent efforts to relate form to mechanical stress can be found in the research of Green and colleagues (summarized in Green, 1986). They have investigated and proposed the concept of hoop reinforcement. This can be visualized as like the straps or hoops around a barrel which, by provision of inward tension, direct the planes of cell division and elongation. Hoop reinforcement can be considered as additional to the higher tension already exerted by the epidermal cell walls and to occur in local regions of epidermal cells. It is proposed that excessive tension in one direction, hoop reinforcement is then changed so that reinforcement occurs in the direction of growth. A reassessment of tensions within the tissue then leads to further modifications of hoop reinforcement. These ideas provide models for understanding the processes of phyllotaxis. The production of form becomes a kind of structural epigenesis and the sensing of tension and compression become critical elements in morphogenesis. Put simply and figuratively, if the apex is squeezed then the new leaf pops out where tension is minimal. But the squeezing is really induced by inner turgor pressure acting upon the surface tension and reinforcement of the epidermal cell walls. Green (1986) has described the detection of reinforcement fields which can induce asymmetries in tension and has shown by computer modelling how local changes in tension lead to much more global ones because of contiguity of the cell wall. Mechanical influences can be modelled through the effects of tension and compression on the cytoskeleton, but the data described here centralize $[Ca^{2+}]_i$ as a critically important molecule in the whole process. $[Ca^{2+}]_i$ is known to alter the gel/sol state of the cytoplasm.

The effects of pressure on shape and form in fungal infection, root formation and polarity

Perhaps the best example that shows pressure to regulate the form and character of cells are the observations that removal of pressure on cambium leads to its conversion to callus (Steeves & Sussex, 1989). In turn, tobacco pith cells subjected to external pressure divided in an orderly fashion to produce radial files resembling cambium.

The notions of signalling through changes in tension and compression are relevant to understanding the responses to fungal attack. Pathogenic fungi can often secrete cell wall hydrolases to aid infection. The subsequent weakening of the wall should signal the infected cell through changes in tension and compression. These signals would complement production of elicitor cell wall fragments which likewise are known to signal the cell and induce defense responses. Furthermore, damage to cells by wounding should involve changes in turgor pressure which may convey information to surrounding areas by a drop in pressure.

The relaxation of cell wall constraint and thus changes in tension and compression may also be sufficient to induce cell division in the pericycle. Lateral roots are initiated by cell divisions in the pericycle and these primordia have to grow and develop inside the tissues of mature roots before eventually breaking through. Unless the pericycle cells acquire a turgor pressure greater than that of surrounding cortical cells, the only simple way in which cell division and the accompanying growth can occur is by weakening the cell wall constraints of the surrounding cell walls. Auxin can induce both lateral and adventitious root formation and also the synthesis and secretion of cell wall hydrolases; cavities have been observed to form in cortical tissue ahead of the growing and developing root primordia (Trewavas, 1979). One interesting possibility is that constraint might simply maintain non-growing pericycle cells in an effectively dormant state and that the direction of constraint relaxation may determine the form and planes of cell division of the new primordium when it grows.

Evidence which relates $[Ca^{2+}]_i$ to the specification of the plane of cell division comes from studies on the *Fucus* zygote. The direction and polarization of the first cell division of the newly formed zygote is believed to be specified by a controlled calcium leak in the nascent rhizoidal tip (Trewavas, 1982). Figure 2 shows a simple axis transposition of a growing *Fucus* zygote which suggests its similarity to the bending cell shown in Fig. 1. It was D'Arcy Thompson (1942) who first initiated the concept of axis transposition for which he is justly famous. The success of axis transposition suggests again the importance of mechanical inputs into the product of form.

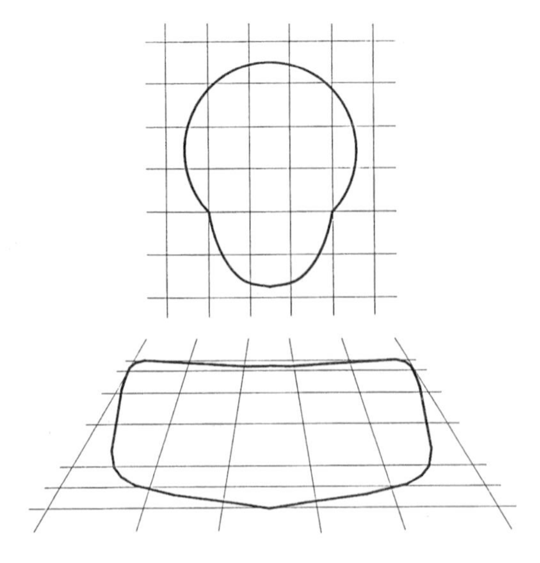

Figure 2 Axis transposition of a growing *Fucus* zygote using the D'Arcy Thompson (1942) procedure. The figure indicates how, by using the axis transposition, a growing *Fucus* zygote can be made to look like a bending cell as drawn in Fig. 1.

CONCLUSION

Mechanical views of shape and form were eclipsed in the later half of this century because of an emphasis instead on specification by positional information. Positional information concepts originated from studies of the development of chick limb and hydra and hinge on assumptions that pattern formation is specified by communication by largely unknown chemicals. The theory has had some popularity in plant terms and growth substances are sometimes posited with the appropriate specific communicating properties, an idea that has been vigorously challenged (Trewavas, 1986a).

Theories that attempt to explain the production of plant shape and form must involve the very specific attributes of plant cells of which the most striking is the presence of the cell wall and the associated turgor pressure. Clear evidence is emerging of the presence of intimate conversations between the wall and the cytoplasm during development and regulation by both chemicals and tension/ compression/pressure would be expected. Cell wall calcium, not considered in this article, will also contribute to the interrelation of tension and response since it helps cross-link and rigidify the walls of growing cells. It would be sterile to argue that form only results from mechanical information or that cellular chemical communication is the ultimate answer. No doubt both are used in different ways and perhaps for different purposes, but any model that fails to take both possibilities into account will only represent a partial description of events.

The data described earlier suggest that $[Ca^{2+}]_i$ acts to transduce physical signals. Other evidence, not considered in any depth here, suggests that chemical signalling may be transduced by the same route. In that case, the necessary integration and interpretation of all signals involved in the production of form may be funnelled through this simple ion. This simplifying hypothesis is adequate justification for a focused research attack which aims to understand in detail the kinetics, dynamics and spatial behaviour of $[Ca^{2+}]_i$.

ACKNOWLEDGEMENTS

Unpublished work was supported by the SERC and the AFRC.

REFERENCES

BRAAM, J. & DAVIS, R.W., 1990. Rain-, wind- and touch-induced expression of calmodulin and calmodulin related genes in *Arabidopsis. Cell, 60:* 357–364.

BUNNING, E., 1941. Uber die Verhinderung des etiolements. *Berichte der Deutschen Botanischen Gesselschaft, 59:* 2–9.

BUNNING, E. & LEMPNNAU, C., 1954. Uber die Wirkung mechnischer und photischer reize auf die gewebe und organbildung von Mimosa pudica. *Berichte der Deutschen Botanischen Gesellschaft, 67:* 10–18.

DARWIN, C., 1887. *The Power of Movement in Plants.* London, U.K.: John Murray.

FRANKLIN-TONG, V.E., RYDE, J.P., READ, N.D., TREWAVAS, A.J. & FRANKLIN, C., 1993. The self-incompatiblity response in *Papaver rhoeas* is mediated by cytosolic free calcium. *Plant Journal 4:* 163–177.

GILROY, S., READ, N.D. & TREWAVAS, A.J., 1990. Elevation of stomatal cytosol calcium using caged calcium and caged inositol phosphate initiates stomatal closure. *Nature, 346:* 769–771.

GRACE, J., 1977. *Plant Response to Wind.* London, U.K.: Academic Press.

GREEN, P. B., 1986. Plasticity in shoot development; a biophysical view. In D.H. Jennings & A.J. Trewavas (eds), *Plasticity in Plants,* pp. 211–233. Cambridge, U.K.: The Company of Biologists.

GURDON, J.B., 1974. *The Control of Gene Expression in Animal Development.* Oxford, U.K.: Clarendon Press.

HANSON, J.B. & TREWAVAS, A.J., 1982. Regulation of plant cell growth; the changing perspective. *New Phytologist, 90:* 1–18.

JAFFE, M.J., 1973. Thigmomorphogenesis; the response of plant growth and development to mechanical perturbation. *Planta, 114:* 143–157.

JAFFE, M.J., HUBERMAN, M., JOHNSON J., & TELEWSKI, F.W., 1985. Thigmomorphogenesis; the induction of callose formation and ethylene evolution by mechanical perturbation in bean stems. *Physiologia Plantarum, 64:* 271–279.

KNIGHT, M., CAMPBELL, A.K., SMITH, S.M: & TREWAVAS, A.J., 1991a. Transgenic plant aequorin reports the effect of touch, cold shock and elicitors on cytoplasmic calcium. *Nature, 352:* 524–526.

KNIGHT, M., CAMPBELL, A., SMITH, S.M. & TREWAVAS, A.J., 1991b. Reconstituted aequorin in *Escherichia coli* reports cytoplasmic calcium. *FEBS Letters, 282:* 405–408.

KNIGHT, M., SMITH, S.M. & TREWAVAS, A.J., 1992. Wind induced plant motion immediately increases cytosolic calcium. *Proceedings of the National Academy of Sciences, USA, 89:* 4967–4972.

KUTSCHERA, U., 1992. The role of the epidermis in the control of elongation growth in stems and coleoptiles. *Botanica Acta, 105:* 246–253.

NEEL, P. & HARRIS, R.W., 1971. Motion-induced inhibition of elongation and induction of dormancy in Liquidamber. *Science, 173:* 58–59.

READ, N.D., ALLAN, W.G.T., KNIGHT, M.R., RUSSEL, A., SHACKLOCK, P. & TREWAVAS, A.J., 1992a. Imaging and measurement of cytosolic free calcium in plant and fungal cells. *Journal of the Microscopical Society, 166:* 57–86.

READ, N.D., KNIGHT, H. & TREWAVAS, A.J., 1992b. Fluorescence ratio imaging and confocal microscopy of calcium in filamentous fungi. *Binary, 4:* 50–52.

SACHS, J., 1887. *Lectures on the Physiology of Plants.* Oxford, U.K.: Clarendon Press.

SHACKLOCK, P., READ, N.D. & TREWAVAS, A.J., 1992. Cytosolic free calcium mediates red light induced photomorphogenesis. *Nature, 358:* 753–755.

SIMON, P., 1992. *The Action Plant.* Edinburgh, U.K.: Blackwells.

SINNOTT, E.W., 1960. *Plant Morphogenesis.* New York, U.S.A.: McGraw Hill Book Co.

STEEVES, T.A. & SUSSEX, I.M., 1989. *Patterns in Plant Development.* Cambridge, U.K.: Cambridge University Press.

THOMPSON, D.W., 1942. *On Growth and Form.* Cambridge, U.K.: Cambridge University Press.

TREWAVAS, A.J., 1979. The dynamics of meristem control by growth substances. In E.C. George (ed.), *Differentiation and the Control of Development in Plants-Potential for Chemical Modification,* British Plant Growth Regulator Group Monograph no. 3, pp. 39–55. Letcombe, Wantage: British Plant Growth Regulator Group.

TREWAVAS, A.J., 1982. Possible control points in plant development. In H. Smith & D. Grierson (eds), *The Molecular Biology of Plant Development,* pp. 10–30. Edinburgh, U.K.: Blackwells.

TREWAVAS, A.J., 1986a. Resource allocation under poor growth conditions. A major role for growth substances in plasticity. In D.H. Jennings & A.J. Trewavas (eds), *Plasticity in Plants,* pp. 31–72. Cambridge, U.K.: Company of Biologists.

TREWAVAS, A.J., 1986b. Timing and memory processes in seed embryo dormancy; a conceptual paradigm for plant development questions. *Bioessays, 6:* 87–93.

TREWAVAS, A.J., 1991. How do plant growth substances work II. *Plant Cell and Environment, 14:* 1–12.

CHAPTER

13

Genes controlling flower development

Genes controlling flower development in *Antirrhinum*

ENRICO COEN, DESMOND BRADLEY, ROBERT ELLIOTT, RÜDIGER SIMON, JOSÉ ROMERO, SABINE HANTKE, SANDRA DOYLE, MARK MOONEY, DA LUO, PAULA McSTEEN, LUCY COPSEY, CORAL ROBINSON & ROSEMARY CARPENTER

Abstract
Following floral induction, plant meristems in certain positions become organized to produce flowers. Flower meristems usually consist of three layers of clonally distinct cells: L-I (epidermis), L-II (subepidermis), L-III (core). The interactions between genes expressed in these layers control the flower development programme. This process is being studied using a combination of genetic, molecular and physiological approaches to *Antirrhinum*. In particular, transposon-induced mutations in genes controlling early switches in floral development are being exploited. Using transposons as tags, several of the affected genes have been isolated and their products characterized. The normal site and timing of gene expression has been determined by *in situ* hybridization to wild-type plants. Some of the regulatory interactions between genes have been revealed by studying *cis*-acting or *trans*-acting mutations which have resulted in ectopic gene expression. Somatic excisions of transposons have also been selected to obtain chimeric plants in which only one meristematic layer expresses a gene. Analysis of these chimeras has shown that some gene interactions are indirect, requiring cell–cell signalling. Many of these experiments have been carried out on plants grown in controlled environmental conditions which promote or inhibit flowering. Taken together, these approaches are providing a framework for understanding how the initiation and development of flowers is controlled.

Editorial note. Dr Coen and his collaborators were unable to produce a complete manuscript. In view of the importance of this topic, the following synopsis of recent research on genes controlling flower development has been written by Dr Andrew Hudson.

Shape and Form in Plants and Fungi
ISBN 0-12-371035-9

Genes controlling flower development

ANDREW HUDSON

CONTENTS

INTRODUCTION

The key to this approach has been the use of two genetically characterized species of dicotyledons. *Antirrhinum majus* and *Arabidopsis thaliana.* Homoeotic mutations which alter the identity of floral meristems or floral organs have been identified in both species and used as the basis for isolation and analysis of the relevant homoeotic genes. Although the ancestors of *Antirrhinum* and *Arabidopsis* probably diverged about 70 million years ago, the structure and functions of their floral homoeotic genes show surprising conservation.

FLORAL ORGAN IDENTITY

The flowers of *Antirrhinum* and *Arabidopsis* are structurally similar, consisting of four concentric whorls of organs. From the outside of the flower inwards, the organs in successive whorls have the identities of sepals, petals, stamens and finally carpels. On the basis of the phenotypes of homoeotic mutants, three similar explanations for the genetic determination of floral organ identity were suggested (Bowman *et al.*, 1991; Carpenter & Coen, 1990; Schwarz-Sommer *et al.*, 1990), and the minor differences between these have subsequently been reconciled to produce a working model applicable to both *Antirrhinum* and *Arabidopsis* (Coen & Meyerowitz, 1991). This proposes that three functions (*a, b* and *c*) are each expressed in domains equivalent to two adjacent whorls of the mature flower (Fig. 1) and that the combination of functions expressed in any whorl is responsible for organ identity (e.g. *a*, alone in whorl 1 specifies sepals, *a* and *b* in whorl 2, petals). Therefore, loss of the *b* function, as in the *Antirrhinum* mutant, *deficiens* (*def*), leads to the replacement of petals and stamens in whorls 2 and 3 with sepals and carpels. However, the phenotypes of mutants that lack either *a* or *c* functions demanded the incorporation of two additional factors into the model. First, that the domains in which *a* and *c* functions become

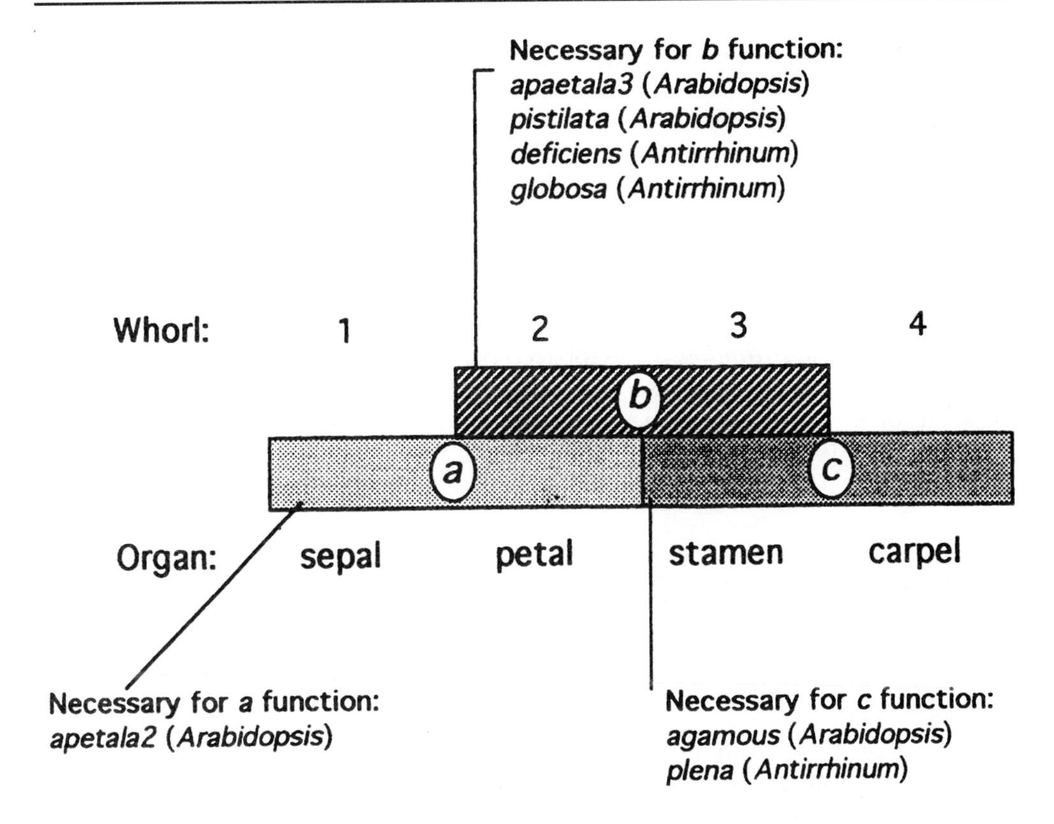

Figure 1 Model for the interaction of three functions, *a*, *b* and *c*, in specifying floral organ identity in *Antirrhinum* and *Arabidopsis*. Of the several genes which may be necessary for each function, only those referred to in the text are shown. Redrawn from Coen & Meyerowitz, 1991.

established are mutually exclusive. Therefore, loss of the *a* function, as in the *Arabidopsis* mutant *apetala2* (*ap2*), allows expression of the *c* function to extend into the two outer whorls of the flower, and accounts for the resulting mirror-image arrangement of organs: carpels, stamens, stamens, carpels. Similarly, loss of the *c* function allows *a* expression to extend inwards. Secondly, it was necessary to assume that the *c* function is needed for determinacy of the floral meristem because its loss, as in the *Arabidopsis* mutant *agamous* (*ag*), not only affects the identity of floral organs, but also leads to the proliferation of additional internal whorls.

This hypothetical model is consistent with the phenotypes of plants which lack more than one function. For example, loss of both the *b* and *c* functions in *Arabidopsis* plants homozygous for *apetala3* (*ap3*) and *ag* mutations, is expected to result in loss of determinacy and expression of the *a* function throughout the flower. As predicted, the flowers of these *ap3*, *ag* double mutants consist of an indeterminate number of sepaloid whorls (Bowman *et al.*, 1991).

Isolation of homoeotic genes necessary for the three functions has allowed their domains of expression to be detected by *in situ* hybridization. The region of the floral meristem in which any gene is expressed at a high level has been found to correspond to the two adjacent whorls of the mature flower predicted by the model. For example, expression of the *Arabidopsis c* function gene, *ag*, is restricted to the central region of the floral meristem destined to form whorls 3 and 4 (Drews *et al.*, 1991). *In situ* hybridization has also confirmed the antagonism between expression of the *a* and *c* functions. In the *ap2* mutant of *Arabidopsis*, which lacks the *a* function, expression of the *ag* gene (*c* function) is no longer restricted to whorls 3 and 4, but can be detected in the outer two whorls of the flower (Drews *et al.*, 1991).

Another important test of the proposed antagonism between expression of the *a* and *c* functions follows from the prediction that a gain of *c* function in whorls 1 and 2 could be equivalent to loss of *a* function, and result in a floral phenotype of carpels, stamens, stamens, carpels. This phenotype has indeed been found to result from constitutive expression of the *c* function gene *ag*, or its homologue from *Brassica napus* in transgenic *Arabidopsis* and tobacco plants, respectively (Mandel *et al.*, 1992; Mizukami & Ma, 1992) and is also seen in *Antirrhinum* plants carrying a semidominant allele of the *c* function gene, *plena*, which is expressed throughout the plant (Bradley *et al.*, 1993).

HOMOEOTIC GENE STRUCTURE AND FUNCTION

All the characterized floral homoeotic genes appear to encode proteins which are members of a larger family of transcription factors (the MADS family) and therefore have the potential to regulate the expression of other genes (Sommer *et al.*, 1990; Yanofsky *et al.*, 1990). Although none of these target genes has as yet been identified, a number of potential candidates may be suggested. First, the homoeotic genes may stimulate their own expression. Such positive feedback might allow domains of expression established in the early floral meristem to be reinforced, and to keep pace with subsequent growth of the meristem by cell division as each cell in a domain passes expression on to its daughters. Secondly, negative regulation of *a* function gene expression by *c* function gene products, and vice versa, would provide a mechanism by which one group of genes could exclude expression of the other from a particular cell. Thirdly, the floral homoeotic genes must regulate, either directly or indirectly, the genes that produce the characteristic shape and structure of the four organ types. Because the expression of the homoeotic genes is not necessarily confined to the early floral meristem, but can in some cases be detected late into development of the organs (e.g. Schwarz-Sommer *et al.*, 1992), the target genes may change with time: from genes directing the patterns of cell division which determine the gross morphology of each organ early in primordial development to those involved in differentiation of specific cell types at a later stage.

Support for this proposal is provided by observations of *Antirrhinum* mutants carrying an unstable, transposon-induced mutation of the *b* function gene, *def* (Carpenter & Coen, 1990). The transposon prevents *def* expression, and organs in the second whorl therefore develop as carpels. Excision of the transposon can, however, restore expression to clones of cells which then develop the characteristic morphology of petal cells within an otherwise sepaloid organ. Restoration of *def* expression early in organ development appears to influence the pattern of cell division, because large revertant clones assume the gross morphology of petals, and outgrow the surrounding sepal tissue to produce a distorted mosaic organ. Late reversion, however, can produce small clones of cells which show the characteristic morphology and pigmentation of petal epidermis cells but which appear not to have undergone a programme of cell divisions any different from that of their sepaloid neighbours.

As well as suggesting that the targets of homoeotic genes may change with development of the organ, the phenotype of this unstable *def* mutant also indicates that early expression of the gene is not necessary for its later expression or function. Similar conclusions have come from studies of temperature-sensitive homoeotic mutants of both *Antirrhinum* and *Arabidopsis* (Bowman *et al.*, 1989; Schwarz-Sommer *et al.*, 1992). In order to produce a nearly normal organ, it was found that a homoeotic gene needs to be functional from a time before organ initiation to shortly before cell differentiation. Loss of function late in development led to the production of intermediate organs with a shape similar to that of the wild-type, but with a cell type characteristic of the mutant.

RELATIONSHIP BETWEEN FUNCTIONS AND HOMOEOTIC GENES

The three functions *a*, *b* and *c*, may each require the expression of more than one homoeotic gene. The *Antirrhinum b* function, for example, consists of at least two homoeotic genes, *def* and *globosa* (*glo*). In this case, expression of the genes appears to be independent, because *def* is expressed normally in *glo* mutants, and *glo* in *def* mutants (Tröbner *et al.*, 1992). The only direct evidence for the nature of the relationship between genes and functions is the finding that the products of the *def* and *glo* genes are able to bind to a potential DNA target sequence only as heterodimers (Tröbner *et al.*, 1992).

HOMOEOTIC GENES AND MERISTEM IDENTITY

A similar molecular genetic approach has been applied to an earlier stage in floral development: the transition between inflorescence and floral meristems. Two *Antirrhinum* genes, *floricaula* (*flo*) and *squamosa* (*squa*) have been identified by mutation as being necessary for the meristem to become floral. The phenotype of *flo* and *squa* mutants is similar. They produce a normal inflorescence axis, but the meristems in the axils of the bracts, which in wild-type develop into flowers, continue to grow as indeterminate, branching inflorescence axes (Carpenter & Coen, 1990; Huijser *et al.*, 1992). Because the mutations cause the replacement of flowers with inflorescence structures, they can be considered homoeotic. Both genes have been isolated and their structures and patterns of expression characterized. The *squa* gene appears to encode a transcription factor of the MADS family (Huijser *et al.*, 1992), and although the product of *flo* shows no homology to previously characterized proteins, it has structural characteristics which suggest that it may also be a transcription factor (Coen *et al.*, 1990). In wild-type plants, *flo* shows a pattern of expression which changes with development of the flower. Expression is detectable first in bract primordia, which is intriguing because the phenotype of *flo* mutants, which bear morphologically normal bracts, suggests that the gene is not necessary for bract development. It raises the possibility that *flo* is involved in signalling between the bract primordium and its axillary meristem which is necessary for the meristem to become floral. From the bract primordium, expression of *flo* moves into the floral meristem, and occurs sequentially in the regions which develop into whorls 1, 2 and 4. This later whorl-specific pattern of expression suggests that *flo* may have a role in specifying floral organ identity in combination with other floral homoeotic genes.

The homologue of *flo* has been isolated from *Arabidopsis* and found to correspond to a locus previously identified by the *leafy* mutation as necessary for the formation of flowers (Weigel *et al.*, 1992). The mechanisms which specify meristem identity, like those that determine the identity of floral organs, therefore appear to have been conserved between *Antirrhinum* and *Arabidopsis.*

GENES DO NOT EXPLAIN EVERYTHING

One obvious feature of homoeotic genes is that their expression is confined to specific regions of the developing plant. Although this suggests that the genes are able to respond to positional information, the mechanisms which might provide the information remain unknown. Two possibilities can, however, be ruled out. First, a sequential mechanism, in which the organs in one floral whorl influence the identity of those in the next whorl, is inconsistent with the phenotypes of floral homoeotic mutants, and therefore appears unlikely. For example, *Antirrhinum* mutants which show ectopic expression of the *c* function gene *plena* have stamens in the third whorl, as in wild-type, although the organs in the first and second whorls are carpels and stamens, respectively (Carpenter & Coen, 1990). Secondly, it appears that the positional information to which homoeotic genes respond is not strictly dependent on cell ancestry, because the domains of homoeotic gene expression include all three layers of the meristem, even though these layers represent three distinct cell lineages established at embryogenesis (e.g. Drews *et al.*, 1991).

EVOLUTIONARY IMPLICATIONS

Although the ancestors of *Antirrhinum* and *Arabidopsis* probably diverged early in the evolution of flowering plants, the mechanisms that determine meristem and floral organ identity in the two species appear homologous. This implies that the floral homoeotic genes were functional in earlier angiosperms. Because most of the homoeotic genes are members of the same MADS gene family, it has been suggested that they arose by duplication of an ancestral gene which originally directed the development of the most primitive flower: one made up of fertile sporophylls. Subsequent divergence of the MADS genes may then have allowed elaboration of the flower into a structure containing additional organ types with specialized protective and attractive roles (Coen, 1991). The homology between *Antirrhinum* and *Arabidopsis* also suggests that the homoeotic mechanisms may be widespread amongst flowering plants. Although the inflorescence architecture and floral structures of *Antirrhinum* and *Arabidopsis* are not typical of all angiosperms, the possibility now exists to examine the role of homoeotic genes in other species, including those with flowers which are not hermaphrodite, or which have variable numbers of organs in a spiral arrangement.

REFERENCES

BOWMAN, J.L., SMYTH, D.R. & MEYEROWITZ, E.M., 1989. Genes directing flower morphogenesis in *Arabidopsis. Plant Cell, 1:* 37–52.

BOWMAN, J.L., SMYTH, D.R. & MEYEROWITZ, E.M., 1991. Genetic interactions among floral homeotic genes of *Arabidopsis. Development, 112:* 1–20.

BRADLEY, D., CARPENTER, R., SOMMER, H., HARTLEY, N. & COEN, E.S., 1993. Complementary floral homeotic phenotypes result from opposite orientations of a transposon at the *plena* locus of Antirrhinum. *Cell, 72:* 85–95.

CARPENTER, R. & COEN, E.S., 1990. Floral homeotic mutations produced by transposon-mutagenesis in *Antirrhinum majus. Genes and Development, 4:* 1483–1493.

COEN, E.S., 1991. The role of homeotic genes in flower development and evolution. *Annual Review of Plant Physiology and Plant Molecular Biology, 42:* 241–279.

COEN, E.S. & MEYEROWITZ, E.M., 1991. The War of the Whorls: genetic interactions controlling flower development. *Nature, 353:* 31–37.

COEN, E.S., ROMERO, J.-M., DOYLE, S., ELLIOTT, R., MURPHY, G. & CARPENTER, R., 1990. *Floricaula:* a homeotic gene required for flower development in *Antirrhinum majus. Cell, 63:* 1311–1322.

DREWS, G.N., BOWMAN, J.L. & MEYEROWITZ, E.M., 1991. Negative regulation of the *Arabidopsis* homeotic gene *AGAMOUS* by the *APETALA2* product. *Cell, 65:* 991–1002.

HUIJSER, P., KLEIN, J., LÖNNIG, W.-E., MEIJER, H., SAEDLER, H. & SOMMER, H., 1992. Bracteomania, an inflorescence anomaly, is caused by the loss of function of the MADS-box gene *squamosa* in *Antirrhinum majus. EMBO Journal, 11:* 1239–1249.

MANDEL, M.A., BOWMAN, J.L., KEMPIN, S.A., MEYEROWITZ, E.M. & YANOFSKY, M.F., 1992. Manipulation of flower structure in transgenic tobacco. *Cell, 71:* 133–143.

MIZUKAMI, Y. & MA, H., 1992. Ectopic expression of the floral homeotic gene *AGAMOUS* in transgenic *Arabidopsis* plants alters floral organ identity. *Cell, 71:* 119–131.

SCHWARZ-SOMMER, Z., HUIJSER, P., NAKEN, W., SAEDLER, H. & SOMMER, H., 1990. Genetic control of flower development by homeotic genes in *Antirrhinum majus. Science, 250:* 931–936.

SCHWARZ-SOMMER, Z., HUE, I., HUIJSER, P., FLOR, P., HANSEN, R., TETENS, F., LÖNNIG, W.-E., SAEDLER, H. & SOMMER, H., 1992. Characterisation of the *Antirrhinum* floral homeotic MADS-box gene *deficiens*: evidence for DNA binding and autoregulation of its persistent expression throughout flower development. *EMBO Journal 11:* 251–263.

SOMMER, H., BELTRÁN, J.P., HUIJSER, P., PAPE, H., LÖNNIG, W.-E., SAEDLER, H. & SCHWARZ-SOMMER, Z., 1990. *Deficiens*, a homeotic gene involved in the control of flower morphogenesis in *Antirrhinum majus*: the protein shows homology to transcription factors. *EMBO Journal 9:* 605–613.

TRÖBNER, W., RAMIREZ, L., MOTTE, P., HUE, I., HUIJSER, P., LÖNNIG, W.-E., SAEDLER, H., SOMMER, H. & SCHWARZ-SOMMER, Z., 1992. *GLOBOSA*: a homeotic gene which interacts with *DEFICIENS* in the control of *Antirrhinum* floral organogenesis. *EMBO Journal, 11:* 4693–4704.

WEIGEL, D., ALVAREZ, J., SMYTH, D.R., YANOFSKY, M.F. & MEYEROWITZ, E.M., 1992. *LEAFY* controls floral meristem identity in *Arabidosis. Cell, 69:* 843–859.

YANOFSKY, M.E., MA, H., BOWMAN, J.L., DREWS, G.N., FELDMAN, K.A. & MEYEROWITZ, E.M., 1990. The protein encoded by the *Arabidopsis* gene *agamous* resembles transcription factors. *Nature, 346:* 35–39.

CHAPTER

14

Genetic and molecular analysis of plant shape and form using *Arabidopsis thaliana* as a model system

NICHOLAS HARBERD, JINRONG PENG & PIERRE CAROL

CONTENTS

Abstract

Arabidopsis is increasingly being used as a model system for the application of genetic and molecular techniques to the study of plant growth and development. This chapter outlines the properties of *Arabidopsis* which make it uniquely suited to this type of approach. The application of several of the methods used in *Arabidopsis* research to the study of the role of the gibberellins in plant growth regulation is described. The gibberellins (GAs) are a group of plant growth regulators which have profound effects on a variety of plant developmental processes. The *gai* mutation of *Arabidopsis* alters the sensitivity of plants to GA, and therefore identifies a gene whose product is

Shape and Form in Plants and Fungi
ISBN 0–12–371035–9

involved with the mediation or modulation of the GA response. Progress towards the genetic characterization and molecular isolation of *gai*, with the eventual aim of understanding how its product affects GA sensitivity is outlined. The genetic and molecular methods described here are generally applicable to the analysis of plant development and to the identification of genes playing important roles in the determination of plant shape and form.

SUITABILITY OF *ARABIDOPSIS* FOR GENETIC RESEARCH

Arabidopsis thaliana is a small cruciferous weed belonging to the economically important Brassicaceae family. It has been used as a model plant in classical genetic studies for over 40 years, and has a number of properties that are convenient for genetic research. The plant is small, has a short generation time, is relatively easy to grow, produces many seeds per parent plant and (although it usually self-pollinates) can be easily cross-pollinated. As a result of many years of mutagenesis experiments an impressive variety of mutant lines exists. These mutants fall into several different classes, including those with affected morphology, plant and seed colour, embryo development, synthesis of, or response to, endogenous plant growth regulators and response to environmental signals which regulate growth and development. Many of these mutations have been mapped via conventional recombination analysis, resulting in a genetic map with five linkage groups which correspond with the five chromosome pairs found in diploid *Arabidopsis* (Koornneef, 1990; Meyerowitz, 1987).

THE *ARABIDOPSIS* GENOME

Two aspects of the *Arabidopsis* genome make it useful in gene cloning experiments. First, it is very small, the haploid nuclear genome size being approximately 100 000 kb pairs. Secondly, the *Arabidopsis* nuclear genome, when compared with that of other plant species, has a very low repetitive DNA content. This makes it possible successively to isolate overlapping genomic clones (a procedure known as chromosome walking), thus obtaining large contiguous segments of chromosomal DNA for further study. Chromosome walking is also facilitated by the availability of restriction fragment length polymorphism (RFLP) linkage maps of the *Arabidopsis* chromosomes (Chang *et al.*, 1988; Nam *et al.*, 1989).

CHROMOSOME WALKING USING YEAST ARTIFICIAL CHROMOSOME (YAC) CLONES

Yeast artificial chromosome (YAC) vectors allow very large (50–700 kb) pieces of DNA to be cloned (Burke *et al.*, 1987). Thus, a representative library of the *Arabidopsis* genome can be contained in a relatively small number of YAC clones. The large inserts contained within YAC clones permit chromosome walking in steps much longer than can be achieved with cosmid or bacteriophage clones. Several *Arabidopsis* DNA YAC libraries are now available (e.g. Grill & Somerville, 1991). The YAC clones can be used to obtain long segments of contiguous DNA from the *Arabidopsis* chromosomes in the following way. Initially, the chromosomal locations of a subset of the YAC clones are known because they hybridize to DNA markers whose chromosomal locations have already been established by RFLP analysis (Chang *et al.*, 1988; Nam *et al.*, 1989). DNA probes from each end of the fragment inserted into these mapped YAC clones are then hybridized back to the YAC library. The YAC clones identified by these probes are then themselves used as a source of end-probes for hybridization with the library. By repeating these procedures, the overlapping inserts of the YAC clones are ordered into large contiguous segments covering the chromosomal region of interest (Gibson & Somerville, 1992; Putterill *et al.*, 1993).

TRANSPOSON MUTAGENESIS IN *ARABIDOPSIS THALIANA*

Transposon mutagenesis is a procedure that has proved highly effective for the isolation of plant genes (e.g. Hake *et al.*, 1989). The procedure relies upon identification of a mutant plant displaying an alteration in phenotype caused by insertion of a transposon (a segment of DNA capable of transposing from one site in the genome to another) within a gene of interest. The transposon sequences are used as hybridization probes to obtain a molecular clone of the mutant allele. DNA flanking the transposon insertion is then used to isolate the wild-type gene. Considerable effort has gone into the development of systems based on maize transposable elements in *Arabidopsis* (Balcells *et al.*, 1991). One example is a two-component system using the elements *Ac* and *Ds* (Swinburne *et al.*, 1992). In this system, the transposing element (*Ds*) is physically distinct from another element (*Ac*-stable) which provides a source of the protein (transposase) which catalyses transposition of *Ds*. The *Ds* element will be stable, and thus will remain at its original location, until plants containing it are crossed with a plant containing *Ac*-stable. Following transposition, the *Ds* can be stabilized at its new location by genetic segregation away from *Ac*-stable.

GIBBERELLIN MUTANTS OF *ARABIDOPSIS THALIANA*

We are using the above technologies in attempts to isolate genes controlling GA biosynthesis and signal transduction. GA-related mutants have been isolated in several species, and are classified according to their sensitivity to exogenously applied GA (Reid, 1986). Recessive, GA-sensitive dwarf mutants are reduced in height and often display impaired seed germination and floral fertility. The phenotype of these mutants can be restored to normal by application of exogenous GA. Mutant alleles of this kind have been isolated at five different loci (*ga1* to *ga5*) in *Arabidopsis* (Koornneef & Der Veen, 1980). Analysis of these dwarf mutants has shown that they contain reduced levels of the GA species thought to be responsible for controlling shoot elongation, suggesting that the wild-type genes are involved with various enzymatic steps in GA biosynthesis (Talon *et al.*, 1990). Another important class of GA-related mutants are the dominant, GA-insensitive dwarf mutants. These mutants have phenotypes closely resembling those of the recessive GA-sensitive dwarfing mutants. However, these mutants are completely unresponsive to exogenously applied GA. A mutant of this kind (*gai*), has been isolated in *Arabidopsis* (Koornneef *et al.*, 1985). This mutant may identify a gene whose product is involved at some stage in GA signal transduction, or in the regulation of tissue sensitivity to GA.

Interestingly, the *gai* and *ga4* loci map to within approximately 0.6 cM of one another (N. Harberd, unpublished observation; Peng & Harberd, 1993) on the top arm of chromosome 1 of *Arabidopsis*. We are focusing on the *gai–ga4* region since the close proximity of these genes will enable us to isolate and study two genes, one involved with GA biosynthesis (*ga4*, this gene is thought to encode a hydroxylase enzyme required for the conversion of GA from a biologically inactive to an active form, Talon *et al.*, 1990) and the other (*gai*) with GA signal transduction/sensitivity.

ISOLATION OF IRRADIATION-INDUCED *gai*-DERIVATIVE ALLELES

gai is a semidominant mutation, a property it shares in common with the GA-insensitive dwarfing mutations of maize (*D8* and *Mpl1*, Harberd & Freeling, 1989), and wheat (*Rht* homoeoallelic series, Gale & Youssefian, 1985). On the assumption that *gai* is a gain-of-function mutation we devised a method for the isolation of potential loss-of-function 'knockout' alleles of *gai*. Approximately 60 000 *gai/gai* seeds were treated with γ-irradiation. The treated seeds were planted out and the (M_1) plants were screened for the presence of bolt stems having internodes longer than that characteristic of plants homozygous for *gai*. Since *gai* is semidominant these bolt stems were potentially composed of tissue that was heterozygous for *gai* and for a novel *gai* derivative allele *gai-d* (Fig. 1). Subsequent genetic analysis of the progeny (obtained via self-pollination) of these stems showed that four of them did indeed contain such

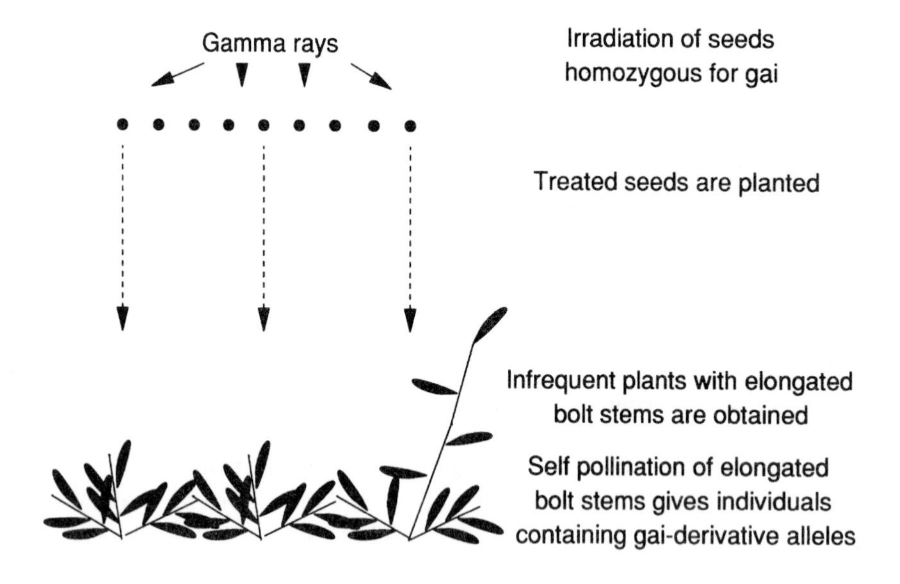

Figure 1 Method for the induction and recovery of *gai* derivative alleles. Seeds homozygous for *gai* are treated with γ-rays and planted out. Plants displaying single elongated stems are observed at very low frequency. These plants are chimeric, the elongated stem contains tissue which contains one copy of the original *gai* mutation. The second copy of the gene has been further mutated by the γ-rays, converting it into a non-functional *gai-d* allele. Thus, the stem displays the elongated phenotype characteristic of tissue heterozygous for *gai*. Seeds containing the *gai-d* allele are recovered by allowing the elongated stem to self-pollinate and harvesting the resulting seeds.

alleles, now designated *gai-d1*, *gai-d2*, *gai-d3* and *gai-d4*. One of these alleles (*gai-d3*) displays abnormal inheritance properties. It is inherited with normal frequency through the egg (female gametophyte), but its transmission through the pollen (male gametophyte) is severely impaired. This behaviour has, in the past, been correlated with the presence of chromosomes carrying deletion mutations (McClintock, 1942). The possibility that the *gai-d* alleles may be deletion or rearrangement mutations is of significance to our attempts to isolate the *gai* locus, since mutations of this kind are easier to identify at the molecular level than nucleotide substitutions.

When homozygous, *gai-d1*, *gai-d2* and *gai-d4* confer a phenotype that is indistinguishable from that of wild-type. Since these alleles are likely to be loss-of-function alleles, this observation suggests that the wild-type gene function is dispensable (Peng & Harberd, 1993). Mutational analyses of genes identified by dominant gain-of-function alleles in other organisms have demonstrated that it is not uncommon for their presumed loss-of-function alleles to confer wild-type phenotypes (Park & Horwitz, 1986).

CHROMOSOME WALKING IN THE *gai–ga4* REGION

The chromosomes of *Arabidopsis* are well-defined by RFLP markers (Chang *et al.*, 1988; Nam *et al.*, 1989). However, the RFLP and classical genetic maps have few markers in common, and this makes accurate cross-reference between these two kinds of map difficult. Thus it is necessary to map accurately a gene with respect to RFLP markers before chromosome walking experiments can be initiated. We have determined the position of the *gai–ga4* region with respect to RFLP markers from the top arm of chromosome 1 of *Arabidopsis*. Plants homozygous for *ga4* (Landsberg *erecta* background) were crossed with wild-type plants of either the Niederzenz or Columbia ecotypes. These ecotypes were used because they differ slightly at the level of DNA sequence from the Landsberg *erecta* strain. These differences result in the occasional loss or gain of sites recognized by restriction endonucleases, resulting in DNA digest fragments of different lengths (diagnostic for each ecotype). *ga4/ga4* homozygotes were obtained in the F_2 generation of the above crosses, and were confirmed by checking that their (F_3) progeny (obtained through self-pollination) were uniformly dwarfed and did not segregate for tall (wild-type) individuals.

Genomic DNA was prepared from the F_3 families and tested for the presence of RFLP alleles associated with the Landsberg *erecta* and Niederzenz genomes. The further apart two loci are on a chromosome, the greater the likelihood that a recombination event (cross-over) will occur between them. Thus, in the *ga4* homozygotes tested in our experiment the closer a DNA sequence is to *ga4* the more likely it is to be Landsberg *erecta* rather than Niederzenz DNA (since *ga4* was isolated from the Landsberg *erecta* background). On this basis we have identified two RFLP markers which are in close proximity to *ga4*, m219 (1.3 cM distal of *ga4*) and g2395 (1.7 cM proximal of *ga4*) as shown in Fig. 2 (see also Peng & Harberd, 1993). Owing to the close proximity of *gai* and *ga4*, it is likely that *gai* also maps to the region spanned by these two markers.

We have also isolated yeast artificial chromosome (YAC) clones containing DNA from the *gai–ga4* region. m219 and g2395 have been used as hybridization probes to identify YACs from the EG YAC library (Grill & Somerville, 1992). m219 identifies

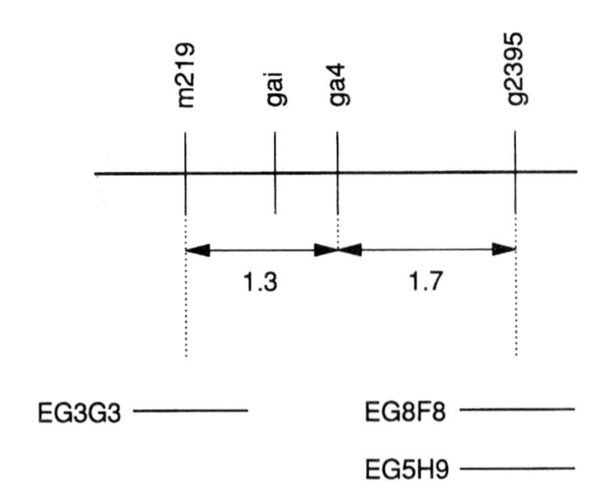

Figure 2 Linkage map of the *gai–ga4* region. m219 and g2395 are restriction fragment length polymorphism (RFLP) markers closely flanking the *gai–ga4* region on chromosome 1 of *Arabidopsis*. Distances given are in centiMorgans (cM). The RFLP markers also identify the yeast artificial chromosome (YAC) clones indicated below the linkage map. The lengths of the *Arabidopsis* DNA fragments cloned within these YACs are not drawn to scale.

YAC EG3G3, and g2395 identifies YACs EG8F8 and EG5H9 (Fig. 2). We are using end-probes derived from these YACs in chromosome walking steps, with the aim of obtaining a contiguous segment of DNA stretching between m219 and g2395. This DNA will contain *gai* and *ga4*. The precise identification of these loci can then be achieved by a number of techniques including complementation of *ga4* (restoration of wild-type phenotype to plants homozygous for *ga4* by introduction into the genome of the cloned wild-type allele) and detailed analysis of the DNA structure of the *gai–ga4* region in the irradiation-induced derivatives.

TRANSPOSON TAGGING OF *gai*

We have also initiated experiments with the aim of isolating *gai* via transposon tagging. *Ac* and *Ds* elements transpose preferentially to positions genetically linked to those from which they excise. We have identified a line containing a *Ds* element approximately 10 cM distal of *gai* (Peng & Harberd, 1993; Peng *et al.*, unpublished; Whitelam *et al.*, 1993). A recombinant chromosome containing both *gai* and the *Ds* was isolated and made homozygous. Recombinant homozygotes were then crossed with plants containing sources of the *Ac* transposase enzyme. We are currently bulking up individuals multiply homozygous for the transposase source, *Ds* and *gai*. From the progeny of these individuals, we aim to isolate plants displaying wild-type phenotype due to the insertion of *Ds* into *gai*. We predict that a transposon-insertion knockout allele of *gai* will confer a wild-type phenotype because the irradiation-induced *gai* derivative alleles confer a phenotype indistinguishable from wild-type. Following *Ds* tagging the *gai* gene will be isolated and sequenced.

CONCLUSIONS

We are applying the genetic and molecular biological resources available in *Arabidopsis* with the aim of isolating and understanding the function of the *gai* and *ga4* loci. The molecular cloning of *ga4* may permit detailed study of an enzyme involved in GA biosynthesis, and an improved understanding of the regulation of GA metabolic pathways. Isolation of *gai* will permit the study of a gene product which is a potential GA signal transducer molecule. The techniques that we have described are, in a general sense, applicable to the analysis of the genetic components determining plant shape and form.

REFERENCES

BALCELLS, L., SWINBURNE, J. & COUPLAND, G., 1991. Transposons as tools for the isolation of plant genes. *Trends in Biotechnology, 9:* 31–37.

BURKE, D.T., CARLE, G.F. & OLSON, M.V., 1987. Cloning of large segments of exogenous DNA into yeast by means of artificial chromosome vectors. *Science, 236:* 806–812.

CHANG, C., BOWMAN, J.L., DeJOHN, A.W., LANDER, E.S. & MEYEROWITZ, E.M., 1988. Restriction fragment length polymorphism map for *Arabidopsis thaliana. Proceedings of the National Academy of Sciences, USA, 85:* 6856–6860.

GALE, M.D. & YOUSSEFIAN, S., 1985. Dwarfing genes in wheat. In G.E. Russell (ed). *Progress in Plant Breeding,* pp. 1–35, London, U.K.: Butterworths.

GIBSON, S. & SOMERVILLE, C., 1992. Chromosome walking in *Arabidopsis thaliana* using yeast artificial chromosomes. In C. Koncz, N-H Chua & J. Schell (eds), *Methods in* Arabidopsis *Research,* pp. 119–143. Singapore: World Scientific.

GRILL, E. & SOMERVILLE, C., 1991. Construction and characterization of a yeast artificial chromosome library of *Arabidopsis* which is suitable for chromosome walking. *Molecular and General Genetics, 226:* 484–490.

HAKE, S., VOLBRECHT, E. & FREELING, M., 1989. Cloning *Knotted* the dominant morphological mutant of maize using *Ds2* as a transposon tag. *European Molecular Biology Organization Journal, 8:* 15–22.

HARBERD, N.P. & FREELING, M., 1989. Genetics of dominant gibberellin-insensitive dwarfism in maize. *Genetics, 121:* 827–838.

KOORNNEEF, M., 1990. Linkage map of *Arabidopsis thaliana* (2n=10) In S.J. O'Brien (ed.), *Genetic Maps,* 5th edn, Book VI, pp. 95–99. New York, U.S.A.: Cold Spring Harbor Press.

KOORNNEEF, M. & van der VEEN J.H., 1980. Induction and analysis of gibberellin sensitive mutants in *Arabidopsis thaliana* (L.) Heynh. *Theoretical and Applied Genetics, 58:* 257–263.

KOORNNEEF, M., ELGERSMA, A., HANHART, C.J., van LOENEN-MARTINET, E.P., van RIGN, L. & ZEEVAART, J.A.D., 1985. A gibberellin insensitive mutant of *Arabidopsis thaliana. Physiologia Plantarum, 65:* 33–39.

McCLINTOCK, B., 1942. Mutable loci in maize. *Carnegie Institute of Washington Year Book, 41:* 181–186.

MEYEROWITZ, E., 1987. *Arabidopsis thaliana. Annual Review of Genetics, 21:* 93–113.

NAM, H.G., GIRAUDAT, J., den BOER, B., MOONAN, F., LOOS W., HAUGE, B. & GOODMAN, H.M., 1989. Restriction fragment length polymorphism linkage map of *Arabidopsis thaliana. Plant Cell, 1:* 699–705.

PARK, E.-C. & HORVITZ, H.R., 1986. Mutations with dominant effects on the behaviour and morphology of the nematode *Caenorhabditis elegans. Genetics, 113:* 821–852.

PENG, J. & HARBERD, N.P., 1993. Derivative alleles of the Arabidopsis gibberellin-insensitive (*gai*) mutation confer a wild-type phenotype. *Plant Cell, 5:* 351–360.

PUTTERILL, J., ROBSON, F., LEE, K. & COUPLAND G., 1993. Chromosome walking with YAC clones in *Arabidopsis*: isolation of 1700 kb of contiguous DNA on chromosome 5, including a 300 kb region containing the flowering-time gene *CO*. *Molecular and General Genetics, 239:* 145–157.

REID, J.B., 1986. Gibberellin mutants. In A.D. Blonstein & P.J. King (eds), *A Genetic Approach to Plant Biochemistry,* pp. 1–34. Vienna, Austria: Springer-Verlag.

SWINBURNE, J., BALCELLS, L., SCOFIELD, S.R., JONES, J.D.G. & COUPLAND, G., 1992. Elevated levels of *Activator* transposase mRNA are associated with high frequencies of *Dissociation* excision in *Arabidopsis. Plant Cell, 4:* 583–595.

TALON, M., KOORNNEEF, M. & ZEEVAART, J.A.D., 1990. Endogenous gibberellins in *Arabidopsis thaliana* and possible steps blocked in the biosynthetic pathways of the semidwarf *ga4* and *ga5* mutants. *Proceedings of the National Academy of Sciences, USA, 87:* 7983–7987.

WHITELAM, G.C., JOHNSON, E., PENG, J., CAROL, P., ANDERSON, M.L., COWL, J.S. & HARBERD, N.P., 1993. Phytochrome A null mutants of Arabidopsis display a wild-type phenotype in white light. *Plant Cell, 5:* 757–768.

Part II

FUNGI

CHAPTER

15

Cellular nature and multicellular morphogenesis of higher fungi

NICK D. READ

CONTENTS

Abstract

An understanding of the morphogenesis of fungi requires an understanding of their life-styles as heterotrophic, sessile organisms. Central to the success of filamentous fungi has been the evolution of hyphae and the developmental plasticity these hyphae possess. Yeasts, in contrast to filamentous fungi, exhibit limited developmental plasticity. The term fungal cell needs to be used rather loosely to cover cellular elements such as hyphae and discrete cells like yeast cells. Five basic cellular types can be recognized: hyphae; yeast cells; spores; conglutinate cells; and a heterogeneous group of other determinate cell types. On a functional level, fungi form true tissues but since multicellular development, involving hyphal aggregation, is undergone in a fundamentally different way, fungal tissues are structurally different from those of plants and animals. Hyphal adhesion plays an important role in the development of multihyphal aggregates, including mycelial cords, rhizomorphs, stromata, sclerotia, lichen thalli and fruitbodies.

Shape and Form in Plants and Fungi
ISBN 0–12–371035–9

FUNGAL LIFE-STYLES AND MORPHOGENESIS

It is important to understand development in fungi within the context of their life-styles. Higher fungi (i.e. the Ascomycotina, Basidiomycotina and Deuteromycotina) are heterotrophic and sessile organisms. These features have undoubtedly been important in influencing their unique modes of differentiation. Heterotrophy necessitates an efficient means for the secretion of extracellular digestive enzymes and absorption of organic materials. Immobility has, perhaps, more profound implications for fungal growth and development. Mycelial fungi typically inhabit heterogeneous, discontinuous and often changing microenvironments. A sessile existence, largely due to the possession of cell walls, imposes an inability to escape environmental variation by motility. Most filamentous fungi overcome this handicap by rapidly changing their patterns of growth and development in response to one or more of a multitude of environmental signals; the growth and development of filamentous fungi is very plastic. The success of filamentous higher fungi as heterotrophic, sessile organisms can largely be attributed to the evolution of hyphae. These cellular elements are, or can readily become, adapted for: exploration and resource capture; nutrient mobilization, uptake and storage; translocation of nutrients and water; defence of occupied substratum; reproduction; and survival (Cooke & Rayner, 1984; Rayner *et al.*, Ch. 17 this volume). Hyphae are also responsible for the enormous diversity in patterns of cellular and multicellular organization found within fungi (Figs 1–9, 11–17).

In contrast, yeasts tend to live in relatively uniform, liquid environments. They respond to fewer environmental signals and do not normally show the high degree of development plasticity typical of filamentous fungi.

Fungi, plants and animals probably parted evolutionary company in the pre-Cambrian era (Burnett, 1987). Since then these three eukaryotic organismal types have evolved their own developmental strategies to cope with characteristically distinct life-styles. These strategies show many important differences although they also exhibit some features in common. To illustrate this, Table 1 compares the development, and factors which influence the development, of filamentous higher fungi, yeasts, angiosperm plants and vertebrate animals. It should be emphasized that some generalizations have had to be made in many of the points of comparison.

THE FUNGAL COLONY

The colony of a filamentous higher fungus is a very repetitive structure. It is composed of a three-dimensional, branched network of interconnecting hyphae and possesses an indefinite number of growing points. Because vegetative hyphae exhibit tip growth (Gooday & Gow, Ch. 19 this volume), the growing part of the mycelium constantly moves away from its original position. Thus the history of a colony's growth and development is more-or-less fixed in place as the colony margin continues to grow outwards. Secondary differentiation of the colony usually occurs within the basic framework laid down during the vegetative growth phase. Initially a colony is trophic but usually changes to having reproductive, dispersal and/or survival functions.

Some hyphae develop into other, often more differentiated, cellular elements (e.g. spores, yeast cells or specialized hyphae such as asci). Hyphae may also aggregate and undergo further morphogenesis within multihyphal systems (e.g. stromata,

Table 1 Comparison of development and features influencing development of filamentous higher fungi, yeasts, angiosperm plants and vertebrate animals

Feature	Filamentous higher fungi	Yeasts	Angiosperm plants	Vertebrate animals
Nutrition	Heterotrophic, absorptive	Heterotrophic, absorptive	Photo-autotrophic	Heterotrophic, phagocytic
Sessile/motile	Sessile	Sessile	Sessile	Motile
Overall structure	Simple, repetitive	Simple, repetitive	Simple, repetitive	Complex, non-repetitive
Developmental hierarchy	Limited	Very limited	Limited	Great
Main cellular structures	Walled hyphae	Walled cells	Walled cells	Wall-less cells
Main determinant of cell shape	Cell wall	Cell wall	Cell wall	Cytoskeleton
Predominate nuclear state of cells	Multinucleate	Uninucleate	Uninucleate	Uninucleate
Type of cell division	Septation	Septation	Cell plate formation	Cell cleavage
Multicellularity	Universal	Very limited	Universal	Universal
Number of cell types	3–10	1–7	10–20	100–200
Cell generation time	Short	Short	Long	Long
Life-cycle length	Short	Short	Long	Long
Developmental plasticity	Great	Very limited	Great	Limited
Developmental autonomy	Great	Limited	Great	Limited
Regenerative prowess	Great	None	Great	Limited
Vegetative reproduction	Present	Present	Present	Absent
Typical number of offspring	Up to many millions	Infinity	Up to many thousands	Up to tens
Reproductive propagules	Present	Present	Present	Absent
Response to adverse conditions	Tolerate or escape	Tolerate or occasionally escape	Tolerate or escape	Usually escape

sclerotia, lichen thalli or fruitbodies; Figs 5–9, 11, 17). Hyphal growth and development vary in different parts of a colony and distinct patterns are often manifested at the macroscopic level (e.g. in the distribution of pigmentation, hyphal aggregates, sporulating structures, aerial mycelium or autolysed hyphae). On heterogeneous substrates and substrata, as usually occur in nature, these patterns may differ significantly from those on homogeneous media, as commonly found in Petri dishes.

Yeast colonies typically lack the variable morphologies of colonies produced by filamentous fungi, although there are exceptions (e.g. in *Candida albicans*, Soll *et al.*, 1993).

THE FUNGAL CELL

Problems are frequently encountered when considering the cellular nature of a fungus. A common view of the cell is of a discrete, uninucleate unit of protoplasm bounded by a plasma membrane and typically by an extracellular matrix (e.g. a cell wall in the case of fungi and plants). The discrete nature of cells is usually taken to mean that cytoplasm and organelles cannot freely interchange between adjacent cells. It is clear that higher fungi do produce cells of this kind. However, they also produce hyphae which often contain extensive regions along which considerable movement of cytoplasm and organelles can occur through septal pores (Gregory, 1984). For convenience it is often easiest to use the term fungal cell rather loosely to cover cellular elements such as hyphae as well as discrete cells such as yeast cells. Semantic problems frequently arise, however, when hyphae or spores are truly multicellular, possessing discrete cell compartments separated by septa with blocked pores. These problems of definition illustrate important differences between fungi and other organisms.

Five basic, morphological types of cellular elements may be recognized in higher fungi: hyphae, yeast cells, spores, conglutinate cells, other determinate cell types (Fig. 1).

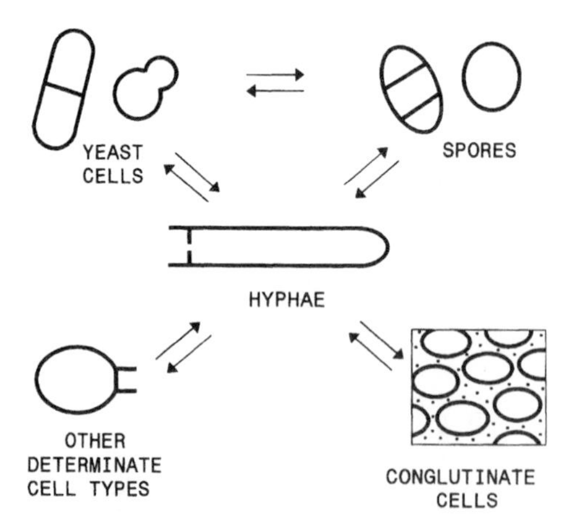

Figure 1 Diagram of the five basic cell types found in higher fungi. Pathways of differentiation or transformation may be in either direction. However, differentiation between different cell types is not commonly reversible (e.g. ascus hypha to ascospores) although examples of reversible transformation are well known (e.g. yeast–hyphal dimorphism).

Hyphae

Hyphae are discrete, tubular, polarized elements exhibiting a pronounced longitudinal growth pattern. The markedly polar nature of their growth is usually primarily a result of tip growth (Gooday & Gow, Ch. 19 this volume) although this is not always the case (e.g. in stipe hyphae of agaric fruitbodies, Mol *et al.*, 1990). They may be relatively non-specialized (e.g. vegetative hyphae) or specialized and sometimes highly differentiated (e.g. asci). Branching and septation are the norm but may be absent in some hyphae such as asci which are true cells. Hyphae are found in a range of microhabitats and are well adapted for efficiently covering, penetrating and ramifying through heterogeneous, solid substrata. The functions of hyphae are very varied. The relatively high surface area to volume ratio of hyphae makes them well adapted for absorption, secretion and excretion but leaves them prone to desiccation. Their tubular nature coupled with their tendency to form branched, anastomosed networks provides an efficient means for the interchange of materials and organelles between different regions of a colony.

Yeast cells

Yeast cells are single cells which reproduce vegetatively by processes involving septation and separation of daughter cells by budding or binary fission. Yeast cells are best adapted for uniform liquid habitats. Being unicells, they possess a higher surface area to volume ratio than hyphae and are capable of higher metabolic rates and more rapid production of biomass in nutrient-rich conditions. Because of their typically passive nature, yeasts lack the exploratory and invasive capabilities of hyphae. Nevertheless, cell separation allows rapid dissemination of propagules within aqueous environments.

Spores

Spores are discrete cellular structures which vary significantly in size, form and origin, and are differentiated in a wide variety of ways (Figs 2, 4, 9, 10). Spores may be either multicellular, due to septation, or unicellular. They can have reproductive, dispersal and/or survival functions. They may be produced mitotically (e.g. conidia) or meiotically (e.g. ascospores or basidiospores) and often in large numbers. Many spores have a low water content, thick walls, and are resistant structures.

Conglutinate cells

Conglutinate cells are elements that arise when the cell walls of adjacent hyphae become firmly cemented together. Conglutinate cells, by definition, exist in groups and tend to form tissues (see later) within multihyphal aggregates (Figs 3–9, 11–17). The extent and strength of adhesion varies within these tissues. Conglutinate cells provide an organ or lichen thallus with mechanical rigidity, particularly if the cells are strongly bonded together. As a result, conglutinate cells can serve to support and protect other components (e.g. developing spores within hyphal aggregates; Fig. 9).

Other determinate cell types

This is a heterogeneous group which include many infection structures (e.g. appressoria, substomatal vesicles, haustorial mother cells and haustoria), basidia and adhesive knobs of nematophagous fungi.

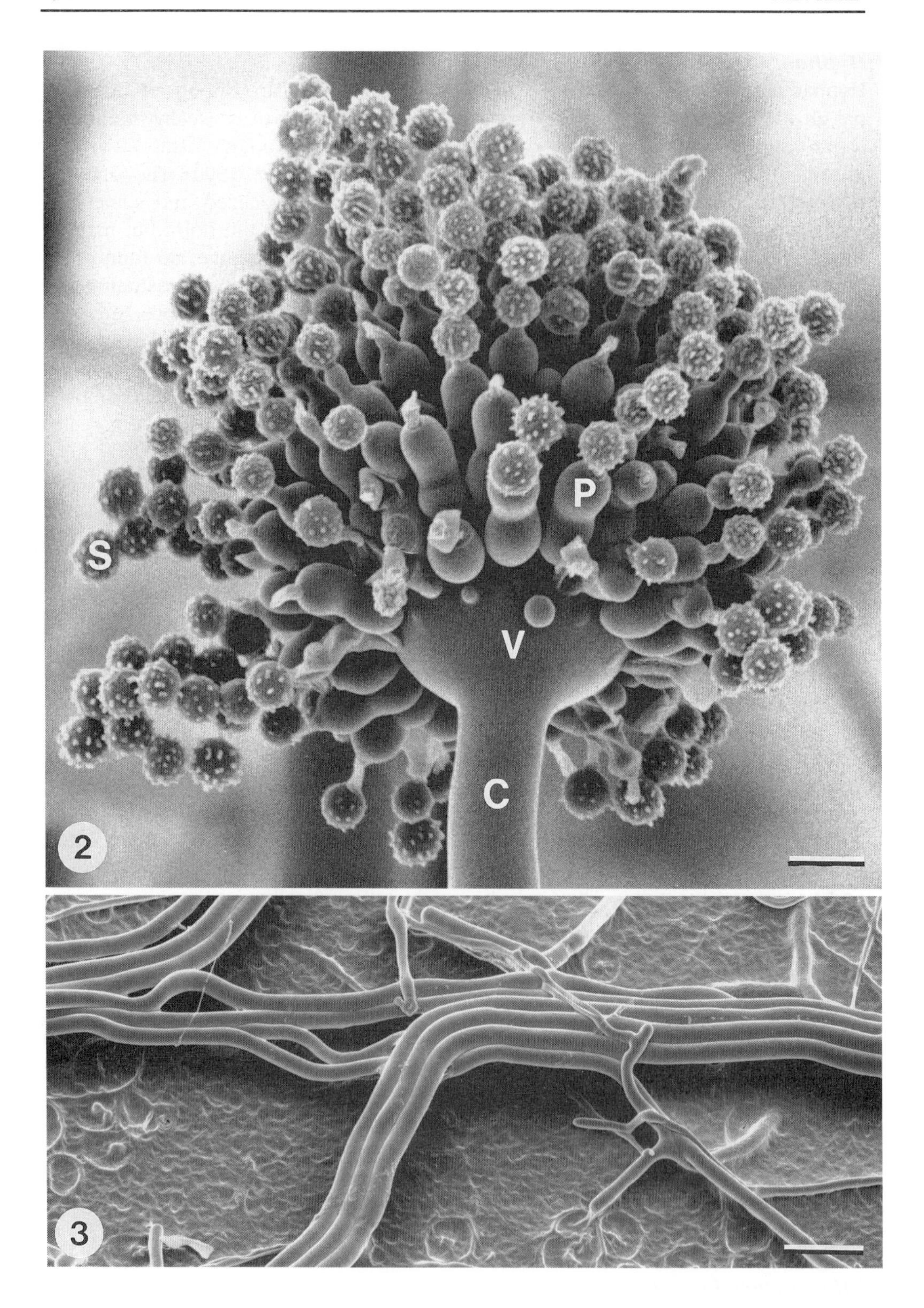
S
P
V
C
2
3

This classification of cellular elements into five basic types should not be too rigid because intermediate states occur. Four examples are as follows. First, during the formation of a pseudomycelium intermediate stages between yeasts and hyphae are formed (Odds, 1985). Secondly, the distinction between yeasts and spores may not be clear cut because they often share common features of proliferation (Cole and Samson, 1979). Thirdly, a whole continuum of cellular types between hyphae and conglutinate cells which have completely lost their hyphal appearance exists (cf. Figs 5–9, 11, 17). Fourthly, the distinction between determinate hyphae and other determinate cell types is not always clear. For example, many of the cells of a toadstool hymenium (e.g. basidia, paraphyses and cystidia, Rosin & Moore, 1985), may often be regarded as determinate hyphae, conglutinate cells or belonging to the group of other determinate cell types. The classification is thus somewhat artificial and primarily represents extremes of cellular type. One of the important features of the classification is that it highlights fundamental ways in which the cellular nature of filamentous fungi has evolved to fulfil basic requirements of the fungal life-style.

FUNGAL TISSUES

Animal and plant tissues usually contain a large number of similarly functioning cells which are bound together in a characteristic arrangement by an extracellular matrix. The tissues of these organisms sometimes contain more than one cell type but generally the cells act in concert to perform a general function. Tissues tend to act as metabolic units and combine together to form organs.

Differentiated multihyphal aggregates are constructed from one or more tissues but these differ from those of animals or plants because the tissues are of hyphal origin. Mycologists often make the distinction between prosenchyma, in which the component hyphae of a fungal tissue still have a hyphal appearance (Figs 5–7), and pseudoparenchyma, in which they do not (Figs 5, 8, 9, 11, 17). Because a fungal tissue is structurally different from that of an animal or plant, the term 'pseudotissue' is sometimes used to describe it. On a functional level, however, fungi clearly form true tissues on a par with animals or plants.

What is actually termed a tissue within a multihyphal aggregate may be debatable, particularly since, as in animals and plants, adjacent tissues often merge into each other (Figs 7, 8, 9, 11, 17). A functional distinction can often be useful here. In this respect, the highly differentiated fruitbody (perithecium) of the ascomycete *Sordaria* may be considered as being composed of three functionally distinct tissues: the centrum (containing asci, ascogenous hyphae and paraphyses); the lower perithecial wall; and the neck (Fig. 9). Their primary functions are to produce spores (centrum),

---

Figure 2 *Aspergillus niger* v. Tieghem (Deuteromycotina; Hyphomycetes). Scanning electron micrograph of the asexual reproductive apparatus. The determinate conidiophore (C) is swollen at its apex into a vesicle (V). From determinate conidiogenous cells (phialides, P), chains of asexual spores (conidia, S) arise. Partially freeze-dried. Scale 10 μm. From Read (1990).

Figure 3 *Gauemannomyces graminis* (Sacc.) v. Arx & Olivier (Ascomycotina; Diaporthales). Scanning electron micrograph of mycelial cords. These are very simple multihyphal aggregates arising from the adhesion of adjacent hyphae. Partially freeze-dried. Scale 10 μm.

protect the developing spores (lower perithecial wall), and direct the spores that are forcibly discharged to a suitable location for germination (neck). If the centrum is thought of as a tissue, it is clear that not all fungal tissues need be composed of conglutinate cells.

MULTICELLULAR DEVELOPMENT

There is a fundamental difference between multicellular development in filamentous fungi and that in animals and plants (with the exception of some filamentous algae), which is that fungi produce hyphae whilst the other organisms do not. In most cases, multicellularity in these fungi is initially achieved by hyphal septation (i.e. cross-wall formation) accompanied by septal pore occlusion. Septa typically grow inwards from the wall across a dividing hypha or cell (Fig. 1). This contrasts with cell plate formation in plants which starts in the centre and develops outwards across the dividing cell, and which can be in more than one plane. Limited multicellularity in fungi may also arise during sporulation (Figs 2, 4, 9, 10). More complex multicellularity requires hyphal aggregation (Figs 3–9, 11–17). The aggregates formed may then undergo further septation, differentiation and morphogenesis. Cell adhesion and differentiation within hyphal aggregates often results in at least some tissues losing their hyphal characteristics as they differentiate into conglutinate cells (Figs 5–9, 11, 17). Some hyphae in these aggregates become considerably specialized in structure and function. Certain specialized hyphae (e.g. asci) or other determinate cells (e.g. basidia) may produce spores (e.g. Fig. 9).

Higher fungi produce a wide range of multihyphal aggregates varying from simple mycelial cords (Fig. 3), more complex conidiophore aggregates (Fig. 4), stromata (Figs 6 and 17), rhizomorphs and sclerotia (Fig. 7), through to the sometimes highly differentiated lichen thalli (Fig. 5) and fruitbodies (Figs 8, 9, 12, 14–16).

Figure 4 *Doratomyces stemonitis* (Pers. ex Steud.) Morton & G. Sm. (Deuteromycotina; Hyphomycetes). Scanning electron micrograph of a coremium. It is composed of numerous, adjacent conidiophores which have adhered together and produced chains of conidia. Partially freeze-dried. Scale bar 25 μm.

Figure 5 *Hypogymnia physodes* (L.) Ach. (Ascomycotina; Lecanorales). Scanning electron micrograph of a vertical fracture through a lichen thallus. The thallus possesses two outer pseudoparenchymatous cortical tissue layers composed of conglutinate cells (C), and an internal prosenchymatous tissue comprising easily recognizable medullary hyphae (M). Algal cells are present in a layer (A) just beneath the upper cortex of the fungal partner. Freeze-fractured, fully frozen-hydrated. Scale bar 25 μm.

Figure 6 *Xylaria hypoxylon* (Fr.) Grev. (Ascomycotina; Xylariales). Scanning electron micrograph of a transversely fractured stroma. The outer layer is composed of a palisade of cells and this surrounds aggregated hyphae which have a parallel arrangement seen here in cross fracture. The latter hyphae are largely surrounded by extracellular matrical material with intermittent air spaces. Freeze-fractured, fully frozen-hydrated. Scale bar 25 μm.

Figure 7 *Sclerotinia sclerotiorum* (Lib.) de Bary (Ascomycotina; Helotiales). Scanning electron micrograph of a fracture through the outer region of a sclerotium. Note the compact outer layer of pseudoparenchymatous rind (R) and cortex (C) composed of conglutinate cells which surround the prosenchymatous medulla (M) tissue comprised of clearly distinguishable interwoven hyphae. Freeze-fractured, fully frozen-hydrated. Scale bar 25 μm.

C
A
M
C
5
4
6
M
C
R
7

Multicellularity serves a number of important roles. The existence of two or more cellular elements establishes a system for intercellular communication and interaction within a colony or between colonies. Fungal cells, like those of other organisms, are social entities; they are engaged in constant communication with their environment and with one another. Multicellularity also provides a means for cell specialization and thus division of labour between different cells within a multicellular element, hyphal aggregate or in the colony as a whole. The physical isolation of adjacent cell compartments is usually a prerequisite for cell specialization and septal pore occlusion is probably important in this (Gull, 1978). It is possible that certain specialized types of occluded septal pores (e.g. those within ascogenous hyphae, Beckett, 1981, and those of basidiomycete dolipore septa, Lu & McLaughlin, 1991) are important for selective cell–cell communication. In this respect, they may have features in common with plasmodesmata and gap junctions in plants and animals, respectively.

The development of the specialized tissues of multicelluar organisms requires the differentiation of distinct cell types and their organization into integrated structures. The fate of a cell may be determined solely by lineage or may also require the reception and interpretation of positional information from its environment. How systems of positional information are generated is one of the most challenging problems of developmental biology. Low-molecular-weight, diffusible, signal molecules (so-called 'morphogens') which establish the spatial pattern of cells during multicellular devel-

Figure 8 *Sordaria macrospora* Auersw. (Ascomycotina; Sordariales). Light micrograph of a longitudinal section through a developing protoperithecium. Note the pseudoparenchymatous protoperithecial wall composed of conglutinate cells devoid of a hyphal appearance. This has an outer region (OW) of thick-walled cells and an inner region (IW) of much more inflated, thin-walled cells. The latter is sometimes termed the centrum pseudoparenchyma. These wall layers surround the developing centrum (C) which at this stage comprises ascogenous hyphae intermixed with, and largely indistinguishable from, paraphyses in this section. Chemically fixed; stained with methylene blue. Scale bar 50 μm.

Figure 9 *Sordaria humana* (Fuckel) Winter (Ascomycotina; Sordariales). Light micrograph of a mature perithecium. This is composed of three functionally distinct tissues: the lower perithecial wall (W), centrum (C) and perithecial neck (N). The pseudoparenchymatous perithecial wall serves to protect the centrum and is composed of conglutinate cells which have lost their hyphal appearance. The inner wall layer (clearly seen in Fig. 8) has become crushed by this stage. The centrum is composed of asci, ascogenous hyphae and paraphyses. The ascogenous hyphae give rise to asci. Meiosis is undergone within the asci and the resultant haploid nuclei are packaged within ascospores. The perithecial neck is composed of fine hyphae (periphyses) lining the neck canal which differentiate the conglutinate cells of the pseudoparenchymatous neck wall (see Fig. 11). The latter lack a hyphal appearance. The neck is positively phototropic which allows the ascus spore guns to be aimed at the light. The ascospore can thus be forcibly discharged away from the dung, in which the fungus resides, and on to surrounding grass which may then be consumed by a herbivore. Eventually the spores are deposited in fresh dung. Freeze-substituted; stained with safranin green and methylene blue. Scale bar 50μm. From Read & Beckett (1985).

Figure 10 *Stigmatomyces diopsis* Thaxter (Ascomycotina; Laboulbeniales). Differential interference contrast light micrograph of a complete thallus. The cell lineage within the thallus is rigidly determined with little developmental plasticity. Note the foot (F) which adheres to the insect host, necked perithecium (P), an ascus (A) released from the distal end of the neck, and appendage (AP) bearing antheridia (AT). Scale bar 25 μm.

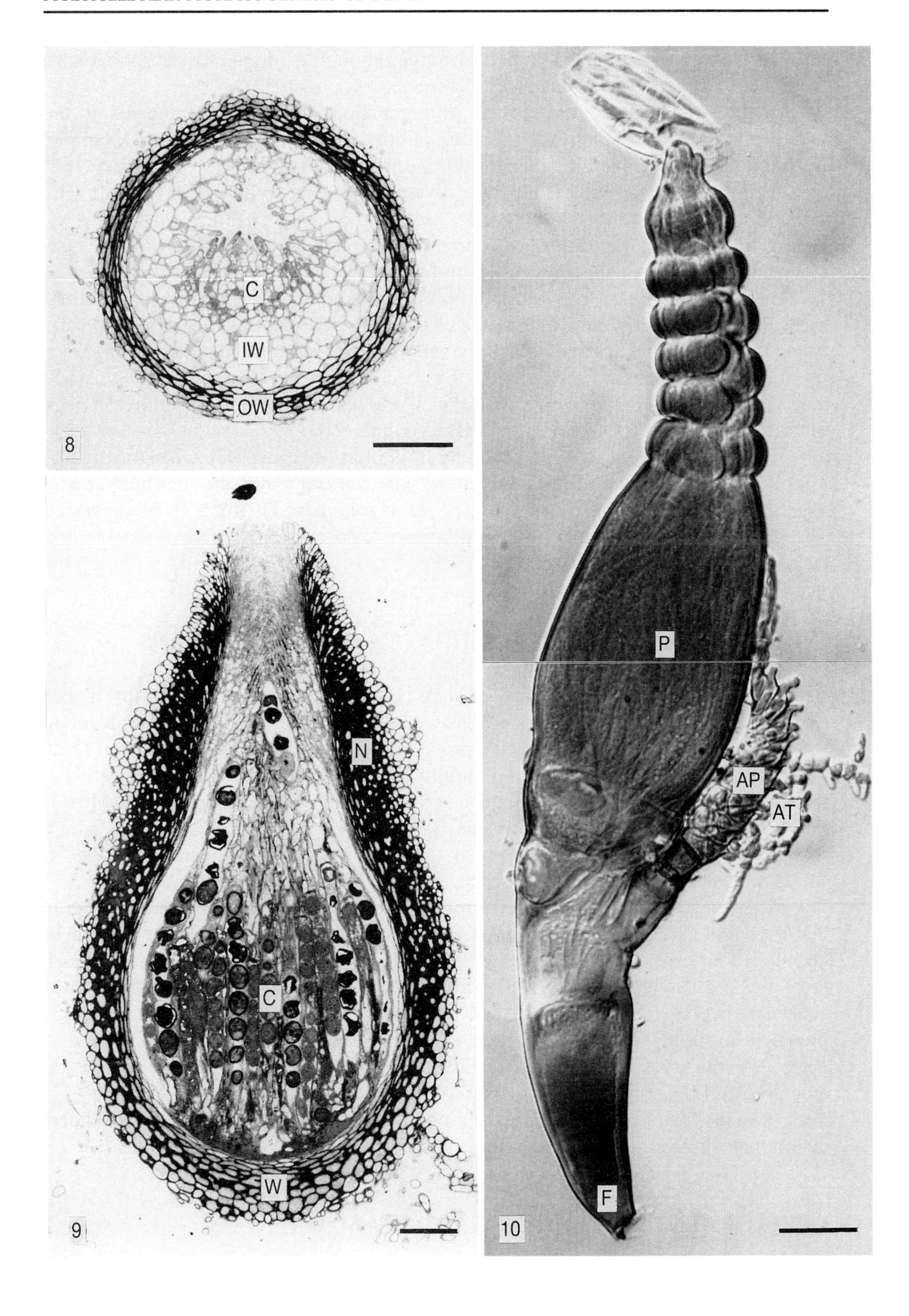
C
IW
OW
8
N
C
W
9
P
AP
AT
F
10

opment have not yet been identified in fungi although evidence for morphogenetic fields within fungal tissues exists (e.g. Horner & Moore, 1987).

Of critical significance during the differentiation of hyphal aggregates is the adhesion of hyphae to each other. During perithecium morphogenesis, for example, hyphal adhesion is initially important during the formation of the conglutinate cells of the protoperithecial wall from interwoven enveloping hyphae (Figs 8, 12). The protoperithecial wall subsequently gives rise to the lower perithecial wall (Fig. 9). Hyphal adhesion is later important during the differentiation of the conglutinate cells of the neck wall from the hyphae (periphyses) which line the neck canal (Figs 9, 11). How hyphal adhesion is brought about is unknown but may involve different mechanisms. Some evidence indicates a role for phenol oxidases (e.g. laccases) catalysing the cross-linking of phenolic polymers between adjacent hyphae. During perithecium morphogenesis, for example, it has been found that laccase activity is associated with the expanding protoperithecial wall and with the emerging perithecial neck of *Sordaria brevicollis* Olive & Fantini (Broxholme *et al.*, 1991). In the last few years a new class of abundant proteins called hydrophobins, because of their highly hydrophobic nature, have been found associated with the cell walls of aerial hyphae and aggregating fruitbody hyphae of *Schizophyllum commune* Fr. It has been suggested that some hydrophobins may, by an unknown mechanism, play a role in binding fruitbody hyphae together (Wessels, 1992).

INDETERMINATE AND DETERMINATE GROWTH PATTERNS

Both indeterminate and determinate growth patterns are common within the higher fungi. Indeterminate (i.e. open-ended) growth is only exhibited at the cellular level by the majority of hyphae; the other cell types all undergo determinate growth (Fig. 1). At the multicellular level, tissues within multihyphal aggregates are commonly determinate and stop growing after reaching a certain fixed size, an example being the lower perithecial wall (Figs 9, 16). Some tissues, however, continue growing indeterminately through the functional life of the hyphal aggregate (e.g. the perithecial neck, Figs 9, 11, 13, 16; Read, 1983; Read & Beckett, 1985).

A characteristic feature of the developmental pattern of many filamentous fungi is that they undergo a switch from indeterminate to determinate growth or vice versa and, in a few cases, this is reversible. Examples of these transitions at the cellular level are: yeast/hyphal dimorphism (Odds, 1985); formation of conidiophores from vegetative hyphae (Cole & Samson, 1979); differentiation of basidia, cystidia and paraphyses from subhymenial hyphae (Rosin & Moore, 1985); differentiation of appressoria from germ tubes (e.g. Bourett & Howard, 1990; Kwon & Hoch, 1991); and the emergence of a germ tube from a spore during germination. The emergence of a neck from the lower bulbous region of a perithecium is an example of a multicellular determinate–indeterminate growth transition (Figs 9, 16).

DEVELOPMENTAL PLASTICITY AND STABILITY

The expression of an individual genotype may be modified by its environment; the amount that it can be modified is termed its plasticity. This phenotypic plasticity can

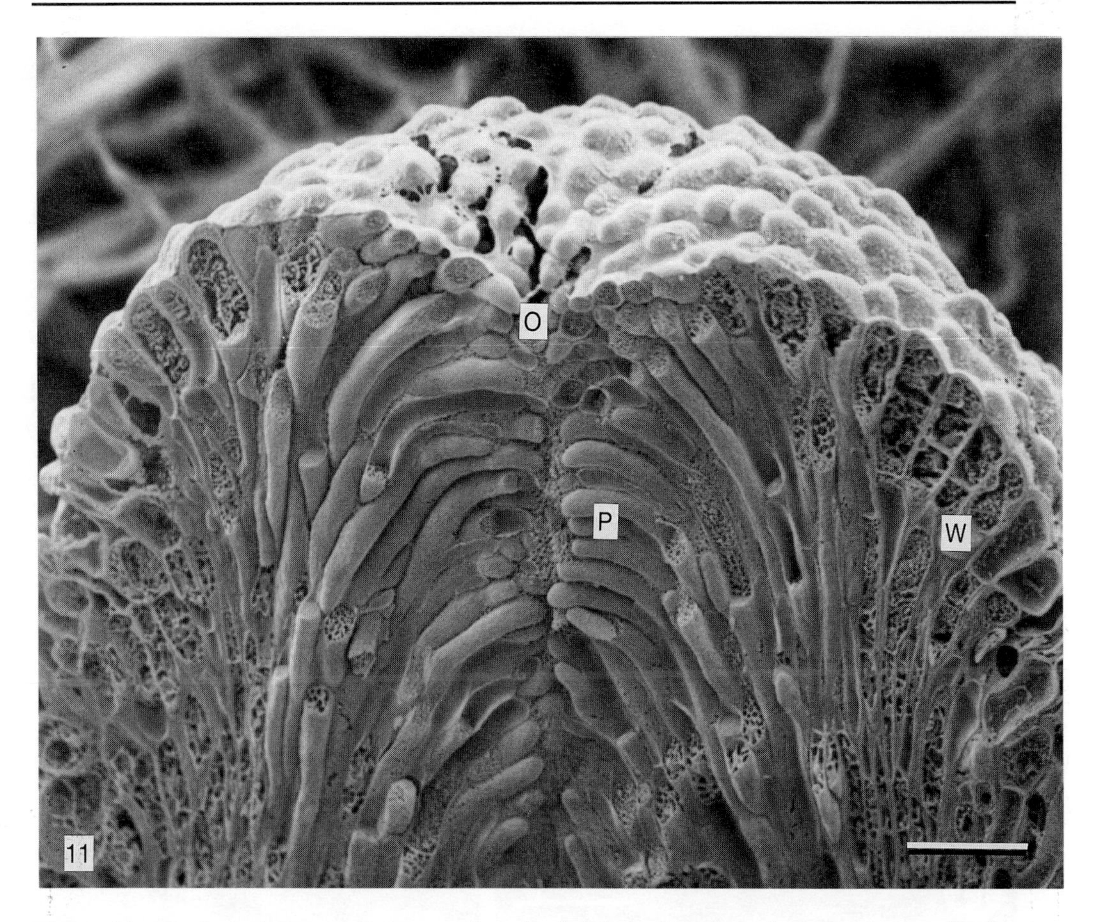

Figure 11 *Sordaria humana* (Ascomycotina; Sordariales). Scanning electron micrograph of a median longitudinal fracture through a growing neck of a mature perithecium (similar to that shown in Fig. 9). The periphyses (P) are fine hyphae which are curved towards the centre of the neck canal in order to grip asci during ascospore discharge. The periphyes grow out of the neck canal over the rim of the opening of the neck (the ostiolar pore, O) and as they pile up on top of each other they adhere together and differentiate into the conglutinate cells of the neck wall (W). At the multicellular level, the perithecium neck primarily exhibits apical growth. Freeze-fractured, fully freeze-dried. Scale bar 10 μm. From Read & Beckett (1985).

be either morphological or physiological (Bradshaw, 1965; Jennings & Trewavas, 1986). In the context of development the term developmental plasticity (syn. developmental versatility) may be used. Filamentous fungi tend to exhibit significantly greater developmental plasticity than yeasts because their microhabitats are less stable.

The mycelial colony and its cellular components contains within its developmental programming certain basic morphogenetic options, allowing or restricting the potential for diversification under episodic selection pressures. Mycelial fungi are able to adapt rapidly to varying conditions by switching between different functional modes, manifest as different growth patterns or developmental pathways. These mode switches include: (1) determinate–indeterminate transitions; (2) alterations in the distance between hyphal branches and in branch angle; (3) aerial versus appressed or submerged growth; (4) compact versus diffuse morphogenesis; and (5) juvenility and senescence (Gregory, 1984; Rayner *et al.*, 1987; Rayner & Coates, 1987). Switches in

12
13
14
15
16

functional mode are accompanied by fundamental changes in hyphal physiology, behaviour and development but the mechanistic basis of mode switching is largely unknown.

The developmental biology of fungi in natural habitats is an important subject which fungal ecologists are increasingly addressing. Intraspecific or interspecific interactions between neighbouring mycelia often result in the modulation of growth or morphogenetic patterns. Within interaction zones, new genes are expressed, particularly those involved in secondary metabolism (e.g. pigment synthesis). However, other changes often occur adjacent to interaction zones. These can include alterations in the mycelial growth pattern or result in the induction of mycelial cords, pseudosclerotial plates, sclerotia or sporophores (Rayner & Coates, 1987; Rayner *et al.*, 1987; Rayner & Boddy, 1988). In natural habitats mycelial interactions may have an important role in influencing growth and development.

Generally, filamentous fungi exhibit more plastic development in their trophic than in their reproductive/survival phase. The unspecialized, iterative, hyphal character of the extending, peripheral parts of fungal colonies and the fact that vegetative hyphae in these zones often contain extensive septate, multinucleate regions along which considerable movement of cytoplasm and organelles can occur, are important for the developmental versatility of the trophic mycelium. Developmental plasticity becomes limited during cellular and multicellular specialization.

Although the fungal colony consists of a complicated, highly cross-linked network of participating components, there is no central control system. Vegetative hyphae exhibit a high degree of autonomy and are very opportunistic (Table 1). This is significant with regard to different parts of a colony. In general, there tends to be significantly less autonomy during the development of hyphal aggregates. However, developmental plasticity is sometimes clearly evident during fruitbody morphogenesis. In *Volvariella bombycina* (Schaeff.:Fr) Sing., for example, fruitbody polymorphisms

---

Figure 12 *Sordaria humana* (Ascomycotina; Sordariales). Scanning electron micrograph of a young protoperithecium. Note the fine and highly branched interwoven enveloping hyphae which differentiate into the conglutinate cells of the protoperithecium wall (see Fig. 8). Partially freeze-dried. Scale bar 10 μm.

Figure 13 *Ceratocystis fimbriata* Ell. & Halsted Pers. (Ascomycotina; Ophiostomatales). Scanning electron micrograph of the tip of the perithecium neck shown in Fig. 16. Note that the neck involves the longitudinal extension of adjacent parallel hyphae. This contrasts with perithecium development in the majority of species which is more akin to that shown in Fig. 11. Partially freeze-dried. Scale bar 10 μm.

Figure 14 *Ascobolus immersus* Pers. (Ascomycotina; Pezizales). Scanning electron micrograph of a young apothecium. The hymenium bearing ascogenous hyphae and paraphyses at this stage is beginning to be exposed as the apothecium opens out. Partially freeze-dried. Scale bar 200 μm.

Figure 15 *Ascobolus immersus.* Scanning electron micrograph of a mature apothecium. The surface of the hymenium is covered with asci and paraphyses. The asci have all exhibited a positive phototropism and in this micrograph are bent towards the left (i.e. the original source of light). Partially freeze-dried. Scale bar 200 μm.

Figure 16 *Ceratocystis fimbriata.* Perithecium (only half of the neck is visible in this micrograph; the tip is shown in Fig. 13). Note the lower bulbous region, within which the asci containing ascospores are differentiated, and the neck which is growing by the coordinated spiral growth of adjacent hyphae. As the asci are continuously produced they are forced up the neck and eventually released passively from its tip, in mucilage. They are dispersed by insects. Partially freeze-dried. Scale bar 50 μm.

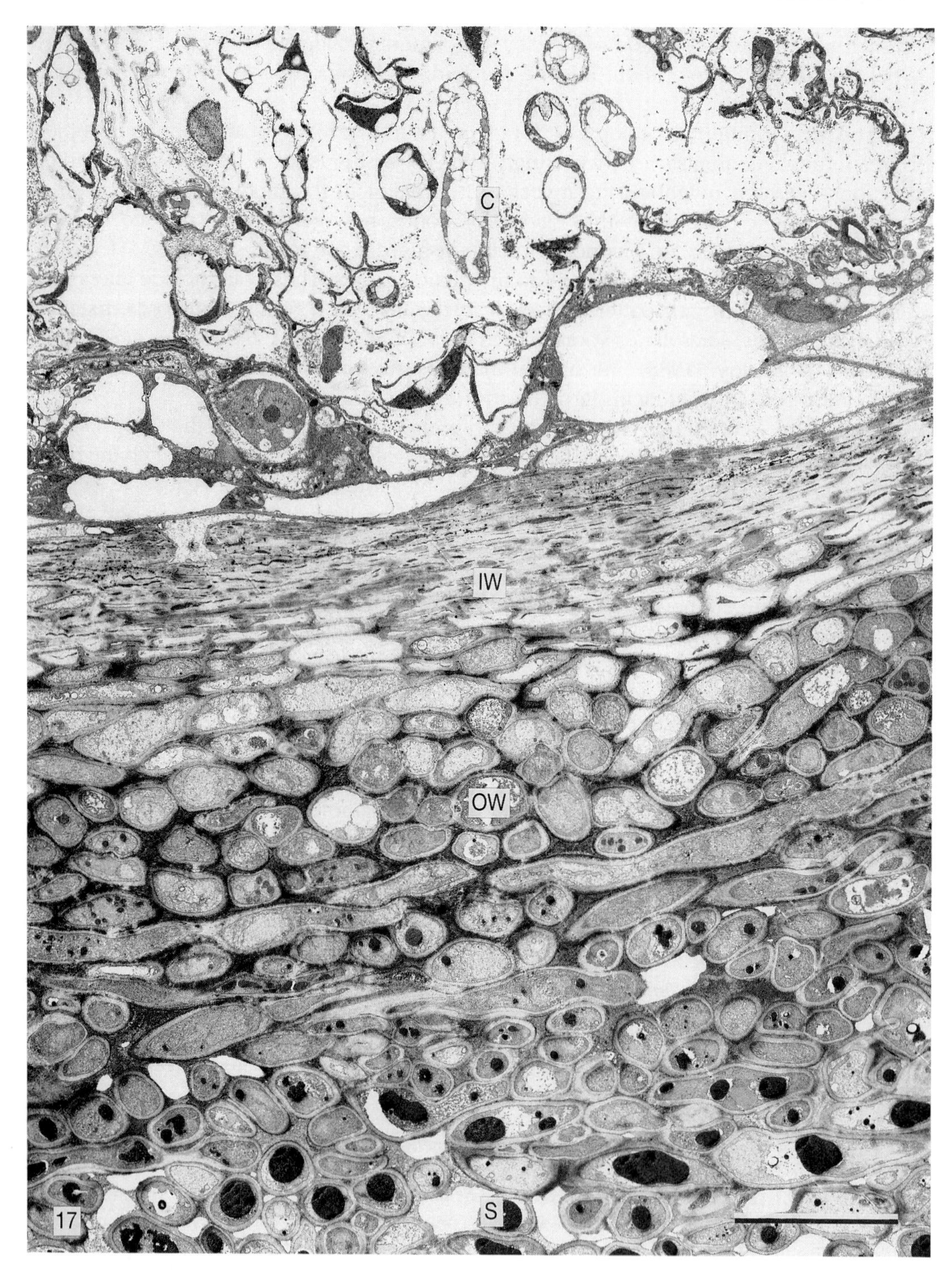

Figure 17 *Xylaria longipes* Nitschke (Ascomycotina; Xylariales). Transmission electron micrograph of a longitudinal section through the base of a perithecium embedded within a stroma (S). In the perithecium a pseudoparenchymatous perithecial wall encloses the centrum (C). The perithecial wall has an outer wall (OW) and inner wall (IW) layer. The centrum contains various cell profiles belonging to ascogenous hyphae, paraphyses and the bases of asci. The perithecial wall and stroma are composed of conglutinate cells which have mostly lost their hyphal appearance. The stroma contains some air spaces (cf. Fig. 6). Note that the stroma, two wall layers of the perithecial wall and the centrum merge into each other. Chemically fixed; stained with uranyl acetate and lead citrate. Scale bar 10 μm (unpublished micrograph kindly provided by A. Beckett).

occur spontaneously. Chiu *et al.* (1989) have suggested that the fruitbody ontogenetic programme may be a sequence of subroutines which can be modulated independently. Coordinated activation of the subroutines specifies the ontogeny and the architecture of a fruiting structure. Developmental plasticity may be an expression of these morphogenetic subroutines in the wrong sequence and/or in the wrong place.

Nutrient limitation imposes less autonomy for growth and differentiation. It often imposes corporate growth with the formation of multihyphal reproductive and survival structures exhibiting markedly more canalized development than the trophic mycelium. A primary reason for most filamentous fungi and plants exhibiting high developmental plasticity is because they are sessile. However, it should be noted that these organisms exhibit aerial and non-aerial, invasive growth which allows them to 'move' parts of themselves from one location to another, a feature that is extremely important for resource capture. Corporate growth involving the formation of mycelial cords (Fig. 3) or rhizomorphs can also be important for moving between isolated nutrient resources (Dowson *et al.*, 1986).

It is clear that many filamentous fungi, as sessile organisms, have been successful in coping with environmental variation because they can rapidly change their growth or developmental pattern in response to one or more environmental cues. Part of this change in developmental pattern may involve spore production. Spore dispersal represents a means of escape from adverse conditions. This method of escape is also exhibited by some plants and contrasts with animals which can simply move to a more favourable habitat (Table 1).

Except for pseudomycelium or spore production, yeasts have limited capabilities for multicellular differentiation. Including ascospores, *Saccharomyces* produces seven different cell types, of which most can only be identified on the basis of the genes they express (Herskowitz, 1989). This compares with less than 10 morphologically distinct cell types differentiated by many filamentous fungi (Table 1). Larger numbers of cell types could undoubtedly be identified in filamentous fungi if classified according to the genes they express. It has been generally assumed that yeasts show very little developmental plasticity. The yeast pathogen *Candida albicans*, however, has been shown to exhibit considerable phenotypic plasticity which may be important in their versatile pathogenic behaviour (Soll *et al.*, 1993).

Not all higher fungi undergoing significant multicellular development exhibit great developmental plasticity. In this respect, the Laboulbeniales deserve special mention. They live in stable environments as ectoparasites of arthropods, often occupying specific positions on their hosts (Hulden, 1983; Tavares, 1984). A complete thallus of one species is shown in Fig. 10. These exotic organisms exhibit rigid development which, in certain respects, is more akin to that of animals than to that of filamentous fungi. Their cell lineages are canalized and the number of uninucleate cells within a thallus is more-or-less predetermined. Unfortunately it has proved impossible to grow a complete thallus of a laboulbeniacious fungus in culture (Tavares, 1984), otherwise this group may have provided some extremely useful experimental systems for developmental biology.

ACKNOWLEDGEMENTS

Thanks are due to: Mr T. Collins for drawing Fig. 1; Dr D. Minter for providing access to the slide collection of the Laboulbeniales, originally prepared by Roland Thaxter in the 1920s and 1930s, and from which N. Read photographed Fig. 10; Dr A. Beckett for supplying Fig. 18; Ms K. Lord for technical assistance; and the Science and Engineering Research Council for grant funding (Grant number GR/D/33472).

REFERENCES

BOURETT, T.M. & HOWARD, R.J., 1990. *In vitro* development of penetration structures in the rice blast fungus *Magnaporthe grisea. Canadian Journal of Botany, 68:* 329–342.

BRADSHAW, A.D., 1965. Evolutionary significance of phenotypic plasticity in plants. *Advances in Genetics, 13:* 115–155.

BROXHOLME, S.J., READ, N.D. & BOND, D.J., 1991. Developmental regulation of proteins during fruit-body morphogenesis in *Sordaria brevicollis. Mycological Research, 95:* 958–969.

BURNETT, J.H., 1987. Aspects of the macro- and micro-evolution of the fungi. In A.D.M. Rayner, C.M. Brasier & D. Moore (eds), *Evolutionary Biology of the Fungi,* pp. 1–15. Cambridge, U.K.: Cambridge University Press.

CHIU, S.W., MOORE, D. & CHANG, S.T., 1989. Basidiome polymorphism in *Volvariella bombycina. Mycological Research, 92:* 69–77.

COLE, G.T. & SAMSON, R.A., 1979. *Patterns of Development in Conidial Fungi.* London, U.K.: Pitman.

COOKE, R.C. & RAYNER, A.D.M., 1984. *Ecology of Saprotrophic Fungi.* London, U.K.: Longman.

DOWSON, C.G., RAYNER, A.D.M. & BODDY, L., 1986. Outgrowth patterns of mycelial cord-forming basidiomycetes from and between woody resource units in soil. *Journal of General Microbiology, 121:* 203–211.

GREGORY, P., 1984. The fungal mycelium: an historical perspective. *Transactions of the British Mycological Society, 9:* 1–11.

GULL, K., 1978. Form and function of septa in filamentous fungi. In J.E. Smith & D.R. Berry (eds), *The Filamentous Fungi,* Vol. 3. *Development Mycology,* pp. 78–93. London, U.K.: Edward Arnold.

HERSKOWITZ, I., 1989. A regulatory hierarchy for cell specialization in yeast. *Nature, 342:* 749–757.

HORNER, J. & MOORE, D., 1987. Cystidial morphogenetic field in the hymenium of *Coprinus cinereus. Transactions of the British Mycological Society, 88:* 479–488.

HULDEN, L., 1983. Laboulbeniales (Ascomycetes) of Finland and adjacent parts of the U.S.S.R. *Karstenia, 23:* 31–136.

JENNINGS, D.H. & TREWAVAS, A.J., 1986. *Plasticity in Plants.* Cambridge, U.K.: The Company of Biologists Ltd.

KWON, Y.H. & HOCH, H.C., 1991. Temporal and spatial dynamics of appressorium formation in *Uromyces appendiculatus. Experimental Mycology, 15:* 116–131.

LU, H. & McLAUGHLIN, D.J., 1991. Ultrastructure of the septal pore apparatus and early septum initiation in *Auricularia auricula-judae. Mycologia, 83:* 322–334.

MOL, P.C., VERMEULEN, C.A. & WESSELS, J.G.H., 1990. Diffuse extension of hyphae in stipes of *Agaricus bisporus* may be based on a unique wall structure. *Mycological Research, 94:* 480–488.

ODDS, F.C., 1985. Morphogenesis in *Candida albicans. Critical Reviews in Microbiology, 12:* 45–93.

RAYNER, A.D.M. & BODDY, L., 1988. *Fungal Decomposition of Wood. Its Biology and Ecology.* Chichester, U.K.: John Wiley.

RAYNER, A.D.M. & COATES, D., 1987. Regulations of mycelial organisation and responses. In A.D.M. Rayner, C.M. Brasier & D. Moore (eds), *Evolutionary Biology of the Fungi,* pp. 115–136. Cambridge, U.K.: Cambridge University Press.

RAYNER, A.D.M., BODDY, L. & DOWSON, C.G., 1987. Genetic interactions and developmental versatility during establishment of decomposer basidiomycetes in wood and tree litter. In M. Fletcher, T.R.G. Gray & J.G. Jones (eds), *Ecology of Microbial Communities,* pp. 83–123. Cambridge, U.K.: Cambridge University Press.

READ, N.D., 1983. A scanning electron microscopic study of the external features of perithecium development in *Sordaria humana. Canadian Journal of Botany, 61:* 3217–3229.

READ, N.D., 1990. Low-temperature scanning electron microscopy of fungi and fungus–plant interactions. In K. Mendgen & D.-E. Lesemann (eds), *Electron Microscopy of Plant Pathogens,* pp. 17–29. Berlin, Germany: Springer-Verlag.

READ, N.D. & BECKETT, A., 1985. The anatomy of the mature perithecium in *Sordaria humana* and its significance for fungal multicellular development. *Canadian Journal of Botany, 63:* 281–296.

ROSIN, I. & MOORE, D., 1985. Differentiation of the hymenium in *Coprinus cinereus. Transactions of the British Mycological Society, 84:* 621–628.

SOLL, D.R., MORROW, B. & SRIKANTHA, T., 1993. High-frequency phenotypic switching in *Candida albicans. Trends in Genetics, 9:* 61–65.

TAVARES, I., 1984. *The Laboulbeniales.* Lehre: J. Cramer.

WESSELS, J.G.H., 1992. Gene expression during fruiting in *Schizophyllum commune. Mycological Research, 96:* 609–620.

CHAPTER

16

Moulding moulds into mushrooms: shape and form in the higher fungi

ROY WATLING & DAVID MOORE

CONTENTS

Abstract

The classification of, and the understanding of relationships between, the larger fungi has been strangled by the rigid adoption in the early part of the 19th century of a system of identification based totally on the appearance of the mature fruit body – viz. the end-product of a multitude of changes of shape and form. It was almost 100 years later before microscopic characters were seriously employed for identification. Out of this sprang the appreciation that all was not well and that these microscopic structures could act as stronger markers of relatedness. Examples of these natural constructions will be given with emphasis on caution not to abandon completely all gross morphology. The classification of most organisms has a degree of developmental study in-built and although studies took place as early as the last decade of the 19th century, the importance of the understanding of the development of the fruit body is only now really percolating into our thoughts. Indeed, revealing studies are still under way. Not surprisingly, these are beginning to suggest that development of

Shape and Form in Plants and Fungi
ISBN 0–12–371035–9

fungal structures depends upon the exercise of relatively simple sets of 'rules' which seem to be organized into 'programming routines' governing the distribution and pattern of cell differentiation in space and time. What distinguishes developmental pathways which lead to grossly different end-product morphologies may be the temporal order of the routines and the spatial position of tissues in which particular routines are invoked. Some of the rules and some of the routines can be identified experimentally.

INTRODUCTION

It is often necessary to remind people that fungi are not plants. There are still a great many people whose education was completed before the revolution in systematics in the mid-1960s and who are firmly convinced that fungi are plants – peculiar plants, perhaps, but plants nevertheless. This notion, of course, is completely wrong. Plants, animals and fungi are now seen to be three quite distinct kingdoms of eukaryotic organisms (Cavalier-Smith, 1981; Margulis, 1974; Whittaker, 1969). This is a systematic arrangement but it is reflected in current ideas about the early evolution of eukaryotes in all of which the major kingdoms are thought to have separated at some protistan level. If this was really the case then plants, animals and fungi became distinct long before the multicellular grade of organization was established in any of them.

The three kingdoms differ from one another in ways that are crucial to determining shape and form. A major aspect of the original definition of the kingdoms (Whittaker, 1969) was nutrition (plants use radiant energy, animals engulf, fungi absorb), and this apparently simple base for separation embraces numerous other correlated differences in structure and life-style strategy that can be catalogued. Once the separation into three distinct eukaryotic kingdoms has been made, though, other non-correlated differences emerge and among these is the way in which multicellular architectures can be organized. A key feature during the embryology of even lower animals is the movement of cells and cell populations, so cell migration (and everything that controls it) plays a central role in animal morphogenesis. Being encased in walls, plant cells have little scope for movement and their changes in shape and form are accommodated by control of the orientation and position of the mitotic division spindle and, consequently, the orientation and position of the daughter cell wall which forms at the spindle equator. Fungi are also encased in walls; but their basic structural unit, the hypha, has two peculiarities which mean that fungal morphogenesis must be totally different from plant morphogenesis. These are that hyphae grow only at their tip and that cross-walls form only at right angles to the long axis of the hypha. The consequence is that fungal morphogenesis depends on the placement of hyphal branches. To proliferate, a hypha must branch; and to form a structure, the position at which the branch emerges and its direction of growth must be controlled.

If it is right that the evolutionary separation between the major kingdoms occurred at a stage before the multicellular grade of organization, then these kingdoms evolved, independently, the mechanisms to organize populations of cells. The fungal hypha differs in so many important respects from animal and plant cells that significant differences in the way cells interact in the construction of organized tissues must be expected. However, comparison can reveal common strategies and conserved pathways as well as alternative approaches, providing insight into the response of very different living organisms to the need to solve the same sorts of morphogenetic control problems. For the mycologist, parallels between fungi and other eukaryotes are worth seeking out so that the conceptual framework that has already been established, in embryology for example, as well as in cell and evolutionary biology, may be used. Rather than repeating past mistakes, lessons should be learned from them.

FUNGAL NAMES

In the study of fungi it has often been claimed that there are not enough characters to attempt a meaningful classification. It is true that the vast array of features offered to the flowering plant specialist are lacking, but careful attention to detail does allow the identification of meaningful traits. It is also true that the microscopic study of larger fungi has lagged behind that of the moulds and similar growths, under the misconception that macroscopic characters were sufficient. Amateur naturalists, and many professionals too, have tended to leave the larger fungi to one side as though everything was settled and even when they have taken them up, workers wax lyrical over such features as the pileus surface, the smell and the taste, so that identification and everything that flows from it becomes almost mystical. These characters, however, are only the final expressions of the truly basic features – the developmental patterns and biochemical pathways that characterize the species.

Perhaps matters were not helped by the 'Father of Mycology', Elias Fries, and his contemporaries all of whom had excellent observational powers and feel for the organism, but in the absence of a concept of evolutionary convergence, shape took an overemphasized part in the understanding of mushrooms and toadstools and their relatives. This fundamental mistake is still apparent.

Fayod (1889) used the microscope to study agaric structures and from his results suggested relationships, but it was not until Patouillard (1900) and his contemporaries that the microscope became a regular tool in identification and classification. Perhaps the necessity of the microscope in identification today has driven the sytem the other way and critical field observations are now less frequently made. It is relevant, however, that the use of scanning and transmission electron microscopes have assisted in the unravelling of many problems.

In very few plants or animals has shape been such a dominant factor in driving 20th century classification. Botanists and zoologists are used to dealing with notions of homology, analogy, neoteny and the like, resolving their differences by application of detailed developmental, structural and anatomical studies. This is not so in many areas of mycology where there is little critical study of anatomy and little consideration of development.

Fries (1821) developed a classification based on the earlier work of Persoon (1801) which was primarily an arrangement founded on the shape and form of the spore-producing tissue, the hymenium, and on the hymenophore, the structure on which the hymenium was borne. Thus, agarics with plates (i.e. gills) beneath an umbrella-shaped pileus, as in the ordinary cultivated mushroom, were brought together in a single group irrespective of spore colour, tissue texture and general aspect (facies). They were contrasted with those fungi that had tubes in a spongy layer beneath the pileus (boletes and bracket fungi) and those with teeth or spines hanging down below (hydnoids). Both of the last groups included taxa which lacked a pileus and were formed as a sheet of fertile tissue on the substrate (resupinate forms). Other major groups included those that bore spores externally on a club-shaped (clavarioid) or coral-like (coralloid) structure and those in which they were borne internally (gasteroid) – a group which was to be expanded many-fold as specimens were sent back for identification from the colonies newly conquered by the European nations.

Although groups such as the boletes with fleshy, putrescent, tube-bearing fruit bodies (basidiomes), but otherwise quite like the mushroom, were separated on morphology alone, it was Fayod's (1889) studies on development and anatomy that forced the realization that the Friesian groupings were artificial. Unfortunately, his observations made no impact on the mycological community until over 40 years later when Roger Heim reintroduced the modern audience to these studies (Heim, 1931), quickly followed and expanded subsequently by Rolf Singer (1936, 1951 *et subseq.*). At the end of the 19th century Fayod's studies were almost heretical and the Friesian classification held sway; indeed it was quite a shock when the conservative British mycologist was exposed by Carleton Rea (1922) not to the traditional approach but the new ideas. This work followed, in part, that of Patouillard (1900) where relationships and groupings not considered before were expressed. The close relationship between some small cup-shaped basidiomycetes and agarics was proposed, the joining together in a single grouping of large woody bracket fungi with those of resupinate structure, etc. In fact this was a preview to the separation of white wood rotters and brown wood rotters, and other similar ecological groupings. At last, after 100 years of stagnation, during which superficial characters were the basis of a classification, the marrying of anatomical and developmental information came about, and when later correlated with chemical data produced suites of characters which led to radical rearrangements. It was not that the early mycologists had not recognized these natural groupings, but rather that they did not have the resources available to them (technical and conceptual) which made the rearrangements possible.

THE SIGNIFICANCE OF THE STRUCTURE OF TISSUES

Taking the structure of the flesh of the basidiome first: it can be shown quite easily that, although the naive view might be that this is fundamentally homogeneous, it differs in detail from one group of larger fungi to another. Fayod (1889) clearly demonstrated this for different groups of agarics; the approach was followed for the bracket fungi (Corner, 1932a,b, 1953; Cunningham, 1965), clavarioid fungi (Corner, 1950) and hydnaceous fungi (Maas Geesteranus, 1971, 1975).

The function of the basidiome is to produce as many basidiospores as the structure will allow, and the structure can be expanded in many ways to give optimum spore production. Thus, in the Russulales this is achieved by columns and rosettes of hyphae expanding in an orchestrated way (Reijnders, 1976), whereas other agarics have simple, gradually elongating and inflating hyphae, accompanied in *Coprinus* by narrow elements (Hammad *et al.*, 1993) which resemble the inducer hyphae in *Russula* and *Lactarius* (Reijnders, 1976; Watling & Nicoll, 1980). In Amanitaceae the formation of the gills is schizohymenial and so differs from other families in the character, but it is correlated with what is called an acrophysalidic tissue (Bas, 1969) by which flesh hyphae through massive inflation of individual cells allow the fruit body to expand. In some members of the Tricholomataceae a further modification is found leading to a two-component flesh termed sarcodimitic (Corner, 1966, 1991); Redhead (1987) believes this is such a fundamental method of basidiome expansion that he has adopted the family Xerulaceae for agarics exhibiting the character.

Bracket fungi achieve massive spore production by increasing the longevity of the basidiome either over several months, as in *Polyporus*, or as a perennial fruit body with resurgence of growth at regular intervals (e.g. *Fomes*). Examination of such flesh reveals an intricate pattern of hyphae which may branch profusely to bind adjacent hyphae together or elongate, thicken and lose their living contents to form strong tubes which then act as structural members in much the same manner as the units in tubular furniture. These hyphae increase in number as the basidiome grows. The major group exhibiting this type of hyphal construction possess woody fruitbodies and a poroid hymenium (polypores), but it can be found in a rudimentary form in some fungi with gills, e.g. *Lentinus*. It is now considered that *Lentinus* is more closely related to the polypores than to the other agarics. In some taxa it may be a complex of two or three hyphal types, viz. dimitic or trimitic. The presence or absence of such mixtures of hyphae agrees with other characters used in the classification of the polyporaceous taxa (Ryvarden, 1992). In general, this approach can be applied to the clavarioid fungi and to the resupinates, both groups of which can then be shown to be heterogeneous groups with some members now considered more advanced than others.

Fayod (1889) also showed that in the homoiomerous agarics (trama composed only of hyphal tissue) the flesh between the hymenial surfaces might be one of four kinds, and although this has had to be modified as more agarics have been examined, the basic idea holds true, viz. that in mature specimens, bilateral (divergent), convergent, regular and irregular patterns of hyphal arrangement can be demonstrated. In each case they correlate with other characters and together are used to define families, a taxonomic rank that appeared late in agaricology. Just as the structure of the flesh is reflected by demonstrable field characters, so can the gill tramal types be recognized.

By careful observation of developing basidiomes, it hs been found, particularly by the critical work of Reijnders (1948; 1963), that there are at least ten different ways by which the familiar mushroom shape can be formed, a shape which is excellently designed to give protection to the developing hymenia in exposed environments (Fig. 1). These vary from those with naked development (Fig. 1A), which includes the majority of the bracket fungi developing from a concentration of tightly bound hyphae forming a rounded structure known to foresters as a 'conk', to those with a complete enclosing tissue (Fig. 1J), or enclosing membrane or membranes which only break before maturity (Fig. 1B–1I). The exciting thing is that the type of development complies with the proposed classification based on other features. Thus, the boletes, with their hymenium exposed at all times during development (gymnocarpic, Fig. 1A) are considered to be close to the Paxillaceae where the pileus gives protection by being pressed against the stipe surface and enclosing the hymenium (pilangiocarpic, Fig. 1G), and Gomphidiaceae, where hyphae from an initially naked primordium envelop the developing hymenium (metavelangiocarpic, Fig. 1E). In addition to this developmental connection there are similarities in morphology of the basidiome, basidiospore structure and chemistry.

The veils which form in the agarics (Fig. 1B–1E) are considered to be protective, allowing the hymenia to develop in a rather well-defined environment. The shape of the gills is strongly tied to the constraints of this environment within the developing basidiome, and has led to the use in identification of a traditional feature, gill attach-

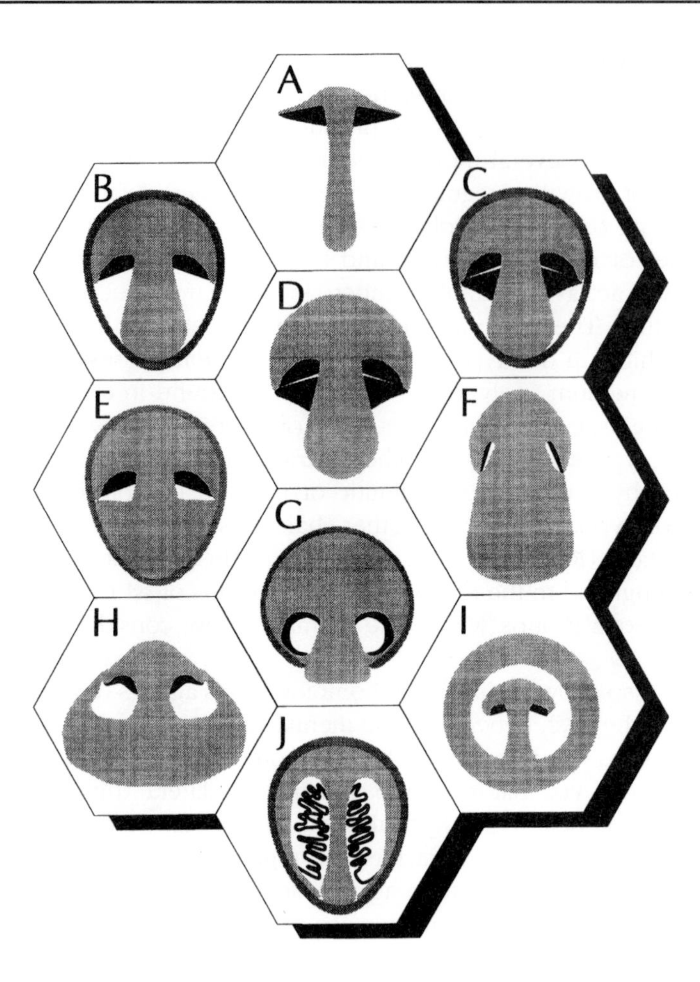

Figure 1 Ten ways to make a mushroom. A montage of diagrammatic sections illustrating the various primordial tissue patterns that eventually mature to form mushroom-like basidiomata; hymenial tissues are shown in black. **A** gymnocarpic, where the hymenium is naked at first appearance and develops to maturity on the fruitbody surface; **B** monovelangiocarpic, with a single (universal) veil enveloping the whole primordium; **C** bivelangiocarpic, in which an inner (partial) veil provides additional protection to the hymenium; **D** paravelangiocarpic, where the veil is reduced and often lost at maturity; **E** metavelangiocarpic, where a union of secondary tissues emerging from the pileus and/or stipe forms an analogue of the universal veil; **F** gymnovelangiocarpic, in which the hymenium is protected by a very reduced veil, seen only at adolescence, formed between the stipe and the closely applied pileus; **G** pilangiocarpic, the hymenium is protected by tissue extending downwards from the margin of the pileus; **H** stipitoangiocarpic, the hymenium is protected by tissue extending upwards from the stipe base, but this does not enclose the primordium; **I** bulbangiocarpic, where the tissue protecting the hymenium is largely derived from the basal bulb of the stipe and initially completely encloses the primordium; **J** endocarpic, where the mature hymenium is enclosed or covered over, just one (the pileate type) of a number of patterns of this gasteromycetous form of fruit body is shown.

ment to the stipe apex. At least eight different types of attachment have been used and they are now being analysed (Pöder, 1990).

There is a large group of fungi which have their basidiospores enclosed in the basidiome, even to maturity (Fig. 1J). They are called the 'gastromycetes', although many of their members do not really follow the original definition. They include puffballs, earth-stars, earth-balls, all 'reduced' to an epigeous or hypogeous sack of spores. In addition, the stink horns are included and these parallel the agarics in gross morphology. It is true that their basidia are stigmatosporic, but the hymenium is designed for insect dispersal as opposed to wind dispersal. Shape and form in this group is highly specialized, and is similar to what may be seen in the colours, bizarre shapes and penetrating smells of insect-attracting flowering plants.

The basidiomycetous hypogeous fungi, as with their ascomycete cousins the truffles, are attractive to animals, often having a distinctive odour which in some taxa resembles the male pheromone of pigs. Many quite unrelated groups have developed hypogeous taxa, which retain the internal form of the ancestor although looking alike from outside, something that the mycological pioneers did not appreciate. This same range of ancestral types is seen in the secotioid fruitbody which resembles a large drumstick. In addition, however, forceful or passive spore dispersal may be found in the species, linking them to their equivalent agaric relative. Thus, instead of a single family (Secotiaceae) there are now up to 20 groupings. These are clearly derived groups and should simply be placed in classification with their undoubted mushroom relatives, although some authorities would argue otherwise (see Singer, 1958, 1986) (Table 1).

Although this paper concentrates on basidiomycetes, it should be emphasized that a similar approach is being applied to the classification of the ascomycetes. The old groups, pyrenomycetes and discomycetes, originally based on the shape of the fruiting body, are now being split in more natural ways; the lichens are being included in their respective groups within the system; the cleistothecium is now dispersed in many families, e.g. Erysiphaceae, Eurotiaceae, Sordariaceae even Ascobolaceae (as *Seliniella* which is really *Ascobolus immersus* (von Arx & Müller, 1955; van Brummelen, 1967)). The yeasts pose rather different questions although they too are no longer classified together simply because they are single-celled. It is clear that yeasts have evolved in several groups, especially amongst the jelly fungi where they may play a part in the life-cycle or may replace the filamentous stage completely.

Amongst the basidiomycetes, there has been some tinkering with the classification and some major alterations have been made, but in the main Fries' classification has stood up well to scrutiny, especially as he did not have the benefit of information on the thousands of taxa throughout the world which is now available. What would Fries have thought of a *Russula* with a ring, as some species have in Africa and South and Central America? Although the facies is dissimilar, at least the heteromerous trama, amyloid basidiospores and presence of macrocystidia show their true relationships. Suites of characters are the key to natural classification and the fungi are no exception. The production of purplish brown or dull pigments, flesh greening in aqueous solutions of ferrous sulphate, with basidiospores which exhibit blunt spines, warts or angled facets are considered fundamental to the delimitation of the Thelephorales, thus bringing together species with teeth (*Hydnellum*), with pores (*Boletopsis*), clavarioid (*Thelephora*), and resupinates (*Tomentella*); i.e. irrespective of the final shape of the basidiome (Table 2). Parallels can be seen in other groups (e.g. Hymenochaetales). Cultural studies have helped to support this, bringing together fungi that were formerly thought dissimilar. The suite of characters including prominent latex channels (lactiferous hyphae ending in macrocystidia on the pileus and stipe surfaces and in the hymenium) has demonstrated a relationship between

Table 1 The Boletales and Russulales illustrate the broad range of hymenophore configurations which members of an order can span; they also have gasteroid relatives – a feature found throughout the agarics

	Hymenogastrales		
	Secotiaceae	Hymenogastraceae and Rhizopogonaceae	Additional hymenophore configurations
Traditional taxa			
Boletales, predominantly form poroid hymenophore			
Hygrophoropsidaceae			Agaricoid
Gyrodontaceae			Agaricoid (?)
Meiorganaceae			Agaricoid
Gomphidiaceae			
Chroogomphus	*Brauniellula*		
Gomphidius	*Gomphogaster*		
Paxillaceae			
Paxillus	*Austrogaster*		Agaricoid
Boletaceae			
Boletus *Suillus* *Leccinum* }	*Gastroboletus*		
Phylloporus	*Paxillogaster* *Gymnopaxillus*		
Chamonixiaceae			
Gyroporus?		*Chamonixia*	
Boletellaceae			Agaricoid
Strobilomycetaceae			
Coniophoraceae			Cantharelloid/hydnoid/merulioid, resupinate
Rhizopogonaceae		*Rhizopogon*	
Corneromycetaceae			Cantharelloid/hydnoid/merulioid, resupinate
Russulales, called the Astrogastraceous series, predominantly with gilled hymenophore			
Russulaceae			
Lactarius	*Arcangeliella*		Agaricoid and poroid
Russula	*Macowanites*		Agaricoid
Elasmomycetaceae	*Elasmomyces*	*Zelleromyces* *Gymnomyces* *Martellia*	

Taxa according to Julich (1981).

Lentinellus and *Clavicorona* made even closer by *Clavicorona* fruiting bodies being formed in cultures of *Lentinellus* (Miller & Stewart, 1971) and species of *Clavicorona* being described as new taxa but in reality being immature *Lentinellus* (Maas Geesteranus, 1971) – parallels to *Seliniella*.

Much still has to be done in the study of the development of the basidiome, and the relationship between pileus, stipe and hymenium. This requires cellular and bio-

Table 2 Comparison of the Ganodermatales, an order with only poroid basidiomes, and the Thelephorales and Hymenochaetales, which each show a wide spectrum of hymenophore configurations (indicated by the extent of the horizontal bars)

	Traditional families							
	Clavariaceae	Thelephoraceae	Cantharellaceae	Hydnaceae		Polyporaceae		Agaricaceae
	Hymenophore configuration							
	Club- to coral-shaped	Smooth to wrinkled	Ridged and veined	Toothed	Tubes resupinate	Tubes Non-resupinate	Tubes Non-resupinate	Gilled
					Annual or perennial (*Poria*)	Annual (*Polyporus*)	Perennial (*Fomes*)	
	Clavarioid	Thelephoroid	Cantharelloid	Hydnoid	Poroid	Poroid	Poroid	Agaricoid
Modern orders and families								
Thelephorales	■	■	■	■		■		■
Thelephoraceae	░	░	░					
Lenzitopsidaceae						░		
Bankeraceae				░				
Boletopsidaceae						░		
Verrucosporaceae*								░
Hymenochaetales	■	■			■	■	■	
Clavariachaetaceae	░							
Hymenochaetaceae		░						
Coltriciaceae						░		
Phellinaceae					░		░	
Ganodermatales							■	
Ganodermataceae							░	
Haddowiaceae							░	

*There is some doubt that *Verrucospora*, the sole representative, belongs to this order.
Taxa according to Julich (1981).

chemical analysis, although Reijnders (1963) has provided a terminology on which such studies can be based. Further studies on the structure of selected components of the hymenia are required. The development and role of brachycystidia (paraphyses) and the cystesium–cystidium combinations have been studied in *Coprinus* (see later); Price (1973) has examined cystidial development in agarics from a purely morphological point of view, as have Eriksson *et al.* (1978) in *Peniophora sect. Peniophora.*

With the plasticity in form of the basidiomycete fruitbody, variation in response to environmental change might be expected, and is indeed found; Bondartsev (1963) was the first to offer a cautionary note to taxonomists for placing too much emphasis on a single character. The taxonomist should not be carried away with the narrow focus of single approaches. Neither microscopic nor macroscopic characters are sufficient alone, each must be measured by its merits. The recent synthesis of fruitbody structures by Reijnders & Stalpers (1992) is a baseline from which to work.

The same range of hymenophoral configurations have now been accepted amongst the distantly related hymenomycetous Heterobasidiae, viz. *Tremella* and *Auricularia* and their allies, with little consternation. Thus the poroid hymenium is found in *Aporium*, hydnoid in *Pseudohydnum*, coralloid in *Tremellodendron*, cupuloid in *Auricularia* and resupinate in *Sebacina*. Even gasteroid heterobasidiae are now known.

VARIATION IN SHAPE AND FORM

Clearly, fungal systematists are now appreciating that fruitbody shape and form should not hold the central position it once did. This is a conceptual point – it is a matter of interpretation of the value of particular features in establishing relationships between a group of fungi, and the belief is growing that fruitbody form is less useful because it is a more flexible feature than has previously been considered. Surprisingly, in biological terms, fruitbody shape and form seems to be equally flexible. From this point this chapter will focus on this variation in fruitbody shape and form. An interpretation of shape and form will be promoted which could have an impact on the understanding of development, evolution and systematics.

The first issue to be addressed is that variation in shape and form occurs at different levels and for different reasons. There is variation (more properly called plasticity) in the shape and form of fruitbodies produced by a particular strain which can be demonstrated by *in vitro* culture; there is variation between collections of what might, on other grounds, be judged the same species; and there is variation at the supraspecific level. It is with the last two categories that taxonomists must be most concerned, but the first category can yield the most valuable information.

Many morphological mutants or variants have been induced or isolated from nature, especially in *Coprinus cinereus* (Schaeff.: Fr.) S.F. Gray and *Schizophyllum commune* Fr. (Kanda & Ishikawa, 1986; Raper & Krongelb, 1958; Takemaru & Kamada, 1972). Such mutants can be instructive in establishing development pathways (Esser *et al.*, 1977; Moore, 1981) and allowing detailed study of particular phenotypes (Kamada & Takemaru, 1977a,b; Kanada *et al.*, 1990; Kanda *et al.*, 1989). Here, more stress will be placed on epigenetic plasticity: instances where, for some reason, the development of a normal genotype is disturbed, but without change to that genotype.

Plasticity in fruiting morphogenesis may be a strategy for adaptation to environmental stress. The 'rose-comb' disease of the cultivated mushroom, *Agaricus bisporus* (Lange) Imbach (syn. *A. brunnescens* Peck.), in which convoluted growths of hymenium develop over the outer surface of the pileus, seems to be caused by mineral oil fumes in mushroom farms (Flegg, 1983; Flegg & Wood, 1985; Lambert, 1930). Viral infections have been involved in some instances, e.g. in *Laccaria, Armillaria* and *Inocybe* (Blattny *et al.*, 1971, 1973), and fungal attack in others. For example, Buller (1922) showed that gill-less fruitbodies of *Lactarius piperatus* (L.:Fr.) S.F. Gray were caused by parasitism by *Hypomyces lactifluorum* (Schw.) Tul. and Watling (1974) showed that primordia of *Entoloma abortivum* (Sprague: Donk) can be converted to a puff-ball structure by interaction with *Armillaria mellea* (Vahl.: Fr.) Kummer.

This sort of fruitbody polymorphism, or developmental plasticity, has been reported in various fungal species (Buller, 1922, 1924; Keyworth, 1942; Singer, 1975), but thorough studies have only been made on *Psilocybe merdaria* (Fr.) Ricken (Reijnders, 1977; Watling, 1971), *Agaricus bisporus* (Atkins, 1950; Flegg & Wood, 1985; Reijnders, 1977; Worsdell, 1915) and *Volvariella bombycina* (Schaeff.: Fr.) Singer (Chiu *et al.*, 1989). In *Agaricus bisporus* the developmental variants reported include: carpophoroids (sterile fruitbodies; Singer, 1975), forking (where a single stipe bears two or more pilei; Atkins, 1950; also called bichotomy by Worsdell, 1915), proliferation (additional secondary pilei arise from pileus tissues; Worsdell, 1915), fasciation (a bundle of conjoined basidiomes; Worsdell, 1915), and supernumerary hymenia (the basidiome has additional hymenia on the upper surface of the pileus; Langeron & Vanbreuseghem, 1965). All of these forms have also been observed in *Volvariella bombycina* (Chiu *et al.*, 1989) and in *Psilocybe merdaria* (Watling, 1971).

In addition, in both *V. bombycina* and *P. merdaria* other fruitbody forms, including morchelloid and gasteromycetoid fruitbodies, arose spontaneously alongside the normal agaric fruitbodies. Thus, an agaric fungus is able to produce both morchelloid (n.b., *Morchella* is an ascomycete) and gasteromycete-like fruitbodies alongside its normal fruitbodies. To put this into the context of an approximate animal counterpart, the parallel would be for cats to be able, quite normally, to give birth to litters containing the odd kitten looking like an aardvark, dolphin or even iguana. Obviously, this assumes that the taxonomic ranks are equivalent; not surprisingly, there is a great debate about taxonomic rankings in fungal systematics!

In *V. bombycina* these teratological forms arose spontaneously, in two different strains, and were found in cultures bearing normal fruitbodies, regardless of the composition of the substrate. Importantly, all hymenia in these forms were functional in the sense that they produced apparently normal basidiospores. The function of the plasticity in fruiting morphogenesis seems to be to maximize spore production and favour dispersal of spores even under environmental stress.

Fruiting is a complex polygenic process in these fungi (Leslie, 1983; Leslie & Leonard, 1984; Meinhardt & Esser, 1983; Prillinger & Six, 1983) further modulated by environmental factors (Leatham & Stahmann, 1987; Manachère, 1985; Manachère *et al.*, 1983; Raudaskoski & Salonen, 1984). There is some genetic evidence that different structures (specifically sclerotia and basidiomes) share a joint initiation pathway (Moore, 1981), and the common growth of basidiomes directly from sclerotia may also suggest morphogenetic similarities. Fruiting in haploid, primary homothallic species, such as *V. bombycina* and *V. volvacea* (Bull.: Fr.) Singer (Chang & Yau, 1971; Chiu & Chang, 1987; Royse *et al.*, 1987), and in homokaryons in heterothallic species

(Dickhardt, 1985; Elliott, 1985; Graham, 1985; Stahl & Esser, 1976; Uno & Ishikawa, 1971) shows that fruiting is independent of the sexual cycle regulated by the incompatibility system in heterothallic species (Kües & Casselton, 1992). It is against this sort of background that basidiome variants must be interpreted.

Transition of the agaric hymenial pattern to the morchelloid one, particularly the position of the hymenium on the upper surface of the pileus, have been suggested to be due to reversion or atavism to a fruitbody organization seen in ascomycetes (Worsdell, 1915). Similarly, the gasteromycetoid forms of *Volvariella bombycina* might, by the same logic, be taken to reflect some phylogenetic relationship with the so-called gasteromycete genus *Brauniella* which Singer (1955, 1963, 1975) has, on other grounds, suggested to be ancestral to some species of *Volvariella.* However, making phylogenetic points on the basis of morphological variants is inherently dangerous. In stressing the value of developmental features in taxonomic and evolutionary interpretations, there is the risk that history will be made to repeat itself and the old notion of ontogeny recapitulating phylogeny might be resurrected in some minds. The zoologists have battled through this stage in the development of evolutionary ideas; mycologists should strive to avoid it. In the first quarter of this century animal evolution was thought of as resulting mainly from modification of adult form and development was seen as a recapitulation of previous mature stages. This was encapsulated in 'the individual in its development recapitulates the development of the race' in MacBride's *Textbook of Embryology* (1914). Walter Garstang's views were diametrically opposed. He coined the term paedomorphosis which he summed up as 'Ontogeny does not recapitulate Phylogeny: it creates it.' He published his views in the usual way (e.g. Garstang, 1922) but most memorably, in verse (Garstang, 1962):

> MacBride was in his garden settling pedigrees,
> There came a baby Woodlouse and climbed upon his knees,
> And said: 'Sir, if our six legs have such an ancient air,
> Shall we be less ancestral when we've grown our mother's pair?'

A MATTER OF ROUTINE

Basidiome developmental variants can be used to comment on the ontogenetic programme. Because they are actually or potentially functional as basidiospore production and dispersal structures, they have been interpreted (Chiu *et al.*, 1989; Moore, 1988) as indicating that normal fruitbody development is comprised of a sequence of independent but coordinated morphogenetic subroutines, each of which can be activated or repressed as a complete entity. For example, there is a 'hymenium subroutine' which, in an agaric, is normally invoked to form the 'epidermal' layer of the hymenophore (gill lamella); but if it is invoked aberrantly and additionally to form the upper epidermis of the pileus, it forms, not a chaotic travesty of a hymenium, but a functional supernumerary hymenium. Similarly, the 'hymenophore subroutine' produces the classic agaric form when invoked on the lower surface of the pileus, but if wrongly invoked on the upper surface, it produces, not a tumorous growth, but a recognizable inverted cap. Thus, development of fungal structures in general is thought to depend upon organized execution of such subroutines; the sequence and location in which they are invoked determining the ontogeny and form of the fruiting structure (Fig. 2). Invocation of these developmental subroutines may be logically equivalent to the 'mode switches' between different mycelial states discussed by Gregory (1984) and Rayner & Coates (1987). Some of the subroutines can be identified with specific structures, such as basal bulb, stipe, pileus, hymenophore, hymenium and veil, but others are rather subtle, affecting positional or mechanical morphogenetic features. One such might be a 'grow to enclose' capability, possibly associated primarily with the veil subroutine but perhaps expressed in the stipe base to generate pilangiocarpic basidiomes.

Essentially the same subroutines could give rise to morphologically very different forms, depending on other circumstances. For example, the agaric gill hymenophore subroutine seems to be expressed with the rule 'where there is space, make gill' (Chiu & Moore, 1990a,b). When this is combined with mechanical anchorages the contortions initially produced by this rule are removed as the gills are stretched along the lines of mechanical stress. Where such anchorages are absent the expansion forces are not communicated through the gills and the labyrinthine structure remains – as in morchelloid forms.

CELL FORM, FUNCTION AND LINEAGE

A cell described as a basidium is quite clearly characterized by karyogamy meiosis and the formation of basidiospores. In other words, application of the nomenclature involves consideration of the past and future behaviour of the cell. Usually, other descriptive terms, like basidiole, paraphysis, sterile element or cystidium, are applied on the basis of the immediate morphology and/or position of the cell without reference to its ontogeny or fate, yet these are important considerations. If the mechanisms of differentiation and morphogenesis are to be understood, the descriptions of developmental pathways must be precise. A tramal hyphal branch which becomes a

Available subroutines

stipe pileus hymenium veil

Possible subroutine activation sequences

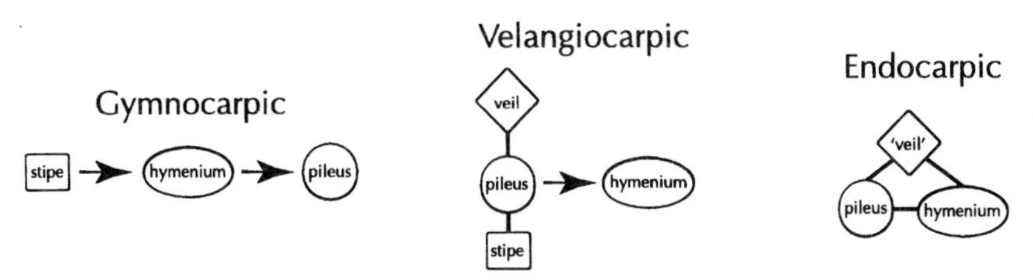

Figure 2 The notion of developmental subroutines illustrated in the form of abbreviated flow diagrams. It is envisaged that development is effectively segmentalized into specific subroutines (four are illustrated here, there are undoubtedly many more) and that different outcomes result from these subroutines being invoked in different sequences.

hymenial cell may be pluripotent initially, but it must follow one developmental pathway and the varied pathways of differentiation open to it involve commitment to expression of particular (probably different) sets of characters at different stages in morphogenesis so that the hymenium is furnished with cells that have appropriate functional characteristics.

Some cell lineages are unable to express morphologies developed by others. In the hymenium of *Agaricus* the 'epidermal pavement' which provides the structural support for basidia is made up of basidioles in an arrested meiotic state. Even after many days of existence, when the fruitbody was close to senescence, 30–70% of the basidioles were in meiotic prophase (Allen *et al.*, 1992). This is not wastage of reproductive potential but use of one differentiation pathway to serve two distinct but essential functions. *Coprinus* illustrates the other extreme in having a highly differentiated cell type, the paraphysis, with which to construct the epidermal pavement. These cells arise after the numerically static basidiole population commits to meiosis, branching from beneath the basidia and forcing their way into the hymenium (Rosin & Moore, 1985). At maturity, individual basidia are surrounded by about five paraphyses; thus, more than 80% of the hymenial cells in *Coprinus* serve a structural function. *Agaricus* and *Coprinus* hymenophore tissues reach essentially the same structural composition by radically different routes.

Some cell lineages reach the same final morphology through different routes. Both *C. cinereus* and *V. bombycina* have facial (pleuro-) and marginal (cheilo-) cystidia. Both types of cystidium in *V. bombycina* are established when the hymenium is first laid down on the folded gills and, apart from location, their differentiation states and ontogeny appear to be identical. Facial cystidia in *C. cinereus* are also established as components of the very first population of dikaryotic hyphal tips which form hymenial tissue (Horner & Moore, 1987; Rosin & Moore, 1985) and are mostly binucleate as a result. Marginal cystidia in *C. cinereus* are the apical cells of branches from the multinucleate gill trama, which become swollen to repair the injury caused when primary gills pull away from the stipe; marginal cystidia retain the multinucleate character of their parental hyphae (Chiu & Moore, 1993).

REGIONAL PATTERNS OF COMMITMENT

The distributions of cystidia and gills in *Coprinus cinereus* have been interpreted as being dependent on interplay between activating and inhibiting factors (Horner & Moore, 1987; Moore, 1988) in a pattern-forming process similar to the model developed by Meinhardt & Gierer (1974, and see Meinhardt, 1984). Successful application of this model to fungi as well as to plants and animals concentrates attention on the fact that the distribution of stomata on a leaf, bristles on an insect and cystidia on a fungal hymenium have a great deal in common at a fundamental mechanistic level. Other similarities emerge when a search for commitment is made.

The classic demonstration of commitment involves transplanting the cell into a new environment; if the transplanted cell continues the developmental pathway characteristic of its origin then it is said to have been committed before transplantation. On the other hand, if the transplanted cell embarks upon the pathway appropriate to its new environment then it was clearly not committed at the time of transplant. Most fungal tissues produce vegetative hyphae very rapidly when disturbed and 'transplanted' to a new 'environment' or medium. This is a regenerative phenomenon which creates the impression that fungal cells express little commitment to their state of differentiation. Very little formal transplantation experimentation has been reported with fungal multicellular structures. The clearest examples of commitment to a developmental pathway has been provided by Bastouill-Descollonges & Manachère (1984) and Chiu & Moore (1988) who demonstrated that basidia of isolated gills of *Coprinus congregatus* Bull.: Fr. and *C. cinereus*, respectively, continued development to spore production if removed to agar medium at early meiotic stages. Other hymenial cells, cystidia, paraphyses and tramal cells, immediately reverted to hyphal growth but this did not often happen to immature basidia. Evidently, basidia are specified irreversibly as meiocytes and they become determined to complete the sporulation programme during meiotic prophase I. Once initiated, the maturation of basidia is an autonomous, endotrophic process that is able to proceed *in vitro*. Clearly, then, even if only to a limited extent, commitment to a pathway of differentiation some time before realization of the differentiated phenotype can occur in these fungi. Although these experiments have been discussed mainly for their value in understanding commitment to the basidium differentiation pathway, it is equally important to stress that other cells of the hymenium showed no commitment; immediately reverting to hyphal growth on explantation as though they have an extremely tenuous grasp on their state of differentiation. That these cells do not default to hyphal growth *in situ* implies that their state of differentiation is somehow continually reinforced by some aspect of the environment of the tissue which they comprise.

FUZZY LOGIC

Discussion of differentiation in fungi often involves use of words like 'switch' in phrases that imply wholesale diversion at some stage between alternative developmental pathways. There are now many examples which suggest that fungal cells behave as though they assume a differentiation state even when all conditions for that state have not been met. Rather than rigidly following a prescribed sequence of steps, differentiation pathways for the sorts of fruitbodies discussed here appear to be based on application of rules that allow considerable latitude in expression; 'decisions' between developmental pathways seem to be made with a degree of uncertainty, as though they are based on probabilities rather than absolutes. For example, facial cystidia of *C. cinereus* are generally binucleate, reflecting their origin and the fact that they are sterile cells, yet occasional examples can be found of cystidia in which karyogamy has occurred (Chiu & Moore, 1993) or of cystidia bearing sterigmata. This suggests that entry to the cystidial pathway of differentiation does not totally preclude expression of at least part of the differentiation pathway characteristic of the basidium (Fig. 3). Equally, the fact that a large fraction of the basidiole population of *A. bisporus* remains in arrested meiosis (Allen *et al.*, 1992) indicates that entry to the meiotic division pathway does not guarantee sporulation. There are many other examples in the literature.

Potential contributions to differentiation

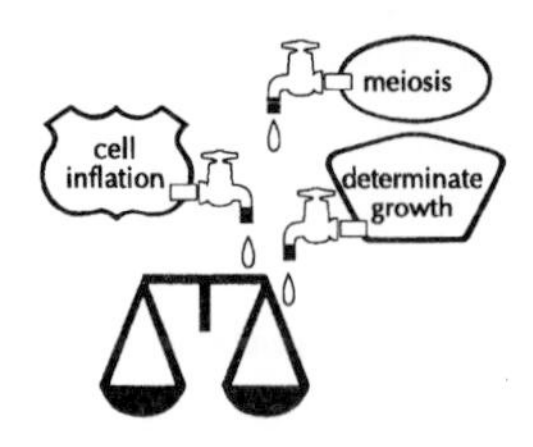

Potential pathways for differentiation

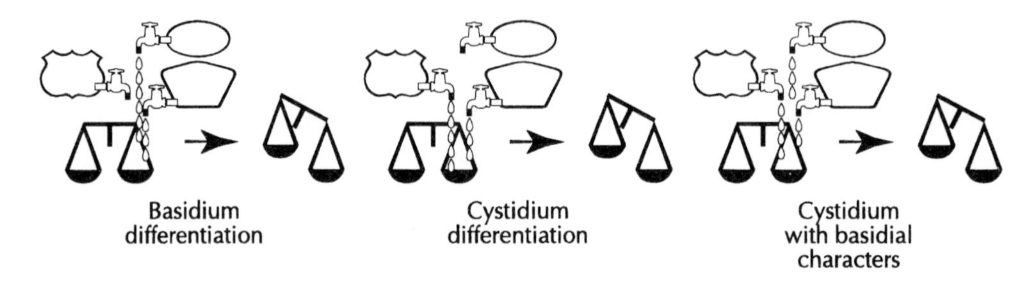

Figure 3 Cartoons illustrating the idea that fungal differentiation involves the cell reaching a state which can be sustained as being different from the vegetative hyphal compartment. The different characters which an undifferentiated cell might express are viewed as making contributions to a balance of differentiation. In most instances the normal state of differentiation results from a specific combination of these contributions. Occasionally, unexpected contributions leads to cells that are unmistakably of one sort, but are expressing characters of another sort. In the example illustrated in the bottom panel, cystidia with basidial characters would result, for example having sterigmata (observed by Watling, 1971) or undergoing karyogamy (Chiu & Moore, 1993).

CONCLUSIONS

The last 25 years have seen the final break away from the vice-like grip of the Friesian system of classification for larger fungi. By the use of developmental and anatomical characters correlated with chemical and microscopic features, natural groupings have been demonstrated. These, however, cut across the old guidelines which were based on the gross morphology of the end-product – be it mushroom, bracket fungus or puff-ball. Throughout, unifying ideas are now becoming evident, allowing worthwhile comparison with related phenomena in the plant and animal kingdoms. The careful examination of teratological forms provides insight into gross developmental patterns, allowing interpretation of developmental strategies. A marriage of information is required and overemphasis of the monstrosities themselves at the expense of using them to understand normal developmental pathways cannot be supported. At the cellular level, too, careful analysis reveals analogies with animal and plant development, giving evidence for varied levels of commitment, for regulation of distinct pathways of differentiation which differ in both time and space, and for pattern-forming processes which might be dependent upon morphogens. On the other hand, differences emerge from these analyses which suggest that analogy with animal and plant development should be limited. Fungal differentiation seems to be far less 'final' than is usually expected of animal and plant cells. Few fungal cells show complete commitment to a pathway of differentiation; except for the meiocytes, even the most highly adapted cells revert readily to the filamentous mode of growth. Also, fungal cells appear to be capable of expressing a state of differentiation even when all of the attributes of that state have not been attained. It seems that steps in fungal developmental pathways depend on balancing probabilities rather than all-or-none switches. These interpretations are important in suggesting the design of experiments but it is important that laboratory data and field observations are given equal weight.

ACKNOWLEDGEMENT

The authors are extremely grateful to Lily Novak Frazer for her constructive comments on the manuscript.

REFERENCES

ALLEN, J.J., MOORE, D. & ELLIOTT, T.J., 1992. Persistent meiotic arrest in basidia of *Agaricus bisporus*. *Mycological Research, 96:* 125–127.

ATKINS, F.C., 1950. *Mushroom Growing Today*. London, U.K.: Faber & Faber.

BAS, C., 1969. Morphology and subdivision of *Amanita* and a monograph of its section Lepidella. *Persoonia, 5:* 285–579.

BASTOUILL-DESCOLLONGES, Y. & MANACHÈRE, G., 1984. Photosporogenesis of *Coprinus congregatus*: correlations between the physiological age of lamellae and the development of their potential for renewed fruiting. *Physiologia Plantarum, 61:* 607–610.

BLATTNY, C., KASALA, B., PILÁT, A., SENTILLIOVA-SVOBODOVÁ, J. & SEMERDZIEVA, M., 1971. Proliferation of *Armillaria mellea* (Vahlin Fl. Dan. ex Fr.) P. Karst. probably caused by a virus. *Ceská Mykologie, 25:* 66–74.

BLATTNY, C., KRALIK, O., VESELSK, J., KASALA, B. & HERZOVA, H., 1973. Particles resembling virions accompanying the proliferation of agaric mushrooms. *Ceská Mykologie, 27:* 1–5.

BONDARTSEV, M.A., 1963. On the anatomical criterion in the taxonomy of Aphyllophorales. *Botanichnyi Zhurnal SSSR, 48:* 362–372.

BULLER, A.H.R., 1922. *Researches on Fungi*, vol. 2. London, U.K.: Longman, Green & Co.

BULLER, A.H.R., 1924. *Researches on Fungi*, vol. 3. London: U.K.: Longman, Green & Co.

CAVALIER-SMITH, T., 1981. Eukaryote Kingdoms: seven or nine? *BioSystems, 14:* 461–481.

CHANG, S.T. & YAU, C.K., 1971. *Volvariella volvacea* and its life history. *American Journal of Botany, 58:* 552–561.

CHIU, S.W. & CHANG, S.T., 1987. *Volvariella bombycina* and its life history. *Mushroom Journal for the Tropics, 7:* 1–12.

CHIU, S.W. & MOORE, D., 1988. Evidence for developmental commitment in the differentiating fruit body of *Coprinus cinereus*. *Transactions of the British Mycological Society, 90:* 247–253.

CHIU, S.W. & MOORE, D., 1990a. A mechanism for gill pattern formation in *Coprinus cinereus*. *Mycological Research, 94:* 320–326.

CHIU, S.W. & MOORE, D., 1990b. Development of the basidiome of *Volvariella bombycina*. *Mycological Research, 94:* 327–337.

CHIU, S.W. & MOORE, D., 1993. Cell form, function and lineage in the hymenia of *Coprinus cinereus* and *Volvariella bombycina*. *Mycological Research, 97:* 221–226.

CHIU, S.W., MOORE, D. & CHANG, S.T., 1989. Basidiome polymorphism in *Volvariella bombycina*. *Mycological Research, 92:* 69–77.

CORNER, E.J.H., 1932a. The fruit-body of *Polystictus xanthopus* Fr. *Annals of Botany, 156:* 71–111.

CORNER, E.J.H., 1932b. A *Fomes* with two systems of hyphae. *Transactions of the British Mycological Society, 17:* 51–81.

CORNER, E.J.H. 1950. *A monograph of* Clavaria *and Allied Genera*. Annals of Botany Memoirs no. 1. London, U.K.: Oxford University Press.

CORNER, E.J.H., 1953. The construction of polypores. *Phytomorphology, 3:* 152–167.

CORNER, E.J.H., 1966. *A Monograph of Cantharelloid Fungi.* Annals of Botany Memoirs no. 2. London, U.K.: Oxford University Press.

CORNER, E.J.H., 1991. *Trogia* (Basidiomycetes. *The Garden's Bulletin, Singapore,* (Suppl. 2), 1–100.

CUNNINGHAM, G.H., 1965. *Polyporaceae of New Zealand.* New Zealand Department of Scientific and Industrial Research, Bulletin no. 164, 303 pp.

DICKHARDT, R., 1985. Homokaryotization of *Agaricus bitorquis* (Querl.) Sacc. and *Agaricus bisporus* (Lange) Imb. *Theoretical and Applied Genetics, 70:* 52–56.

ELLIOTT, T.J., 1985. Developmental genetics – from spore to sporophore. In D. Moore, L.A. Casselton, D.A. Wood & J.C. Frankland (eds), *Development Biology of Higher Fungi,* pp. 451–465. Cambridge, U.K.: Cambridge University Press.

ERIKSSON, J., HJORTSTAM, K. & RYVARDEN, L., 1978. The Corticiaceae of North Europe, *Fungiflora, 5:* 889–1047.

ESSER, K., STAHL, U. & MEINHARDT, F., 1977. Genetic aspects of differentiation in fungi. In J. Meyrath & J.D. Bu'lock (eds), *Biotechnology and Fungal Differentiation,* pp. 67–75. London, U.K.: Academic Press.

FAYOD, M.V., 1889. Histoire naturelle des Agaricinés. *Annales des Sciences Naturells, 9:* 181–411.

FLEGG, P.B., 1983. Response of the sporophores of the cultivated mushroom (*Agaricus bisporus*) to volatile substances. *Scientia Horticulturae, 21:* 301–310.

FLEGG, P.B. & WOOD, D.A., 1985. Growth and fruiting: In P.B. Flegg, D.M.Spencer & D.A. Wood (eds), *The Biology and Technology of the Cultivated Mushroom,* pp. 141–178. Chichester, U.K.: John Wiley.

FRIES, E., 1821. *Systema Mycologicum,* vol.I. Mauritius: Gryphiswald.

GARSTANG, W., 1922. The theory of recapitulation: a critical re-statement of the biogenetic law. *Linnean Society of London, Zoological Journal, 35:* 81–101.

GARSTANG, W., 1962. *Larval Forms with other Zoological Verses* (with an introduction by A.C Hardy). Oxford, U.K.: Blackwell.

GRAHAM, K.M., 1985. Mating type of progeny from haploid sporocarps of *Pleurotus flabellatus* (Berk. et Br.) Sacc. *Malaysian Applied Biology, 14:* 104–106.

GREGORY, P.H., 1984. The First Benefactors' Lecture. The fungal mycelium: an historical perspective. *Transactions of the British Mycological Society, 82:* 1–11.

HAMMAD, F., WATLING, R. & MOORE, D., 1993. Cell population dynamics in *Coprinus cinereus*: narrow and inflated hyphae in the fruit body stem. *Mycological Research 97:* 275–282.

HEIM, R.J., 1931. Le genre Inocybe. *Encyclopédie Mycologique,* vol. I, pp. 1–429. Paris, France. Lechevalier et Fils.

HORNER, J. & MOORE, D., 1987. Cystidial morphogenetic field in the hymenium of *Coprinus cinereus. Transactions of the British Mycological Society, 88,* 479–488.

JULICH, W., 1981. Higher taxa of Basidiomycetes. *Bibliotheca Mycologia, 85:* 1–485.

KAMADA, T. & TAKEMARU, T., 1977a. Stipe elongation during basidiocarp maturation in *Coprinus macrorhizus:* mechanical properties of stipe cell wall. *Plant & Cell Physiology, 18:* 831–840.

KAMADA, T. & TAKEMARU, T., 1977b. Stipe elongation during basidiocarp maturation in *Coprinus macrorhizus:* changes in polysaccharide composition of stipe cell wall during elongation. *Plant & Cell Physiology, 18:* 1291–1300.

KANADA, T., ARAKAWA, H., YASUDA, Y. & TAKEMARU, T., 1990. Basidiospore formation in a mutant of incompatibility factors and in mutants that arrest at meta-anaphase I in *Coprinus cinereus. Experimental Mycology, 14:* 218–226.

KANDA, T. & ISHIKAWA, T., 1986. Isolation of recessive developmental mutants in *Coprinus cinereus. Journal of General and Applied Microbiology, 32:* 541–543.

KANDA, T., GOTO, A., SAWA, K., ARAKAWA, H., YASUDA, Y. & TAKEMARU, T., 1989. Isolation and characterization of recessive sporeless mutants in the basidiomycete *Coprinus cinereus*. *Molecular and General Genetics, 216:* 526–529.

KEYWORTH, W.G. 1942. The occurrence of tremelloid outgrowths on the pilei of *Coprinus ephemerus*. *Transactions of the British Mycological Society, 25:* 307–310.

KÜES, U. & CASSELTON, L.A., 1992. Fungal mating type genes – regulators of sexual development. *Mycological Research, 96:* 993–1006.

LAMBERT, E.B., 1930. Two new diseases of cultivated mushrooms. *Phytopathology, 20:* 917–919.

LANGERON, M. & VANBREUSEGHEM, R., 1965. *Outline of Mycology,* vol. 2, 2nd edn, translated from the French by J. Wilkinson, London, U.K.: Pitman.

LEATHAM, G.F. & STAHMANN, M.N., 1987. Effect of light and aeration on fruiting of *Lentinula edodes*. *Transactions of the British Mycological Society, 88:* 9–20.

LESLIE, J.F., 1983. Initiation of monokaryotic fruiting in *Schizophyllum commune:* multiple stimuli, multiple genes. *Abstracts, Third International Mycological Congress, Tokyo,* p. 163.

LESLIE, J.F. & LEONARD, T.J., 1984. Nuclear control of monokaryotic fruiting in *Schizophyllum commune*. *Mycologia, 76:* 760–763.

MAAS GEESTERANUS, R.A., 1971. *Hydnaceous Fungi of the Eastern Old World.* Amsterdam, The Netherlands: North-Holland.

MAAS GEESTERANUS, R.A., 1975. *The Terrestrial Hydnums of Europe.* Amsterdam, The Netherlands: North-Holland.

MACBRIDE, E.W., 1914. *Textbook of Embryology,* vol. I. *Invertebrata,* W. Heape (ed.). London, U.K.: Macmillan.

MANACHÈRE, G., 1985. Sporophore differentiation of higher fungi: a survey of some actual problems. *Physiologie Vegétale, 23:* 221–230.

MANACHÈRE, G., ROBERT, J.C., DURAND, R., BRET, J.P. & FÈVRE, M., 1983. Differentiation in the Basidiomycetes. In J.E. Smith (ed.), *Fungal Differentiation: a Contemporary Synthesis,* pp. 481–514. New York, U.S.A.: Marcel Dekker.

MARGULIS, L., 1974. Five-Kingdom classification and the origin and evolution of cells. *Evolutionary Biology, 7:* 45–78.

MEINHARDT, H., 1984. Models of pattern formation and their application to plant develoment. In P.W. Barlow & D.J. Carr (eds), *Positional Controls in Plant Development,* pp. 1–32. Cambridge, U.K.: Cambridge University Press.

MEINHARDT, F. & ESSER, K., 1983. Genetic aspects of sexual differentiation in fungi. In J.E. Smith (ed.), *Fungal Differentiation: a Contemporary Synthesis,* pp. 537–557. New York, U.S.A.: Marcel Dekker.

MEINHARDT, H. & GIERER, A. 1974. Applications of a theory of biological pattern formation based on lateral inhibition. *Journal of Cell Science, 15:* 321–346.

MILLER, O.K. & STEWART, L., 1971. The genus *Lentinellus*. *Mycologia, 63:* 333–369.

MOORE, D., 1981. Developmental genetics of *Coprinus cinereus:* genetic evidence that carpophores and sclerotia share a common pathway of initiation. *Current Genetics, 3:* 145–150.

MOORE, D., 1988. Recent developments in morphogenetic studies of higher fungi. *Mushroom Journal for the Tropics, 8:* 109–128.

PATOUILLARD, N., 1900. *Essai Taxonomique sur les familles et les genres des Hyménomycètes.* Lons-Le-Saunier, France: Lucien Declume.

PERSOON, D.C.H., 1801. *Synopsis methodica Fungorum.* Gottingae: Henricum Dieterich.

PÖDER, R., 1990. Phylogenetical aspects of gill development and proportions in basidiocarps. In A. Reisinger & A. Bresinsky (eds), *Abstracts, 4th International Mycological Congress, Regensberg,* p. 89, abstract IB-89/4. Regensburg: IMC4.

PRICE, I.P., 1973. A study of cystidia in effused Aphyllophorales. *Nova Hedwigia, 24:* 515–618.

PRILLINGER, H. & SIX, W., 1983. Genetic analysis of fruiting and speciation of basidiomycetes: genetic control of fruiting in *Polyporus ciliatus*. *Plant Systematics and Evolution, 141:* 341–371.

RAPER, J.R. & KRONGELB, G.S., 1958. Genetic and environmental aspects of fruiting in *Schizophyllum commune* Fr. *Mycologia, 50:* 707–740.

RAUDASKOSKI, M. & SALONEN, M., 1984. Interrelationships between vegetative development and basidiocarp initiation. In D.H. Jennings & A.D.M. Rayner (eds), *The Ecology and Physiology of the Fungal Mycelium*, pp. 291–322. Cambridge, U.K.: Cambridge University Press.

REA, C., 1922. *British Basidiomycetae.* Cambridge, U.K.: Cambridge University Press.

RAYNER, A.D.M. & COATES, D., 1987. Regulation of mycelial organisation and responses. In A.D.M. Rayner, C.M. Brasier & D. Moore (eds), *Evolutionary Biology of the Fungi.* pp. 115–136. Cambridge, U.K.: Cambridge University Press.

REDHEAD, S.A., 1987. The Xerulaceae (Basidiomycetes), a family with sarcodimitic tissues. *Canadian Journal of Botany, 65:* 1551–1562.

REIJNDERS, A.F.M., 1948. *Études sur les développement et l'organisation histologique des carpophores dans les Agaricales.* Gouda: N.V. Drukkerig v/H Koch & Knuttel.

REIJNDERS, A.F.M., 1963. *Les problèmes du développement des carpophores des Agaricales et de quelques groupes voisins.* The Hague: Dr W. Junk.

REIJNDERS, A.F.M., 1976. Recherches sur le développement et l'histogénèse dans les Asterosporales. *Persoonia, 9:* 65–83.

REIJNDERS, A.F.M., 1977. The histogenesis of bulb- and trama tissue of the higher Basidiomycetes and its phylogenetic implications. *Persoonia, 9:* 329–361.

REIJNDERS, A.F.M. & STALPERS, J.A., 1992. *The Development of the Hymenophoral Trama in the Aphyllophorales and the Agaricales.* Studies in Mycology, no. 34. Baarn, The Netherlands: Centraalbureau voor Schimmelcultures.

ROSIN, I.V. & MOORE, D., 1985. Differentiation of the hymenium in *Coprinus cinereus. Transactions of the British Mycological Society, 84:* 621–628.

ROYSE, D.J., JODON, M.H., ANTONIO, G.G. & MAY, B.P., 1987. Confirmation of intraspecific crossing and single and joint segregation of biochemical loci of *Volvariella volvacea. Experimental Mycology, 11:* 11–18.

RYVARDEN, L., 1992. Genera of polypores, nomenclature and taxonomy. *Synopsis Fungorum, 5:* 1–363.

SINGER, R., 1936. Studien zur Systematik der Basidiomyceten. *Beihefte zum Botanischen Centralblatt, 56B:* 137–174.

SINGER, R., 1951. The Agaricales in modern taxonomy. *Lilloa, 22:* (1949): 1–830.

SINGER, R., 1955. New and interesting species of Basidiomycetes IV. *Mycologia, 47:* 763–777.

SINGER, R., 1958. The meaning of the affinity of the Secotiaceae with the Agaricales. *Sydowia, 12:* 1–43.

SINGER, R., 1963. Notes on secotiaceous fungi: *Galeropsis* and *Brauniella. Koninklijke Nederlandse Akademie van Wetenschappen – Proceedings Series C: Biological and Medical Sciences, 66:* 106–117.

SINGER, R., 1975. *The Agaricales in Modern Taxonomy,* 3rd edn. Vaduz: Cramer.

SINGER, R., 1986. *The Agaricales in Modern Taxonomy,* 4th edn. Koenigstern: Koeltz.

STAHL, U. & ESSER, K., 1976. Genetics of fruit-body production in higher Basidiomycetes I. Monokaryotic fruiting and its correlation with dikaryotic fruiting in *Polyporus ciliatus. Molecular and General Genetics, 148:* 183–197.

TAKEMARU, T. & KAMADA, T., 1972. Basidiocarp development in *Coprinus macrorhizus* I. Induction of developmental variations. *Botanical Magazine (Tokyo), 85:* 51–57.

UNO, I. & ISHIKAWA, T., 1971. Chemical and genetic control of induction of monokaryotic fruiting bodies in *Coprinus macrorhizus. Molecular and General Genetics, 113:* 228–239.

VAN BRUMMELEN, J., 1967. A world monograph of the genera *Ascobolus* and *Saccobolus* (Ascomycetes, Pezizales). *Persoonia* (Suppl. 1), 1–260.

VON ARX, J.A. & MÜLLER, E., 1955. Über die Gattungen *Selinia* Karst. und *Seliniella* Nov. Gen. und ihre phylogenetische Bedeutung. *Acta Botanica Neerlandica, 4:* 116–125.

WATLING, R., 1971. Polymorphism in *Psilocybe merdaria. New Phytologist, 70:* 307–326.

WATLING, R., 1974. Dimorphism in *Entoloma abortivum. Bulletin bimensuel de la Société Linnéenne de Lyon (numéro spécial dédiés à R. Kühner*): 449–470.

WATLING, R. & NICOLL, H., 1980. Sphaerocysts in *Lactarius rufus. Transactions of the British Mycological Society, 75:* 331–333.

WHITTAKER, R.H., 1969. New concepts of kingdoms of organisms. *Science, 163:* 150–160.

WORSDELL, W.C., 1915. *The Principles of Plant Teratology,* vol. 1. London, U.K.: The Ray Society.

CHAPTER

17

Differential insulation and the generation of mycelial patterns

A.D.M. RAYNER, G.S. GRIFFITH & H.G. WILDMAN

CONTENTS

Abstract

Developmentally indeterminate systems generate diverse patterns by automatically shifting the dynamic balance between energy-capturing, conserving and distributing processes according to circumstances. This balance depends critically on the degree of insulation, the resistance to deformation and penetration, of the system's boundary. The latter regulates the translation of pressure, generated by active assimilation of energy-yielding resources, into thrust.

Fungal mycelia epitomize such patterns, being capable of shifting, in a versatile

Shape and Form in Plants and Fungi
ISBN 0–12–371035–9

manner, between a variety of developmental modes. Correspondingly, hyphae may interconvert with unicellular states, change their frequency and direction of branching, and undergo septation, anastomosis and protoplasmic degeneration. Moreover they may aggregate or diverge, and switch into or out of non-assimilative modes with protective, redistributive and reproductive properties.

In such ways, mycelial systems are able to respond fluidly to the opportunities and constraints that they encounter as they traverse spatiotemporally heterogeneous environments. A possible explanation of this ability may lie in metabolic feedback processes that control the hydrodynamic properties of mycelia via the differential insulation of hyphal boundaries with water-resistant substances. The latter include certain polypeptides and terpenoid and aromatic compounds, sensitive to polymerization and conversion to free radicals by phenol-oxidizing enzymes, whose synthesis is enhanced when the use of mitochondrially generated ATP to fuel active transport is diverted or prevented.

INDETERMINACY AND FLUID DYNAMICS

According to Sun Tzu (500 BC, cited by Liddell Hart, 1967), 'military tactics are like unto water; for water in its natural course runs away from high places and hastens downwards. So in war, the way to avoid what is strong is to strike what is weak. Water shapes its own course according to the ground over which it flows; the soldier works out his victory in relation to the foe he is facing.'

According to the argument to be developed later, this description of circumstance-driven, self-organizing, distributive patterns has extraordinary parallels with the developmental indeterminacy which characterizes fungal mycelia. A mycelium can indeed be likened to an army, or to any other social grouping that possesses some form of collective, intercommunicative organization within a deformable boundary. Correspondingly, its behaviour is essentially fluid-dynamical, with expansive (dissociative) processes being counteracted by constraining (associative) processes. The pattern in which it distributes itself depends on the way, and the degree to which, it polarizes pressure, generated by active uptake of energy-yielding resources from the environment, into thrust. Thrust, in its turn, depends on how strongly or weakly lateral boundaries are insulated, i.e. their resistance to deformation and penetration, and on the disposition of internal barriers to communication.

It will be suggested that the latter properties can be varied by means of a delicate interplay between external conditions and internal genetic information, mediated by metabolic feedback. This facility allows mycelia to respond in a sensitive, versatile and energy-efficient manner to the opportunities and constraints of an uncertain life in spatiotemporally heterogeneous conditions.

On the other hand, an inherent feature of such creatively unstable dynamic systems is their susceptibility, locally or generally, to disharmony and consequent degeneracy. It is important to understand how such degeneracy can be regulated so as not to be catastrophic, and to realize that in an indeterminate system it can have positive as well as negative evolutionary effects, enabling the redistribution and sequestration of resources.

MYCELIA AS DYNAMIC SYSTEMS

A prevalent view would seem to be that fungal mycelia only exhibit growth, not development, the latter term being restricted largely to the formation of determinate offshoots such as fruitbodies (cf. Moore *et al.*, 1985). This attitude corresponds with an idealized approximation of growing mycelium to an unsophisticated, homogeneous, purely absorptive structure which expands in direct proportion to the amount that it assimilates in discrete hyphal growth units (cf. Prosser, 1991).

The quest of experimentalists and industrialists to obtain fully reproducible, and therefore predictable, data and yields has only served to reinforce this way of thinking. Correspondingly, in focusing on the behaviour of individual, well-characterized strains maintained as pure cultures in uniform regimes, fungal physiology has departed progressively from the conditions in which fungi thrive in their natural habitats. These real-life conditions are liable to be heterogeneous, and therefore locally unpredicatable, both in space and time. The mycelium responds to their challenge as a sophisticated, indeterminate or open-ended dynamic structure that varies its developmental pattern with circumstances and not in any preset sequence (e.g. Andrews, 1992; Gregory, 1984; Rayner, 1991; Stenlid & Rayner, 1989).

Pattern-switching

Dynamic stages in the development of the mycelium begin when a spore germinates, with a phase of isotropic expansion or swelling. This is usually followed by symmetry breaking, which initiates the indeterminate development of one or more polarized hyphal tubes. However, even at this early stage, options between indeterminate and determinate patterns can occur, resulting in mycelial–yeast dimorphisms and sporulation. Moreover, the interior of the emerging hyphal tubes may or may not, depending partly on organism and partly on circumstance, be partitioned by septa into uninucleate, binucleate or multinucleate compartments (e.g. Ainsworth & Rayner, 1991; Boddy & Rayner, 1983; Fig. 1).

Branching of the hyphal tubes is often spatiotemporally closely correlated with septation (Trinci, 1978; Fig. 1) and generates a radiating structure whose fractal dimension (space-filling capacity) progressively increases (Ritz & Crawford, 1990; Crawford & Ritz, Ch. 18 this volume). In higher fungi, branching is followed by the process of hyphal fusion or anastomosis which has the critically important effect of establishing an interior communications network behind the expansive colony margin (Rayner, 1991).

As branches emerge, they may do so with variable frequency and at varying angles, representing different degrees of commitment to radial (explorative) or tangential (exploitative/consolidative) vectors. The higher the ratio between the radial and tangential vectors, the lower will be the fractal dimension of the structure and the greater will be its polarity, a situation reminiscent of the operation of the transmission system in a motor vehicle (Rayner & Coates, 1987).

Such changes in branching pattern, when coordinated, give rise to changes in the organization and extension rate of the mycelial margin. Where these changes occur abruptly, they result in what have been termed slow-dense/fast-effuse dimorphisms (Rayner & Coates, 1987), and are often manifested as sectoring or 'point growth' phenomena (Coggins *et al.*, 1980).

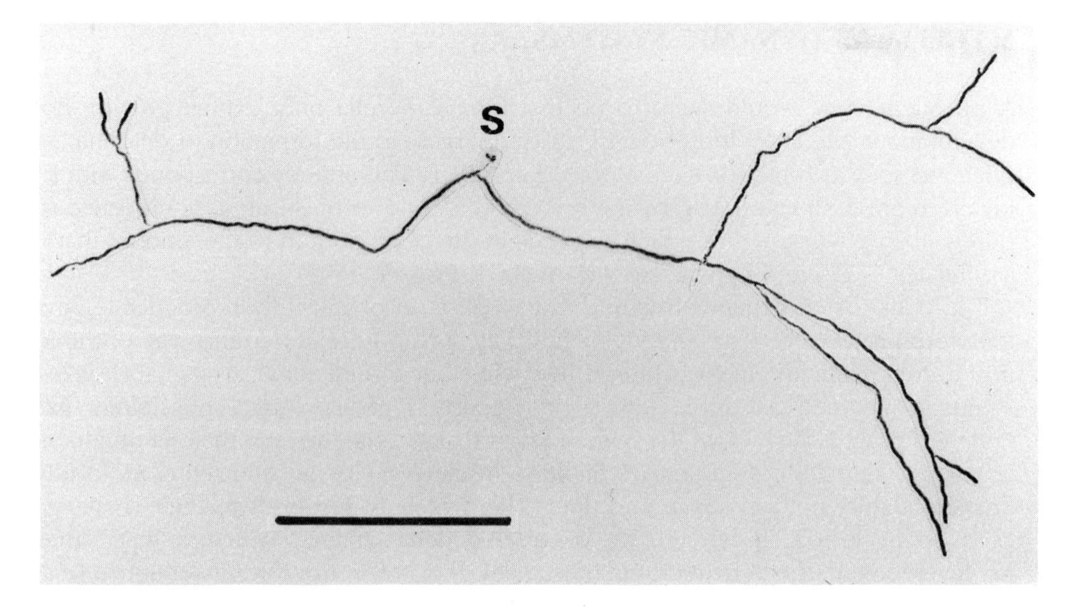

Figure 1 Coenocytic mycelium of *Phanerochaete magnoliae* originating from a basidiospore (S). Branches have only formed terminally, resulting in a structure with a very low fractal dimension (close to 1). Once the colony was about 1.5 cm diameter, septa, lateral branches and anastomoses formed in rapid succession, converting the mycelium into a partial network and associated with expression of laccase enzyme activity. Before septation, protoplasmic streaming at rates up to 50 μm s^{-1} was observed (Ainsworth & Rayner, 1991).

Another important aspect of mycelial coordination depends on the capacity of hyphae to develop either diffusely or in compact associations. The latter may have migratory or connective roles in the case of rhizomorphs and mycelial cords (which are often capable of far faster extension than individual hyphae), protective roles in sclerotia and pseudosclerotia, and reproductive roles in stromata and fruitbodies.

Such roles also depend on another fundamental property of hyphal systems, the capacity to shift between assimilative and non-assimilative – acquisitive and distributive or conservative – states. The existence of this capacity is at odds with the notion that mycelia are, actually or operationally, uniformly absorptive structures. However, it is arguable that without it mycelia would logistically be unable to extend their domain at a steady rate in more than one spatial dimension (see later). More particularly, it enables mycelial systems to penetrate into and ameliorate physicochemically adverse domains and to forage efficiently (i.e. with minimum dissipation appropriate to resource availability) between discontinuous resource supplies (Dowson *et al.*, 1986, 1988a, 1989a; Fig. 2).

As shown in Fig. 2, efficient foraging may also often require redistribution of resources from redundant to active mycelial phases. This may be accomplished by degenerative or cell death processes, some of which have been referred to as 'senescence' (e.g. Kück, 1989), that temporarily or persistently, locally or generally, result in autolysis or discontinuation of expansion. Such redistributive processes may establish powerful, self-sustaining, source-to-sink fluxes that can supply emerging sporophores (e.g. Ruiters & Wessels, 1989; Sietsma & Wessels, 1979; Watkinson, 1977) as well as explorative mycelial fronts. For example, in the leaf-litter decomposing

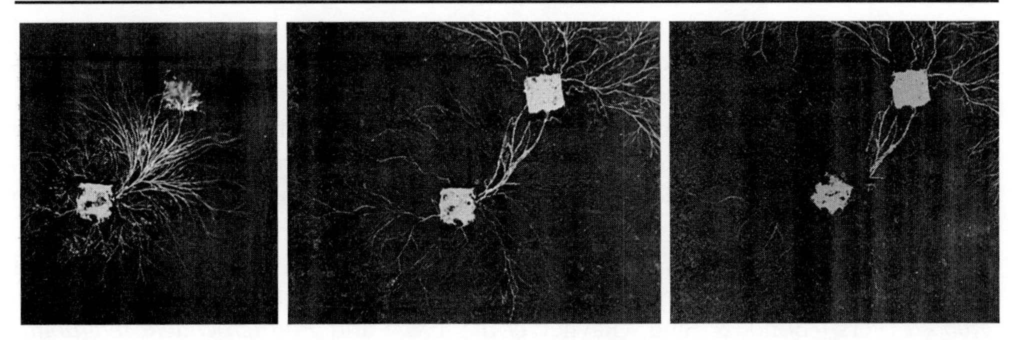

Figure 2 Stages (left to right) in development of a foraging system of *Steccherinum fimbriatum* (Pers.: Fr.) Erikss. between an inoculum beechwood block (8 cm^3) and an uncolonized beechwood block bait respectively placed in the centre and to one side of a tray of unsterile soil. The rapidly extending sector, thickening of connective mycelial cords and regression of non-connective cords as the mycelium re-emerges from the bait all demonstrate the redistributive nature of the system. From Dowson *et al.* (1988a).

basidiomycete, *Clitocybe nebularis* (Batsch: Fr.) Kummer, the mycelium consists of a 'fairy ring' with exploratory mycelial cords at its outer margin and degenerating mycelium at its trailing edge. Segments cut from the annulus extend with conserved rates and polarity if transplanted inside or outside the ring, but degenerate if reoriented before being reinserted within the annulus, resulting in a permanent gap in the mycelial front (Dowson *et al.*, 1989b). Another basidiomycete, *Peniophora lycii* (Pers.) v. Höhn. & Litsch., exhibits 'polar growth' along twigs of ash (*Fraxinus excelsior* L.), such that it cannot be reisolated from locations through which its growth front has already passed (Griffith & Boddy, 1991).

In addition to resource redistribution, degenerative processes may limit damage by sealing off boundaries within and around mycelial systems. At the same time, the existence of these processes also implies that mycelia, as part of their developmental repertoire, contain the seeds of their own destruction. Given suitable kinds of interference from outside, these seeds may be activated to propagate disharmony and ultimate collapse of the system.

Responses to neighbours

The operation, in concert, of the varied operational modes just described can often be observed directly when mycelia are challenged by a particularly important type of environmental heterogeneity: that which results from an encounter with a neighbouring mycelium belonging either to the same or to a different species. Such encounters occur frequently in natural populations and communities of fungi inhabiting relatively undisturbed, stress-free habitats (Cooke & Rayner, 1984).

Where the neighbour is of an unrelated species, there is normally little likelihood of hyphal fusion leading to protoplasmic continuity between opposing colonies (e.g. Gregory, 1984). Under these circumstances, the outcome seems to depend first on the sensitivity of individual hyphae to protoplasmic degeneration at long or short range from opposing hyphae, and secondly on the corporate capacity of mycelia to produce non-assimilative phases.

Hyphal degeneration at long range is generally attributed to diffusible or volatile factors or antibiotics. Degeneration following contact or very close proximity is described as hyphal interference, and is usually thought to be particularly characteristic of interactions involving basidiomycetes (Ikediugwu & Webster, 1970). It may be analogous to the hypersensitive resistance reactions exhibited by plants when they are invaded by incompatible fungal parasites (Rayner, 1986). Correspondingly, an absence or delay in expression of hyphal interference reactions may permit invasion by another mycelium.

Three wood-inhabiting basidiomycetes which are able by such means to take over domain from specific former residents are *Lenzites betulina* (Fr.) Fr., *Pseudotrametes gibbosa* (Pers.) Bond. & Sing. (Rayner *et al.*, 1987) and *Phanerochaete magnoliae* (Berk. & Curt.) Burds. (Ainsworth & Rayner, 1991). These examples highlight both the protective effect of degenerative processes, and the way that they may be subverted. *L. betulina* and *P. gibbosa* respectively replace *Coriolus* Quél. and *Bjerkandera* Karst. species by means of invasive, mycoparasitic mycelial fronts which elicit no immediate response by the host species. By contrast, *P. magnoliae* possesses a 'guerilla strategy' (cf. Lovett-Doust, 1981) whereby rapidly extending, sparsely distributed hyphae penetrate deep into the interior of *Datronia mollis* (Somf.: Fr.) Donk colonies, initiating widespread hyphal interference reactions from which it eventually emerges as sole survivor (Figs 3, 4).

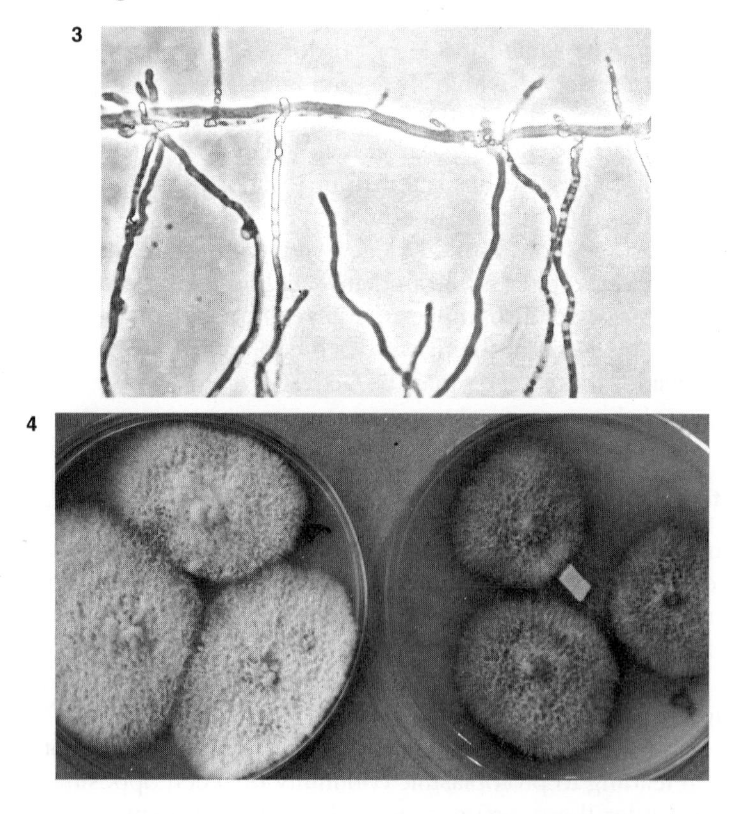

Figures 3 and 4 Interactions between *Phanerochaete magnoliae* and *Datronia mollis*. **3**. Protoplasmic degeneration in *D. mollis* hyphae (vertically aligned) following contact with a hypha (horizontally oriented, 6 μm diameter) of *P. magnoliae*. **4**. Colonies of *D. mollis* growing in two 9 cm Petri dishes containing 2% malt agar. The dish on the right has been inoculated centrally with *P. magnoliae*, whilst the other dish has not. Note the restricted extension and pigmentation of *D. mollis* colonies in contact with *P. magnoliae*. From Ainsworth & Rayner (1991).

Assemblages of emergent, presumably non-assimilative hyphae that either accumulate at or invade across mycelial interfaces are critical to the outcome of interactions involving many basidiomycetes and some ascomycetes inhabiting non-ephemeral substrata such as decaying wood. Observations of the manner and patterns of production of these assemblages on laboratory media are strongly reminiscent on the one hand of the disposition of troops on a battlefield, and on the other hand of the effects of driving a non-compressible fluid into or through a resistance (Figs 5–7). Such observations clearly indicate the corporate ability of the mycelium to seal its boundaries, redistribute its resources and coordinate its responses to external events.

When a neighbouring mycelium belongs to the same species, then, at least in ascomycetous and basidiomycetous fungi, hyphal fusion can occur, potentiating physiological unification and genetic invasion or exchange. When the neighbour is geneti-

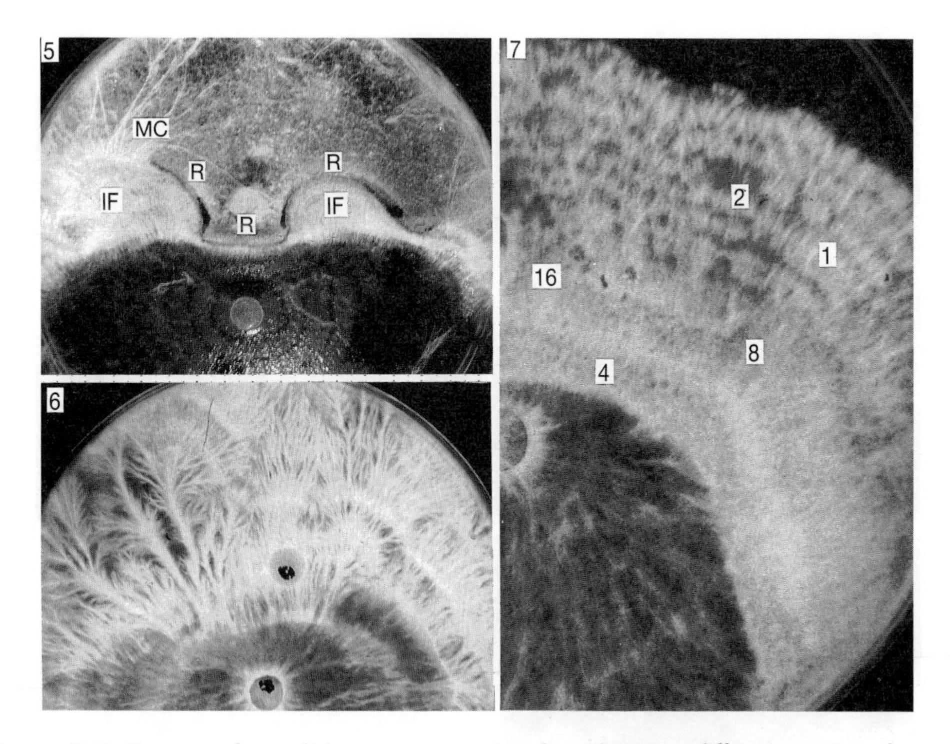

Figures 5–7 Patterns of mycelial emergence at interfaces between different species of wood-inhabiting basidiomycetes paired in 9 cm Petri dishes containing 3% malt agar and incubated in darkness at 20°C. The strong directionality of invasive systems across interaction interfaces, and localization of non-invasive systems to interaction interfaces are noteworthy. **5**. *Phanerochaete velutina* (Fr.) Karst. (bottom) against *Phlebia radiata*. Periodically extending invasion fronts (IF) produced by *P. velutina* have halted where *P. radiata* has produced ridges (R) of salmon pink aerial mycelium, but these ridges have subsequently been broached by penetrative mycelial cords (MC). **6**. *Hypholoma fasciculare* (bottom) against *Phlebia rufa* (Fr.) M.P. Christ. *H. fasciculare* has invaded *P. rufa* both in the form of dendritically branched mycelial cords (left) and as more diffuse bands exhibiting periodic shifts in hyphal density. **7**. Part of an interaction between *H. fasciculare* (bottom) and *Peniophora lycii*, showing that major bands of equal width within a replacement front produced by *H. fasciculare* appear to contain **1**, **2**, **4**, **8** and possibly **16** narrower minor bands of dense emergent mycelium alternating with zones of sparser mycelium. Such a pattern is reminiscent of the 'period doubling' phenomena exhibited by non-linear dynamic systems that are driven beyond their carrying capacity (e.g. Schaffer, 1987).

cally identical, it is usually accepted as self, such that somatic integration is achieved: in heterogeneous environments this can lead to marked redistributional effects (Fig. 8). By contrast, when the neighbour is sufficiently genetically different to be 'recognized' as non-self, then two seemingly opposite outcomes are possible (e.g. Rayner, 1991). Somatic incompatibility, or 'rejection', maintains individual physiological and genetical integrity via a degenerative response in fused hyphal segments (Figs 9, 10). Non-self 'acceptance' allows the temporary or persistent formation of heteroplasmons, heterokaryons or, exceptionally, allodiploids and in basidiomycetes is a fundamental requirement for sexual outcrossing. The fact that both these outcomes can sometimes be expressed in individual combinations of strains (Fig. 10) may have relevance to the origins of genetic and epigenetic stability and instability in mycelial systems (see later).

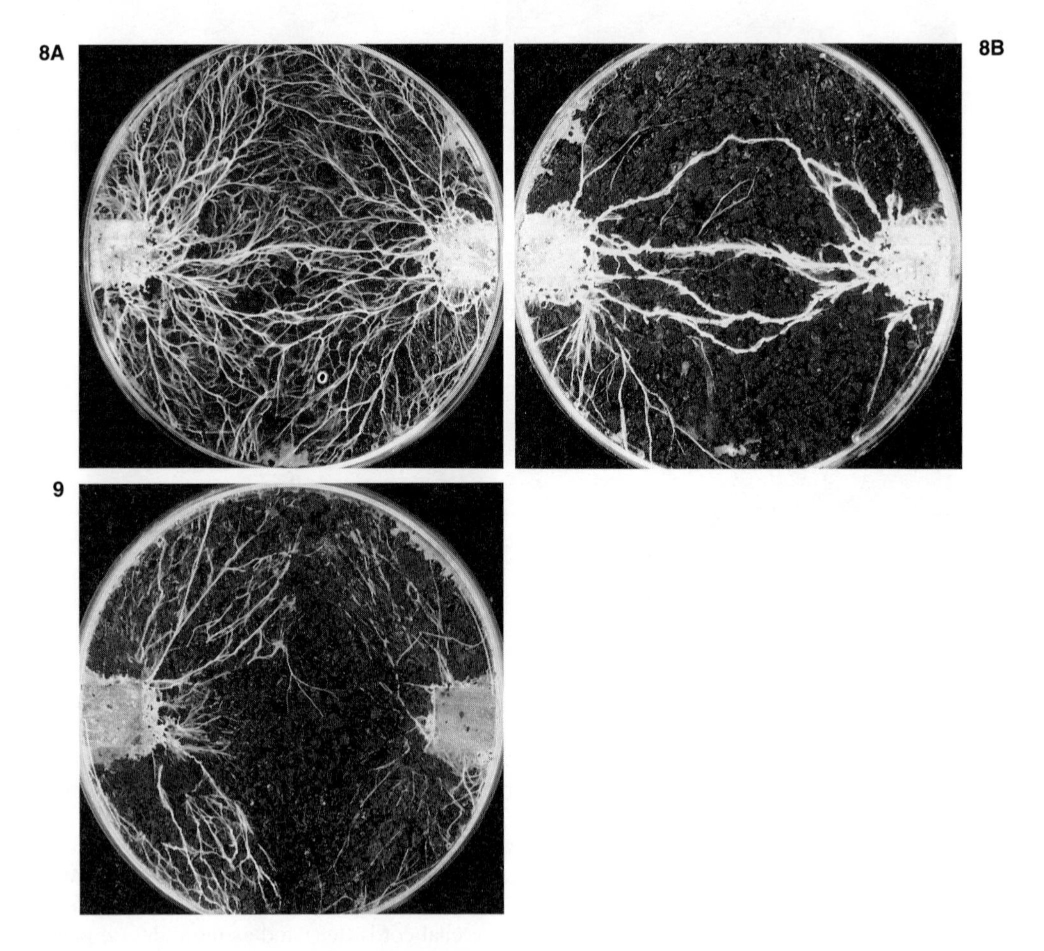

Figures 8 and 9 Consequences of anastomosis between mycelial cord systems of *Phanerochaete velutina* growing out from beechwood blocks inoculated into 14 cm diameter Petri dishes containing non-sterile soil. **8** Early (A) and late (B) stages of interaction betweeen heterokaryotic mycelia of like genotype, showing initial anastomosis followed by regression of all but a few, thickened, connective cords. **9** Development of an unoccupied zone between persistent, somatically incompatible, heterokaryotic mycelia of unlike genotype. From Dowson, Rayner & Boddy, 1988b.

Figure 10 Interaction on 2% malt agar between single basidiospore isolates (homokaryons) of *Stereum hirsutum* from a fruitbody from the Ukraine. The two plates on the right contain mating type-compatible combinations where a localized secondary mycelium (heterokaryon) has emerged in a bow-shaped region, bounded by pigmented somatic rejection zones. In the centre is a mating type-compatible combination giving rise to uniform emergence of secondary mycelium across the plate. The two plates on the left show mating-type incompatible combinations exhibiting strong and weak rejection responses. Courtesy of Dr A.M. Ainsworth.

REGULATING MYCELIAL PATTERNS BY MEANS OF VARIABLE THRUST

Efforts to explain mycelial patterns have traditionally tended both to be teleological and to focus from outside in: that is, they have sought to identify what the mycelium needs to do, in order to be adaptive, at the beck and call of its environment. The consequences of a mycelial organization may then be perceived as its functions in the same way that a river might be thought to flow downhill to reach the sea. For example, apical extension and branching have been explained as the means by which a mycelium maximizes energy capture along nutrient or oxygen concentration gradients (e.g. Deacon, 1984). More generally, the formation of extrahyphal gradients of stimulatory or inhibitory factors has been pursued as the primary determinant of mycelial patterns (e.g. Prosser, 1991), notwithstanding that these gradients may be just as much a consequence as a cause of such patterns.

Once a mycelium's appropriate responses to its needs have been characterized in this way, the next stage is a quest for the mechanisms: the gadgetry of signal transmitters, receptors, transducers and their genetic specifications, by which such responses, both structural and physiological, are achieved. The mycelium is approached as a machine, designed by natural selection and with specific tasks to fulfil, which can be understood if only its components can be identified and their functions defined.

The approach to be followed later contrasts with and complements the above in that it focuses on the intrinsic properties that a mycelium possesses as a direct consequence of its manner of organization as a dynamic system. In this approach, which is perhaps more apposite to understanding indeterminate development, process is a more primary concern than mechanism, and how a pattern can be generated is more primary than how it is adaptive (cf. Gould & Lewontin, 1979). Moreover, since the focus is from inside the system out, it is more relevant to ask how the mycelium distributes to rather than is drawn to its environment. Having identified the basic pattern-generating processes in mycelia, questions about how they can be regulated by diverse mechanisms, and how they may be optimized to meet particular functional requirements can then be pursued as important, but secondary issues.

The starting point for defining the generative processes underlying mycelial patterns is to recognize that, due to active uptake of solutes and consequent osmosis across their boundaries, mycelia are pressurized systems. This situation, either tacitly assumed or explicitly stated, has been basic to a number of models of hyphal and mycelial growth (e.g. Prosser, 1991). These models have viewed pressure both as a driving force (Bartnicki-Garcia, 1973; Wessels, 1986) and as a determinant of cell shape via changes in conformation of the hyphal wall to minimize surface stress (Koch, 1982; Saunders & Trinci, 1979). However, they have fallen short of addressing the *systemic* significance of pressure as a means of communication and determinant of pattern. Even those studies that have pointed to the importance of mass flow in translocation seem to have focused more on the way pressure is generated at source than on the way it is transmitted to outlets at the sytem's boundary (e.g. Jennings, 1984, 1987). That it is the latter process that is critical to pattern formation is, however, evident from comparative studies of indeterminate dynamics in driven systems of all kinds.

The central issue then, is not just how pressure is generated, but the degree to which it can be transformed into a vectorial or hydraulic parameter, thrust, both at the threshold when a system breaks spherical symmetry, and when it branches. In other words, it is necessary to understand how the relationship between energy uptake and throughput across and within the system's boundary is regulated. It is important to note here that thrust does not imply that a pressure gradient exists within the system – only across its boundaries (cf. Money, 1990). The development of thrust is therefore not equivalent to establishment of the turgor gradients that are said to cause translocation in mycelia (Jennings, 1984, 1987), in fact the converse. It may also be arguable whether turgor, as a hydrostatic potential, usefully relates to the hydrodynamic properties of systems with expanding boundaries.

Essentially, the more effectively sealed (resistant to deformation and leakage) are an elongated system's lateral boundaries relative to its axial boundaries, the greater the thrust (throughput) can be for a given rate of uptake and cross-sectional diameter. Moreover, the greater the sustainable thrust, the more efficient can be the long-range distribution pattern, with branches being both fewer and more acutely angled (cf. Prosser, 1991, who suggests that such patterns in mycelia may be due to the effects of diffusible inhibitors).

However, whilst boundary sealing conserves energy, it also resists further uptake and expansion, and so the boundary must always be partly open (penetrable and/or deformable) if the system is not to stagnate. This provides the basis for the fundamental non-linearity (counteractivity), and potential instability, of driven systems. The dynamic balance struck between boundary opening and boundary sealing (neither of which can be absolute) dictates the maximum sustainable thrust (equivalent to carrying capacity) and therefore the distribution pattern and outward form adopted by these systems. If this balance can be varied according to circumstances, so too can be the pattern. Moreover, situations can arise when overall uptake is increased above a threshold (equivalent to that determining the onset of turbulence) where it cannot, because of boundary constraints (e.g. tube diameter) logistically be converted into a constant throughput. On such occasions, the system will respond by leaking, bursting, deforming, branching and/or displaying simple to complex oscillations, depending on the degree of the overshoot and the inherent properties of the boundary. One such occasion would be an encounter with a nutrient-rich medium. Another would be an encounter with a neighbouring mycelium which acts as a resistance to axial deformation and hence throughput (cf. Figs. 5–7).

Some of the ways in which thrust could be changed in mycelial systems by means of varying the resistance to deformation and penetration of hyphal walls and the disposition of internal communication barriers due to septal sealing and protoplasmic degeneration, are illustrated in Fig. 11.

Materials

One aspect of the regulation of thrust, the deformability of hyphal boundaries, has long been a primary concern of those seeking to explain fungal growth, and has been widely recognized to be a function of the disposition of microfibrillar hyphal wall components, such as chitin. In particular, attempts to explain apical hyphal growth have sought to account for the rigidity of the wall around the sides of hyphae as compared with the extensibility of the wall at the tip. Two models have received most attention. The 'balanced lysis' model of Bartnicki-Garcia (1973) suggests that the apical wall is rendered plastic by the action of lytic enzymes which continually loosen the microfibrillar framework as it is being assembled by synthetic enzymes. The 'steady state' model of Wessels (1986) suggests that the apical wall is synthesized as a viscoelastic material that becomes progressively rigidified by means of chemical cross-linking.

Both the above models, in effect, envisage a dynamic balance between counteractive processes resulting in extensibility and rigidity. So long as this balance is sustained, the hypha will maintain a constant diameter and extension rate, appropriate to the amount of thrust. If it is not sustained, then the tip may become more plastic, increase in diameter and in extreme circumstances, burst, or it may rigidify more quickly, leading to narrowing or total inextensibility – features which are regularly observed in fungal colonies.

Mathematical models of apical extension have not to date incorporated this intrinsic non-linearity, even though it provides an obvious route to heterogeneity in mycelia. They have instead largely assumed, implicitly or explicitly, that extension rate is purely a function of the rate of supply of apical vesicles (Prosser, 1991), notwithstanding that the latter rate may itself be a function of thrust and the diameter of the tube.

Whereas components affecting the strength properties of hyphal walls have received much attention, though perhaps not in the terms just outlined, components affecting an equally important thrust-regulating property, penetrability, seem to have been neglected. Of particular interest are water-resistant materials that coat, impregnate or line hyphal walls. Ostensibly, the production of such materials could have considerable importance in transitions between assimilative and non-assimilative states. Current candidates for such a role include two main categories of compounds.

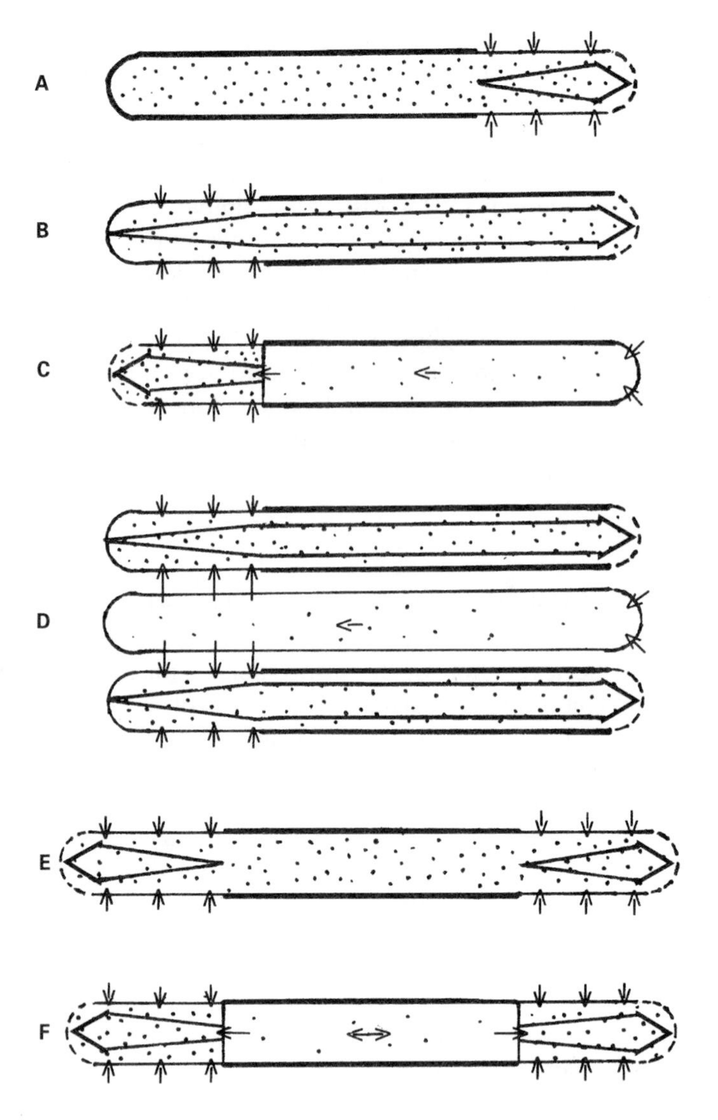

Figure 11 Diagrams illustrating how patterns of uptake (narrow arrows) and development of thrust (broad/tapering arrows) in axially elongated metabolic systems may be regulated by internal partitioning (membrane discontinuity) and boundary properties. Rigid boundaries are represented by a solid line; water-resistant boundaries are indicated by a thickened line; deformable boundaries are indicated by a broken line. Stippling density indicates solute concentration. External water potential is assumed to be uniformly high, but evapotranspirative or other mechanisms leading to extraction or discharge of water will enhance or suppress the fluxes according to circumstances, as will the formation of storage compounds such as glycogen. **A** A single protoplasmic continuum, where the deformable region encounters energy-rich domain, and thrust builds as it extends, increasing the surface through which net uptake occurs; e.g. a germinating spore or migratory mycelium encountering a food base. When the uptaking surface exceeds the critical value (which may be prevented if boundary-sealing keeps pace with extension) necessary to sustain constant maximum thrust (equivalent to linear extension), the system will be prone to branch. **B** A continuum where the deformable region enters non-nutritive domain, such that lateral boundaries are insulated, as in a hypha or hyphal system emerging from a food base. Thrust, developed in the assimilative region is sustained to an extent that depends on the efficacy of lateral sealing.

In the first category are a group of water-repellent polypeptides, which have been referred to as 'hydrophobins' (Wessels, 1991, 1992). These polypeptides probably correspond with the 'rodlets' which have been seen, using electron microscopy, to coat the surface of hyphae and spores of a wide range of fungi (e.g. Cole *et al.*, 1979) and probably contain strongly conserved sequences of amino acids in both ascomycetes and basidiomycetes (Stringer *et al.*, 1991). In *Schizophyllum commune* Fr. their spatiotemporal distribution coincides precisely with what might be expected of insulating compounds; they are bound to the walls of emergent aerial hyphae and fruitbody initials, but are released into the medium from submerged hyphae (Wessels, 1991, 1992).

With regard to the consumption of nitrogen, the use of polypeptides as insulators is, however, very costly, such that there may be a requirement to have access to sufficient nitrogen before emergent mycelial phases can be produced. This could, for example, set limits to the amount of biomass production necessary before a fungus becomes competent to fruit. Moreover, limitations on nitrogen supply may result in considerable competition between alternative non-assimilative phases, a possibility that would account for numerous observations of reciprocal relationships between such phases. All these limitations would be particularly acute in habitats where nitrogen availability is intrinsically low, for example, in wood.

An alternative, less nitrogen costly, means of insulation could be provided by aromatic and terpenoid products from the acetate, polyketide and shikimate pathways (notably absent from animals with determinate development, but widespread in plants and fungi). Many such compounds may be capable of conversion to free radicals when acted upon by phenol-oxidizing enzymes (laccases, tyrosinases and peroxidases), leading in turn both to polymerization and depolymerization. Such processes acting outside the plasmalemma would affect the insulating properties of hyphal walls (as well as the degradation of external substrates such as lignocellulose), whilst inside the plasmalemma they would instigate cell death. Phenol-oxidizing activity would also be expected to be substrate-induced and to affect polypeptides.

It is therefore of interest that changes in phenoloxidase activity have been linked with a variety of developmental landmarks, including fruitbody initiation, rhizomorph production and expression of somatic incompatibility (e.g. Leatham & Stahmann,

---

Figure 11 continued **C** Protoplasmic discontinuum, in which the assimilative region is separated from the non-assimilative region by a complete membrane, such as would occur at a sealed septum (septa, in that they are usually diaphragms with a central communication channel, may act as valves, allowing protoplasmic communication between hyphal compartments when open, but not when blocked or sealed). Here, water and solutes would be drawn from and through the less metabolically active non-assimilative component into the more competitive (in terms of active transport) assimilative component. This mechanism would allow redistribution from autolysing/degenerative regions, for example during foraging and following anastomosis (cf. Figs 2 & 8). **D** Coupled system, in which a central freely permeable, metabolically inactive system is surrounded by highly insulated hyphae. The central system provides a channel through which water and solutes absorbed passively at the mycelial margin is distributed to assimilative components, as could apply to foraging mycelial cords emerging from a food base. **E** Linked system, with assimilative sites at each end. The connection will remain intact as long as it does not become isolated, e.g. a system bridging two assimilative sites as a result of foraging (Fig. 2) or self-fusion (Fig. 8). **F** Linked system in which the connection becomes isolated, degenerates, and loses resources to assimilative phases, as between two somatically incompatible mycelia (Fig. 9).

1981; Li, 1981; Ross, 1985; Worrall *et al.*, 1986). There is evidence that enhancement of phenol-oxidizing activity is associated with production of emergent mycelial phases at interspecific interaction interfaces (as in Figs 5–7) (White & Boddy, 1992; Griffith, Rayner & Wildman, unpublished observation). Moreover, there may be a reciprocal relationship between laccase activity, associated with repression of emergence of aerial mycelium, and tyrosinase and peroxidase activity associated with promotion of the latter. In *Peniophora lycii*, autolytic zones associated with polar growth correspond, on media amended with L-cysteine (a free radical scavenger) with the deposition of purple intrahyphal pigment – perhaps implicating free radical production in the degenerative process (cf. Frese & Stahl, 1990).

These and other observations suggest that material means to bring about degenerative processes and to insulate and deinsulate hyphal systems are available to fungal mycelia. This raises the question as to how these processes may be initiated at appropriate times and places.

Local feedback

In outline, a promising approach to this question may be possible through knowledge of feedback relationships between what have been termed primary and secondary metabolism but which might better be thought of as inductive and transductive metabolism. As active transport depletes resources from the external medium, so increasing the pressure difference across the hyphal boundary, a switchover occurs from inductive to transductive processes (differentiation) yielding hydrophobic compounds. This switchover would accomplish an efficient transition from energy assimilation to energy conservation and distribution. An appropriate cue would be a drop in energy charge (Bushell, 1989a, b), brought about externally by some environmental signal, internally by mitochondrial dysfunction, or by some combination of both. By the same token, an increase in energy charge, following arrival and reinitiation of assimilation in nutrient-rich domain, would provide the basis for dedifferentiation.

EXPERIMENTAL AND CONCEPTUAL APPROACHES TO TEST AND ENHANCE UNDERSTANDING

The aim so far has been to demonstrate the potential value of a systemic approach based on an appreciation of how interactions between fundamentally simple processes operating at mycelial boundaries can generate complex and apparently sophisticated patterns of behaviour. Inevitably there are many ways in which the argument, and/or the information on which it is based, is incomplete; however, further progress may be possible through finding answers to four basic types of questions. First, what water-resistant compounds are present in mycelial/hyphal boundaries, how are they distributed, are they correlated with developmental mode and how much do they affect permeability? Secondly, what part, if any, does mitochondrial dysfunction play in the induction of developmental shifts and degenerative processes? Thirdly, what are the metabolic, genetic and phenotypic consequences of combining disparate nuclear and mitochondrial genomes in the same protoplasm? Finally, can a logistic (non-linear) mathematical model of mycelial dynamics be constructed that successfully reproduces observed patterns of development?

A brief discussion of these questions, and description of relevant findings from some of the authors' studies now follows.

Hydrophobic extracellular chemistry

The possible relevance of the distribution of hydrophobins to assimilative–non-assimilative transitions has already been mentioned. The authors have used, in their research, high-pressure liquid chromatography and other techniques to provide metabolic fingerprints of extractable hydrophobic compounds that are produced, released or sequestered at key developmental stages by pure and interacting cultures of wood-inhabiting basidiomycetes. The results so far are at least consistent with an autonomous extracellular polymerization/depolymerization chemistry involving hydrophobic compounds and their interaction with light and phenol-oxidizing enzymes in the determination of mycelial patterns. Specifically:

1. there are reproducible changes in the spectrum of compounds extractable from monocultures at different developmental stages, especially after arrest of extension of colony margins;
2. some of these changes will occur in cell-free extracts, especially following exposure to light, and certain irreversible changes are associated with crystallization or precipitation of specific compounds;
3. in *Phlebia radiata*, four phenolic/quinonic compounds are associated with the septate mycelial phase which follows behind the coenocytic growth front, and form a precipitate similar in colour to that of aerial mycelium;
4. interspecific interactions, associated with enhanced mycelial emergence and changes in phenoloxidase activity, lead to reduced incidence of particular hydrophobic compounds;
5. addition of the laccase inhibitor, cetyl trimethyl ammonium bromide (CET) to the culture medium significantly increases the incidence of hydrophobic compounds;
6. addition of the metabolic inhibitor, 2,4-dinitrophenol to the medium has similar effects to interspecific interactions on metabolite profiles.

Mitochondrial dysfunction, developmental shifts and degenerative processes

The possible role of mitochondria in the determination of mycelial developmental patterns is indicated by a variety of observations (Rayner & Ross, 1991). These include the possible relation between a mitochondrial intron, 'senescence', laccase and free radical generation in *Podospora pauciseta* (Ces.) Traverso (Frese & Stahl, 1990; Küch, 1989), the reciprocal relation between oxidative phosphorylation and phenoloxidase activity (e.g. Lyr, 1958, 1963) and the relation between phenoloxidase activity and developmental landmarks (see earlier). The pivotal position of mitochondria in metabolic networks and the probable relationship between primary and secondary metabolic processes and energy charge (see earlier) all add weight to the conclusion that mitochondria are suitably placed for an executive role in indeterminate development.

One approach to investigating this role would be to examine the effects on mycelial morphogenesis of agents known to affect mitochondrial functioning in a variety of ways. Correspondingly, it may be significant that from a range of metabolic inhibitors, the uncoupling agent, 2,4-dinitrophenol, brought about responses in mycelia of wood-inhabiting basidiomycetes most strikingly similar to those elicited by interaction with a neighbour (Figs 12, 13). The similarity between the effects on extracellular metabolite profiles of this agent and interspecific interactions (see earlier) suggests that endogenously produced uncoupling agents, e.g. certain phenolics, could play a role in developmental regulation. Any such agents would, through their influence on energy charge, be likely to feedback positively on their own biosynthesis, as well as eliciting phenoloxidase activity.

A problem with this approach, however, is the difficulty in distinguishing changes due to effects on mitochondria from those due to action of the agent on other cell components. This difficulty is compounded by the need to apply the agents from outside the system, at concentrations and under circumstances that may have little direct relation to conditions where genes are being expressed within hyphal interiors. A more systemic approach may, however be available through addressing the next issue to be discussed.

Effects of combining disparate nuclear and mitochondrial genomes in the same protoplasm

In keeping with general developmental biological theory (e.g. Pritchard, 1986), it has been assumed, so far, that changes in genetic information content *per se* are not involved in pattern shifts. It has also been implied that changes in gene expression may play a following rather than leading role in ordering indeterminate development.

This is not to say that changing the genomic context within systems that are intrinsically finely balanced between competing but interdependent expansive and constraining processes cannot potentiate even greater instability. In particular, the coexistence of conflicting sets of information in disparate nuclear and mitochondrial

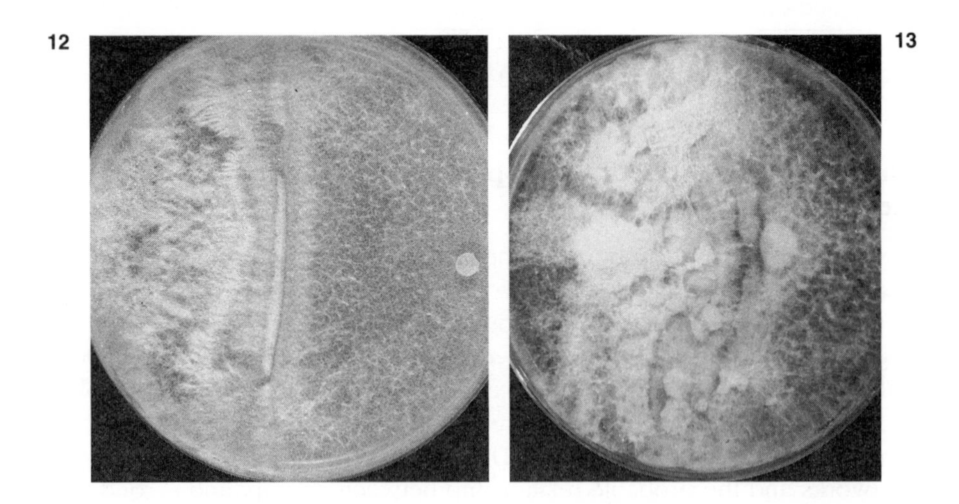

Figures 12 and 13 Effects of interspecific interactions and 125×10^{-5} g 2,4-dinitrophenol (DNP) on mycelial emergence patterns in *Phlebia radiata* grown on 3% malt agar. **12** Production, by *P. radiata* (inoculated right), of dense bands of emergent, salmon pink bands of mycelium when inoculated facing a DNP-impregnated assay disc. **13** Production of similar bands when facing a mycelium of *Bjerkandera fumosa* (Pers.: Fr.) Karst.

genomes is a potentially major source of instability. Such instability may be fundamental to the true nature of incompatibility and speciation processes.

Basidiomycete mating involves reciprocal exchange of numerous nuclei, but not mitochondria, between fused homokaryotic hyphal systems. The process of associating two types of nuclei within cytoplasmic backgrounds corresponding with two parents is therefore reiterated many times. This enables any potential instability in the interaction between any particular combination of genomes to be expressed and characterized (Rayner, 1991).

Correspondingly, recent studies of mating between strains of *Stereum hirsutum* (Willd.: Fr.) S.F. Gray have demonstrated that, in addition to the uniform outgrowth of secondary mycelium between compatible strains from like locations, both degenerative and highly complex patterns of phenotypic output occur between strains from disparate locations (Ainsworth *et al.*, 1990a, b; 1992).

These patterns sometimes are and sometimes are not accompanied by changes in nuclear and mitochondrial genomic organization, evident in DNA fingerprinting profiles, as well as accumulation of pigments and crystalline outgrowths of the sesquiterpene, (+)-torreyol (Figs 14, 15). Moreover, there is evidence for the wholesale silencing and replacement of nuclear genomes within particular cytoplasmic backgrounds. Much further research is needed for a full understanding of the mechanisms underlying these outcomes, but the latter do point to the occurrence of a dynamic relationship between genomic composition and expression, with hydrophobic metabolites and phenol-oxidizing enzymes playing a key role.

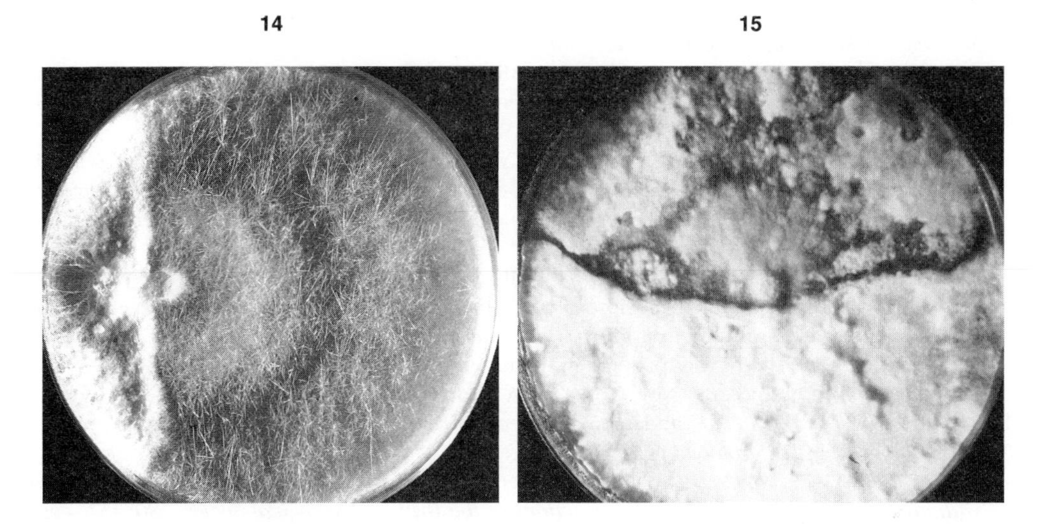

Figures 14 and 15 Degenerative and complex outcomes of reciprocal nuclear exchange between homokaryotic strains of *Stereum*. **14** Alternating dense and sparse zones of crystalline aggregates of (+)-torreyol emerging from a degenerative Australian strain of *S. hirsutum* following interaction with an English strain (inoculated left). From Ainsworth *et al.* (1990b). **15** Pairing between a strain of *S. complicatum* Fr. from the U.S.A. (lowermost) and a strain of *S. hirsutum* from what at the time was the U.S.S.R. Whereas the U.S.A. strain has retained a uniform morphology, like that of its progenitor, the U.S.S.R. strain has broken down into numerous subdomains of variable morphology and separated by pigmented zones. From Ainsworth *et al.* (1992).

Non-linear mathematical models

The essentially linear mathematical models currently favoured (e.g. Prosser, 1991), correspond with some aspects of mycelial growth, namely early exponential development from a germinating spore and the linear radial increment of the peripheral growth zone of mature colonies. The associated concept of a hyphal growth unit which is regularly duplicated and its association with septation, also accord with some observations of branching patterns. However, these models do not in themselves elucidate many important issues, e.g. how a constant mass density is conserved in peripheral growth zones expanding linearly in more than one spatial dimension; why branching occurs in both septate and coenocytic systems, and how heterogeneity originates. These characteristics, as well as those that can as a special case be modelled linearly, may on the other hand be explicable in terms of non-linear models that reflect the relation between uptake and throughput processes in partially anastomosed systems.

ACKNOWLEDGEMENT

We thank Glaxo Group Research Ltd for financial support.

REFERENCES

AINSWORTH, A.M. & RAYNER, A.D.M., 1991. Ontogenetic stages from coenocyte to basidiome and their relation to phenoloxidase activity and colonization processes in *Phanerochaete magnoliae*. *Mycological Research, 95:* 1414–1422.

AINSWORTH, A.M., RAYNER, A.D.M., BROXHOLME, S.J. & BEECHING, J.R., 1990a. Occurrence of unilateral genetic transfer and genomic replacement between strains of *Sterum hirsutum* from non-outcrossing and outcrossing populations. *New Phytologist, 115:* 119–128.

AINSWORTH, A.M., RAYNER, A.D.M., BROXHOLME, S.J., BEECHING, J.R., PRYKE, J.A., SCARD, P.T., BERRIMAN, J., POWELL, K.A., FLOYD, A.J. & BRANCH, S.K., 1990b. Production and properties of the sesquiterpene, (+)-torreyol, in degenerative mycelial interactions between strains of *Stereum*. *Mycological Research, 94:* 799–809.

AINSWORTH, A.M., BEECHING, J.R., BROXHOLME, S.J., HUNT, B.A., RAYNER, A.D.M. & SCARD, P.T., 1992. Complex outcome of reciprocal exchange of nuclear DNA between two members of the basidiomycete genus *Stereum*. *Journal of General Microbiology, 138:* 1147–1157.

ANDREWS, J.H., 1992. Life history patterns in the fungi. In G.C. Carroll & D.T. Wicklow (eds), *The Fungal Community*, 2nd edn. New York, U.S.A.: Marcel Dekker.

BARTNICKI-GARCIA, S., 1973. Fundamental aspects of hyphal morphogenesis. *Symposia of the Society for General Microbiology, 23:* 245–267.

BODDY, L. & RAYNER, A.D.M., 1983. Mycelial interactions, morphogenesis and ecology of *Phlebia radiata* and *Phlebia rufa* in oak. *Transactions of the British Mycological Society, 80:* 437–448.

BUSHELL, M.E., 1989a. The process physiology of secondary metabolite production. *Symposia of the Society for General Microbiology, 44:* 95–120.

BUSHELL, M.E., 1989b. Biowars in the bioreactor. *New Scientist, 124 (1685):* 42–45.

COGGINS, C.R., HORNUNG, U., JENNINGS, D.H. & VELTKAMP, C.J., 1980. The phenomenon of 'point growth', and its relation to flushing and strand formation in the mycelium of *Serpula lacrimans*. *Transactions of the British Mycological Society, 75:* 69–76.

COLE, G.T., SEKIYA, M., KASAI, R., YOKOYAMA, T. & NOZAWA, Y., 1979. Surface ultrastructure and chemical composition of the cell walls of conidial fungi. *Experimental Mycology, 3:* 132–156.

COOKE, R.C. & RAYNER, A.D.M., 1984. *Ecology of Saprotrophic Fungi*. London, U.K.: Longman.

DEACON, J.W. 1984. *Introduction to Modern Mycology,* 2nd edn. Oxford: Blackwell Scientific.

DOWSON, C.G., RAYNER, A.D.M. & BODDY, L., 1986. Outgrowth patterns of mycelial-cord-forming basidiomycetes from and between woody resource units in soil. *Journal of General Microbiology, 132:* 203–211.

DOWSON, C.G., RAYNER, A.D.M. & BODDY, L., 1988a. Foraging patterns of *Phallus impudicus, Phanerochaete laevis* and *Steccherinum fimbriatum* between discontinuous resource units in soil. *FEMS Microbiology Ecology, 53:* 291–298.

DOWSON, C.G., RAYNER, A.D.M. & BODDY, L., 1988b. The form and outcome of mycelial interactions involving cord-forming decomposer basidiomycetes in homogeneous and heterogeneous environments. *New Phytologist, 109:* 423–432.

DOWSON, C.G., SPRINGHAM, P., RAYNER, A.D.M. & BODDY, L., 1989a. Resource relationships of foraging mycelial systems of *Phanerochaete velutina* and *Hypholoma fasciculare* in soil. *New Phytologist, 111:* 501–509.

DOWSON, C.G., RAYNER, A.D.M. & BODDY, L., 1989b. Spatial dynamics and interactions of the woodland fairy ring fungus, *Clitocybe nebularis. New Phytologist, 111:* 699–705.

FRESE, D. & STAHL, U., 1990. Ageing in *Podospora anserina* — a consequence of alternative respiration? In A. Reisinger & A. Bresinsky (eds), *Fourth International Mycological Congress IMC4 Abstracts:* 184: University of Regensburg.

GOULD, S.J. & LEWONTIN, R.C., 1979. The spandrels of San Marco and the Panglossian paradigm: a critique of the adaptionist programme. *Proceedings of the Royal Society, London, B, 205:* 581–598.

GREGORY, P.H., 1984. The fungal mycelium: an historical perspective. *Transactions of the British Mycological Society, 82:* 1–11.

GRIFFITH, G.S. & BODDY, L., 1991. Fungal decomposition of attached angiosperm twigs. IV. Effect of water potential on interactions between fungi on agar and in wood. *New Phytologist, 117:* 633–641.

IKEDIUGWU, F.E.O. & WEBSTER, J., 1970. Antagonism between *Coprinus heptemerus* and other coprophilous fungi. *Transactions of the British Mycological Society, 54:* 181–204.

JENNINGS, D.H., 1984. Water flow through mycelia. In D.H. Jennings & A.D.M. Rayner (eds), *The Ecology and Physiology of the Fungal Mycelium,* pp. 143–164. Cambridge, U.K.: Cambridge University Press.

JENNINGS, D.H., 1987. Translocation of solutes in fungi. *Biological Reviews, 62:* 215–243.

KOCH, A.L., 1982. The shape of the hyphal tips of fungi. *Journal of General Microbiology, 128:* 947–951.

KÜCK, U., 1989. Mitochondrial DNA rearrangements in *Podospora anserina. Experimental Mycology, 13:* 111–120.

LEATHAM, G.F. & STAHMANN, M.A., 1981. Studies on the laccase of *Lentinus edodes*: specificity, localization and association with developing fruit bodies. *Journal of General Microbiology, 125:* 147–157.

LI, C.Y., 1981. Phenoloxidase and peroxidase activities in zone lines of *Phellinus weirii. Mycologia, 73:* 811–821.

LIDDELL HART, B.H., 1967. *Strategy.* London, U.K.: Faber & Faber.

LOVETT-DOUST, L., 1981. Population dynamics and local specialization in a clonal perennial (*Ranunculus repens*). I. The dynamics of ramets in contrasting habitats. *Journal of Ecology, 69:* 743–755.

LYR, H., 1958. Die Induktion der Laccase-Bildung bei *Collybia velutipes* Curt. *Archiv für Mikrobiologie, 28:* 310–324.

LYR, H., 1963. Enzymatisches Detoxifikation cholorierter Phenole. *Phytopathologie Zeitschrift, 38:* 342–354.

MONEY, N.P., 1990. Measurement of hyphal turgor. *Experimental Mycology, 14:* 416–425.

MOORE, D., CASSELTON, L.A., WOOD, D.A. & FRANKLAND, J.F: (eds), 1985. *Developmental Biology of the Higher Fungi.* Cambridge, U.K.: Cambridge University Press.

PRITCHARD, D.J., 1986. *Foundations of Developmental Genetics.* London, U.K.: Taylor & Francis.

PROSSER, J.I., 1991. Mathematical modelling of vegetative growth of filamentous fungi. In D.H.

Arora, B. Rai, K.G. Mikerji & G.R. Knudsen (eds), *Handbook of Applied Mycology,* pp. 591–623. New York, U.S.A.: Marcel Dekker.

RAYNER, A.D.M., 1986. Mycelial interactions — genetic aspects. In G.W. Gould, M.E. Rhodes Roberts, A.K. Charnley, R.M. Cooper & R.G. Board (eds), *Natural Antimicrobial Systems,* pp. 277–296. Bath, U.K.: Bath University Press.

RAYNER, A.D.M., 1991. The challenge of the individualistic mycelium. *Mycologia, 83:* 48–71.

RAYNER, A.D.M. & COATES, D. 1987. Regulation of mycelial organisation and responses. In A.D.M. Rayner, C.M. Brasier & D. Moore (eds) *Evolutionary Biology of the Fungi,* pp. 115–136. Cambridge, U.K.: Cambridge University Press.

RAYNER, A.D.M. & ROSS, I.K., 1991. Sexual politics in the cell. *New Scientist, 129 (1762):* 30–33.

RAYNER, A.D.M., BODDY, L. & DOWSON, C.G., 1987. Genetic interactions and developmental versatility during establishment of decomposer basidiomycetes in wood and tree litter. *Symposia of the Society for General Microbiology, 41:* 83–123.

RITZ, K. & CRAWFORD, J., 1990. Quantification of the fractal nature of colonies of *Trichoderma viride. Mycological Research, 94:* 1138–1141.

ROSS, I.K., 1985. Determination of the initial steps in differentiation in *Coprinus congregatus.* In D. Moore, L.A. Casselton, D.A. Wood & J.F. Frankland (eds), *Developmental Biology of the Higher Fungi,* pp. 353–373. Cambridge, U.K.: Cambridge University Press.

RUITERS, M.H.J. & WESSELS, J.G.H., 1989. *In situ* localization of specific RNAs in whole fruiting colonies of *Schizophyllum commune. Journal of General Microbiology, 135*: 1747–1754.

SAUNDERS, P.T. & TRINCI, A.P.J., 1979. Determination of tip shape in fungal hyphae. *Journal of General Microbiology, 110:* 469–473.

SCHAFFER, W.M., 1987. Chaos in ecology and epidemiology. In H. Degn, A.V. Holden & L.F. Olsen (eds), *Chaos in Biological Systems,* pp. 233–248. New York, U.S.A.: Plenum Press.

SIETSMA, J.H. & WESSELS, J.G.H., 1979. Evidence for covalent linkages between chitin and β-glucan in a fungal wall. *Journal of General Microbiology, 114:* 99–108.

STENLID, J. & RAYNER, A.D.M., 1989. Environmental and endogenous controls of developmental pathways: variation and its significance in the forest pathogen, *Heterobasidion annosum. New Phytologist, 113:* 245–258.

STRINGER, M.A., DEAN, R.A., SEWALL, T.C. & TIMBERLAKE, W.E., 1991. *Rodletless,* a new *Aspergillus* developmental mutant induced by directed gene inactivation. *Genes and Development, 5:* 1161–1171.

TRINCI, A.P.J., 1978. The duplication cycle and vegetative development in moulds. In J.E. Smith & D.R. Berry (eds), *The Filamentous Fungi,* vol. 3, *Developmental Mycology,* pp. 132–163. London, U.K.: Edward Arnold.

WATKINSON, S.C., 1977. Effect of amino acids on coremium development in *Penicillium claviforme. Journal of General Microbiology, 101:* 269–275.

WESSELS, J.G.H., 1986. Cell wall synthesis in apical hyphal growth. *International Review of Cytology, 104:* 37–79.

WESSELS, J.G.H., 1991. Fungal growth and development: a molecular perspective. In D.L. Hawksworth (ed.), *Frontiers in Mycology,* pp. 27–48. Kew, Surrey, U.K.: CAB International.

WESSELS, J.G.H., 1992. Gene expression during fruiting in *Schizophyllum commune. Mycological Research, 96:* 609–620.

WHITE, N.A. & BODDY, L., 1992. Extracellular enzyme localization during interspecific fungal interactions. *FEMS Microbiology Letters, 98:* 75–80.

WORRALL, J.J., CHET, I. & HÜTTERMANN, A., 1986. Association of rhizomorph formation with laccase activity in *Armillaria* spp. *Journal of General Microbiology, 132:* 2527–2533.

CHAPTER

18

Origin and consequences of colony form in fungi: a reaction–diffusion mechanism for morphogenesis

JOHN W. CRAWFORD & KARL RITZ

CONTENTS

Abstract

Shape and form of various structures in organisms have historically been exploited to provide the bases for a descriptive taxonomy. Such shapes and forms arise as a consequence of the interaction between genotype and environment, and are useful diagnostics of the processes linking the genetic code to phenotypic expression. Studies on the relationships between genotype and phenotype have been slow in part because of the lack of a suitable theoretical framework linking complex geometries to dynamical growth models. Application of fractal geometry is yielding promising developments towards providing that framework. In this chapter these developments are discussed in relation to the growth of fungal colonies. The measurement and interpretation of the fractal dimension of hyphal distribution in space is discussed.

Shape and Form in Plants and Fungi
ISBN 0–12–371035–9

Colony form in fungi is heterogeneous though highly organized, with fractal branching patterns implying spatially correlated growth. Such ordered growth is shown to result in an optimal configuration for a heterogeneous environment where resource distribution is likely to be patchy. It is shown how a quantification of shape and form can lead to this interpretation in terms of competition between concurrent growth strategies, viz. those oriented toward spatial exploration or exploitation. A non-linear growth model is presented whose testable assumptions may yield insight into the relative importance of environmental influences and self-organization in large- and small-scale branch patterning in fungi. The basis of the model is a reaction–diffusion mechanism which governs the spatial distribution of calcium in the cytoplasm. Such a mechanism results in continuous extension and localized branching events which occur subapically. Large-scale patterning in the form of growth rings and spirals can also be reproduced.

INTRODUCTION

Rutherford may only have been half serious when he remarked that 'all science [was] either physics or stamp collecting', but the point is made that taxonomy by itself is not a means to understanding in biology. Rather, the hope is that classification schemes are meaningful, and that they are an aid to identifying processes which give rise to the essential differences they purport to express. A taxonomy based on shape and form alone may hide the fact that the great diversity perceived is actually a consequence of some continuum in realization of the same, essentially simple, biophysical or biochemical phenomena. The final goal of studies of shape and form is to reach an understanding of how the 'physical laws', hidden in the genetic code, interact with the environment and physicochemical processes in organisms, to result in the observed morphology. The aim of this chapter is to show how quantitative and objective measurements of colony form in fungi constrain theories of morphogenesis by which such form is created. Fungi are convenient models for this type of study because they share properties of both single-cell and multicellular organisms and thus such studies may provide insight into the problem of scaling to more complex organisms. Furthermore, the complexity and diversity of fungal form presents both a challenge to any theory of morphogenesis and a wealth of opportunity for testing alternative mechanisms.

OBSERVATIONS

The morphology of eucarpic fungal colonies is governed by the processes of hyphal elongation and branching. These are essentially distinct processes; elongation being continuous in time, whereas branch initiation represents a discrete event in space and time. If form is random, in the sense that the spatial distribution of hyphae is disordered, then the two processes may be governed by independent mechanisms. If, however, the distribution is ordered, and the processes governing form are oriented towards generating some optimal configuration for a particular environment, then the two processes cannot act independently. They are either governed by the same mechanism, or they are a consequence of a chain of mechanisms that are coordinated in a manner which is influenced by the external environment. The question of whether the colony is ordered in some optimal way is clearly important for developing any theory of morphogenesis.

Despite their complex mycelial patterns, hyphal colonies do exhibit some highly ordered growth properties. Colonies growing under optimal conditions expand radially at a constant rate (Trinci, 1978). This is thought to be evidence that hyphal tip growth is supported by only a limited length of subtending mycelium and that more distal parts of the colony do not contribute to growth, i.e. the peripheral growth zone hypothesis (Trinci, 1971). This is an extremely important conclusion since, if correct, it implies that growth of spatially separated parts of the colony should be independent or uncorrelated. However, constant linear growth rates are well known in other systems where cell extension is limited to some zone of constant width (e.g. plant leaf tissue). In these systems, expansion rates are thought to be controlled by the physical properties of the cell boundary (Fry, 1989; Passioura & Fry, 1992) rather than by the rate of biomass production. Evidence that this may be also the case in fungi is suggested in the results of Schmid & Harold (1988), where elongation rates were observed to be far more affected than biomass production in conditions of low extracellular Ca^{2+} concentration. Lockart (1965) has demonstrated that longitudinal extension will be exponential when the entire length of the growing tube is extensible, and linear when only some fixed length can extend (cf. Passioura & Fry, 1992). Constant linear growth rate is more probably a consequence of the fact that only the tip regions of hyphae are capable of extension (Wessels, 1990) rather than by supply limitations of wall precursors.

Growth rings provide another example of order in the growth of fungal colonies (Chevaugeon & Van Huong, 1969). These modulations in hyphal density and sporulation can either be in the form of concentric rings (Lysek, 1978; Fig. 1) or spirals (Bourret *et al.*, 1969; Fig. 2), while some species seem to be capable of producing both (Fig. 3A, B). As illustrated in Fig. 1, initially independent colonies of the same

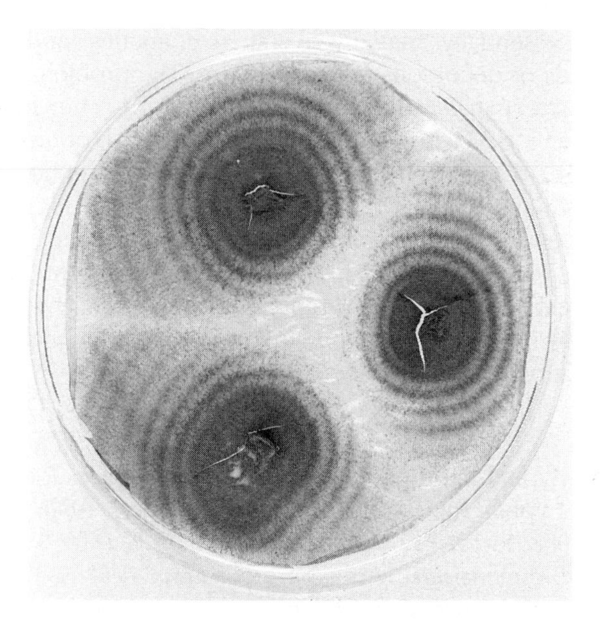

Figure 1 Three colonies of *Cylindrocarpon destructans* (Zinssm.) Scholten growing on potato-dextrose agar under conditions of continuous darkness. Original diameter of dish 9 cm.

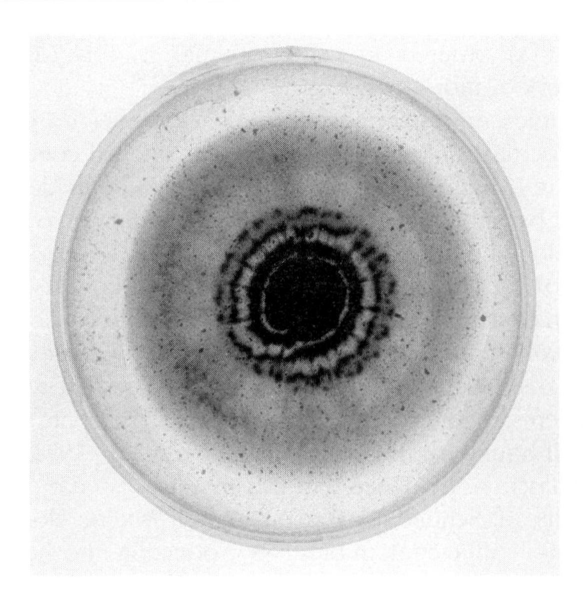

Figure 2 Single colony of *Penicillium claviforme* Bainier growing on malt extract agar under conditions of continuous darkness. Original diameter of dish 9 cm.

species grown in the same environmental conditions show rings with similar regular spacing and density profiles. Suggested mechanisms for these phenomena include circadian rhythms and feedback between hyphal proliferation and nutrient depletion (Edelstein & Segel, 1983; Watkinson, 1975). However, not all rhythms can have a circadian origin (Bourret, *et al.*, 1969).

A frequent way of describing the morphology of a fungal colony and its growth dynamics is by deriving parameters such as number of tips per unit area, or hyphal density. These are essentially spatially averaged quantities, and some models are based on these kinds of definitions (Prosser, 1992). The problem with using specific quantities is that such spatial averages are ill-defined on a heterogeneous structure such as a hyphal network. Figure 4 shows a schematic plot of total hyphal length as a function of the area over which total length is calculated. A homogeneous colony would produce a straight line on this type of plot, i.e. total length per unit area is constant and independent of the spatial scale of measurement. For a real colony, however, the hyphal density defined in this way is not constant but depends on the scale of measurement (Fig. 4). A ln:ln transformation of total hyphal length vs area renders the relationship linear. Assuming hyphal mass/unit hyphal length is constant, this implies the relation:

$$M(R) \propto R^D \tag{1}$$

where $M(R)$ is the hyphal mass enclosed in a circular region R and D is the fractal dimension of the hyphal (mass) distribution (Falconer, 1990; Mandelbrot, 1983; Peitgen & Saupe, 1988; Ritz & Crawford, 1990). Values of D for a colony growing on the surface of a two-dimensional surface are constrained to lie between 1 and 2. A colony of dimension 2 occupies the surface area homogeneously, whereas dimensions less than 2 imply a structure which has an increasingly 'clumpy' spatial distribution in two dimensions. Fractal growth as implied by Equation (1) implies that the

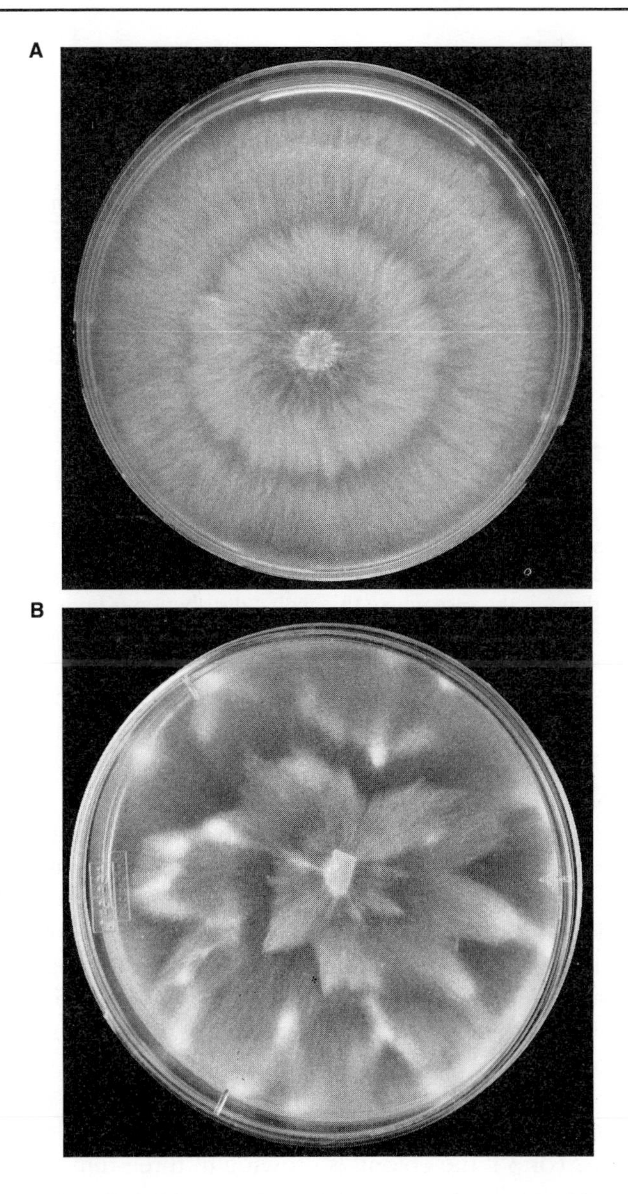

Figure 3 Colonies of *Mortierella elongata* Linnem. growing on potato-dextrose agar. **A** Colony grown in continuous light; **B** colony grown in continuous darkness. Original diameter of dish 9 cm.

structure cannot be disordered, i.e. the distribution of hyphae is spatially correlated and the structure at large scales is not independent of structure at small scales.

The growth of fungal colonies is in fact far more ordered than Equation (1) implies. Measurements of fractal dimension made during the development of colonies of *Trichoderma viride* Pers. constrained to grow on the surface of two-dimensional agar plates are time-dependent (Ritz & Crawford, 1990). We have made similar measurements on colonies of *Alternaria alternata* (Fr.) Keissler and *Rhizoctonia solani* Kuhn which confirm that this phenomenon is not confined to a single species. Thus, spatially correlated growth persists during colony development, although the form of growth (described by *D*) may vary.

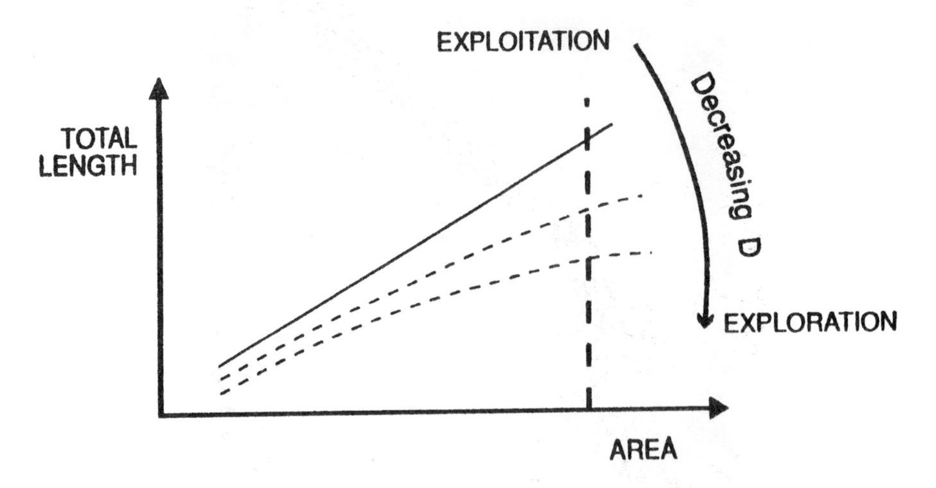

Figure 4 Schematic plot of total hyphal length versus area over which total length is measured, for a colony growing across a two-dimensional surface. The slope of the curves gives the hyphal density. The solid line denotes the case where the hyphal distribution is homogeneous. The broken curves illustrate the behaviour of a fractal distribution with heterogeneity increasing as the fractal dimension, *D*, decreases in the direction of the arrow as indicated. The vertical broken line intersects the curves at the points which indicate the total hyphal length invested in filling a given area.

EXPLORATION VERSUS EXPLOITATION

Fungi absorb growth agents, primarily through hyphal tips, from the medium in which they grow. Effective exploitation of the medium involves investment of resource in building a structure capable of acting as a diffusive sink, of detecting sites of high nutrient concentration and of growing to encounter such sites. Any resultant growth morphology must therefore be a consequence of these competing growth strategies of exploration versus exploitation (Ritz & Crawford, 1990), corresponding to non-assimilative and assimilative growth phases (Rayner, 1991). As shown in Fig. 4, a colony growing homogeneously must invest a relatively large amount of material in occupying a given region of space, and proportionately more as it continues to grow. A colony growing heterogeneously with a mass distribution described by some fractal dimension less than 2 (or 3 if the colony is growing in three-dimensional space) has, in contrast, to invest a smaller hyphal mass in occupying the same region of space, and progressively less as it continues to grow (Fig. 4). If the colony encounters a resource, nutrient is less limiting and hyphal mass will concentrate in that region to maximize the surface area available for resource capture. If, however, resource is limiting, colonies which are more economical in their expenditure of mass will perform more effectively. Through control of branching and elongation, the fractal dimension is lowered and therefore less mass is committed to filling the area. When nutrition is severely limiting, branching is greatly reduced (Parkinson *et al.*, 1989), the colony fills the area very inefficiently and uses a minimum amount of mass to explore as large an area as possible (Ritz & Crawford, 1990). Fractal growth therefore appears to be an extremely effective mode for facilitating plastic adaptation to growth where resources are distributed heterogeneously, i.e. in most natural environments. The observations described above greatly constrain hypotheses relating to the underlying mechanisms responsible for hyphal growth and colony morphogenesis.

CALCIUM AS A MORPHOGENIC AGENT

Calcium has been invoked as an important morphogenic agent in a number of biological systems (Goodwin & Trainor, 1985; Rasmussen & Barrett, 1984; Trewavas and Knight, Ch. 12 this volume). Little is known about the spatial and temporal dynamics of calcium in fungi (Gooday & Gow, 1990) primarily because of the difficulty in imaging the element, especially in a cytosolic free form. Fungal hyphae require calcium in the external medium to lie within an optimal concentration, around 10^{-3} M in *Neurospora crassa* Shear & Dodge (Schmid & Harold, 1988). Below this concentration, elongation rate and to a lesser extent biomass production becomes inhibited until at around 10^{-7} M; some cells lose the capacity for polarized growth altogether (Schmid & Harold, 1988). Fungal branching is also influenced by calcium (Harold & Harold, 1986), and there is evidence that such effects may be mediated via cAMP (Reissig & Kinney, 1983). Links between low concentrations of cytosolic cAMP and branching have been observed in mutants of *Neurospora* (Pall *et al.*, 1981) and high concentrations of calcium are known to inhibit production of cAMP (Rasmussen & Goodman, 1977). Schmid & Harold (1988) demonstrated distinct gradients in intracellular calcium, showing maxima towards the tips, in growing hyphae. However, these gradients were revealed by CTC fluorescence, which only visualizes membrane-bound calcium.

A number of observations in systems analogous to hyphae may also be pertinent to an understanding of the role of calcium in fungal growth. Pollen tube elongation rates are affected by external calcium in a similar manner to hyphae, with optimal growth occurring at concentrations of around 10^{-3} M, with little extension below 10^{-5} M or above 10^{-2} M (Steer & Steer, 1989). Order-of-magnitude gradients in calcium concentration ([Ca]) have been detected in pollen tubes from tip to base (Herth *et al.*, 1990). There is also a suggestion that high [Ca] may destabilize cross-linking in the cytoskeleton and result in weakening of the mechanical resistance of the cell wall to turgor (Steer, 1990; Steer & Steer, 1989). This mechanism may explain directed growth of hyphal tips, but cannot explain production of a subapical branch, since the main support against turgor in mature wall is the secondary wall which must first be plasticized (Bartnicki-Garcia, 1990; Wessels, 1990). Calcium plays a role in governing the mobility of vesicles in the cytoplasm of algal germ tubes (Kohno & Shimmen, 1988a, b). Approximately 75% of vesicles in characaean algae moved at internal concentrations below 10^{-7} M; this reduced to only 10% at 10^{-6} M and at 4×10^{-6} M only 1% of vesicles were translocated. Calcium also appears to be an important agent in facilitating vesicle fusion with the cell wall in animal cells via calcium stimulation of a G-protein system (Trifaro *et al.*, 1984). At concentrations below about 10^{-8} M, vesicle fusion to the cell wall is inhibited, whereas at concentrations above 10^{-6} M, fusion is enhanced.

These observations on the influence of [Ca] on organelle dynamics suggest that the observed gradients in calcium could lead to accumulation and enhanced absorption of vesicles near the tips of extending tubes. The relationships between calcium gradients and polarized growth remains controversial, with some workers claiming gradients are a consequence rather than a cause of polarized growth (Schmid & Harold, 1988; Steer & Steer, 1989). Certainly, polarized growth is not necessary for calcium gradients to become established, since localized enhancement of internal [Ca] occurs in germinating spores, before the emergence of a germ tube (Reiss *et al.*, 1985). An important question relating to the morphogenic role of calcium is therefore whether calcium gradients are a consequence or cause of polarized growth.

EXISTING THEORIES OF COLONY GROWTH

An influencial theory relating to colony growth is the peripheral growth zone hypothesis (Trinci, 1971). An explanation for constant radial growth rate is sought in terms of the specific growth rate of the colony, under the assumption that it is biomass production which drives hyphal extension. Under this assumption, it is shown that linear growth rate can only be achieved if a peripheral zone of constant width contributes to colony expansion. A prediction of this hyphothesis is that the colony extension rate K_r should be related to the specific growth rate μ according to

$$K_r = \mu w$$

where w, the width of the peripheral growth zone, is a constant (Prosser, 1990; Trinci, 1971). Experiments show that K_r/μ is indeed approximately constant for a species growing in similar environments and that it defines a length scale of the order of the length of the apical compartment (Trinci, 1971). This is held in support of the validity of the peripheral growth zone hypothesis and its underlying assumptions. As pointed out earlier, however, constant linear growth rate is more likely to be a consequence of the physical characteristics associated with the cell wall, than of biomass limitations. Making the assumption that extension rate K_r is constant, then the rate of increase in total hyphal length L is given by

$$\frac{dL}{dt} = bK_r$$

where b is the total number of hyphal tips at time t. If n is the mean number of intercalary compartments per unit length of hypha, and c is the expectation for the number of branches per compartment, then if colony mass M is proportional to L

$$\frac{dM}{dt} = cnK_rM$$

$$\Rightarrow K_r = \frac{1}{nc}\mu$$

where μ is the colony specific growth rate. This result implies that a constant hyphal extension rate will lead to a linear relation between K_r and μ, if c and n are constant over time. The constant of proportionality is equal to the average interval between branches which, if anastomoses are negligible, is the hyphal growth unit, G (Prosser, 1990). This predicted relation between K_r and μ in terms of the hyphal growth unit is similar to the one observed by Morrison & Righelato (1974) for *Penicillium chrysogenum* except for two essential differences. First, they used values for μ and G which were derived from colonies grown in submerged culture, whereas K_r was derived from colonies grown on the surface of a different agar-based medium, and secondly an additional constant multiplier was required on the right-hand side of the expression to achieve agreement with the data. It is conceivable that if K_r, μ and G had been measured from the same colony grown on the same medium, that the multiplicative constant would have been redundant. However, by the nature of the assumptions we have used, the relation derived above must only be considered approximate. Nevertheless, a plausible alternative to the theory that colony growth rate is governed by biomass production in a peripheral growth zone exists.

Whilst the peripheral growth zone hypothesis is the major theory in place for understanding colony extension, there are many models for branching in fungi. Prosser and Trinci (1979) build on the assumptions of the peripheral growth zone hypothesis, and base their model on detailed hypotheses relating to spore germination, production rates and transport of vesicles, and branching. Following in the vein of the peripheral growth zone hypothesis, adjacent intercalary compartments are assumed to be independent of one another from the point of view of growth dynamics, by a suitable choice of parameterization. Branching is assumed to occur when the vesicle concentration in an intercalary compartment rises above some hypothetical threshold. Because of the choice of parameters, the model results depend rather sensitively on the assumptions (which are largely empirical) relating to the creation of septa. As such it addresses the question of branching by replacing it with a new question concerning the mechanisms of septum formation. The model in its presented form also cannot lead to an understanding of branching in coenocytes. Furthermore, the assumption that adjacent compartments are independent means that correlated growth, including rings, cannot be accounted for.

On the basis of this discussion, it is premature to conclude that growth and branching are governed by vesicle flow alone, and that tip growth is controlled only by immediately adjacent hyphal compartments.

Bartnicki-Garcia (1990) proposed the existence of a vesicle supply centre (corresponding to the Spitzenkörper) as a central point source of vesicles which supplies wall precursors to the adjacent tip, hence determining tip shape and growth dynamics. The Prosser and Trinci mechanism is invoked to govern the position and frequency of these new centres and as such the model therefore fails for the reasons outlined earlier.

Edelstein (1982) and Edelstein and Segel (1983) proposed a model of colony growth based on the dynamics of tip production and loss, and on the feedback between colony growth and nutrient depletion. Their model includes the effects of tip death, anastomosis, tip bifurcation, lateral branching and hyphal death on the mass and tip production dynamics. A phenomenological expression is used to relate tip production to hyphal and tip densities which takes into account these tip–hyphal interactions. Branches are assumed to be created at a constant rate and hyphae extend at a constant linear rate. As such, colony morphology is governed by localized events such as the tip–hyphal interactions listed earlier, and by tip and hyphal death rates. Therefore, the model cannot account for the observed spatial correlations resulting from fractal structure. It is capable of reproducing ring structures, although these are rather weak and sensitive to model parameters.

A REACTION–DIFFUSION MECHANISM FOR SELF-REINFORCED POLARIZED GROWTH AND BRANCHING

The contrast between external optimal calcium concentrations ($\sim 10^{-3}$ M) and internal optimal free cytosolic calcium levels ($\sim 10^{-7}$ M) suggests that calcium levels in hyphae are strongly regulated. Turing (1952) showed that non-linear regulation of one chemical species by another could lead to the formation of stable spatial patterns driven by diffusion. This idea has been explored in a number of biological arenas to explain phenomena ranging from shell patterning to the transmission of visual information to the brain (Murray, 1989). It has also been suggested as a mechanism for whorl morphogenesis in the algal cell *Acetabularia mediterranea* Lamouroux (Goodwin & Trainor, 1985; Murray, 1989).

Considering two-species reaction kinetics, the general form for the equations governing the spatiotemporal dynamics of calcium and an agent, A, responsible for calcium depletion in the cytoplasm, may be written thus

$$\frac{\partial[\mathrm{Ca}]}{\partial t} = f([\mathrm{Ca}],\mathrm{A}) + D_{\mathrm{Ca}}\nabla^2[\mathrm{Ca}]$$

$$\frac{\partial \mathrm{A}}{\partial t} = g([\mathrm{Ca}],\mathrm{A}) + D_{\mathrm{A}}\nabla^2\mathrm{A} \qquad (2)$$

where D_A and D_{Ca} are the diffusion coefficients for A and calcium in the cytoplasm respectively, and f and g are two functions describing the reaction kinetics between A and Ca. Before gradients in [Ca] may be supported, a number of conditions must be met: (i) $D_{Ca} < D_A$, otherwise changes in the distribution of A cannot occur sufficiently quickly to accommodate spatial confinement of calcium; (ii) the reaction kinetics f and g must be such that in a well-mixed system (where diffusion-driven gradients cannot become established) [Ca] and A reach stable steady-state values; (iii) this steady-state is unstable to spatial disturbances. These last two conditions mean that for any particular set of reaction kinetics, there exist ranges in the rate coefficients outside which stable patterns cannot be supported (Murray, 1982). The conditions on f, g, D_A and D_{Ca} for spatial patterning are summarized in Murray (1989). Given that f and g satisfy the constraints outlined above, and pattern initiation is possible, the form of the resulting pattern is sensitive primarily to the size and shape of the boundary enclosing the spatial domain confining the reagents, and to the capacity of the hyphal network to mediate transport of the reagents across it. In general, as the domain increases in volume, more complex patterns result. The scale of the resulting pattern depends on the form of the reaction kinetics, the rate coefficients and the value of D_A/D_{Ca}. In particular, they define a minimum length scale for the spatial domain below which spatial patterns will not be supported. In the absence of a detailed knowledge of these, the analysis is limited to a study of the feasibility of producing spatial and temporal calcium dynamics consistent with the observed properties of hyphal growth. Many constraints on the proposed mechanisms are detailed in the second section, and these constrain even phenomenological theories of growth and branching.

As previously discussed, all models for colony development and branching require the pre-existence of a hyphal tip. Here no such assumption is necessary and the model begins with an expanding spore. After an initial period of swelling, the spore reaches a critical diameter above which diffusion-driven instabilities can support a pattern in [Ca] being produced via the above mechanism. The form of the patterns which may be excited, are described by the solutions of the eigenvalue equation (Murray, 1989) with k a constant

$$\nabla^2[\mathrm{Ca}] + k[\mathrm{Ca}] = 0 \qquad (3)$$

the angular parts of which are, assuming spherical geometry, the spherical harmonics (Abramovitch & Stegun, 1965). Three distinct phases of pattern development ensue, as follows.

1. The first pattern to form which could result in asymmetrical radial growth is of the form

$$[\mathrm{Ca}](\theta,\phi) \propto \cos\theta$$

where ϕ and θ are the polar angles of azimuth and elevation respectively. This will result in a 'hotspot' of calcium concentration on one side of the expanding spore. The enhanced concentration there will, according to the discussion in the previous section, result in reduced mobility of vesicles as well as increased vesicle fusion thus depositing wall plasticizers and precursors. This, in turn, will result in localized enhanced growth at this point, and the consequent emergence of a germ tube. Because the emergence of the tube will not initially result in a significant distortion of the boundary geometry which largely dictates the form of the pattern, this peak in calcium will remain associated with the newly formed tip.
2. The geometry of the spatial domain is subsequently dominated by the hypha, i.e. a cylinder. The spatial pattern to develop in this domain will initially be a [Ca] gradient with a maximum at the tip (Fig. 5). This will result in enhanced vesicle movement in basal parts of the hypha and reduced mobility at the tip, causing net apical movement of vesicles. Enhanced fusion driven by a higher [Ca] environment will continue to sustain polarized growth at a rate ultimately limited by the capacity of the cytoplasm to accommodate and continue to support tip extension. As extension continues, new modes for calcium distribution become possible, all of which maintain a concentration at the tip.

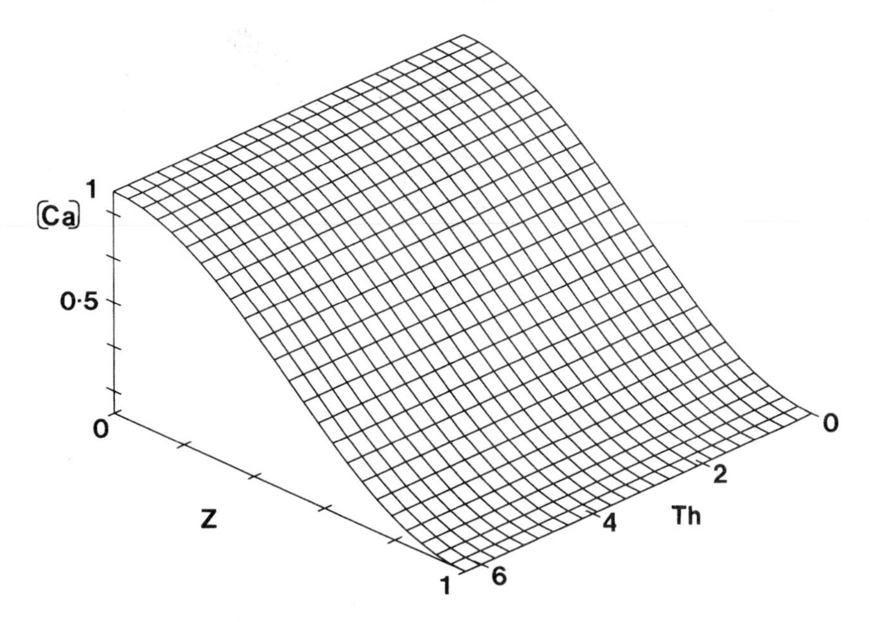

Figure 5 Plot of calcium concentration [Ca] (arbitrary units) versus distance from hyphal tip, z, versus polar angle, Th, on the surface of a germ tube of unit length assuming cylindrical geometry. For a tube of length greater than the circumference of its circular cross-section, the first potentially excitable pattern is shown as derived from Equation (3).

3. As hyphal length reaches a critical threshold, an asymmetric mode is excited (Fig. 6). This results in calcium being concentrated at the tip and half-way down the tube, but to one side. The latter concentration will result in localized accumulation of vesicles and enhanced fusion, resulting in the generation of a branch. Note that cytoskeletal loosening alone, while offering a mechanism for oriented tip growth (Steer & Steer, 1989) cannot explain branch formation, since the strongly cross-linked wall must first be loosened by wall plasticizers, contained in vesicles (Bartnicki-Garcia, 1990). The critical hyphal length at which a branch is produced is controlled during the development of a colony by the binding and release rates of calcium and the kinetics of production of A, implicit in the functions f and g in Equations (2).

If the structure is to develop a morphology consistent with the nutrient status of the environment, then these rate coefficients must be linked to metabolic activity. Since calcium and cAMP are known to antagonistically regulate one another, and that branching has been linked to low intracellular cAMP level (Reissig & Kinney, 1983) we can postulate that cAMP could be the agent A. High metabolic activity may then impinge on the rate coefficients via enhanced cAMP production which should lead to shortening of the interval between [Ca] maxima if increased branching is to result.

The reaction–diffusion type mechanism outlined above could lead to continuous self-reinforced polarized extension as well as discrete branching events. Such a mechanism, by the dependence for pattern formation on the geometry of the colony and a diffusion-based phenomenon, must also give rise to spatially correlated distribution of calcium and hence hyphae. The scale of the correlations depends upon the extent to which septa, if they occur, impede the movement of calcium through the hyphal network. The solution to Equations (2) with a geometrically complex and dynamic boundary such as a mycelium is a formidable task, and only worthwhile if experimen-

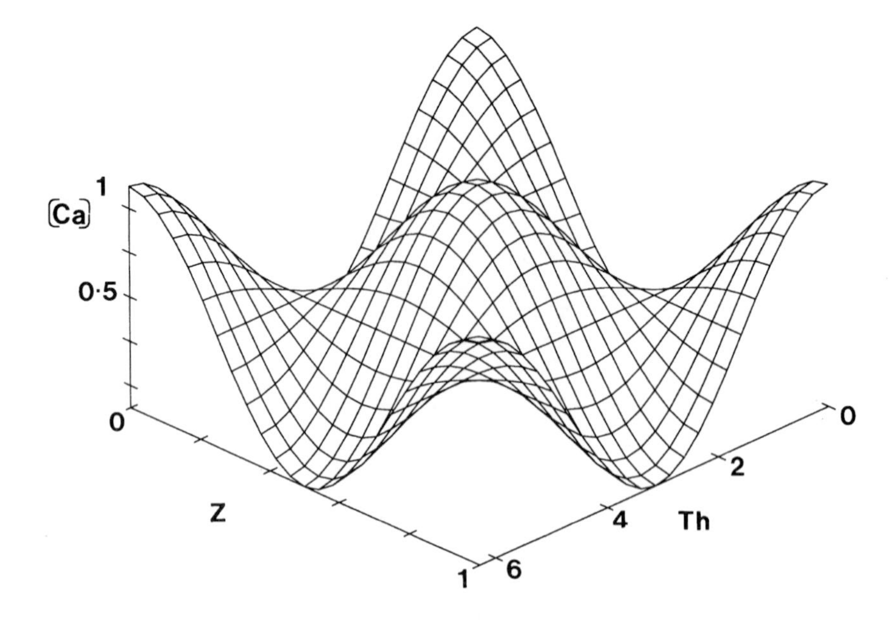

Figure 6 Plot with axes as described in caption to Fig. 5, showing the distribution of calcium corresponding to the initiation of a subapical branch.

tal support for the mechanism is found as well as some reasonable form for the reaction kinetics. It is therefore not at present possible to verify if a reaction–diffusion type mechanism can result in fractal structure. However, the possibility of reproducing large-scale correlated patterns such as rings and spirals can more readily be examined.

If it is assumed that a colony has formed according to the above mechanism, that it is fractal, and that some degree of connectivity (anastomosis) exists between hyphae, then large-scale patterns in calcium distribution will occur which are described by the eigenvalue equation

$$\nabla.(Br^{-\theta}\nabla[\mathrm{Ca}]) + k[\mathrm{Ca}] = 0 \tag{4}$$

where k and B are constants and $\theta = 2[(D/\bar{\bar{d}}) - 1]$, D is the Hausdorff dimension of the hyphal distribution, and $\bar{\bar{d}}$ is the fracton (or spectral) dimension (Crawford *et al.*, 1993; Orbach, 1986). If the colony is constrained to grow across a surface, then $0 \leq \theta \leq 2$, and analytic solutions can be found from the special cases $\theta = 0$, 1 and 2. If the colony is homogeneous and highly connected then $\theta = 0$, in contrast with the case $\theta = 2$ which describes a colony that is not well connected. The radially symmetric solutions of Equation (4) for each case are:

$$\begin{aligned}[\mathrm{Ca}](r) &= J_0\left[\left(\frac{k}{B}\right)^{1/2} r\right] && \theta = 0 \\ &= Ai\left[-\left(\frac{k}{B}\right)^{1/3} r\right] && \theta = 1 \\ &= rJ_{1/2}\left(\frac{k}{4B}r^2\right) && \theta = 2\end{aligned}$$

where J_n and Ai denote the Bessel function of order n and the Airy function respectively. The solutions are displayed graphically in Fig. 7 and indicate that radial oscillations are predicted. Both the spacing between the rings and their radial density profiles depend on the degree of heterogeneity and connectivity of the underlying hyphal network. Where [Ca] is highest, vesicle accumulation and fusion will be concomitantly high and growth of hyphae will be enhanced as a consequence. For consistency with the required branching response to nutrient status, large-scale pattern formation must be confined to zones of low metabolic activity in the colony, i.e. the older (and usually central) parts.

Reaction–diffusion provides a mechanism for orchestrating the continuous process of extension and discontinuous process of branching via calcium-regulated vesicle dynamics, possibly augmented by control of cytoskeletal resistance to turgor-driven expansion. The mechanism implies that growth must be spatially correlated, although fractal structure has yet to be established. Large-scale patterning results as a natural consequence of this mechanism. Although analogous results will not be derived here, reaction–diffusion type mechanisms are also capable of producing spiral patterns in the distribution of the reactants (Murray, 1989).

Reaction–diffusion is by no means the only mechanism by which excitable dynamics of the type discussed may be produced. Indeed there are real problems with assuming that the patterns are stable to cytoplasmic streaming (Goodwin & Trainor, 1985) and that modes become de-excited and excited in the manner suggested as the boundary extends. Nevertheless, it remains a viable and testable theory.

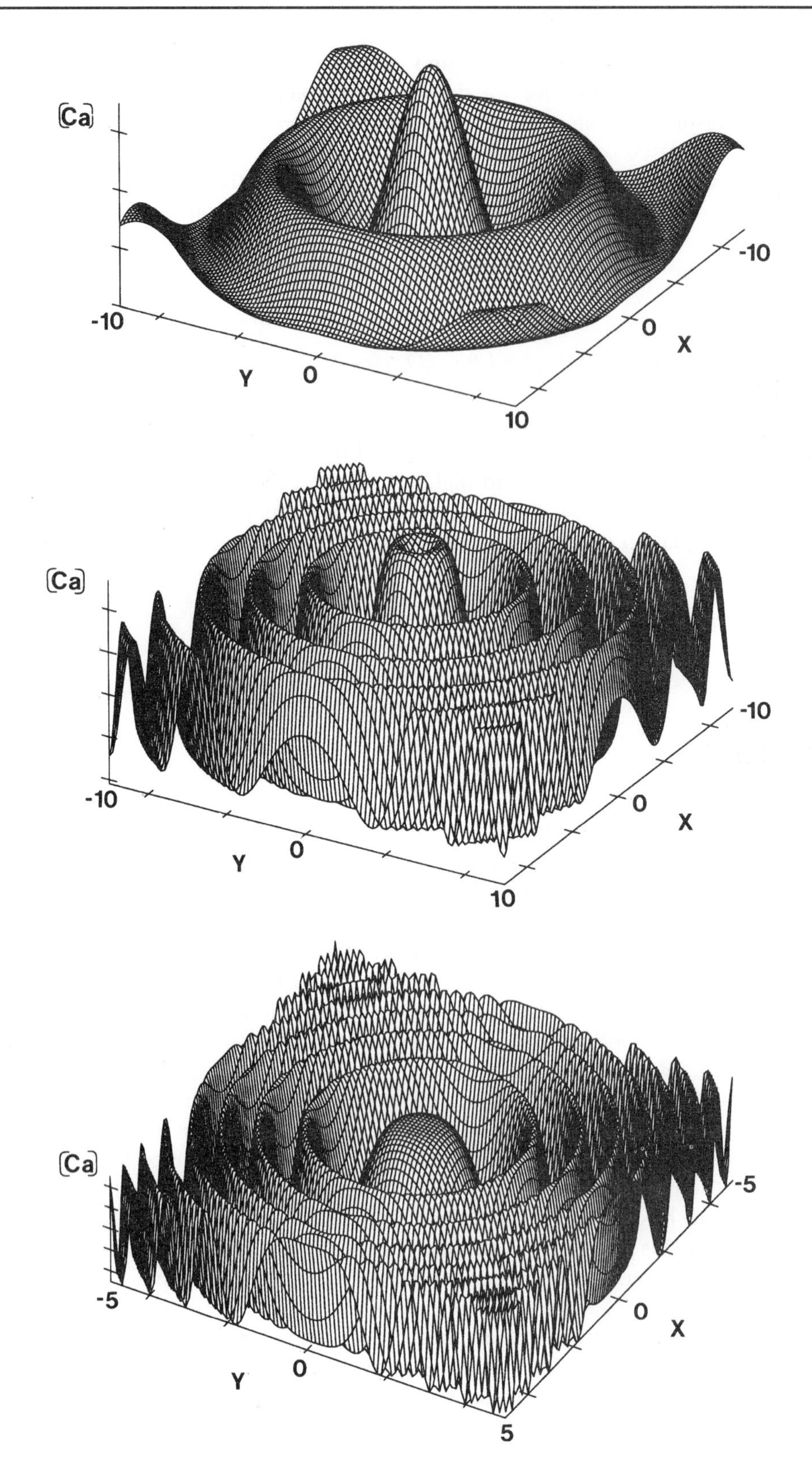

Figure 7 Plot of [Ca] as a function of position on the surface of a two-dimensional colony (all quantities have arbitrary units). Curves correspond to solutions of Equation (4) for $\theta = 0$ (upper curve); $\theta = 1$ (middle curve) and $\theta = 2$ (lower curve).

Goodwin and Trainor (1985) and Briere and Goodwin (1988) propose a mechanism for tip and whorl morphogenesis in the marine alga *Acetabularia mediterranea* with predictions which agree with some aspects of observation (see Goodwin, Ch. 2 this volume). A calcium regulated strain field in the cytogel is assumed with coupling of the cytogel with the cell wall via a non-linear dependence of wall elasticity on cytogel strain. The mechanism can produce spatially localized distortion of the cell wall consistent with observed tip generation, and reproduce some of the properties of growth dynamics. However, the equations were designed to be applied to regeneration of an excised tip in the organism, and therefore only apply to the primary wall. Indeed, it is assumed that the secondary wall provides an inert and rigid support for the expanding apex. In its present form, the theory does not carry over to the generation of branches from the subapical wall of hyphae or to the generation of rings. However, married to some hypothesized process for the coordination of localized vesicle fusion, such as the one presented here, it may provide a mechanism for calcium-controlled branching and extension in hyphae.

CONCLUSIONS

Observation of fungal growth and form have been reviewed and demonstrated to place severe constraints on the viability of proposed theories of morphogenesis. None of the existing models for generation of colony form suggests mechanisms which lead to agreement between observation and theory. There are still very large gaps in the understanding of morphogenesis as it relates to fungi, and indeed to other single and multicellular organisms.

The predictions of the reaction–diffusion theory proposed here could be verified experimentally through imaging of cytosolic free calcium which should show spatial patterning before any morphogenic event. However, the theory can only be complete when suitable reaction kinetics for calcium release and binding can be derived. This will only be possible if a suitable morphogenic partner responsible for reducing [Ca] in the cytoplasm can be identified.

ACKNOWLEDGEMENTS

The authors would like to thank Drs Alan Rayner, Bruce Marshall and Prof. B.D. Sleeman for interesting discussions and Ms Susan Verral for technical assistance. This research is supported by the Scottish Office Agriculture and Fisheries Department.

REFERENCES

ABRAMOWITZ, M. & STEGUN, I.A., 1965. *Handbook of Mathematical Functions.* New York, U.S.A.: Dover.

BARTNICKI-GARCIA, S., 1990. Role of vesicles in apical growth and a new mathematical model of hyphal morphogenesis. In I.B. Heath (ed.), *Tip Growth in Plant and Fungal Cells*, pp. 211–232. London, U.K.: Academic Press.

BOURRET, J.A., LINCOLN, R.G. & CARPENTER, B.H., 1969. Fungal endogenous rhythms expressed by spiral figures. *Science, 156:* 763–764.

BRIERE, C. & GOODWIN, B.C., 1988. Geometry and dynamics of tip morphogenesis in *Acetabularia*. *Journal of Theoretical Biology, 131:* 461–475.
CHEVAUGEON, J. & VAN HUONG, N., 1969. Internal determinism of hyphal growth rhythms. *Transactions of the British Mycological Society, 53:* 1–14.
CRAWFORD, J.W., RITZ, K. & YOUNG, I.M., 1993. Quantification of fungal morphology, gaseous transport and microbial dynamics in soil: an integrated framework utilising fractal geometry. *Geoderma, 56:* 157–172.
EDELSTEIN, L., 1982. The propagation of fungal colonies: a model for tissue growth. *Journal of Theoretical Biology, 98:* 679–701.
EDELSTEIN, L. & SEGEL, L.A., 1983. Growth and metabolism in mycelial fungi. *Journal of Theoretical Biology, 104:* 187–210.
FALCONER K., 1990. *Fractal Geometry*. New York, U.S.A.: J. Wiley.
FRY, S.C., 1989. Cellulases, hemicelluloses and auxin-stimulated growth: a possible relationship. *Physiologia Plantarum, 75:* 532–536.
GOODAY, G.W. & GOW, A.R., 1990. Enzymology of tip growth in fungi. In I.B. Heath (ed.), *Tip Growth in Plant and Fungal Cells*, pp. 31–58. London, U.K.: Academic Press.
GOODWIN, B.C. & TRAINOR, L.E.H., 1985. Tip and whorl morphogenesis in *Acetabularia* by calcium-regulated strain fields. *Journal of Theoretical Biology, 117:*, 79–106.
HAROLD, R.L. & HAROLD, F.M., 1986. Ionophores and cytochalasins modulate branching in *Achlya bisexualis*. *Journal of General Microbiology, 132:* 213–219.
HERTH, W., REISS, H.-D. & HARTMANN, E., 1990. Role of calcium ions in tip growth of pollen tubes and moss protonema cells. In I.B. Heath (ed.), *Tip Growth in Plant and Fungal Cells*, pp. 91–118. London, U.K.: Academic Press.
KOHNO, T. & SHIMMEN, T., 1988a. Accelerated sliding of pollen tube organelles along *Characeae* actin bundles regulated by Ca^{2+}. *Journal of Cell Biology, 106:* 1539–1543.
KOHNO, T. & SHIMMEN, T., 1988b. Mechanism of Ca^{2+} inhibition of cytoplasmic steaming in lily pollen tubes. *Journal of Cell Science, 91:* 501–509.
LOCKHART, J.A. 1965. Cell extension. In J. Bonner & J.E. Varner (eds), *Plant Biochemistry*, pp. 826–849. New York, U.S.A.: Academic Press.
LYSEK, G., 1978. Circadian rhythms. In J.E. Smith & D.R. Berry (eds), *The Filamentous Fungi*, Vol. 3, pp. 376–388. London, U.K.: Edward Arnold.
MANDELBROT, B.B., 1983. *The Fractal Geometry of Nature*. New York, U.S.A.: W.H. Freeman.
MORRISON, K.B. & RIGHELATO, R.C., 1974. The relationship between hyphal branching, specific growth rate and colony radial growth rate in *Penicillium chrysogenum*. *Journal of General Microbiology, 81:* 517–520.
MURRAY, J.D., 1982. Parameter space of the Turing instability in reaction diffusion mechanisms: a comparison of models. *Journal of Theoretical Biology, 98:* 143–163.
MURRAY, J.D., 1989. *Mathematical Biology*. Berlin, Germany: Springer-Verlag.
ORBACH, R., 1986. Dynamics of fractal networks. *Science, 231:* 814–819.
PALL, M.L., TREVILLYAN, J.M. & HINMAN, N., 1981. Deficient cyclic adenosine 3′,5′-monophosphate control in mutants of two genes of *Neurospora crassa*. *Molecular and Cellular Biology, 1:* 1–8.
PARKINSON, S.M. WAINWRIGHT, M. & KILLHAM, K., 1989. Observations on oligotrophic growth of fungi on silica gel. *Mycological Research, 93:* 529–534.
PASSIOURA, J.B. & FRY, S.C., 1992. Turgor and cell expansion: beyond the Lockhart equation. *Australian Journal of Plant Physiology, 19:* 565–576.
PEITGEN, H. & SAUPE, D., 1988. *The Science of Fractal Images*. Berlin, Germany: Springer-Verlag.
PROSSER, J.I., 1990. Comparison of tip growth in prokaryotic and eukaryotic filamentous microorganisms. In I.B. Heath (ed.), *Tip Growth in Plant and Fungal Cells*, pp. 233–259. London, U.K.: Academic Press.
PROSSER, J.I., 1992. Mathematical modeling of vegetative growth of filamentous fungi. In D.K. Arora, B. Rai, K.G. Mukerji & G.R. Knudsen (eds), *Handbook of Applied Mycology*, pp. 591–623. New York, U.S.A.: Marcel Dekker.
PROSSER, J.I. & TRINCI, A.P.J., 1979. A model for hyphal growth and branching. *Journal of General Microbiology, 111:* 153–164.

RASSMUSSEN, H. & BARRETT, P.Q., 1984. Calcium messenger system — an integrated view. *Physiological Reviews, 64:* 938–984.

RASSMUSSEN, H. & GOODMAN, D.B.P. 1977. Relationships between calcium and cyclic nucleotides in cell activation. *Physiological Reviews, 57:* 421–509.

RAYNER, A.D.M., 1991. The challenge of the individualistic mycelium. *Mycologia, 83:* 48–71.

REISS, H.-D., HERTH, W. & NOBILING, R., 1985. Development of membrane- and calcium-gradients during pollen germination of *Lilium longiflorum. Planta, 163:* 84–90.

REISSIG, J.L. & KINNEY, S.G., 1983. Calcium as a branching signal in *Neurospora crassa. Journal of Bacteriology, 154:* 1397–1402.

RITZ, K. & CRAWFORD, J., 1990. Quantification of the fractal nature of colonies of *Trichoderma viride. Mycological Research, 94:* 1138–1152.

SCHMID, J. & HAROLD, F.M., 1988. Dual roles for calcium ions in apical growth of *Neurospora crassa. Journal of General Microbiology, 134:* 2623–2631.

STEER, M.W., 1990. Role of actin in tip growth. In I.B. Heath (ed.), *Tip Growth in Plant and Fungal Cells,* pp. 119–145, London, U.K.: Academic Press.

STEER, M.W. & STEER, J.M., 1989. Pollen tube tip growth. *New Phytologist, 111:* 323–358.

TRIFARO, J.-M., BADER, M.-F. & DOUCET, J.-P., 1984. Chromaffin cell cytoskeleton: its possible role in secretion. *Canadian Journal of Biochemical Cell Biology, 63:* 661–679.

TRINCI, A.P.J., 1971. Influence of the width of the peripheral growth zone on the radial growth rate of fungal colonies on solid media. *Journal of General Microbiology, 67:* 325–344.

TRINCI, A.P.J., 1978. The duplication cycle and development in moulds. In J.E. Smith & D.R. Berry (eds), *The Filamentous Fungi,* vol. 3, pp. 132–163. London, U.K.: Edward Arnold.

TURING, A.M., 1952. The chemical basis of morphogenesis. *Philosophical Transactions of the Royal Society of London, B:* 37–72.

WATKINSON, S.C., 1975. Regulation of coremium morphogenesis in *Penicillium claviforme. Journal of General Microbiology, 87:* 292–300.

WESSELS, J.G.H., 1990. Role of cell wall architecture in fungal tip growth generation. In I.B. Heath (ed.), *Tip Growth in Plant and Fungal Cells,* pp.1–29. London, U.K.: Academic Press.

CHAPTER

19

Shape determination and polarity in fungal cells

GRAHAM W. GOODAY & NEIL A.R. GOW

CONTENTS

Abstract

The varieties of shapes of fungal cells are the results of different patterns of assembly of wall material. The key structural component of most fungal walls is chitin. It is synthesized by several chitin synthases, apparently each with a different role in the cell. The site of wall assembly is dictated by endogenous or exogenous signals and is characterized by its being the focus for the delivery of microvesicles containing enzymes such as chitin synthases and autolysins, wall matrix materials and plasma membrane. This delivery is coordinated by cytoskeletal structures, which may in turn

Shape and Form in Plants and Fungi
ISBN 0–12–371035–9

have been guided by interactions with membrane proteins, and controlled by ion channels in the plasma membrane. Patterns of wall deposition in hyphal tips, branches, septa, intercalary extension and dimorphic growth are discussed in the light of these processes. Responses of hyphae to applied electric fields are discussed and models are suggested to explain these.

INTRODUCTION

Inside their walls fungal cells differ little from other eukaryotic cells. It is the formation of their rigid walls that gives them their distinctive fungal character: as hyphae, growing as ramifying mycelia through substrata; or as yeast cells in colonies on damp nutritious surfaces or as individual cells in liquids. The many different forms found in fungi result from different patterns of polarized deposition of cell wall materials. The cytoplasmic turgor pressure exerts mechanical stresses on the wall to give the rigidity required to form structures such as sporangiophores, conidiophores and multicellular fruiting bodies. The wall, however, is not an inert extracellular framework; it is metabolically active to allow changes of form with which we are familiar: spore germination, hyphal branching, anastomoses of vegetative and mating hyphae, spore formation by hyphae and tissues. This account concerns aspects of the studies of the authors of basic mechanisms of polarization of cell wall formation in fungal cells, and discusses these in the light of current advances in fungal cell biology.

CHITIN AS A SHAPE-DETERMINING MOLECULE

Much of our work has concerned the polysaccharide chitin, the characteristic structural component of the walls of most fungal cells. Its chemical nature, as a β(1–4)-linked linear homopolymer of *N*-acetylglucosamine, gives it great strength, since the molecules are held rigid by hydrogen bonds along chains, and adjacent chains are cross-linked by more hydrogen bonds to give microfibrils (Gooday, 1979). Covalent cross-links are formed between chitin chains and other wall components, notably glucans (Wessels, 1990). These properties explain the widespread occurrence of chitin as a component of structures requiring mechanical strength, not only in fungi but also in protozoa and invertebrates. The shape-determining role of chitin in fungal cells is shown by chemical and/or enzymic removal of all other components, to leave chitinous ghosts retaining the shape of the original cell.

APICAL GROWTH

Growth of vegetative hyphae of fungi is strictly polarized. Most of the synthesis of structural wall polymers takes place in the apical dome, as shown for example by light microscopic autoradiography of incorporation of tritiated precursors (Bartnicki-Garcia & Lippman, 1969; Gooday, 1971). The wall at the apex is plastic, but becomes rigid as it progressively forms the tubular hyphal wall. This process of rigidification has been investigated in particular by Wessels and co-workers, who describe two major processes, progressive crystallization of chitin and glucan and the covalent cross-linking of these components (Wessels, 1986, 1990). Inhibition of chitin synthesis by antibiotics such as nikkomycins and polyoxins or of glucan synthesis by echinocandin or cilofungin leads to bursting of hyphal apices (Gooday, 1990; Zhu & Gooday, 1992). Cytology of hyphal tips has shown that they are characterized by accumulations of a variety of microvesicles and cytoskeletal structures (Girbardt, 1969; Grove & Bracker, 1970; Howard, 1981; Roberson & Fuller, 1988). The assemblage of microvesicles forms a structure visible in the light microscope, the Spitzenkörper. The microvesicles are formed in the cytoplasm in subapical regions of hyphae and transported to the advancing apices (Fig. 1). The major locomotive system probably involves actin microfilaments, perhaps also with some role for microtubules (Gow, 1989a; Heath, 1990; Hoch *et al.*, 1987; McKerracher & Heath, 1987).

Bartnicki-Garcia *et al.* (1989) describe a mathematical model demonstrating that the rate and organization of the flow of microvesicles can account for the observed kinetics of hyphal growth. The vesicles for each tip are formed in the peripheral growth zone, the length of hypha supporting the growth of one tip (Gooday & Trinci, 1980). Most of the microvesicles are probably formed in Golgi bodies. Some are chitosomes, rich in zymogenic chitin synthases (Kamada *et al.*, 1991a). The vesicles fuse to the plasma membrane in the apical dome, adding membrane-bound enzymes such as chitin synthases to the membrane, and releasing contents such as mannoproteins and lytic enzymes into the matrix of the wall. The wall at the apex is plastic, but does not burst despite the unavoidable mechanical stress from the cell's turgor pressure. There is increasing evidence that its integrity is maintained by a cytoskeletal internal dynamic scaffold on which the wall is built (cf. Jackson & Heath, 1990;

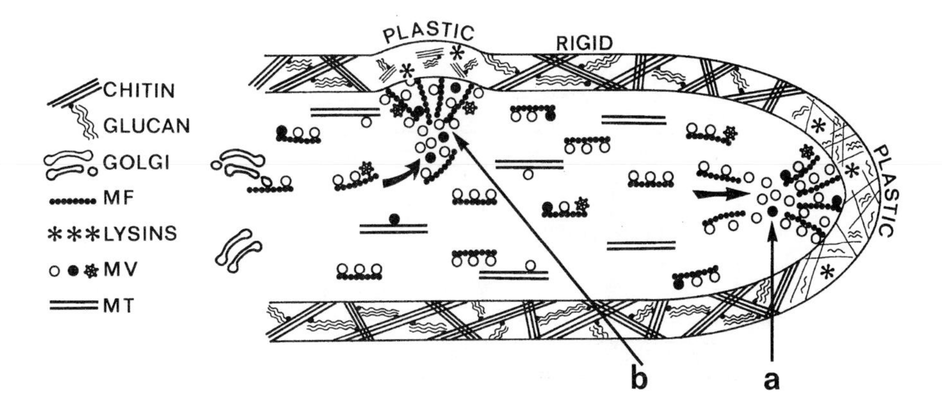

Figure 1 Model of some features of apical growth and branching of a hypha. Microvesicles of a variety of types (MV) are shown being produced by the Golgi bodies and being transported to two sites, the Spitzenkörper of the apex **(a)** and of the new branch **(b)**. The microvesicles are associated with microfilaments (MF) or microtubules (MT) which have each been implicated in vesicle transport. The plastic wall at the tip is shown with nascent fibrils of chitin (straight lines) and glucan (wavy lines). The wall becomes rigidified by their crystallization, and cross-linking by covalent bonds (black dots). At the branch site the wall has been made plastic again by the action of lytic enzymes (stars). It is likely that lytic enzymes are also secreted at the apex, but their role in apical wall growth is unclear.

Wessels, 1986, 1990). Fungi in the exponential phase of growth produce lytic enzymes, such as chitinases (Rast *et al.*, 1991), but their role in apical extension remains unclear (Gooday *et al.*, 1992; Rast *et al.*, 1991).

With sufficient nutrients, fungal mycelia grow exponentially by regular subapical branching, so as to exploit the available substrate. Little is known of the mechanism of choice of site of branching but this is clearly a major factor in the development of a colony. In *Candida albicans*, branches are formed from subapical cells that have accumulated sufficient protoplasm to initiate cytokinesis, and are sited at the proximal ends of the cells, just behind the septa (Gow *et al.*, 1986; Gow & Gooday, 1987a). This pattern of branching just behind septa is seen in many other fungi, and may be thought of as a continuation of the original polarity of the hypha. Although direct evidence is lacking, it is likely that the initial event in the siting of a branch is a signal to accumulate microvesicles, including those containing wall lytic enzymes such as chitinases (Fig. 1). When these are released into the wall, they will weaken it to produce a 'soft spot', which will bulge under the stress of the turgor pressure. With the formation of a new Spitzenkörper, the branch grows as a new tip. There is evidence for the direct involvement of cellulase in the formation of antheridial branches by male hyphae of *Achlya ambisexualis* in response to the sex hormone, antheridiol. Addition of antheridiol results in a large increase in cellulase activity and formation of 'soft spots' in the wall, the presence of which is indicated by sudden popping out of branches (Gow & Gooday, 1987b). Electron micrographs show accumulations of microvesicles, some probably containing cellulase, just under the plasma membrane at these sites (Mullins & Ellis, 1974).

The plasma membrane in the region of such a 'soft spot' will be stretched, and mechanical stress exerted on it by the turgor pressure could play a role in the localized activation of chitin synthases and other synthetic enzymes required for new tip wall synthesis (Fig. 2). There is preliminary evidence that such a mechanism may occur,

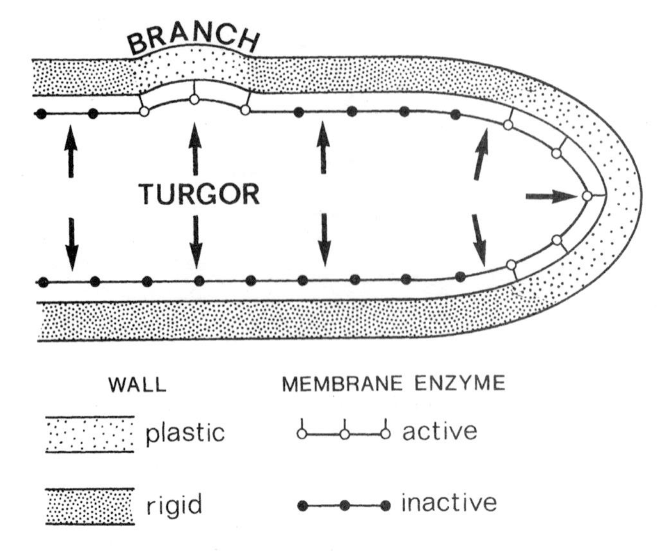

Figure 2 Model of activation of enzymes and/or ion channels by stress on the plasma membrane. Turgor pressure exerts an outward force on the wall, which yields at the apex and branch, stretching the plasma membrane and changing the state of some enzymes from inactive to active.

as enzyme preparations made from fungal protoplasts and cells subjected to hypo-osmotic stress show higher native chitin synthase activities than those made from unstressed cells (D. Schofield & G.W. Gooday, unpublished results). Other membrane proteins may behave in the same way, for example Ca^{2+}-stretch-activated channels (Garrill *et al.*, 1992; Zhou *et al.*, 1991).

SEPTA

Hyphae of most filamentous fungi have septa at regular or irregular intervals. They are formed by centripetal growth of wall material. Consequences of their formation are axial strengthening of the hyphal wall and partial or complete compartmentation of the hypha. Most septa have central pores large enough to allow flow of organelles including nuclei. In *C. albicans* the septum has a central 25 nm micropore, too small to allow passage of organelles (Gow *et al.*, 1980). Plasmodesmata have been reported in some complete septa such as those delimiting gametangia of Zygomycetes (Hawker *et al.*, 1966). As for plant cells, these would effectively seal the cell as an individual but allow transport of nutrients from neighbouring cells. In Ascomycetes and Deuteromycetes with septa with central pores, the septa can seal hyphae in which apical cells are disrupted, as the resultant mass outflows of cytoplasm are stemmed by occlusion of the septal pores by Woronin bodies, proteinaceous crystalline organelles characteristic of these fungi (Markham & Collinge, 1987).

Septa are typically rich in chitin. This is demonstrated by their dense staining with the fluorescent brightener, Calcofluor white, and with fluorescein-labelled wheatgerm agglutinin, which have high affinities for chitin, and by their dense labelling in autoradiographs of hyphae fed with tritiated *N*-acetyl glucosamine (Hunsley & Gooday, 1974). In the plane of the septal plate, the microfibrils are arranged in tangential fashion (Gow & Gooday, 1983; Hunsley & Gooday, 1974). The hyphal septum of *C. albicans* is composed of two chitin-rich septal plates, staining densely in transmission electron micrographs, separated by a thin translucent layer (Gooday & Gow, 1983). As this appearance is typical for fungal septa it is probable that many septa have this double structure. Septa of *Neurospora crassa* (Hunsley & Gooday, 1974; Mahadevan & Tatum, 1967) and *Schizophyllum commune* (van der Valk *et al.*, 1977) show peripheral 'septal rings', triangular in section, apparently rich in chitin but resistant to treatment with chitinase, perhaps because they are highly crystalline and highly cross-linked with glucans.

Septa of hyphae of Homobasidiomycetes typically have a toroidal central swelling, the dolipore, with a membrane-bound organelle, the parenthesome, on either side (Gull, 1978; Moore & McClear, 1962). This complex structure is not a barrier to protoplasmic streaming and the passage of small organelles (Bracker & Butler, 1964) but is a barrier to nuclear migration. The septa of monokaryotic hyphae of many Basidiomycetes, such as *S. commune*, are digested by autolysins during dikaryotization. Studies with isolated wall preparations have shown that this process requires the activities of chitinases and glucanases (Wessels & Marchant, 1974). The growth of the resulting dikaryon involves the formation of clamp connections, in which backward-facing lateral branches bend to fuse with the lateral hyphal walls. Following nuclear division and migration, two septa are formed. Working with *Trametes (Coriolus) versicolor*, Girbardt (1979) described the formation of a belt of 4–7 nm microfilaments, which he suggested were actin, next to the plasma membrane. This belt predicted the site of septal formation. Girbardt suggested that, at an early stage of mitosis, the nucleus signals its position to the plasma membrane. The microfilamentous belt is formed at that point and is the irreversible site of the ensuing septum formation. An invagination of the plasma membrane develops and the septum is formed. The microfilamentous belt becomes positioned at either side of the ingrowing plasma membrane. In *Neozygites* sp., a Zygomycetous fungus which grows as a fission yeast, Butt & Heath (1988) describe the formation of a transient belt of actin microfibrils that forms following mitosis at the site where the septum will later be made. This actin belt is absent during septum formation, but there are actin plaques associated with the developing septum.

During budding of *Saccharomyces cerevisiae*, a ring of chitin is made at the base of the emerging bud. When budding is complete, this ring grows inwards to form the chitin-rich primary septum on the mother cell wall (Cabib, 1987; Cabib *et al.*, 1982, 1992). Release of the daughter cell involves the lytic action of chitinase. Cells of *S. cerevisiae* in which the chitinase gene has been disrupted accumulate as clumps, the daughter cells being unable to separate (Kuranda & Robbins, 1991). Such clumps also accumulate when *S. cerevisiae* or yeast cells of *C. albicans* are grown in the presence of allosamidin or demethylallosamidin, antibiotics that specifically inhibit chitinase (Gooday *et al.*, 1992; Sakuda *et al.*, 1990). Mitchell & Soll (1979) compare septation during budding and hyphal growth in *C. albicans*, and show that they are distinct processes. During budding, the septum is formed at the site of budding, whereas in the developing hyphae it is about 2 μm along the germ tube. In cells of *S. cerevisiae* and *C. albicans* a ring of membrane-associated 10 nm microfilaments forms at the site of septum formation (Byers & Goetsch, 1976; Kim *et al.*, 1991; Soll & Mitchell, 1983). Results of experiments with cell cycle mutants of *S. cerevisiae* led Kim *et al.* (1991) to suggest that these microfilaments are necessary for the normal positioning of the chitin ring, but not for its formation. They suggest that these 10 nm microfilaments are composed of several proteins and are novel components of the eukaryotic cytoskeleton, quite distinct from actin. It may be that they are also components of the microfilamentous belt predicting the site of septation in hyphae of *T. versicolor*, which Girbardt (1979) ascribed to actin.

INTERCALARY GROWTH OF FUNGAL HYPHAE

Some hyphae elongate by intercalary growth. A notable example is that of stipe cells of basidiocarps of *Coprinus cinereus*, in which chitin is synthesized throughout cell elongation in a uniform intercalary fashion (Gooday, 1979, 1982). In contrast to the isotropic arrangement of chitin microfibrils of most vegetative hyphal walls, those of these stipe cells are predominantly transverse in arrangement, being in shallow helices, which may be right- or left-handed (Gooday, 1979; Kamada *et al.*, 1991b). The walls of these long vertical cylindrical cells are good examples of stress-bearing structures. The turgor pressure of each cell is contained by the stress-bearing fabric of its wall, resulting in the multicellular structure having great strength. The fruitbodies of *Coprinus* species are well-known for their ability to push up rapidly through obstacles, such as asphalt paths. In a cylindrical cell, the circumferential surface stress is twice that of the axial stress, Pr as opposed to $Pr/2$ (P, turgor pressure; r, radius) (Koch, 1988). Kamada *et al.* (1991b) suggest a model for the uniform insertion of

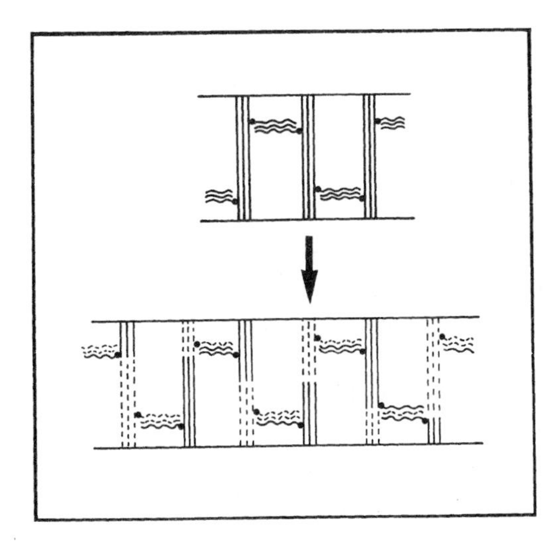

Figure 3 Model of intercalary elongation of stipe cell wall of *Coprinus cinereus*. Shallow helices of chitin microfibrils (straight lines) are shown cross-linked by glucan chains (wavy lines) via covalent bonds (black dots). The wall elongates by uniform intussception of new chitin microbrils (dashed lines) between existing ones and formation of new glucan chains.

new chitin microfibrils throughout the length of the cells. They suggest that the structural integrity of the wall is provided by cross-links between chitin chains and glucan chains (cf. Wessels, 1990) and that controlled breakage of these cross-links, insertion of new microfibrils and formation of new cross-links would lead to a uniform elongation of the cylindrical wall, with maintenance of both wall diameter and shallow helical pitch of the microfibrils (Fig. 3).

YEAST–HYPHAL DIMORPHISM

All fungi are pleomorphic, growing in different forms in different environmental conditions. The term dimorphic, however, is commonly used for fungi that can grow as yeast cells or hyphae. This has special significance for medical mycologists, as most of the major human pathogens show this phenomenon. Working with *C. albicans*, the authors have shown that the hyphal form, implicated in tissue invasion, is the migratory form of the fungus, exhibited when it is under some forms of stress. Outgrowth of a hypha from a yeast cell involves relatively little synthesis of new cytoplasm, but instead is accompanied by massive vacuolation of the mother yeast cell and migration of a slug of nucleated cytoplasm in the apex of a hypha extending at a steady rate (Gow & Gooday, 1982a,b). As the tip extends, it leaves behind highly vacuolated hyphal compartments, each with a nucleus, some of which will branch after accumulating sufficient reserves to allow cytokinesis (Gow *et al.*, 1986; Gow & Gooday, 1987a). The tip shows exploratory behaviour when growing on surfaces. It grows in a helical fashion, presumably reflecting a radial asymmetry in some aspect of wall assembly, and it senses contact with the surface, so that it enters pores that it encounters (Sherwood *et al.*, 1992). Contact guidance is a well-established phenomenon for hyphae of some plant pathogens suych as *Uromyces* species (Read *et al.*,

1992). These properties are probably common to hyphal tips of many fungi and may be thought of as behaviour to exploit available substrates to the fullest.

CHITIN METABOLISM

Chitin is synthesized by intrinsic plasma membrane proteins, the chitin synthases. These enzymes catalyse glycosidic bond formation from the nucleotide sugar substrate, uridine diphospho-*N*-acetylglucosamine:

$$2\text{UDP-GlcNAc} + (\text{GlcNAc})_n \rightarrow (\text{GlcNAc})_{n+2} + \text{UDP}$$

A phospholipid environment and divalent cations such as magnesium, manganese or cobalt are required for activity. The UDP product is inhibitory to activity (Gooday, 1979).

Chitin synthesis in *S. cerevisiae* involves at least three chitin synthases, I, II and III (Table 1). These three enzymes have different properties. Chitin synthases I and II are zymogenic, i.e. produced as proenzymes which are activated by protease activity. Chitin synthase III does not require proteolysis for activity. Chitin synthase I is at least an order of magnitude more sensitive to inhibition by 0.5 M NaCl, polyoxin D and nikkomycins X and Z than chitin synthase II (Cabib, 1991; Sburlati & Cabib, 1986). Chitin synthase I has a pH optimum at about 6.5, chitin synthase II at 7.5–8.0. Chitin synthase I is inhibited by cobalt ions, while chitin synthase II is stimulated (Sburlati & Cabib, 1986). Chitin synthases I and II have similar K_m values for UDP-GlcNAc, of about 0.8–0.9 mM.

Chitin synthase I zymogen is the most abundant of the three enzymes in the yeast cell. Disruption of its gene *CHS1*, does not affect chitin synthesis during normal growth. The only apparent deficiency of these mutants is a tendency to leak cell contents from the septum following release of the daughter cells when growing in media of low pH. As described earlier, daughter cell release is the result of chitinase activity. Cabib *et al.* (1992) suggest that the leakage of cell contents is the result of overaction of chitinase, favoured by acidic pH, which in wild-type cells would be corrected by the synthesis of chitin via chitin synthase I; i.e. this is a 'repair enzyme'.

Chitin synthase II is responsible for chitin synthesis in the primary septa, as mutants of *CHS2* do not form them, and show clumpy growth with aberrant cell shapes and

Table 1 Chitin synthases of *Saccharomyces cerevisiae*

Chitin synthase	Gene	% Chitin loss in mutant	Zymogenicity	Proposed role
I	*CHS1*	< 10	Yes	Repair
II	*CHS2*	< 10	Yes	Primary septum
III	*CSD2**	> 90	No	Growth, mating, sporulation

*Requires other gene products for activity; allelic with *CAL1*, *CAL4*, *DIT101*. Table based on compilation by Bulawa (1992).

sizes (Silverman *et al.*, 1988). Double mutants of *CHS1* and *CHS2*, however, have no detectable chitin deficiency *in vivo* (Bulawa and Osmond, 1990).

The structural genes for chitin synthases I and II are well-characterized, but the situation with chitin synthase III is more complex, as more than one gene appears to be involved in its activity (Bulawa, 1992; Shaw *et al.*, 1991; Valdivieso *et al.*, 1991). Several mutants are deficient in chitin synthase III. These include *Cal1* and *Cal4*, resistant to Calcofluor, the fluorescent brightener which binds to nascent chitin, preventing its orderly deposition (Shaw *et al.*, 1991; Valdivieso *et al.*, 1991) and *CSD2* (chitin synthesis deficient) which is allelic to *CAL1* and *CAL4* (Bulawa, 1992). Chitosan, formed by deacetylation of chitin, is a major component of spore walls of *S. cerevisiae*, occurring together with dityrosine. *CSD2* mutants produce no chitosan or dityrosine layers in their spore walls. The mutant *dit101*, identified by lack of dityrosinein in its spore walls, has proved to be allelic to *CSD2* (Bulawa, 1992). Mutants of *CAL1* produce thin chitinous septa but no chitinous ring and no chitin in their cell walls (Shaw *et al.*, 1991). Assessment of mutants deficient in chitin synthase III activity shows that this enzyme is responsible for synthesis of chitin during growth, maturing and sporulation (Bulawa, 1992). A mating yeast cell, a shmoo, has an elevated chitin content in the walls of its conjugation tube. Expression of *CSD2* is required for synthesis of most of this chitin, but there appears to be no increase in activity of chitin synthase III (Orlean, 1987). The observed increase in chitin synthesis thus appears to be the result of increased flux through the chitin synthesis pathway, with no change in enzyme activity.

Mutants in any one of these three genes are not lethal; the cell has sufficient backup activity from the other enzymes to be able to grow. A triple mutant, however, was inviable. It could be rescued by a plasmid containing the *CHS2* gene under the control of a *GAL1* promoter. Transfer of the mutant from galactose to glucose as carbon source resulted in cell division arrest followed by cell death (Shaw *et al.*, 1991).

Two genes have been identified from *C. albicans* that encode chitin synthases, designated *CHS1* and *CHS2* (Au-Young & Robbins, 1990; Chen-Wu *et al.*, 1992). Sequence comparisons suggest that the *CHS1* gene of *C. albicans* is more similar to the *CHS2* of *S. cerevisiae* than is the *Candida CHS1* (Bowen *et al.*, 1992). Both enzymes are zymogenic, *CHS1* having a pH optimum of 6.5. Northern analyses of mRNA expression showed a marked difference between the two genes, with very much higher levels of *CHS2* message during hyphal outgrowth than yeast growth. The *CHS2* gene of *C. albicans* has been disrupted but as for similar experiments with *S. cerevisiae*, its loss led to a 50% reduction of the normal chitin content of the hyphal wall and no effect on the yeast chitin content, but had little effect on growth or development (N.A.R. Gow, unpublished observation; Gow *et al.*, 1993).

A chitin synthase gene, designated *CHS1*, has been identified in *N. crassa* that has homologies with *CHS1* and *CHS2* of *S. cerevisiae* (Yarden & Yanofsky, 1991). Disruption of this gene led to a sparsely growing mutant strain with much lower chitin synthase activity than wild-type, with abnormal hyphal swellings, but with septa that stained with Calcofluor as strongly as those of wild-type hyphae. Mutant colonies were much more sensitive to inhibition by nikkomycin Z than wild-type colonies.

Using polymerase chain reactions (PCR) with oligomeric DNA primers designed from conserved sequences in *S. cerevisiae CHS1* and *CHS2* and *C. albicans CHS1* genes, 32 genes for chitin synthases have been identified from a further 13 species of fungi (Bowen *et al.*, 1992). Their sequences fall into three classes, but with *S. cerevisiae CHS1* being separate.

Borgia (1992) characterized chitin-deficient, temperature-sensitive mutants of *Aspergillus nidulans, orlA* (osmotically remedial lysis) and *tsE*, and showed that these two genes are necessary for production of L-glutamine: fructose-6-phosphate amidotransferase, the key enzyme on the pathway to UDP-*N*-acetylglucosamine, the precursor of chitin and glycoproteins. Growth of the mutants was remedied by osmotic stabilizers and by *N*-acetylglucosamine.

IMPOSING POLARITY WITH APPLIED ELECTRIC FIELDS

For a range of fungi, applied electric fields result in oriented growth (Gow, 1987). Examples include outgrowth of hyphae during regeneration of protoplasts of *S. commune* (De Vries & Wessels, 1982); hyphae of *N. crassa, A. nidulans, Mucor mucedo, Trichoderma harzianum* and *Achlya bisexualis* (Cho *et al.*, 1991; McGillivray & Gow, 1986); germinating spores of *Phycomyces blakesleeanus* (Van Laere, 1988); rhizoids and hyphae of *Allomyces macrogynus* (De Silva *et al.*, 1992; Youatt *et al.*, 1988); buds, germ tubes and hyphae of *C. albicans* (Crombie *et al.*, 1990; Gow *et al.*, 1991). Applied electric fields, however, did not affect the phenomenon of yeast–hyphal dimorphism in *C. albicans* (Crombie *et al.*, 1990). In contrast electrical fields promoted a yeast to hyphal transition in *Mycotypha africana* (Wittekindt *et al.*, 1989). Depending on fungus and conditions, initial growth can be anodotropic or cathodotropic. In *N. crassa*, the direction of growth of hyphae was dependent on pH; at pH 4 they were cathodotropic, at pH 8 they were anodotropic; in *C. cinereus* and *A. nidulans*, however, hyphae became increasingly cathodotropic at high pH values (Lever, Robertson, Buchan, Gooday & Gow, unpublished observations). In *A. bisexualis*, hyphae growing at pH 7.0 grew towards the anode or the cathode in different growth media (Cho *et al.*, 1991). In *N. crassa* and some other fungi prolonged growth of germ tubes in an electric field resulted in hyphae reorientating themselves from parallel to perpendicular growth in the field (McGillivray & Gow, 1986). Hyphae of *A. macrogynus* showed a similar response (De Silva *et al.*, 1992). Hyphae of *C. albicans* did not show this reorientation (Crombie *et al.*, 1990). Experiments with *C. albicans* showed that cells had a memory of a previous field. When yeast cells were exposed to electric fields and then these were switched off, germ tubes formed later, in the absence of electrical fields, showed polarized growth (Crombie *et al.*, 1990). For example, in conditions when germ tubes started to emerge after 90 min of incubation, cells exposed to a field of 28 mV per cell diameter for only the first 30 min were polarized when germ tubes emerged 60 min later.

These galvanotropic responses are apparently unconnected with the endogenous electric currents detected in fungal hyphae, which are epiphenomena reflecting net anisotropic exchanges of ions with the medium along the hyphae (Cho *et al.*, 1991; De Silva *et al.*, 1992; Gow, 1989b).

In *C. albicans*, there is evidence that galvanotropic responses require calcium ions: they are inhibited by EGTA, a chelating agent, and by cobalt and lanthanum ions, which block calcium ion channels (Buchan, Gooday & Gow, unpublished observations). Similarly, EGTA suppressed galvanotropism of *M. africana* (Wittekindt *et al.*, 1989). In *A. macrogynus*, in contrast, cobalt and lanthanum ions did not inhibit

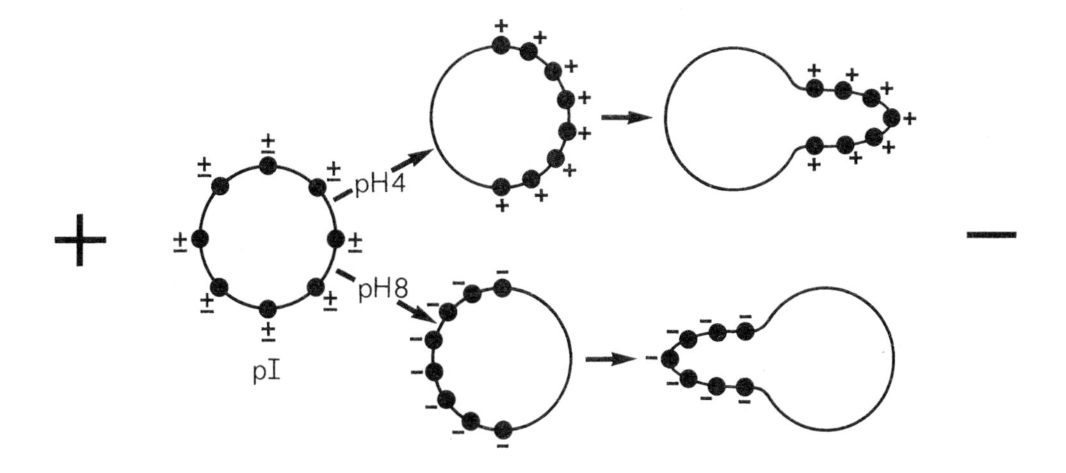

Figure 4 Model of electrophoresis of membrane proteins during galvanotropism. At their pI, the proteins will not move; at pH values below their pI they will move towards the cathode; at values above, towards the anode.

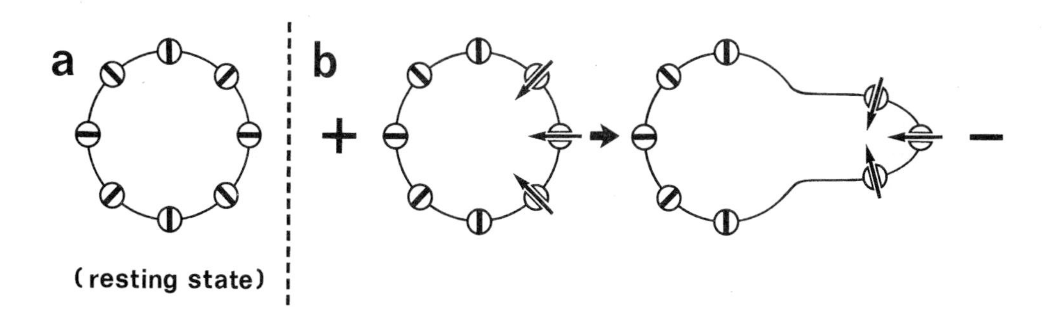

Figure 5 Model of an effect of an applied electric field on voltage-sensitive ion channels during galvanotropism. In the resting state, i.e. with no applied electric field, the channels are closed **(a)**. In an applied field, when the membrane is depolarized at the cathode-facing end of the cell, voltage-sensitive channels will open **(b)**.

galvanotropism, and even low concentrations of EGTA rendered the cells sensitive to inhibition by the applied electric fields (De Silva *et al.*, 1992).

The variety of galvanotropic responses shown by fungal cells may be the result of interacting processes involving a variety of membrane proteins. One of these is electrophoresis and/or electro-osmosis of intrinsic membrane proteins along the membranes by their interaction with the surface charge of the membrane. Electrophoresis, for example, is consistent with the effect of pH on galvanotropism of *N. crassa*: at low pH there would be a net positive charge on ionic groups exposed at the surface of the membrane; at high pH there would be a net negative charge. These proteins would move laterally in the membrane, towards cathode or anode, respectively (Fig. 4). Contenders for being among such mobile proteins are stretch-activated and voltage-gated ion channels and microfilament-associated membrane proteins. For example, stretch-activated ion channels, including ones activated by calcium, have been described in the membranes of protoplasts derived from hyphal tips of *Uromyces appendiculatus* (Zhou *et al.*, 1991) and *Saprolegnia ferax* (Garrill *et al.*, 1992). Heath (1990) discusses the possible role of ion channels and membrane-bound actin-binding proteins in polarized growth. The asymmetric distribution of such proteins could play a key role in maintaining polarity of a growing hypha, and applied electric fields could impose such an asymmetric distribution.

Another process which may be involved is non-uniform perturbation of the membrane potential. The cathodotropic responses of hyphae of *C. cinereus* and *A. nidulans* could reflect their possession of voltage-sensitive ion channels, such as Ca^{2+} channels that are opened by membrane depolarization and closed by hyperpolarization (Fig. 5). Although evidence for this in fungi is indirect, electric fields have been demonstrated recently to stimulate tip growth and cause membrane depolarization and calcium uptake at the cathodal ends of nerve cells (Bedlack *et al.*, 1992; Davenport & Kater, 1992). The more alkaline the medium, the more prone the channels would be to membrane depolarization. As suggested by McGillivray & Gow (1986), the reorientation of hyphae from being parallel to being perpendicular to the field may be a response to an effect on membrane potential, which would be increasingly depolarized at the cathodic ends and hyperpolarized at the anodic ends of hyphae extending parallel to the electric field. Crombie *et al.* (1990) suggest that the lack of this reorientation by hyphae of *C. albicans* could reflect their cellular organization, being composed of compartments separated from each other by septa. As discussed earlier, the septum of *C. albicans* has a membrane-lined micropore, which may electrically insulate each compartment. This is in contrast with the structures of hyphae showing reorientation, which are either coenocytic (e.g. *A. macrogynus*) or have irregularly spaced septa with relatively large pores (e.g. *N. crassa*).

CONCLUSIONS

Some aspects of the mechanisms of generating, maintaining and changing polarity in a fungal cell have been discusssed. The polarity of a fungal cell is transformed into shape by the moulding of its cell wall. There is increasing evidence that this involves membrane-associated cytoskeletal assemblies guiding the wall-synthesizing activities. This organization of cytoskeleton in turn would be regulated by signals, which may be internal, as in tip growth, septum formation and stipe elongation; or external, as in some branching and in vegetative and sexual anastomoses. Strong contenders for involvement in primary signalling are plasma membrane ion channels, particularly for Ca^{2+}. These areas are under active investigation with a variety of fungi in several laboratories and exciting results can be expected in the next few years.

ACKNOWLEDGEMENTS

The authors thank Ann Hoyle for the artwork, and their students and collaborators for contributions to the work described here.

REFERENCES

AU-YOUNG, J. & ROBBINS, P.W., 1990. Isolation of a chitin synthase gene *(CHS1)* from *Candida albicans*, by expression in *Saccharomyces cerevisiae. Molecular Microbiology, 4:* 197–207.

BARTNICKI-GARCIA, S. & LIPPMAN, E., 1969. Fungal morphogenesis: cell wall construction in *Mucor rouxii, Science, 165:* 302–303.

BARTNICKI-GARCIA, S., HERGERT, F. & GIERZ, G., 1989. Computer simulation of fungal morphogenesis and the mathematical basis for hyphal (tip) growth. *Protoplasma, 153:* 46–57.

BEDLACK, R.S., WEI, M-D, & LOEW, L.M. 1992. Localized membrane depolarizations and localized calcium influx during electrical field-guided neurite growth. *Neuron, 9:* 393–403.

BORGIA, P., 1992. Roles of the *orlA, tsE,* and *bimG* genes of *Aspergillus nidulans* in chitin synthesis. *Journal of Bacteriology, 174:* 384–389.

BOWEN, A.R., CHEN-WU, J.L., MOMANY, M., SZANISZLO, P.J. & ROBBINS, P.W., 1992. Classification of fungal chitin synthases. *Proceedings of the National Academy of Sciences, USA, 89:* 519–523.

BRACKER, C.E. & BUTLER, E.E., 1964. Function of septal pore apparatus of *Rhizoctonia solani* during protoplasmic streaming. *Journal of Cell Biology, 21:* 152–157.

BULAWA, C.E., 1992. *CSD2, CSD3,* and *CSD4,* genes required for chitin synthesis in *Saccharomyces cerevisiae:* the *CSD2* gene product is related to chitin synthases and to developmentally regulated proteins in *Rhizobium* species and *Xenopus laevis. Molecular and Cellular Biology, 12:* 1764–1776.

BULAWA, C.E. & OSMOND, B.C., 1990. Chitin synthase I and chitin synthase II are not required for chitin synthesis in vivo in *Saccharomyces cerevisiae. Proceedings of the National Academy of Sciences, USA, 87:* 7422–7428.

BUTT, T.M. & HEATH, I.B., 1988. The changing distribution of actin and nuclear behaviour during the cell cycle of the mite-pathogenic fungus *Neozygites* sp. *European Journal of Cell Biology, 46:* 499–505.

BYERS, B. & GOETSCH, L., 1976. A highly ordered ring of membrane-associated filaments in budding yeast. *Journal of Cell Biology, 69:* 712–721.

CABIB, E., 1987. The synthesis and degradation of chitin. *Advances in Enzymology, 59:* 59–101.

CABIB, E., 1991. Differential inhibition of chitin synthetases 1 and 2 from *Saccharomyces cerevisiae* by polyoxin D and nikkomycins. *Antimicrobial Agents and Chemotherapy, 35:* 170–173.

CABIB, E., ROBERTS, R. & BOWERS, B. 1982. Synthesis of the yeast cell wall and its regulation. *Annual Review of Biochemistry, 51:* 763–793.

CABIB, E., SILVERMAN, S.J. & SHAW, J.A., 1992. Chitinase and chitin synthase I: counter balancing activities in cell separation of *Saccharomyces cerevisiae. Journal of General Microbiology, 138:* 97–102.

CHEN-WU, J.L., ZWICKER, J., BOWEN, A.R. & ROBBINS, P.W., 1992. Expression of chitin synthase genes during yeast and hyphal growth phases of *Candida albicans. Molecular Microbiology, 6:* 497–502.

CHO, C-W, HAROLD, F.M. & SCHREURS, W.J.A., 1991. Electrical and ionic dimensions of apical growth in *Achlya* hyphae. *Experimental Mycology, 15:* 34–43.

CROMBIE, T., GOW, N.A.R. & GOODAY, G.W., 1990. Influence of applied electric fields on yeast and hyphal growth of *Candida albicans. Journal of General Microbiology, 136:* 311–317.

DAVENPORT, R. & KATER, S.B., 1992. Local increases in intracellular calcium elicit local filopodial responses in helisoma neuronal growth cones. *Neuron, 9:* 405–416.

DE SILVA, L.R., YOUATT, J., GOODAY, G.W. & GOW, N.A.R., 1992. Inwardly directed ionic currents of *Allomyces macrogynus* and other water moulds indicate sites of proton-driven nutrient transport but are incidental to tip growth. *Mycological Research, 96:* 925–931.

DE VRIES, S.C. & WESSELS, J.G.H., 1982. Polarized outgrowth of hyphae by constant electrical fields during reversion of *Schizophyllum commune* protoplasts. *Experimental Mycology, 6:* 95–98.

GARRILL, A., LEW, R.R. & HEATH, I.B., 1992. Stretch-activated Ca^{2+} and Ca^{2+}-activated K^+ channels in the hyphal tip plasma membrane of the Oomycete *Saprolegnia ferax. Journal of Cell Science, 101:* 721–730.

GIRBARDT, M., 1969. Die Ultrastruktur der Apikalregion von Pilzhyphen. *Protoplasma, 67:* 413–441.

GIRBARDT, M., 1979. A microfilamentous septal belt (FSB) during induction of cytokinesis in *Trametes versicolor* (L. ex Fr.). *Experimental Mycology, 3:* 215–228.

GOODAY, G.W., 1971. An autoradiographic study of hyphal growth of some fungi. *Journal of General Microbiology, 67:* 125–133.

GOODAY. G.W. 1979. Chitin synthesis and differentiation in *Coprinus cinereus.* In J.H. Burnett & A.P.J. Trinci (eds), *Fungal Walls and Hyphal Growth*, pp. 203–223. Cambridge, U.K.: Cambridge University Press.

GOODAY, G.W., 1982. Metabolic control of fruitbody morphogenesis in *Coprinus cinereus.* In K. Wells & E.K. Wells (eds), *Basidium and Basidiocarp, Evolution, Cytology, Function and Development.* pp. 157–173. Berlin, Germany: Springer-Verlag.

GOODAY, G.W., 1990. Inhibition of chitin metabolism. In P.J. Kuhn, A.P.J. Trinci, M.J. Jung, M.W. Goosey & L.G. Copping (eds), *The Biochemistry of Cell Walls and Membranes in Fungi*, pp. 61–79. Berlin, Germany: Springer-Verlag.

GOODAY, G.W. & GOW, N.A.R, 1983. A model of the hyphal septum of *Candida albicans. Experimental Mycology, 7:* 370–373.

GOODAY, G.W. & TRINCI, A.P.J., 1980. Wall structure and biosynthesis in fungi. In G.W. Gooday, D. Lloyd & A.P.J. Trinci (eds), *The Eukaryotic Microbial Cell, Society for General Microbiology Symposium,* vol. 30, pp. 207–251. Cambridge, U.K.: Cambridge University Press.

GOODAY, G.W., ZHU, W-Y. & O'DONNELL, R.W., 1992. What are the roles of chitinases in the growing fungus? *FEMS Microbiology Letters, 100:* 387–392.

GOW, N.A.R., 1987. Polarity and branching in fungi induced by electric fields. In R.K. Poole & A.P.J. Trinci (eds), *Spatial Organization in Eukaryotic Microbes.* Oxford, U.K.: IRL Press.

GOW, N.A.R., 1989a. Control of extension of the hyphal apex. *Current Topics in Medical Mycology. 3:* 109–152.

GOW, N.A.R., 1989b. Circulating ionic currents in microorganisms. *Advances in Microbial Physiology, 30:* 89–123.

GOW, N.A.R. & GOODAY, G.W., 1982a. Growth kinetics and morphology of colonies of the filamentous form of *Candida albicans. Journal of General Microbiology, 128:* 2187–2194.

GOW, N.A.R. & GOODAY, G.W., 1982b. Vacuolation, branch production and linear growth of germ tubes of *Candida albicans. Journal of General Microbiology, 128:* 2195–2198.

GOW, N.A.R. & GOODAY, G.W., 1983. Ultrastructure of chitin in hyphae of *Candida albicans* and other dimorphic and mycelial fungi. *Protoplasma, 115:* 52–58.

GOW, N.A.R. & GOODAY, G.W., 1987a. Cytological aspects of dimorphism in *Candida albicans. CRC Critical Reviews in Microbiology, 15:* 73–78.

GOW, N.A.R. & GOODAY, G.W.,, 1987b. Effects of antheridiol on growth, branching and electric currents of hyphae of *Achyla ambisexualis. Journal of General Microbiology, 133:* 3531–3535.

GOW, N.A.R., GOODAY, G.W., NEWSAM, R.J. & GULL, K., 1980. Ultrastructure of the septum in *Candida albicans. Current Microbiology, 4:* 357–359.

GOW, N.A.R., HENDERSON, G. & GOODAY, G.W., 1986. Cytological interrelationships between the cell cycle and duplication cycle of *Candida albicans. Microbios, 47:* 97–105.

GOW, N.A.R., CROMBIE, T. & GOODAY, G.W., 1991. Polarised morphogenesis of *Candida albicans*: cytology, induction and control. In E. Tumbay, M.P.R. Seeliger & O.Ang (eds), *Candida and Candidamycosis,* pp. 13–19. New York, U.S.A.: Plenum Press.

GOW, N.A.R., SWOBODA, R., BERTRAM, G., GOODAY, G.W. & BROWN, A.P.J., 1993. Key genes in the regulation of dimorphism of *Candida albicans.* In H. Vanden Bossche, F.C. Odds & D. Kerridge (eds), *Fungal Dimorphism*, New York, U.S.A.: Plenum. pp. 61–71.

GROVES, S.M. & BRACKER, C.E., 1970. Protoplasmic organization of hyphal tips among fungi: vesicles and Spitzenkorper. *Journal of Bacteriology, 104:* 989–1009.

GULL, K., 1978. Form and function of septa in filamentous fungi. In J.E.. Smith & D.R. Berry (eds), *The Filamentous Fungi,* vol. 3, pp. 78–93. London, U.K.: Arnold.

HAWKER, L.E., GOODAY, M.A. & BRACKER, C.E., 1966. Plasmodesmata in fungal cell walls. *Nature, 212:* 635.

HEATH, I.B., 1990. The roles of actin in tip growth in fungi. *International Review of Cytology, 123:* 95–127.

HOCH, H.C., TUCKER, B.E. & STAPLES, R.C. 1987. An intact microtubule system is necessary for mediation of the signal cell differentiation in Uromyces. *European Journal of Cell Biology, 45:* 209–218.

HOWARD, R.J. 1981. Ultrastructural analysis of hyphal tip cell growth in fungi: Spitzenkorper, cytoskeleton and endomembranes after freeze-substitution. *Journal of Cell Science, 48:* 89–103.

HUNSLEY, D. & GOODAY, G.W., 1974. The structure and development of septa in *Neurospora crassa*. *Protoplasma, 82:* 125–146.

JACKSON, S.L. & HEATH, I.B., 1990. Evidence that actin reinforces the extensible hyphal apex of the Oomycete *Saprolegnia ferax*. *Protoplasma, 157:* 144–153.

KAMADA, T., BRACKER, C.E. & BARTNICKI-GARCIA, S., 1991a. Chitosomes and chitin synthetase in the asexual life cycle of *Mucor rouxii*. *Journal of General Microbiology, 137:* 1241–1252.

KAMADA, T., TAKEMARU, T., PROSSER, J.I. & GOODAY, G.W., 1991b. Right and left handed helicity of chitin microfibrils in stipe cells of *Coprinus cinereus*. *Protoplasma, 165:* 64–70.

KIM, H.B., HAARER, R.K. & PRINGLE, J.R., 1991. Cellular morphogenesis in the *Saccharomyces cerevisiae* cell cycle: localization of the *CDC3* gene product and the timing of events at the budding site. *Journal of Cell Biology, 112:* 535–544.

KOCH, A.L., 1988. Biophysics of bacterial cell walls reviewed as stress-bearing fabric. *Microbiological Reviews, 52:* 337–353.

KURANDA, M.J. & ROBBINS, P.W., 1991. Chitinase is required for cell separation during growth of *Saccharomyces cerevisiae*. *Journal of Biological Chemistry, 266:* 19758–19767.

McGILLIVRAY, A.M. & GOW, N.A.R., 1986. Applied electric fields polarise the growth of mycelial fungi. *Journal of General Microbiology, 132:* 2515–2525.

McKERRACHER, L.J. & HEATH, I.B., 1987. Cytoplasmic migration and intracellular movements of organelles during tip growth of fungal hyphae. *Experimental Mycology, 11:* 79–100.

MAHADEVAN, P.R. & TATUM, E.L., 1967. Localization of structural polymers in the cell walls of *Neurospora crassa*. *Journal of Cell Biology, 35:* 295–302.

MARKHAM, P. & COLLINGE, A.J., 1987. Woronin bodies of filamentous fungi. *FEMS Microbiology Reviews, 46:* 1–11.

MITCHELL, L.H. & SOLL, D.R., 1979. Temporal and spatial differences in septation during synchronous mycelium and bud formation by *C. albicans*. *Experimental Mycology, 3:* 298–309.

MOORE, R.T. & McCLEAR, J.H., 1962. Fine structure of mycota 7. Observations on septa of ascomycetes and basidiomycetes. *American Journal of Botany, 62:* 86–94.

MULLINS, J.T. & ELLIS, E.A., 1974. Sexual morphogenesis in *Achlya:* ultrastructural basis for the hormone induction of antheridial hyphae. *Proceedings of the National Academy of Sciences, USA, 71:* 1347–1350.

ORLEAN, P., 1987. Two chitin synthases in *Saccharomyces cerevisiae*. *Journal of Biological Chemistry, 262:* 5732–5739.

RAST, D.M., HORSCH, M., FURTER, R. & GOODAY, G.W., 1991. A complex chitinolytic system in exponentially growing mycelium of *Mucor rouxii*: properties and function. *Journal of General Microbiology, 137:* 2797–2810.

READ, N.D., KELLOCK, L.J., KINGHT, H. & TREWAVAS, A.J. (1992). Contact sensing during infection by fungal pathogens. In J.A. Callow & J.R. Green (eds), *Perspectives in Plant Cell Recognition*, pp. 137–172. Society for Experimental Biology Seminar Series, vol, 48. Cambridge, U.K.: Cambridge University Press.

ROBERSON, R.W. & FULLER, M.S., 1988. Ultrastructural aspects of the hyphal tip of *Sclerotium rolfsii* preserved by freeze substitution. *Protoplasma, 146:* 143–149.

SAKUDA, S., NISHIMOTO, Y., OHI, M., WATANABE, M., TAKAYAMA, J., ISOGAI, A. & YAMADA, Y., 1990. Effects of demethylallosamidin, a potent yeast chitinase inhibitor, on cell division in yeast. *Agricultural and Biological Chemistry, 54:* 1333–1335.

SBURLATI, A. & CABIB, E. (1986). Chitin synthetase 2, a presumptive participant in septum formation in *Saccharomyces cerevisiae*. *Journal of Biological Chemistry, 261:* 15147–15152.

SHAW, J.A., MOL, P.C., BOWERS, B., SILVERMAN, S.J., VALDIVIESO, M.H., DURAN, A. & CABIB, E., 1991. The function of chitin synthases 2 and 3 in the *Saccharomyces cerevisiae* cell cycle. *Journal of Cell Biology, 114:* 111–123.

SHERWOOD, J., GREGORY, D., GOW, N.A.R., GOODAY, G.W. & MARSHALL, D., 1992. Contact sensing in *Candida albicans. Journal of Medical and Veterinary Mycology, 30:* 461–469.

SILVERMAN, S.J., SBURLATI, A., SLATER, M.L. & CABIB, E., 1988. Chitin synthase 2 is essential for septum formation and cell division in *Saccharomyces cerevisiae. Proceedings of the National Academy of Sciences, USA, 85:* 4735–4739.

SOLL, D.R. & MITCHELL, L.H., 1983. Filament ring formation in the dimorphic yeast *Candida albicans, Journal of Cell Biology, 96:* 486–493.

VALDIVIESO, M.H., MOL, P.C., SHAW, J.A., CABIB, E.C. & DURAN, A., 1991. *CAL1,* a gene required for activity of chitin synthase 3 in *Saccharomyces cerevisiae. Journal of Cell Biology, 114:* 101–109.

VALK, P. VAN DER, MARCHANT, R. & WESSELS, J.G.H., 1977. Ultrastructural localization of polysaccharides in the wall and septum of the basidiomycete *Schizophyllum commune. Experimental Mycology, 1:* 69–82.

VAN LAERE, A., 1988. Effects of electric fields on polar growth of *Phycomyces blakesleeanus. FEMS Microbiology Letters, 49:* 111–116.

WESSELS, J.G.H., 1986. Cell wall synthesis in apical hyphal growth. *International Review of Cytology, 104:* 37–79.

WESSELS, J.G.H., 1990. Role of cell wall architecture in fungal tip growth generation. In I.B. Heath (ed.), *Tip Growth in Plant and Fungal Cells,* 1–29. New York, U.S.A.: Academic Press.

WESSELS, J.G.H. & MARCHANT, R. (1974). Enzymatic degradation of septa in hyphal wall preparations from a monokaryon and a dikaryon of *Schizophyllum commune. Journal of General Microbiology, 83:* 359–368.

WITTEKINDT, E., LAMPRECHT, I. & KRAEPELIN, G. 1989. DC electrical fields induce polarization effects in the dimorphic fungus *Mycotypha africana. Endocytobiosis and Cell Research, 6:* 41–56.

YARDEN, O. & YANOFSKY, C., 1991. Chitin synthase 1 plays a major role in cell wall biogenesis in *Neurospora crassa. Genes and Development, 5:* 2420–2430.

YOUATT, J., GOW, N.A.R. & GOODAY, G.W., 1988. Bioelectric and biosynthetic aspects of cell polarity in *Allomyces macrogynus. Protoplasma, 146:* 118–126.

ZHOU, X-L., STUMPF, M.A., HOCH, H.C. & KUNG, C. 1991. A mechanosensitive channel in whole cells and in membrane patches of the fungus. *Uromyces. Science, 253:* 1415–1417.

ZHU, W-Y. & GOODAY, G.W., 1992. Effects of nikkomycin and echinocandin on differentiated and undifferentiated mycelia of *Botrytis cinerea* and *Mucor rouxii. Mycological Research, 96:* 371–377.

CHAPTER

20

Molecular tools to study sexual development in fungi

ALISON M. ASHBY & KEITH JOHNSTONE

CONTENTS

Abstract

In response to environmental signals, fungal hyphae differentiate to form a variety of new shapes. Current knowledge suggests that ultimate shape and form are determined by a combination of hyphal growth, physical forces, and triggers and physical constraints imposed by the environment. Fungal development may be studied in two ways: (1) the 'outside to inside' approach which allows the analysis of differentiation of fungal hyphae in response to changes in the immediate environment surrounding the hyphae; and (2) the 'inside to outside' approach which focuses on the molecular events that originate from within the cell and govern the determination of shape and form through a series of developmental cascades. The light leaf spot pathogen

Shape and Form in Plants and Fungi
ISBN 0–12–371035–9

Pyrenopeziza brassicae, which can undergo mutually exclusive pathways of asexual and sexual development, has been chosen as a model system in which to study the generation of shape and form in fungi. Using the 'outside to inside' strategy we have identified low-molecular-weight lipoidal compounds or sex factors (SF) which control switching from asexual to sexual reproduction in *P. brassicae*. The contents of a molecular toolbox to allow use of the 'inside to outside' approach to identify genes that are expressed during sexual morphogenesis and in response to SF are described. The potential of reporter genes such as β-glucuronidase (GUS) to study the spatial and temporal contribution of differentially regulated genes to the generation of shape and form is established. Finally, the use of GUS as a reporter to mark single mating type isolates to allow analysis of the contribution made by each mating type to fruiting body formation is described.

INTRODUCTION

An integrated approach to the study of fungal sex, involving both biochemical and molecular strategies, may lead to a greater understanding of both the determination of shape and form and of sexual morphogenesis in fungi. This approach, which has been adopted for analysis of development of the ascomycete fungus *Pyrenopeziza brassicae*, is described and constitutes a model system for analysis of fungal sexual development.

In response to a combination of environmental, genetic and hormonal signals fungal hyphae can differentiate to form a variety of new shapes and forms, including yeast-like cells, infection or resting structures and the specialized hyphae which comprise the sexual fruiting body. During sexual reproduction, fungi produce a remarkable array of new shapes and forms, sizes and colours which add a new dimension to the fungal kingdom. Current knowledge suggests that the ultimate shape and form of fungi is determined by a combination of factors including programmed hyphal growth, differentiation, branching and fusion; the new physical forces created by the combination of the internal osmotic pressures of the hyphae, hyphal extension and friction; and the triggers and physical constraints imposed by the environment. The molecular contribution of individual genes and their products to these processes is at present unknown.

Approaches to the study of sexual development in fungi

Fungal sexual development can be investigated in two ways (Ashby & Dyer, 1992) as illustrated in Fig. 1. First, the 'inside to outside' approach concentrates on analysis of the molecular events which originate within the cell nucleus and are primarily controlled by the mating type idiomorphs. The resulting cascades of gene expression govern the commitment to sexual development and the eventual determination of shape and form through the formation of the sexual fruiting body. This approach was adopted to clone the mating type idiomorphs from *Neurospora crassa, Podospora anserina, Schizophyllum commune, Saccharomyces cerevisiae, Ustilago maydis* and *Coprinus cinerea* (Astell *et al.*, 1981; Bolker *et al.*, 1992; Froelinger & Leong, 1991;

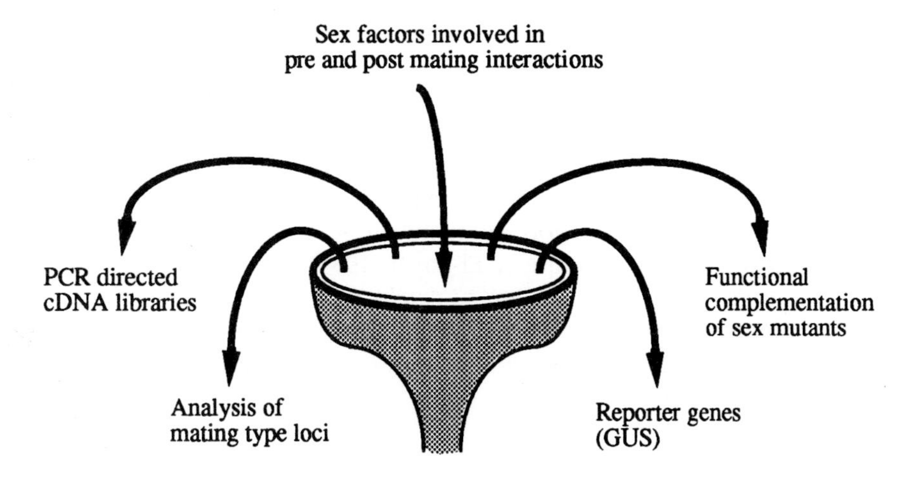

Figure 1 The 'inside-outside' and 'outside-inside' approaches to the analysis of fungal sexual development, involving both biochemical and molecular strategies.

Giasson *et al.*, 1989; Glass *et al.*, 1990; Kronstad & Leong, 1989; Mutasa *et al.*, 1990; Picard *et al.*, 1991; Staben & Yanofsky, 1990) and has subsequently involved the analysis of the mating type loci and genes which are closely linked to the mating type regions. Secondly, the 'outside to inside' approach concentrates on the analysis of differentiation of fungal hyphae in response to changes in the immediate chemical environment.

Sex factors in fungal sexual development

A multitude of sex hormones coordinate sexual development in fungi. The first group, termed pheromones, is found in the heterothallic fungi and coordinates the initial recognition and attraction of compatible mating type partners, as well as stimulating their initial differentiation into male or female gametes. In one of the first studies of the fungal sex factors, Raper (1952) discovered that female hyphae of the water mould *Achyla* release a sterol attractant, antheridiol, which stimulates the male hyphae to differentiate short branches bearing antheridia. These differentiated hyphae then respond to produce a complementary hormone, oogoniol, a sterol ester, which induces further differentiation in female hyphae. In other studies, terpenoid trisporic acids were found to suppress the formation of asexual spores and to stimulate the formation of sexual zygospores in the mucorales (pin moulds), with compatible partners having incomplete yet complementary biosynthetic pathways for trisporic acid synthesis (Gooday, 1974, 1978; Van den Ende, 1976). More recently, the complementary sex hormones parisin and sirenin, produced by the male and female gametes of the water mould *Allomyces*, have been identified and characterized (Machlis, 1966, 1972; Pommerville & Olson, 1987).

The second group of fungal sex hormones triggers and coordinates sexual development once compatible hyphae have fused. Such hormones, often termed 'sex morphogens', may act by triggering sexual development as a whole, in homothallic species, or by specifically inducing fruiting body formation independently of 'true sex' in heterothallic species. These compounds are of particular interest because they appear to control the transition from a relatively undifferentiated form to a well-ordered three-dimensional structure. Understanding the processes whereby such sex morphogens stimulate morphological differentiation may give new insights into how shape and form are determined during the development of more complex multicellular organisms. Champe and colleagues discovered that a series of 'precocious sexual inducing' (Psi) factors produced by sexually reproducing cultures of *Aspergillus nidulans* were able to suppress asexual reproduction and to initiate sexual reproduction when added back to vegetative cultures (Champe & El-Zayat, 1989). Similarly, Kawai and Ikeda (1985) purified a group of lipids known as cerebrosides which trigger fruiting body formation in *Schizophyllum commune*. There are several other chemical and proteinaceous sex factors which have been found to stimulate fruiting body formation in fungi (Dyer *et al.*, 1992; Gooday, 1974).

Pyrenopeziza brassicae as a model for study of the determination of shape and form

P. brassicae has been chosen as a model system to study sexual development and the determination of shape and form in fungi. It is a haploid member of the Ascomycotina and is the causal agent of light leaf spot disease of oilseed rape. The fungus is heterothallic, with two mating types determined by two alleles at a single locus designated *MAT1-1* and *MAT1-2* (Courtice & Ingram, 1987). *P. brassicae* can undergo two mutually exclusive pathways of development (Fig. 2). In the absence of a compatible mating type partner, spores of the fungus will germinate to produce hyphae

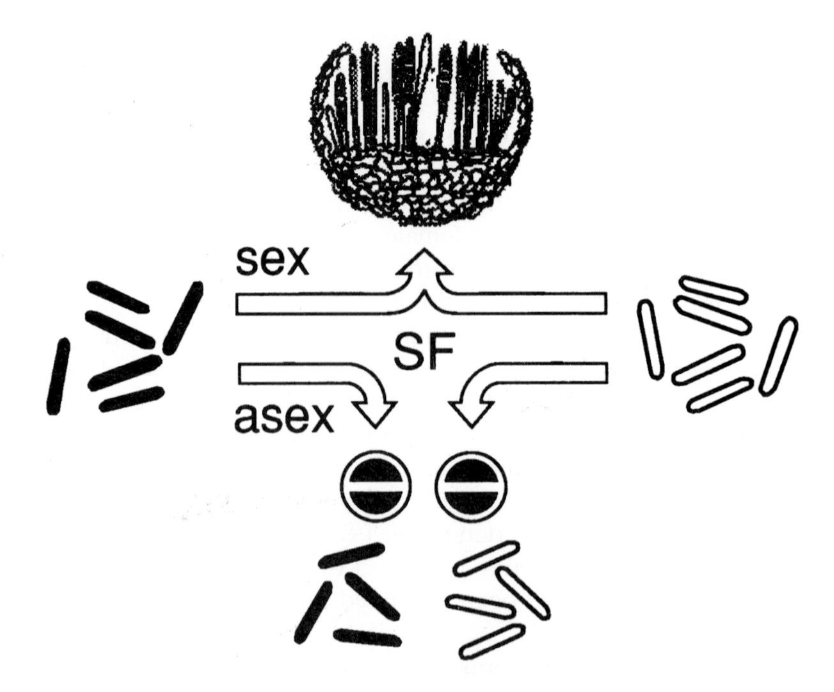

Figure 2 Mutually exclusive pathways of development in *Pyrenopeziza brassicae* and the role of the sex factor (SF) in controlling these pathways. White and black conidiospores indicate the two different mating types of the fungus. The sex factor (SF) inhibits asexual sporulation, resulting in all available resource being diverted into sexual reproduction.

which differentiate by a process of enteroblastic conidiation to form asexual conidia. Conversely, in the presence of a compatible partner, asexual sporulation is suppressed and a complex, but coordinated pathway of development culminates in the production of cup-like fruiting bodies or apothecia.

What follows describes how the 'outside to inside' approach has been used to identify and purify an organic sex factor termed SF which controls the switching from asexual to sexual morphogenesis in *P. brassicae* by inhibiting asexual sporulation. The use of this sex factor as a probe in the molecular analysis of sexual development following the 'inside to outside' approach is then discussed.

THE 'OUTSIDE TO INSIDE' APPROACH

When compatible mating type spores of *P. brassicae* are mixed in equal proportions and incubated for 4 weeks on CMM medium (Rainey, 1989), fertile apothecia develop. A methanolic extract of cultures made at this stage of development was termed SF (Siddiq *et al.*, 1989, 1992). Addition of the crude methanolic extract to single mating type cultures of *P. brassicae* inhibited asexual sporulation and stimulated morphological differentiation characteristic of the initial stages of ascocarp development. When added to mated cultures, the effect of SF was to increase the number of fertile ascocarps and to reduce the time in which they were produced. Crude methanolic extracts from single isolates at the same stage of development had no effects on either asexual or sexual development when added to single and mated cultures of *P. brassicae* (Siddiq *et al.*, 1989, 1992).

Preliminary analysis demonstrated that crude SF is lipophilic, with the active component having a molecular mass of less than 5000 and present as a minor proportion of the crude extract (Siddiq *et al.*, 1989).

Purification of the active component of SF

The active component from crude SF was first partially purified by rotary evaporation of the methanolic extract, followed by resuspension of the residue in water and centrifugation at 15 000**g** for 1 h to remove water-insoluble components. The supernatant was extracted in ethyl acetate and the ethyl acetate-soluble fraction recovered by rotary evaporation and analysed by normal-phase HPLC on a lichrosorb diol column (0.8 cm × 15 cm, Hichrom) eluted with a hexane:propan-2-ol gradient. The resulting fractions were bioassayed as follows for their ability to inhibit asexual sporulation of single mating type isolates of *P. brassicae* after 15 days of incubation at 18°C in the dark. Aliquots (50 μl) of a final concentration of between 0.5 and 1.0 mg ml^{-1} of each fraction in methanol were pipetted individually on to the surface of 1 ml of 3% MA (3% malt extract, 1.2% oxoid agar) in the wells of a 24-well cell culture plate (Bibby, Sterilin), the solvent allowed to evaporate and then each well inoculated with 25 μl of a 1×10^{6} ml^{-1} suspension of conidia of a single isolate of *P. brassicae* (Fig. 3A). Changes in colony morphology were noted at 15 days. Further purification of the bioactive fractions from normal-phase separation was achieved by reverse-phase HPLC on a C18 column (0.8 cm × 15 cm, Spherisorb ODS) eluted with a water : acetonitrile gradient. This purification procedure identified a single bioactive peak (Fig. 3B). SF represents only a minor portion of the crude mixture and the

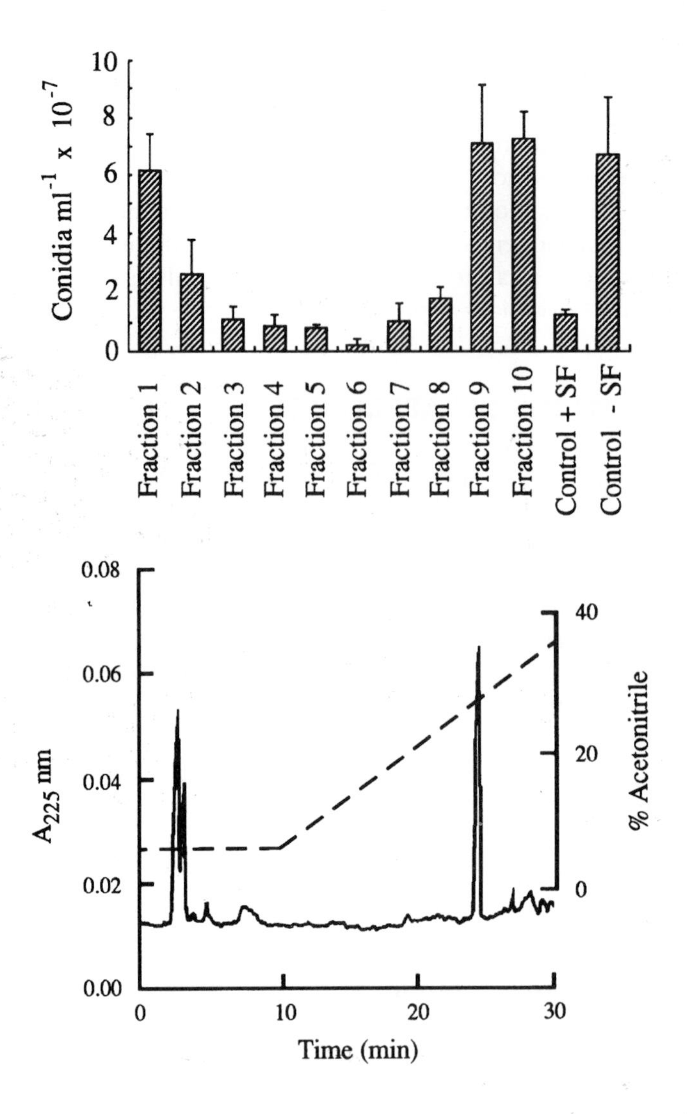

Figure 3 Bioassay and purification of the sex factor from *Pyrenopeziza brassicae:* **A** results of the bioassay of sequential one minute fractions of crude sex factor (SF) separated by normal phase HPLC as described in the text; **B** further purification of the normal-phase active fraction 6 by reverse-phase HPLC to yield a single bioactive peak as described in the text.

extraction procedure is currently being scaled up to enable structural analysis of this active component.

These results provide evidence that one component of SF is an inhibitor of asexual sporulation which, when present in mated cultures, results in all of the available resources being diverted into sexual morphogenesis (Fig. 2).

THE 'INSIDE TO OUTSIDE' APPROACH

The identification of a single fraction displaying high activity in the bioassay will make it possible to identify genes that are differentially regulated in response to SF in single mating type isolates. This analysis comprises part of the 'inside to outside' approach.

The contents of a molecular toolbox

Before an analysis of the regulation of genes by SF it was first essential to develop a molecular 'toolbox' as follows:

Extraction of fungal DNA

The ability to extract high-molecular-weight DNA is an essential prerequisite for cloning fungal genes. Using the method of Raeder and Broda (1985) yields of up to 1 μg DNA were obtained from 1 mg dry weight of tissue of *P. brassicae* with a molecular weight of at least 30 kb. The mycelium was grown in potato dextrose broth (PDB) for 4 weeks and then washed and lyopholized before extraction of the DNA. Reasonable yields of genomic DNA have also been obtained from mycelial colonies grown on solid MA medium.

Vectors for transformation

A transformation vector is an extremely useful tool for the introduction of cloned DNA into a fungus. There are several such vectors available, containing auxotrophic as well as antibiotic markers. The vectors most suitable for *P. brassicae* were from the pAN7 series (Punt *et al.*, 1987). Both pAN 7-1 and pAN 7-2 are shuttle vectors which encode resistance to ampicillin for expression in *E. coli* and carry the hygromycin β-phosphotransferase gene flanked by the *gpd* promotor and *trp* C terminator sequences from *Aspergillus* for expression of resistance to hygromycin in the fungus. pAN 7-2 is a cosmid vector in that it carries a *cos* site for the packaging *in vitro* of large fragments of genomic DNA for library preparation. Since *P. brassicae* is highly sensitive to hygromycin (growth is inhibited at concentrations above 5 μg ml^{-1}), these vectors were useful in establishing an efficient transformation procedure.

Transformation

Fungal protoplasts were prepared by a method based on that of Yelton *et al.* (1984), and transformed using a modified method of Vollmer and Yanofsky (1986) as described by Ball *et al.* (1991a). Variable transformation frequencies were obtained in different experiments, but an average of up to 50 transformants per μg of transforming DNA were obtained routinely.

Generation and complementation of mutants

Mutants may be generated by ultraviolet irradiation (UV) or by exposure to chemical mutagens and screened for deficiencies in a number of phenotypes including the ability to complete the sexual cycle or to respond to specific sex factors. Using the procedures described earlier, auxotrophic mutants of *P. brassicae* have been complemented to restore prototrophy (Ball *et al.*, 1991a), and a protease-negative nonpathogenic mutant has been restored to protease production and pathogenicity (Ball *et al.*, 1991b).

Transformation of sexually deficient mutants with a genomic library of fungal DNA and the screening of the transformants for restored wild-type phenotypes allows key developmentally regulated genes to be targeted. A UV-derived mutant of *P. brassicae* has been isolated which is unable to complete the sexual cycle when crossed with its compatible mating type and shows no response to the sex factor SF (Siddiq *et al.*, 1989). Complementation of this mutation by transformation with a cosmid from a genomic library from *P. brassicae* resulted in a transformant that had restored fertility and responsiveness to SF (Fig. 4). When total genomic DNA was extracted from the transformant and digested with an enzyme that does not cut the transforming cosmid, and the digested DNA probed with the original parent cosmid, a band of approximately 19 kb was identified, of which 10 kb represents flanking genomic DNA (Fig. 5). Digestion of the total genomic DNA from the transformant with an enzyme which cuts once within the cosmid, followed by probing with the parent cosmid, revealed two hybridizing bands. This demonstrates that a single integration of the cosmid into the genome of the mutant had restored the required phenotype. Such results are typical when analysing *P. brassicae* transformed with homologous DNA sequences.

Once a mutation has been functionally complemented, the cloned sequences

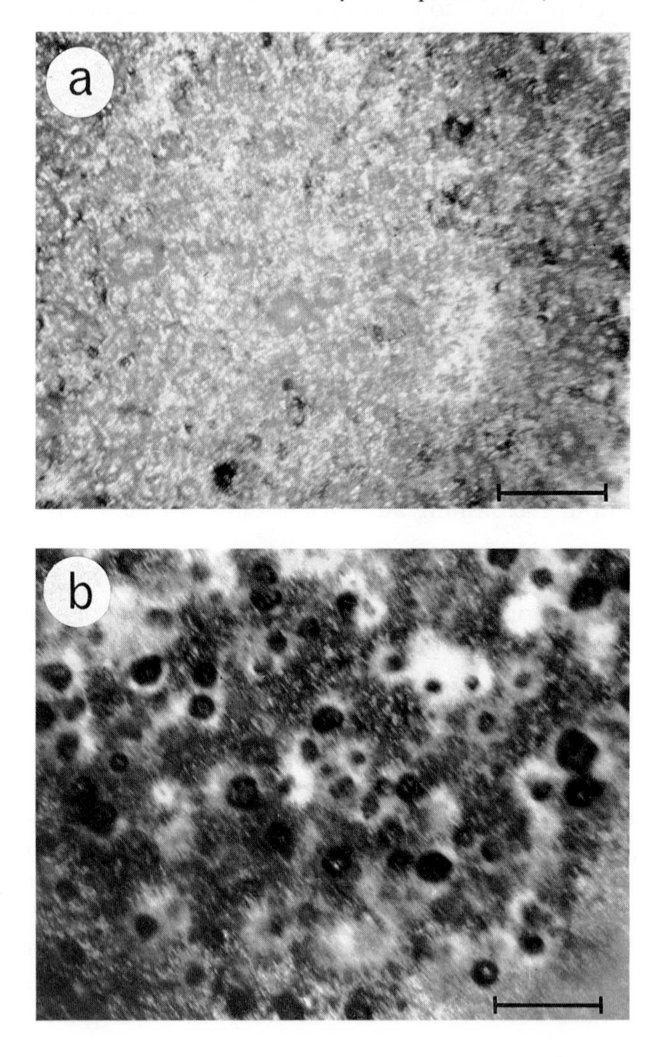

Figure 4 Complementation of the *Pyrenopeziza brassicae* developmental mutant JH26 973 with a wild-type genomic library which resulted in transformant JH26 973(T) with restored fertility and responsiveness to SF. (a) Light micrograph of JH26 973 (*MAT1-2*) crossed with wild-type NH10 (*MAT1-1*) after 14 days on CMM medium in which development of asexual spores but not of sexual structures has occurred (bar 5 mm). (b) Light micrograph of JH26 973(T) crossed with wild-type NH10 after 14 days on CMM medium in which very little asexual sporulation and abundant fertile apothecia are evident (scale bar 5 mm).

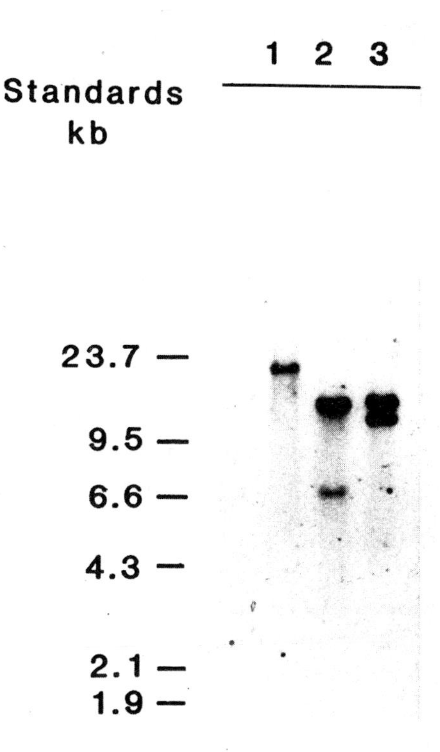

Figure 5 Southern blot of DNA from *Pyrenopeziza brassicae* sex transformant JH26 973(T). Genomic DNA was restricted with *Cla*I (which does not cut within the parent cosmid; track 1), and *Sac*I and *Nco*I (which cut once within the parent cosmid; tracks 2 and 3 respectively) was probed with the parent cosmid pAN7-2.

resulting in the acquired phenotype may be recovered by a technique termed cosmid rescue (Perucho *et al.*, 1980). The technique allows the rescue of the original transforming cosmid vector together with flanking genomic DNA of interest. This technique can now be used to clone the flanking region which may contain a gene involved in sexual morphogenesis (e.g an SF receptor, an SF transport protein or even the *MAT* locus itself). If a gene of interest is not present in the flanking region, this region can be used as a probe in chromosome walking, eventually making it possible to target the gene by probing clones from the genomic library.

With the development of the molecular tool box it is possible to study fungal sexual processes using the 'inside to outside' approach. First, the mating type idiomorphs may be cloned and expression from the *MAT* locus analysed, as has been performed in the case of the ascomycetes *Neurospora crassa* (Glass *et al.*, 1990; Staben & Yanofsky, 1990), *Podospora anserina* (Picard *et al.*, 1991), *Cochliobolus heterostrophus* (Yoder *et al.*, 1989) as well as in the basidiomycetes *Coprinus cinerea* (Mutasa *et al.*, 1990) and *Schizophyllum commune* (Giasson *et al.*, 1989). Two parallel approaches are currently being used to clone the mating type idiomorphs of *Pyrenopeziza brassicae*. These are (1) transformation with a genomic library from the opposite mating type and (2) heterologous probing with cloned mating type loci from *Neurospora crassa*. Secondly, sexuality mutants may be generated, complemented and the genes of interest cloned. Thirdly, polymerase chain reaction (PCR)-directed cDNA libraries may be constructed from single developing fruiting bodies to allow identification of differentially regulated genes. Fourthly, reporter genes such as β-glucuronidase (GUS) may be used to study the spatial and temporal expression of developmentally related genes during sexual morphogenesis.

Generation of PCR-directed cDNA libraries

The technique of PCR allows the amplification of an unknown DNA sequence from between two known primer sequences using a special heat-stable TAQ DNA polymerase from *Thermus aquaticus*. This technique has been well established by Gurr *et al.* (1991) for the identification of genes expressed at the host–pathogen interface. Here it is shown how PCR can be used to target genes that are specifically induced in spores of *P. brassicae* by SF. The following paragraphs describe how developmentally regulated genes may be targeted using PCR to generate cDNA libraries from small numbers of spores of *P. brassicae*.

Induction of RNA synthesis and extraction of total RNA

SF may be added to a spore suspension of *P. brassicae* at time zero (T_0) and total RNA extracted at different time points (T_x). Both messenger RNA and other RNA species will be present in the extract at this stage. The mRNA will include genes that are differentially expressed in response to SF (Fig. 6a). First strand cDNA synthesis from total RNA is directed by a general oligo(dT) primer complementary to the poly(A) tail of the mRNA which, in the presence of reverse transcriptase, generates a single-stranded cDNA template (Fig. 6b). A homopolymer tail (poly(dG)) is then added to the 3′ end of the first strand cDNA catalysed by the enzyme terminal transferase (Fig. 6b). The cDNA sequences of interest will then be flanked by known primer sequences and thus the template can be used for successive rounds of PCR driven by oligo(dC) and oligo(dT) primers in the presence of TAQ polymerase and the four dNTPs (Fig. 6c).

The PCR reaction is performed by a succession of cycles of: denaturation (95°C), where double-stranded DNA is separated into two single-stranded templates; annealing of primers (37–60°C), where small oligonucleotide primers which flank the sequence of interest bind to the 3′ end of the DNA strand; and polymerization (72°C), where DNA is synthesized in the 5′ to 3′ direction by TAQ polymerase in the presence of the four dNTPs (dA, dG, dC, dT) (Fig. 6c). After about 20 successive cycles the DNA sequence is amplified by over a million-fold. The technique of PCR may therefore be valuable in the analysis of gene expression from a single spore or single fruiting body.

Library construction and screening

The amplified cDNAs may be ligated into a suitable vector such as λ ZAPII (Stratagene) and screened by the process of differential hybridization using DNA derived from the original uninduced and induced mRNA as probes to identify genes specifically induced by the presence of SF after various time intervals (Fig. 6d). The technique could also be used to target genes at specific stages in the developmental cycle by the analysis of single developing fruiting bodies. Such experiments remain to be done.

(a)

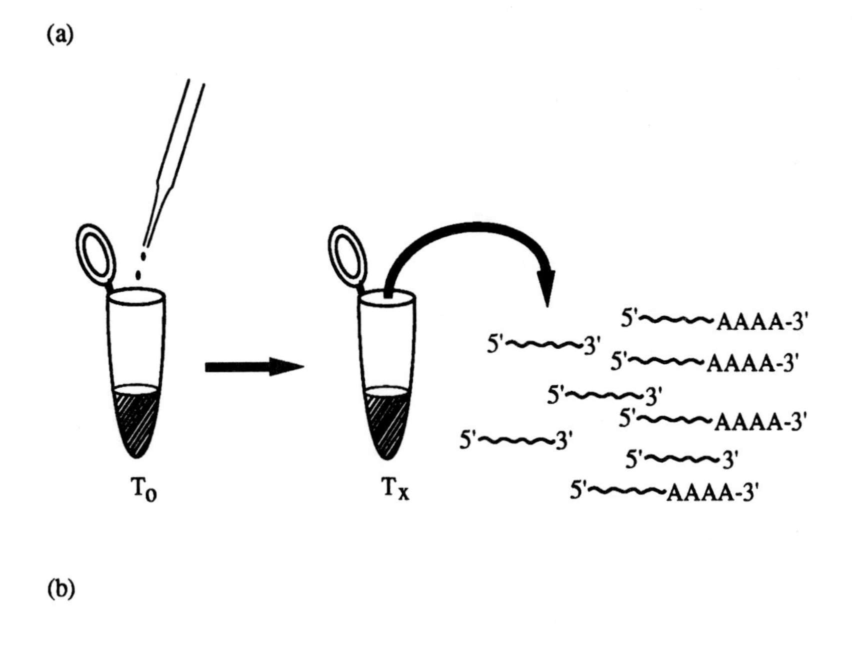

(b)

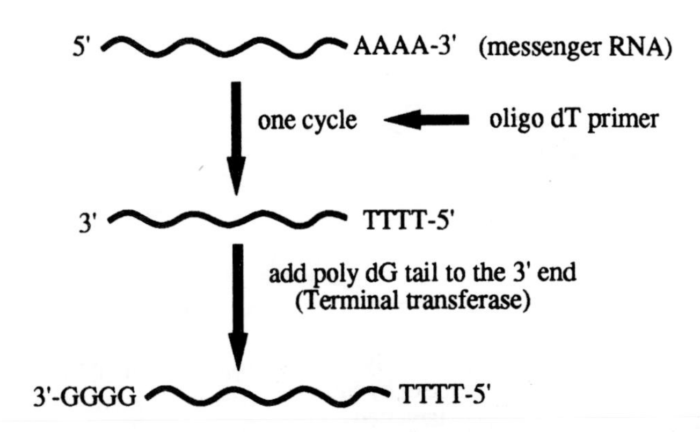

Figure 6 A scheme for using PCR-directed cDNA libraries to target genes that are differentially expressed in response to the sex factor SF. (a) Extraction of total RNA at defined time points (T_x) after induction of spores with the SF. (b) First strand cDNA synthesis from total RNA using an oligo(dT) primer followed by the addition of a poly(dG) tail to the 3′ end of the first strand cDNA. (c) Amplification of cDNA by successive rounds of PCR using oligo(dC) and oligo(dT) primers. (d) Differential hybridization of cDNA clones to target genes which are differentially expressed in the presence of SF.

USE OF REPORTER GENES IN STUDYING DEVELOPMENTAL INTERACTIONS

Reporter genes such as those encoding the synthesis of β-galactosidase, β-glucuronidase (GUS) and luciferase are widely used to study plants and bacteria, but are relatively new to mycology. The enzymes encoded by the reporter genes hydrolyse specific chromogenic or fluorogenic substrates, liberating a specific colour or fluorescence

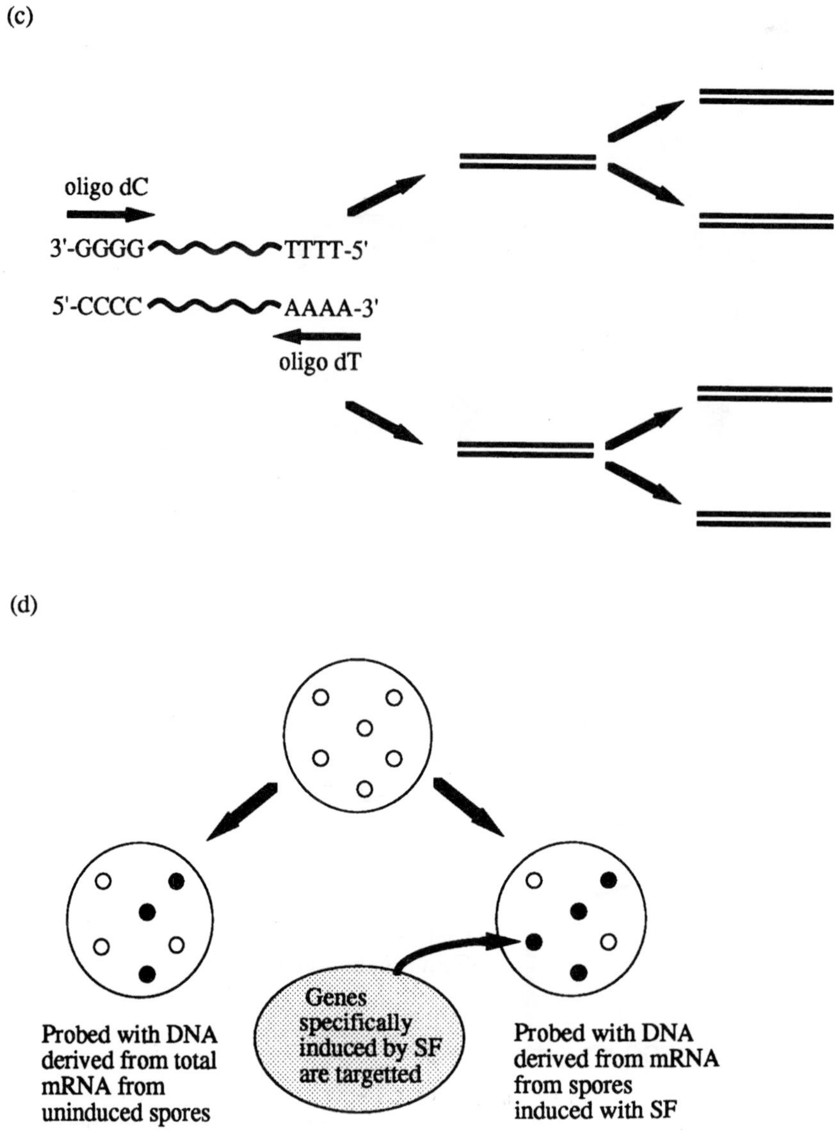

(c)

oligo dC

3'-GGGG~~~~TTTT-5'

5'-CCCC~~~~AAAA-3'

oligo dT

(d)

Probed with DNA derived from total mRNA from uninduced spores

Genes specifically induced by SF are targetted

Probed with DNA derived from mRNA from spores induced with SF

Figure 6 Continued

indicative of reporter gene activity. Such reporter genes can be used for analysis of developmental processes in fungi in two ways. First, if a fungus is transformed with a vector which can constitutively express GUS, this technique can be used to mark the fungus. This is of value in the analysis of otherwise indistinguishable hyphal interactions such as those occurring during complex sexual processes where it may be useful to analyse the contribution to fruiting body formation made by each mating type in a heterothallic interaction. In addition, such a procedure may facilitate the analysis of complex mycoparasitic interactions. Secondly, once key developmentally regulated genes have been targeted, the physiological role of these genes may be

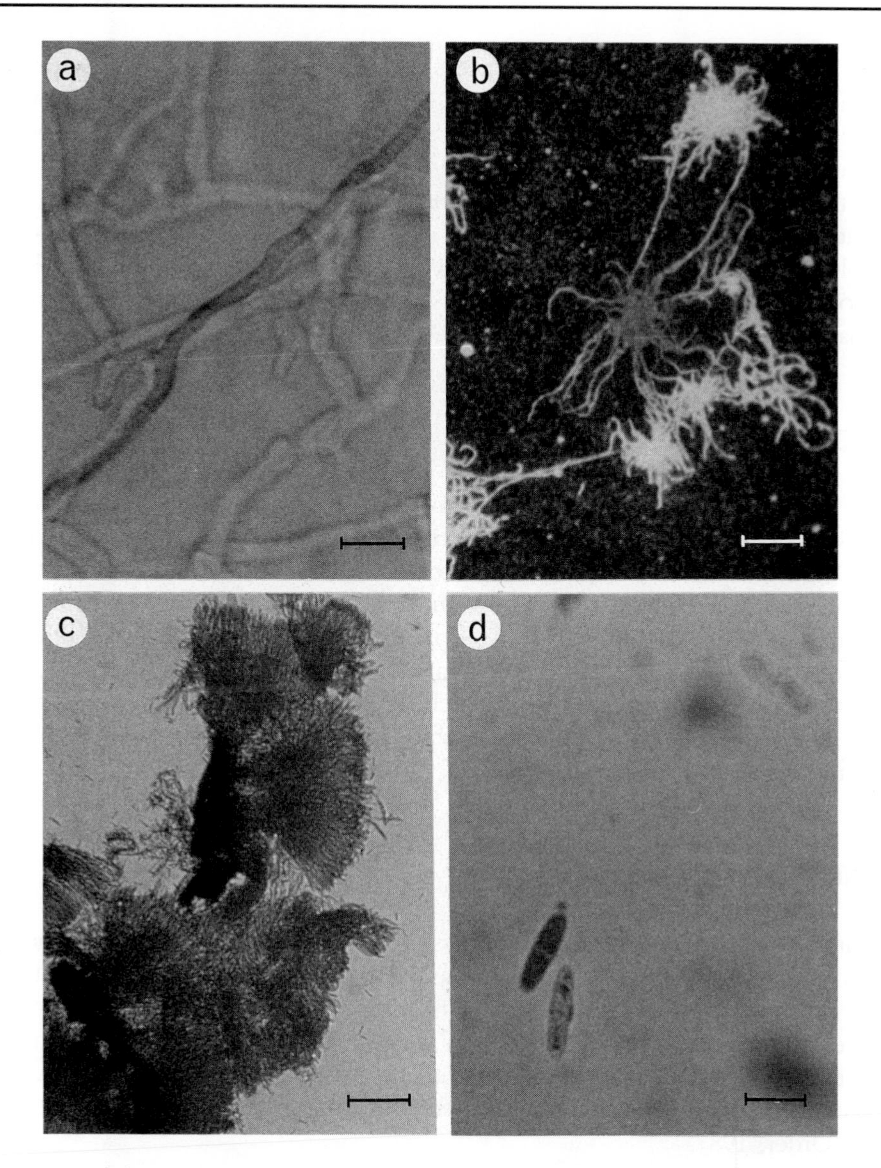

Figure 7 Use of β-glucuronidase (GUS) as a reporter to study complex developmental interactions in *Pyrenopeziza brassicae* by light microscopy. (a) Mycelium of a GUS transformant of *P. brassicae* NH10 (NH10pNOM102/1) crossed with wild-type JH26, 7 days after crossing stained with magenta-β-D-glucuronide (scale bar 25 μm). (b) Mycelium of NH10pNOM102/18 interacting with wild-type JH26 mycelium after 4 days on CMM medium and stained with X-gluc (scale bar 1 mm). (c) Crushed apothecia from a cross of *P. brassicae* NH10pNOM102/18 and wild-type JH26 and stained with X-gluc (scale bar 250 μm). (d) Septate ascospores resulting from a cross of *P. brassicae* NH10pNOM102/18 and wild-type JH26 stained with X-gluc (scale bar 25 μm). The frontispiece shows a coloured version of this figure.

studied by linking the promoter of interest to the reporter gene and by analysing where and when reporter activity is expressed.

The value of using GUS as a reporter gene to study developmental interactions has been assessed by Ashby and Johnstone (1993), with a view to using the technique to study the spatial and temporal expression of developmentally regulated genes during apothecial formation in *P. brassicae*. Both mating types of *P. brassicae* can be transformed with a vector constitutively expressing GUS activity. When crosses are performed on GUS *MAT1-1* × wt *MAT1-2* and GUS *MAT1-2* × wt *MAT1-1*, an analysis may be made of the contribution each mating type makes to fruiting body development (Fig. 7a–d). No endogenous GUS activity is evident in untransformed cells and the product resulting from the hydrolysis of the chromogenic substrate is stable, making histochemical analysis relatively easy. A range of other chromogenic substrates is available which does not appear to affect the developmental interaction between the two mating types. Thus direct analysis can be made on fresh tissue in real time and space.

CONCLUSIONS

Within the last decade a number of tools have been developed for the molecular analysis of filamentous fungi, including several vectors for transformation, efficient and reliable transformation systems, protocols for the extraction of good yields of DNA and, more recently, techniques such as PCR. Such tools, in combination, will allow analysis of how differential gene expression governs the ultimate determination of shape and form in fungi. Molecular analysis, in parallel with biochemical, physiological and cytological approaches, on such amenable systems as the fungi will inevitably lead to rapid growth of the understanding of the determination of biological shape and form.

ACKNOWLEDGEMENTS

The authors thank Sándor Értz for discussion and technical assistance in the purification of SF, David Ingram, John Friend and Maria Chamberlain for the contribution of ideas relating to sexual development in fungi, the AFRC for a project grant (KJ) and the Royal Society for a University Research Fellowship (AMA). The work described was performed under the provisions of license number PHF 174A/91 (28) issued by the Ministry of Agriculture, Fisheries and Food under the Plant Health (Great Britain) Order, 1987.

REFERENCES

ASHBY, A.M. & DYER, P.S., 1992. The secret sex lives of fungi. *New Scientist, 133:* 32–35.

ASHBY, A.M. & JOHNSTONE, K., 1993. Expression of the *E. coli* β-glucuronidase gene in the light leaf spot pathogen *Pyrenopeziza brassicae* and its use as a reporter gene to study developmental interactions in fungi. *Mycological Research, 97:* 575–581.

ASTELL, C.R., AHLSTROM-JONASSON, L. & SMITH, M., 1981. The sequence of the DNAs coding for the mating-type loci of *Saccharomyces cerevisiae*. *Cell, 27:* 15–23.

BALL, A.M., SAWCZYC, M.C., ASHBY, A.M., INGRAM, D.S. & JOHNSTONE, K., 1991a. Transformation of the fungus *Pyrenopeziza brassicae*, cause of light leaf spot of Brassicas, and complementation of mutants using a genomic library. *Experimental Mycology, 15:* 243–254.

BALL, A.M., ASHBY, A.M., DANIELS, M.J., INGRAM, D.S. & JOHNSTONE, K., 1991b. Evidence for the requirement of extracellular protease in the pathogenic interaction of *Pyrenopeziza brassicae* with oilseed rape. *Physiological and Molecular Plant Pathology, 38:* 147–161.

BOLKER, M., URBAN, M. & KAHMANN, R., 1992. The *a* mating type locus of *U. maydis* specifies cell signalling components. *Cell, 68:* 441–450.

CHAMPE, S.P. & EL-ZAYAT, A.A.E., 1989. Isolation of a sexual sporulation hormone from *Aspergillus nidulans. Journal of Bacteriology, 171:* 3982–3988.

COURTICE, G.R.M. & INGRAM, D.S., 1987. Isolation of auxotrophic mutants of the hemibiotrophic Ascomycete pathogen of brassicas, *Pyrenopeziza brassicae. Transactions of the British Mycological Society 89:* 301–306.

DYER, P.S., INGRAM, D.S. & JOHNSTONE, K., 1992. The control of sexual morphogenesis in the Ascomycotina. *Biological Reviews, 67:* 421-458.

FROELINGER, E.H. & LEONG, S.A., 1991. The *a* mating type genes of *Ustilago maydis* are idiomorphs. *Gene, 100:* 113–122.

GIASSON, L., SPECHT, C.A., MILGRIM, C., NOVOTNY, C.P. & ULLRICH, R.C., 1989. Cloning and comparison of *Aa* mating-type alleles of the Basidiomycete *Schizophyllum commune. Molecular and General Genetics, 218:* 72–77.

GLASS, N.L., GROTELUESCHEN, J. & METZENBERG, R.L., 1990. *Neurospora crassa A* mating-type region. *Proceedings of the National Academy of Sciences, USA, 87:* 4912–4916.

GOODAY, G.W., 1974. Fungal sex hormones. *Annual Review of Biochemistry, 43:* 35–49.

GOODAY, G.W., 1978. Functions of trisporic acid. *Philosophical Transactions of the Royal Society of London, B, 284:* 509–520.

GURR, S.J., MCPHERSON, M.J., SCOLLAN, C., ATKINSON, H.J. & BOWLES, D.J., 1991. Gene expression in nematode-infected plant roots. *Molecular and General Genetics, 226:* 361–366.

KAWAI, G. & IKEDA, Y., 1985. Structure of biologically active and inactive cerebrosides prepared from *Schizophyllum commune. Journal of Lipid Research, 26:* 338–343.

KRONSTAD, J.W. & LEONG, S.A., 1989. Isolation of two alleles of the *b* locus of *Ustilago maydis. Proceedings of the National Academy of Sciences, USA, 86:* 978–982.

MACHLIS, L., 1966. Sex hormones in fungi. In G.C. Ainsworth & A.S. Sussmann (eds), *The fungi,* vol. 2, pp. 415–433. London, U.K.: Academic Press.

MACHLIS, L., 1972. The coming of age of sex hormones in plants. *Mycologica, 64:* 234–247.

MUTASA, E.S., TYMON, A.M., GOTTGENS, B., MELLON, F.M., LITTLE, P.F.R. & CASSELTON, L.A., 1990. Molecular organization of an A-mating type factor of the Basidiomycete fungus *Coprinus cinereus. Current Genetics, 18:* 223–229.

PERUCHO, M., HANAHAN, D., LIPSICH, L. & WIGLER, M., 1980. Isolation of the chicken thymidine kinase gene by plasmid rescue. *Nature, 285:* 207–210.

PICARD, M., DEBUCHY, R. & COPPIN, E., 1991. Cloning the mating types of the heterothallic fungus *Podospora anserina:* developmental features of haploid transformants carrying both mating types. *Genetics, 128:* 539–547.

POMMERVILLE, J. & OLSON, L.W., 1987. Evidence for a male-produced pheromone in *Allomyces macrogynus. Experimental Mycology, 11:* 245–248.

PUNT, P.J., OLIVER, R.P., DINGEMANSE, M.A., POUWELS, P.H. & VAN DEN HONDEL, C.A.M.J.J., 1987. Transformation of *Aspergillus* based on the hygromycin resistance marker from *E. coli. Gene, 56:* 117–124.

RAEDER, U. & BRODA, P., 1985. Rapid preparation of DNA from filamentous fungi. *Letters in Applied Microbiology, 1:* 17–20.

RAINEY, P.B., 1989. A new laboratory medium for the cultivation of *Agaricus bisporus. New Zealand Natural Sciences, 16:* 109–112.

RAPER, J.R., 1952. Chemical regulation of sexual processes in Thallophycetes. *Botanical Reviews, 18:* 447–545.

SIDDIQ, A.A., INGRAM, D.S., JOHNSTONE, K., FRIEND, J. & ASHBY, A.M., 1989. The control of asexual and sexual development by morphogens in fungal pathogens. *Aspects of Applied Biology, 23:* 417–426.

SIDDIQ, A.A., JOHNSTONE, K. & INGRAM, D.S., 1992. Evidence for the production during mating of factors involved in suppression of asexual sporulation and the induction of ascocarp formation in *Pyrenopeziza brassicae. Mycological Research, 96:* 757–765.

STABEN, C. & YANOFSKY, C., 1990. *Neurospora crassa a* mating type region. *Proceedings of the National Academy of Sciences, USA, 87:* 4917–4921.

VAN DEN ENDE, H., 1976. *Sexual Interactions in Plants: The role of specific substances in sexual reproduction*. London, U.K.: Academic Press.
VOLLMER, S.J. & YANOFSKY, C., 1986. Efficient cloning of genes of *Neurospora crassa*. *Proceedings of the National Academy of Sciences, USA, 83:* 4869–4873.
YELTON, M.M., HAMER, J.E. & TIMBERLAKE, W.E. (1984) Transformation of *Aspergillus nidulans* by using a *Trp C* plasmid. *Proceedings of the National Academy of Sciences, USA, 81:* 1470–1474.
YODER, O.C., TURGEON, B.G., SCHAFER, W., CIUFFETTI, L., BOHLMAN, H. & VAN ETTEN, H.D., 1989. Molecular analysis of mating type and expression of a foreign pathogenicity gene in *Cochliobolus heterostrophus*. In H. Nevalainen & M. Penttila (eds), *Molecular Biology of Filamentous Fungi,* pp. 189–196. Helsinki: Foundation for Biotechnical and Industrial Fermentation Research.

INDEX

Shape and Form in Plants and Fungi
ISBN 0–12–371035–9

Linnean Society Symposium Series

Number 1. **The Evolutionary Significance of the Exine**, edited by I.K. Ferguson and J. Muller, 1976, xii + 592pp., 0.12.253650.9.

Number 2. **Tropical Trees—Variation, Breeding and Conservation**, edited by J. Burley and B.T. Styles, 1976, xvi + 244pp., 0.12.145150.X

Number 3. **Morphology and Biology of Reptiles**, edited by A. d'A. Bellairs and C. Barry Cox, 1976, xvi + 290pp., 0.12.085850.9

Number 4. **Problems in Vertebrate Evolution**, edited by S. Mahala Andrews, R.S. Miles and A.D. Walker, 1977, xii + 412pp. 0.12.059950.3

Number 5. **Ecological Effects of Pesticides**, edited by F.H. Perring and K. Mellanby, 1977, xii + 194 pp., 0.12.551350.X

Number 6. *Botanical Society of the British Isles Conference Report No. 16*, **The Pollination of Flowers by Insects**, edited by A.J. Richards, 1977, xii + 214pp., 0.12.587460.X

Number 7. **The Biology and Taxonomy of the Solanaceae**, edited by J.G. Hawkes, R.N. Lester and A.D. Skelding, 1979, xviii + 738pp. 0.12.333150.1

Number 8. **Petaloid Monocotyledons—Horticultural and Botanical Research**, edited by C.D. Brickell, D.F. Cutler and Mary Gregory, 1980, xii + 222pp., 0.12.133950.5

Number 9. **The Skin of Vertebrates**, edited by R.I.C. Spearman and P.A. Riley, 1980, xiv + 322pp., 0.12.656950.9

Number 10. **The Plant Cuticle**, edited by D.F. Cutler, K.L. Alvin and C.E. Price, 1982, x + 462pp., 0.12.199920.3

Number 11. **Ecology and Genetics of Host–Parasite Interactions**, edited by D. Rollinson and R.M. Anderson, 1985, xii + 266pp., 0.12.593690.7

Number 12. **Pollen and Spores: Form and Function**, edited by S. Blackmore and I.K. Ferguson, 1986, xvi + 443pp., 0.12.103460.7

Number 13. **Desertified Grasslands: Their Biology and Management**, edited by G.P. Chapman, 1992, xii + 360pp., 0.12.168570.5

Number 14. **Evolutionary Patterns and Processes**, edited by D.R. Lees and D. Edwards, 1993, xii + 326pp., 0.12.440895.8

Number 15. **The Biology of Lemmings**, edited by N.C. Stenseth and R.A. Ims, 1994, xvi + 683pp., 0.12.666020.4

Number 16. **Shape and Form in Plants and Fungi**, edited by D.S. Ingram and A. Hudson, 1994, x + 381pp., 0.12.371035.9